Handbook of
Safety Assessment of Nanomaterials

Pan Stanford Series on Biomedical Nanotechnology

Series Editors

Vladimir Torchilin and Mansoor Amiji

Titles in the Series

Vol. 1

Handbook of Materials for Nanomedicine

Vladimir Torchilin and Mansoor Amiji, eds.

2010

978-981-4267-55-7 (Hardcover)
978-981-4267-58-8 (eBook)

Vol. 2

Nanoimaging

Beth A. Goins and William T. Phillips, eds.

2011

978-981-4267-09-0 (Hardcover)
978-981-4267-91-5 (eBook)

Vol. 3

Biomedical Nanosensors

Joseph Irudayaraj, ed.

2013

978-981-4303-03-3 (Hardcover)
978-981-4303-04-0 (eBook)

Vol. 4

Nanotechnology for Delivery of Therapeutic Nucleic Acids

Dan Peer, ed.
2013

978-981-4411-04-2 (Hardcover)
978-981-4411-05-9 (eBook)

Vol. 5

Handbook of Safety Assessment of Nanomaterials: From Toxicological Testing to Personalized Medicine

Bengt Fadeel, ed.

2014

978-981-4463-36-2 (Hardcover)
978-981-4463-37-9 (eBook)

Forthcoming

Vol. 6

Inorganic Nanomedicine

Bhupinder Singh Sekhon, ed.

Vol. 7

Nanotechnology for Personalized Cancer Treatment

Julia Ljubimova, ed.

Vol. 8

Translation Industrial Nanotechnology

Thomas Redelmeier, ed.

Handbook of Safety Assessment of Nanomaterials

From Toxicological Testing to Personalized Medicine

edited by
Bengt Fadeel

PAN STANFORD PUBLISHING

Published by

Pan Stanford Publishing Pte. Ltd.
Penthouse Level, Suntec Tower 3
8 Temasek Boulevard
Singapore 038988

Email: editorial@panstanford.com
Web: www.panstanford.com

British Library Cataloguing-in-Publication Data
A catalogue record for this book is available from the British Library.

Handbook of Safety Assessment of Nanomaterials:
From Toxicological Testing to Personalized Medicine

ISBN 978-981-4463-36-2 (Hardcover)
ISBN 978-981-4463-37-9 (eBook)

Printed in the USA

Contents

Preface

". . . for I was never so small as this before, never!"
Lewis Carroll, *Alice's Adventures in Wonderland* (1865)

Nanomedicine is the application of nanobiotechnology in clinical medicine. For instance, nanotechnologies offer exciting opportunities for targeted drug delivery, thus bringing to life the concept of a "magic bullet" imagined by Paul Ehrlich a century ago. Nevertheless, understanding whether such nanoscale objects per se exert adverse effects in a biological system is of critical importance. Nanotoxicology, in turn, may be viewed as the study of the undesirable interference between man-made nanomaterials and cellular nanostructures. In this handbook, included in the Pan Stanford series on *biomedical nanotechnology*, we attempt to bridge nanotoxicology and nanomedicine by applying the lessons learned from toxicological testing of manufactured nanomaterials to the field of nanomedicine.

The present volume opens with a historical perspective on the development of nanomedicine, written by Dr. Duncan, a pioneer in the field. Dr. Duncan points out that a balanced discussion of the risks and benefits of nanotechnologies is critically important to ensure the speedy and safe realization of the promises of nanomedicine. Indeed, this is the underlying motivation for the entire volume. Then, Dr. Stone et al. discuss the basic principles of nanotoxicology, highlighting progress in the field in recent years; the authors also provide recommendations for the proper design of experiments to assess nanomaterial hazards. Drs. Warheit and Sayes touch on the need for robust physicochemical characterization of nanomaterials for toxicity testing, and Drs. Fadeel and Parak discuss the biological "identity" of nanomaterials.

These introductory chapters are followed by a series of chapters on different approaches to nanomaterial testing: Dr. Hartung makes the case for in vitro tests, while Drs. Lai and Warheit argue that short-term in vivo (animal) studies are needed. Dr. Burello adds an important perspective on mathematical modeling of quantitative structure–activity relationships (QSARs) for nanomaterials, pointing toward a predictive nanotoxicology. Finally, Dr. Riviere explores the use of physiologically based nanomaterial pharmacokinetic models, or PBNPKs, with which to describe nanomaterial distribution and fate in vivo.

Our immune system serves as the first line of defense against foreign intrusion, and it is therefore of key importance to understand nanomaterial interactions with the immune system, not only from a toxicological point of view, but also if we are to develop nanocarriers for targeted drug delivery or imaging. Three chapters are devoted to immune interactions of nanomaterials: Dr. Moghimi et al. discuss factors that regulate nanomaterial interactions with the innate and adaptive immune system, leading to immunostimulation or immunosuppression, while Dr. Szebeni focuses on complement activation by nanomaterials. Dr. Kostarelos et al. discuss a special case of immune cell interactions with nanomaterials, namely, the biodegradation of carbon-based nanomaterials by enzymes expressed in innate immune cells (or in plants).

Next, we find a comprehensive chapter devoted to genotoxicity and carcinogenicity of nanomaterials (Dr. Woei Ng et al.) and a series of chapters on nanomaterial toxicity affecting specific organs, including chapters on pulmonary and cardiovascular toxicity (Drs. Cassee and Castranova), neurotoxicity (Drs. Sharma and Sharma), dermatotoxicity (Drs. Monteiro-Riviere and Riviere), and reproductive toxicity (Dr. Saunders et al.). The chapter on pulmonary and cardiovascular toxicity focuses on two commercially relevant nanomaterials, titanium dioxide and carbon nanotubes, and on the inhalation route of exposure of particular relevance for occupational exposure. These findings may nevertheless inform us on mechanisms of relevance for nanomedicine. Similarly, the chapter on neurotoxicity takes as its starting point accidental exposure to various types of nanoparticles, but the authors add an exciting perspective on the use of nanomaterials for neuroprotection. The chapter on dermal effects of nanoparticles offers an overview of current literature, and the discussion is of equal relevance from pharmacological (i.e., topical application of drugs, vaccines) and toxicological points of view. The potential for nanoparticles to exert adverse effects on the male or female reproductive systems remains poorly understood, but this is of particular importance not only to understand occupational/environmental exposure but also in the context of the deliberate administration of nanoscale objects in patients.

Finally, a perspective on ethical aspects of nanomedicine is provided. Here, Dr. Kuiken argues that there may be nothing new in terms of the ethical questions that arise as we are confronted with nanomedicines; the question is how much risk we are willing to accept with a new technology before it is proven effective and "safe." This will become even more evident as personalized medicine is enabled, in part, through nanomedicine. This, then, brings us full circle: medicine, and nanomedicine, is essentially the art and science of risk-benefit assessment. Nanotoxicology provides the tools to deal with the "risk."

The book closes with a personal view of the future of (nano) medicine, written by Dr. Hunziker, president of the European Society of Nanomedicine (ESNAM).

I wish to thank the authors who contributed their valuable time and expertise toward the preparation of this book. I hope that the present volume will serve as a useful manual for students and scientists interested in the safe development of nanomedicines.

Bengt Fadeel

Stockholm, July 2014

Chapter 1

Nanomedicine(s) and Their Regulation

Ruth Duncan
Polymer Therapeutics Laboratory, Centro de Investigacion Príncipe Felipe, Valencia, Spain
profruthduncan@btinternet.com

1.1 Background: A Decade of "Nano"; Where Are We Now?

The historical development of, and emerging opportunities for, nanopharmaceuticals and nanoimaging agents, that is, "nanomedicine(s)," has been reviewed in depth elsewhere [1]. The question of safety is critically important. We pointed out that "some seem unaware of the historical background and in their enthusiasm over exaggerate (hype) the potential benefits, giving the impression that their proposed new technologies are already reality today. Others express disproportionate concern as to the possible risks. Objective nanomedicinista (nanomedicine researchers) understand that the truth lies firmly between these two extremes and that the risk-benefit-promise of each technology should be clearly presented on its own merits" [1]. A balanced discussion of all the arguments is critically important in the context of nanomedicine development

Handbook of Safety Assessment of Nanomaterials: From Toxicological Testing to Personalized Medicine
Edited by Bengt Fadeel

ISBN 978-981-4463-36-2 (Hardcover), 978-981-4463-37-9 (eBook)
www.panstanford.com

and safety to ensure speedy, and safe, realization of healthcare benefits. To understand how studies on nanomaterial toxicology might relate to nanomedicine safety it is important to consider (i) the current regulatory framework (legally binding requirements and guidelines) in place for all medicine evaluations [2], (ii) the scientific basis (quality safety and efficacy) and experience gained during the transfer of first-generation nanomedicines into routine clinical use and the clinical adverse effects already documented for these products, and (iii) the likely relevance of in vitro and in vivo nanomaterial safety studies to specifically designed nanomedicine products, considering that administration will be as a sterile pharmaceutical formulation, given via a specific route of administration, using a particular dose and dosing schedule. The goal of nanomedicine is improved diagnosis and treatment of life-threatening and debilitating diseases for which there are currently no adequate therapies. No medicinal product is free of risk. The remit of regulatory agencies is to ensure medicinal products exhibit an acceptable risk versus benefit (Fig. 1.1a). With these issues in mind, this chapter reviews the background, some of the lessons learned during transfer of first-generation nanomedicines into the clinic, and areas where interdisciplinary collaboration between nanomedicinista and nanotoxicologists might aid the early identification of potential risks of those nanomaterials currently being proposed for use in nanomedicines. Nanomaterial toxicity testing is dealt with in the following chapters in this volume.

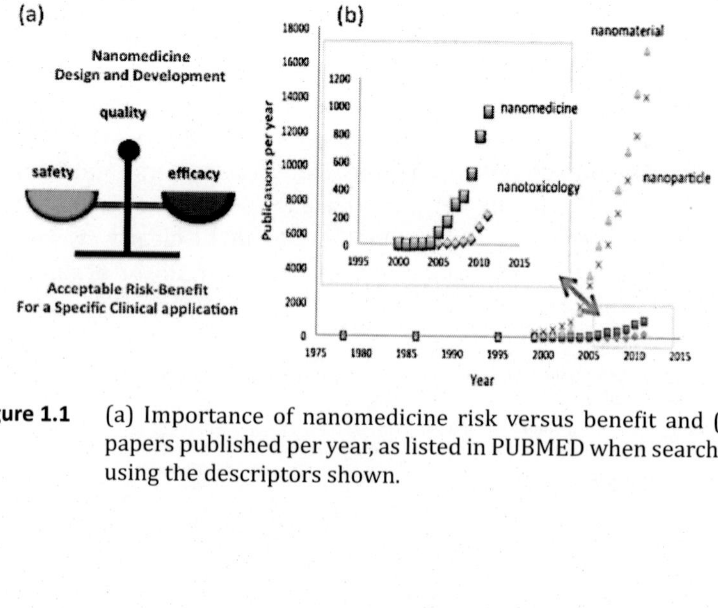

Figure 1.1 (a) Importance of nanomedicine risk versus benefit and (b) papers published per year, as listed in PUBMED when searched using the descriptors shown.

1.1.1 Convergence of Scientific Disciplines: Old Ideas, New Terminology?

The last decade has witnessed the convergence of many diverse scientific disciplines keen to exploit advances in nanoscience to develop novel nanotechnologies for medical applications, that is, for "nanomedicine" [1,3]. In parallel, there has been a growing realization that "ultrafine" nanoparticles in the environment and/or the workplace are particularly hazardous during prolonged accidental human exposure [4]. This fact, together with the increasing use of nanomaterials in a broad range of applications (from construction to aerospace materials, from environmental applications and electronic components to cosmetics and consumer products, etc.), has led to the birth of the field now termed "nanotoxicology" [4–6], which is overviewed elsewhere in this volume. Although the number of publications using the descriptors either "nanomedicine" or "nanotoxicology" has increased rapidly over the last decade (Fig. 1.1b), compared to the terms "nanoparticles" (first used in pharmaceutical sciences since the 1970s) and "nanomaterials" (first used in materials science since the 1980s) and "ultrafine particle toxicology" (used since the 1980s), the number of papers published is still significantly lower. It important to be aware of all current and historical literature relating to the toxicological properties of nanosized materials in relation to environmental, accidental human, and purposeful (nanomedicines) human exposure.

For the reader new to the area of "nanomedicine(s)" it is also important to point out that the overall field of nanomedicine encompasses many different potential healthcare applications of nanotechnologies:

(i) Diagnostics, devices/biosensors, and surgical tools usually *used outside the patient*

(ii) Diagnostics and theranostics (a combination of therapeutic and diagnostic functions) designed for *patient administration*

(iii) Nanostructured biomaterials (often combined with cell therapy) for in vitro or *in-patient* biomedical applications

(iv) Nanomedicines (sometimes called nanopharmaceuticals) *given to patients* by a variety of routes of administration (includes also diagnostics)

This chapter focuses on the safety issues relating to nanomedicines/nanopharmaceuticals. Nanomedicines include inherently active therapeutics, advanced drug delivery systems (e.g., for low-molecular-weight drugs, proteins, genes, aptamers, and small interfering RNA [siRNA]), and vaccine delivery systems (discussed in Ref. [1]). They are often hybrid multicomponent, nanosized structures, and there is growing ambition to combine diagnostic and therapeutic capacity in a single system, that is, to create what has been termed a nanotheranostic. All nanomedicines are carefully designed to exhibit therapeutic benefit against a particular disease (stage of the disease) in a particular patient population using a particular route for administration. They are carefully engineered nanosized constructs in the 1–1000 nm size range and can be made using either top-down or bottom-up manufacturing techniques (discussed at length in Ref. [1]).

Why the 1–1000 nm size range? The need for a scientifically justified, legally binding definition of a "nanomaterial" applicable for safety regulation across all sectors is evident. It has, however, quickly become apparent that it is difficult to find an overarching definition suitable to catch all nanomaterials (in relation to a specific size range having unique physicochemical/toxicology properties) or indeed could include all types of nanomedicines. Many scientific experts have acknowledged that none of the popular size demarcation thresholds (often 1–100 nm) are scientifically applicable across "all" nanomaterials [7]. Moreover it is agreed by many that "catch-all" size descriptors are more likely to create regulatory gaps rather than close them [8, 9]—a dangerous game in the context of nanomedicines. Although a large number of learned societies, governmental agencies, and also the International Organization for Standardization (ISO) have invested much committee time elaborating libraries of "nanodefinitions" (e.g., [10]), it is wise to embrace them with caution, as many proposed definitions are still viewed as controversial or simply wrong. The terminology used in this chapter is defined in Box 1.1.

1.1.2 Medicine Regulation: Evolution, Not Revolution

The last 60 years have seen ~1,200 new medicines enter clinical use [13]. The majority have been low-molecular-weight chemical entities,

Box 1.1 Terminology

†**The field of nanomedicine** uses nanosized tools for the diagnosis, prevention, and treatment of disease and to gain an increased understanding of the complex underlying pathophysiology of diseases.

†**Nanopharmaceuticals/nanomedicine(s)**: The terms are used interchangeably. They can be developed either as drug delivery systems or as biologically active drug products, and they usually consist of at least two components, one of which is the biologically active ingredient. The term encompasses nanometer-size-scale (1–1000 nm) complex systems. Biotechnology products developed as drugs (e.g., proteins, antibodies) would not be included per se, but the term does include nanosized drug crystal technology engineered to increase drug solubility/bioavailability. Nanomedicines are typically multicomponent nanosized systems.

††**A nanomaterial** is a material with at least one dimension in the nanoscale (usually viewed as between 1 nm and 100 nm). Nanomaterials can be particulate in nature (e.g., titanium dioxide, iron oxide), fiber-like, (e.g., carbon nanotubes), or sheet-like (e.g., graphene). Nanomaterials can also be defined in terms of their functionality, as opposed to relying strictly on their size alone.

††**Nanotechnologies** are technologies that either incorporate or employ nanomaterials.

†after [11]; ††after [12]

but over recent decades there has been a rising number of protein therapeutics (now >100) and a growing number of nanomedicines (>40, discussed in detail in Ref. [1]). Modern regulatory practice has continually evolved to address emerging needs as, and when, they have arisen, for example, biotechnology products, gene therapies, cell therapies, and other advanced medicinal products [14–16]. The sulfanilamide tragedy in the United States in the 1930s [14] and the thalidomide tragedy in Europe in the late 1950s [15] highlighted early on the need to understand the fundamental scientific basis of a drug's mechanism of action (safety and efficacy) and physicochemical characteristics controlling the therapeutic index in order to assess risk versus benefit. Proactive risk management is essential.

Although a detailed description of the medicine regulatory process is beyond the scope of this article, it is important to have some knowledge of the current practice applied to all new medicines [2] if one is to consider future additional studies that may be needed to establish the toxicity profile of emerging nanomedicines. Regulatory evaluation includes an integrated assessment of quality (see Section 1.3 for a discussion of "quality"), safety, and efficacy in order to estimate risk versus benefit in relation to the proposed clinical use, that is, target disease, target population, route, frequency of administration, etc. This approach was used to safely introduce first-generation nanomedicines into routine clinical practice (see Section 1.1.3). Typical regulatory considerations applied to all medicinal products are listed in Box 1.2.

As explained by the UK Medicines and Health Regulatory Agency (MHRA), "Decisions will be based on good science and robust methodology. Judgements on safety, quality and performance will be informed by all available, relevant and reliable evidence, with the burden of proof often resting on companies. A cautious approach to a decision will be needed where there is scientific uncertainty about the existence or extent of a risk but reasonable grounds for anticipating the possibility of severe adverse effects" [16]. Important regulatory checkpoints for any novel medicine include the moment at which permission is given to enter into first-in-man studies, the time of market authorization, and not least review of information captured by postmarket surveillance.

At all stages of preclinical and clinical testing it is essential to use well-characterized materials if the results obtained are to be meaningful. Scientific facts/advances are always used to guide the evolution of the medicines regulatory framework. Indeed the recent introduction of European Medicines Agency (EMA) guidance relating to the development of pediatric and advanced therapies medicinal products (ATMPs) has been suggested as a model to pave the way for a regulatory approach in nanomedicines in the European Union (EU) [17]. Over the years a question frequently asked has been, Is there always timely introduction of new concepts [14]? Indeed, the robustness of proactive risk management for nanomedicine regulation is widely debated. It is fair to say that opinions differ. Some categorically state that "regulatory guidance in this area

Box 1.2 Typical regulatory considerations from active substance to medicinal products

- Manufacture and characterization of the active "drug substance"
  - characterization of the active substance
  - impurities
  - definition of an acceptable specification definable by validated techniques (in relation to safety/efficacy profile, route of administration, and proposed clinical use)
- Nonclinical pharmacology and pharmacokinetics
- Nonclinical safety
  - Good laboratory practice (GLP) toxicology evaluation, single dose, repeated dose
  - toxicokinetics
  - genotoxicity
  - carcinogenicity
  - reproductive and developmental toxicology
  - local toxicity
  - immunotoxcity
  - hematoxicity
- Pharmaceutical development of the "drug product"
  - excipients
  - containers
  - sterilization
  - characterization and determination of impurities
  - stability/shelf life
- Packaging
- Clinical development
  - dose for first-in-man studies
  - clinical trial design (including patient population)
  - drug interactions
  - clinical pharmacokinetics/pharmacology
  - fixed combination medicinal products
- Environmental risk assessment
- Pharmacovigilance
- Pharmacoeconomics

is generally lacking" (e.g., [18]); others suggest that the "current dynamic regulatory structure maybe robust enough to adapt to the technological challenges posed by nanomedicines" but that it is not evolving fast enough (e.g., [19]). Balanced discussion requires understanding of the regulatory process in place today. This was used guide the introduction of 1st-generation nanomedicines into routine human use, with the first market entry being more than two decades ago (see Section 1.1.3), coupled to the vision to objectively highlight potential "regulatory gaps" that may need action, if safe introduction of 2nd-generation nanomedicines is to be achieved [20, 21].

Another question often asked is, Is there a need for nanomedicine-specific regulation? The general consensus is no. Nanomedicines should continue to be assessed using the existing guidelines because (i) this process has a successful track record for first-generation nanomedicines and (ii) it entails thorough product-specific evaluation in the context of its proposed clinical use (Box 1.2). However, aware of the need to be vigilant, the US Food and Drug Administration (FDA) has published draft guidance documents relating to nanomaterials and cosmetics and food [22]. The EMA [23] and the MHRA [24] have also published position papers on nanomedicine and nanotechnology regulation. All parties agree that it is essential to avoid complacency and that regular review is needed to identify potential regulatory gaps as new nanomaterials/nanomedicines emerge in order to ensure that these can be swiftly addressed and closed. Early international harmonization of the regulatory landscape would be of benefit to all. The following are particularly important:

(i) Keeping abreast of all advances in nanoscience to enable assessment of the potentially unique risks of each specific novel nanomaterial/nanomedicine as it arises and to enable development of a robust methodology for assessment of toxicity during preclinical development (see Section 1.3.).

(ii) Ensuring that nanomedicines are subjected to regulatory evaluation as "medicines" and not as "nanodevices." The latter are governed by alterative regulatory paths, for example, the FDA's Center for Devices and Radiological Health (CDRH) and EC Medical Devices Directives. A CE marking should not be seen as a quick route to product approval or as a means to bypass more stringent clinical safety assessment. It is also important to ascertain which nanomedicines/theranostics should be considered as combination products. Complex hybrid products are most likely to fall within a "regulatory gap."

(iii) Ensuring rigorous postmarket surveillance, that is, "pharmacovigilance," to quickly identify any problems that may emerge once a nanomedicine is widely used in the general population. Global use may expose the technology to diverse patient cohorts of different age, ethnic, environmental, and genetic backgrounds.

The terminology widely used in relation to nanomedicine toxicity can also be confusing. We have mentioned before [25] that overenthusiastic researchers often make erroneous statements when describing the potential "toxicity" of their latest invention. For example, "it is nontoxic," "it is biocompatible," or it is GRAS (FDA term: *generally recognized as safe*)! The phrases are frequently used without any further qualification, even though the terms "toxicity," "biocompatibility," and "GRAS" can only be applied to substances of precisely defined product specification and in the context of the intended conditions of use (e.g., route of administration, dose, frequency of dosing). Moreover, should nanomaterials/nanomedicines be discussed in terms of their "biocompatibility or toxicity"? What do the words really mean? The pharmaceutical industry refers to a drug's adverse effects in terms of its "toxicity." This is a measure of the nonspecific, unwanted harm that it may elicit toward cells, organs, or indeed the patient as a multiorgan system (Fig. 1.2). "Biocompatibility," on the other hand, was introduced by the field of implantable biomedical materials. This term was defined many years ago (1986) by consensus at a conference of the European Society for Biomaterials [26] as "the ability of a material to perform with an appropriate host response in a specific application." This definition underlines the potential detrimental effect of a material in the body (i.e., toxicity) but also highlights the potential benefits of material biointeraction (interaction with proteins, cells, and tissues at the biointerface) in order to improve performance (Fig. 1.2). This definition also emphasizes material "biocompatibility" in the precise context of its use. "Toxicity" rather than "biocompatibility" is more appropriate in relation to nanomedicines.

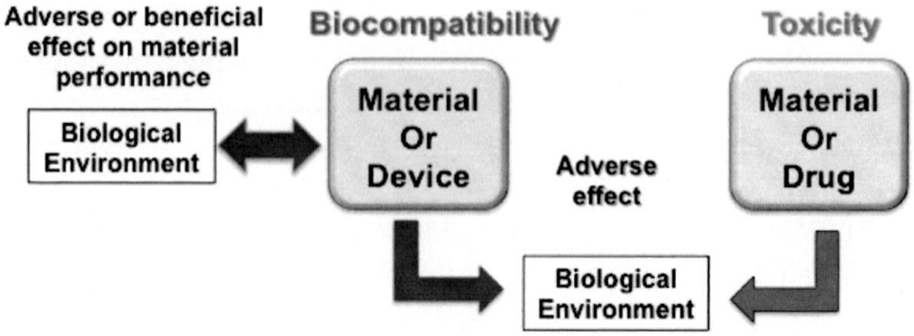

Figure 1.2 Distinction between biocompatibility and toxicity.

1.1.3 Lessons Learned from >40 Years of Clinical Experience with Nanomedicines

1.1.3.1 Products in routine use and clinical development

Nanomedicines are not new. The first advanced drug delivery system (DDS), now discussed as first-generation nanomedicines, entered clinical development more than three decades ago, and they make an important contribution to the modern therapeutic arsenal [1]. (DDS is a long-established term and was used by the pharmaceutical industry and pharmaceutical scientists before "nanomedicines" became a popular descriptor.) The goals for design continue to be improved transport of drugs across biological barriers, for example, transport across the blood–brain barrier (BBB), improved oral bioavailability of poorly soluble drugs, provision of more convenient routes of administration (e.g., oral, transdermal, or pulmonary medicines), improved disease-specific targeting, and/or controlled release. Some technologies are designed to improve drug targeting while at the same time controlling the release of the bioactive substance such that it remains in its therapeutic window for the required duration of therapy and to maximize therapeutic efficacy and/or patient convenience. In the 1980s, advances in the biotech industry began to highlight the urgent need for vectors able to protect emerging "molecular medicines" (especially genes, antisense, etc.) during transit and promote their efficient delivery to the desired intracellular compartment on arrival at target cells. There has been a recent resurgence of interest in such vectors due to the need to promote cytosolic delivery of siRNA and also for intracellular organelle-specific targeting, but it is fair to say that realization of these technologies to marketed products has so far eluded those who have tried.

The main classes of first-generation nanomedicines are shown schematically in Fig. 1.3. Typical product examples are listed in Table 1.1, with a comprehensive listing of all products in Ref. [1]. They include the iron nanoparticle–based products first introduced in the 1930s and used as treatments for anemia and more recently for patient imaging. Products commercialized since the 1990s include liposomal and lipidic complexes, antibody-based drug conjugates and radiopharmaceuticals, and various polymer therapeutics, including protein conjugates. Over the last decade drug a growing number of nanocrystal technology-based products entered routine clinical use as a means of enabling commercialization of poorly soluble drugs. The first true "nanoparticle" (Abraxane) product also entered routine clinical use.

The tendency to call all nanomedicines "nanoparticles" is unfortunate. In many cases this is neither scientifically accurate (e.g., for dendrimers, which are macromolecules/polymers), nor is it helpful as it overlooks the unique features of each class of nanomedicine, for example, liposomes are lipidic vesicles in various forms, not nanoparticles. Most are not round, and they may have a well-structured (hard) surface or a dynamic, hydrated, or mobile polymeric (soft) surface. Cartoons of nanomedicine structure do little justice to the temporal mobility and heterogeneity that interfaces with the biological environment. Each individual nanosized construct in a batch can have different dimensions and a different surface composition. Given the average composition, each product in each class displays typical size ranges in the nanoscale (Table 1.1). The critical design features (often called critical product attributes) that govern pharmacokinetics, safety, and efficacy are carefully optimized (one may use the term "engineered") during lead candidate design, often followed by subsequent refinement after first clinical testing. Several of the first-generation nanomedicinal products are nearing the end of their patent life, so "follow-on" (cf generic) versions are already being proposed for the market. This has led regulatory agencies to reflect carefully on the procedures needed to ensure comparability of identity and performance (both safety and efficacy) of such complex follow-on products when compared to an originator product [27, 28].

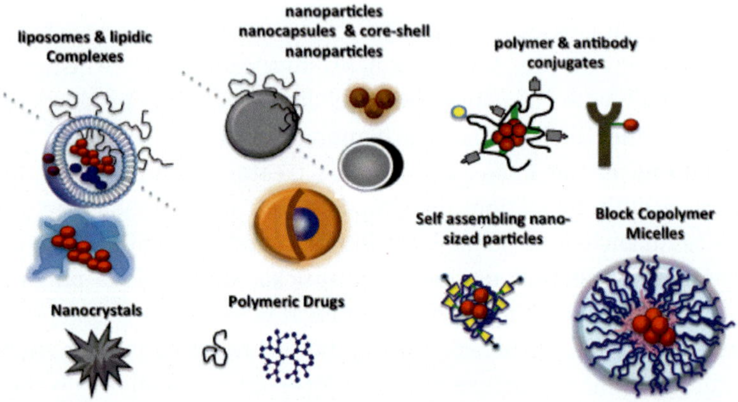

Figure 1.3 Schematic showing classes of first-generation nanomedicines in routine clinical use or clinical development. The possibility for liposomes and nanoparticles to have a covalent coating is shown and for liposomal products to contain two drugs (blue/red) also indicated. Polymer–drug conjugates containing an imaging agent as well as targeting residues have also been tested clinically. (Not to scale for size ranges see Table 1.1.)

Review of the growing historical preclinical and clinical database for these first-generation products is enabling a greater appreciation of species differences in the preclinical pharmacokinetics and mechanisms of toxicity observed [29]. This is also helping to identify those preclinical models that give the best in vitro–in vivo–clinical correlation in respect of biological behavior. Many nanomedicines include coatings. They may be noncovalently bound, for example, iron colloids, or covalently bound, for example, polyethyleneglyocol (PEG)-coated liposomes. Others include rationally designed linking chemistry that can be used to impart stability (PEG proteins) and/or facilitate site- and rate-specific triggered drug release (e.g., antibody- and polymer–drug conjugates) (reviewed in Ref. [1]). Thus interpretation of any toxicity seen for a nanomedicinal product can be particularly complex. It is essential to consider the dose of all components (not just the labeled bioactive), the stability of the entire construct in the biological milieu relative to the route of

Table 1.1 Classes of nanomedicines

	Class	Typical size (nm)	Typical admin. route	Use	Product examples
Products approved for the market	Coated iron colloidals	10–100	i.v.	Treatment of anaemia MRI imaging agents	Feridex, Venofer
	Liposomes and lipidic drug carriers	60–80 nm–>1 μM aggregates	i.v.	Anticancer therapies treatment of opportunist infections	Doxil/Caekyx Ambisome, Ablecet
	Drug nanocrystals	50–1000	oral	Immunosuppressive anti emetic	Rapamune, Emend
	Antibody targeted radiotherapy	<20	i.v.	Cancer	Zevalin, Bexxar
	Antibody-drug conjugate	<20	i.v.	Anticancer therapy	Adcetris, Kadcyla
	Nanoparticles		i.v.	Anticancer therapy	Abraxane
	Polymer therapeutics • polymeric drugs and sequestrants	NA	i.m. oral	Multiple sclerosis renal failure, diabetes	Copaxone Renagel, Welchol
	• polymer protein conjugates	10–20		Cancer, infectious diseases	Neulasta, PEGasys

(*Continued*)

Table 1.1 (*Continued*)

	Class	Typical size (nm)	Typical admin. route	Use	Product examples
Products in clinical development	Polymer therapeutics • polymer-drug conjugates	5–20	i.v. oral	Anticancer therapies	Opaxio, Prolindac NK150, SP1049
	• block copolymer micelles	20–100	i.v.	Anticancer therapies	
	• self assembled polymer conjugate nanoparticles	70	i.v.	Anticancer siRNA delivery	CALAA01
	Liposomal combination therapy		i.v.	Anticancer therapy	CPX-351
	Hollow gold nanoparticles	150	i.v.	Anticancer therapy	Aurolase
	Silica nanoparticles		i.v.	Anticancer therapies	
	Iron oxide nanoparticles	15	†i.t.	Anticancer thermal ablation	Nanotherm
	Silver nanoparticles	1–100	topical	Anticancer/ dermatitis	Nucryst
	Hafnium oxide crystals		†i.t.	Tumour radiotherapy	NBTXR3

†intra-tumoural; i.t.

administration and the pharmacokinetics, and the mechanism and route of elimination of all components.

1.1.3.2 Clinically documented adverse reactions

While there has been a recent call to "pursue and publish clinical trials in nanomedicine" with the observation that "many individuals who specialise in material science, analytical chemistry, or physics have not had prior experience with the regulatory process for human therapeutics" [30]. These comments underline the need to connect all the disciplines working on nanomedicines lab to clinic. Incomers to the field must quickly become familiar with the relevant background literature, even if it is in pharmaceutical sciences or clinical pharmacology journals. Although academic nanomedicine papers rarely cite these benchmarks, there are many primary publications describing the outcomes of clinical trials (successes and failures), GLP preclinical toxicology testing, and the techniques used for validated characterization and/or formulation development. There are also a growing number of publications describing the results of pharmacovigilance studies and pharmacoenconomic evaluation.

For the newer nanomedicines currently in good clinical practice (GCP)-regulated clinical trials, the clinical trial design and its goals are freely accessible (this is an ethical requirement) through clinical trial databases (e.g., [31, 32]). In addition ongoing clinical trial results are routinely presented as abstracts at clinical pharmacology meetings related to the disease setting under study.

The product information guides available for first-generation nanomedicines provide a rich source of information regarding clinically documented toxicities for a specific product, together with their frequency of occurrence in the overall patient population receiving that medicine. Interpretation of the toxicity observed, and the underlying mechanisms, in terms of "nanotoxicology" is, however, very complex. The toxicological profile reflects the biological behavior of the nanomedicinal product in its entirety. Toxicity may be (sometimes strongly) influenced by the nature of any bioactive substance present as a payload (e.g., a cytotoxic drug, a protein, or siRNA), but excipients in the formulation and the formulation characteristics can also play a major role. The pharmacokinetics, body distribution, nanomedicine stability, and the metabolic fate of all components impact the overall toxicological profile seen. Certain adverse reactions seen clinically for first-generation nanomedicines may be generally relevant to the design and use of future nanomedicines, and examples include the following:

- Infusion reactions: These have been documented for a number of nanomedicinal products administered intravenously, for example, for PEGylated liposomes [33], iron infusion solutions [34], and certain antitumor nanoparticles [35] and polymer–drug conjugates [36]. The phenomenon can often be mitigated clinically by slowing the infusion rate and/or by patient premedication, making the risk versus benefit acceptable. It is important to note that not all patients exhibit such reactions, for example, in the case of Doxil the incidence is $<10\%$; therefore $>90\%$ of patients do not show any reaction.
- Hypersensitivity reactions: Severe hypersensitivity reactions can be life threatening and can (albeit rarely) occur during intravenous administration of nanomedicines, for example, iron oxide colloidal solutions (incidence relates to the nature of the coating material [37]), liposomal products [33], and antitumor nanoparticles [38]. Monoclonal antibodies and immunoconjugates also show these effects (reviewed in [39]). Of course allergic reactions are not unique to nanomedicines, for example, antibiotic allergy. Each product in any particular class must be considered independently, for example, for iron infusion solutions the early, high-molecular-weight dextran-coated iron preparations carried a greater risk of anaphylaxis than the more recently introduced products. Low-molecular-weight dextran (Cosmofer) and iron sucrose (Venofer) have a significantly lower risk and do not require a patient test dose before therapy begins. Even so, the mechanisms underlying the acute and the long-term safety of the newer iron preparations are still not well established.
- Oxidative stress: The possibility that nanomaterials induce oxidation stress is currently widely discussed. Clinically used iron solutions have the potential to induce acute and chronic oxidative stress, kidney injury, and accelerated atherosclerosis. Released iron can be redox active, and it has been suggested that intravenous iron sucrose may induce

acute as well as chronic toxicity, as oxidative stress has been reported at a time much earlier than peak concentration of catalytically active iron, so a direct effect of iron sucrose has been proposed [40]. However, the nature of the coating can also play a role as it influences colloidal stability as well as the body distribution and intracellular fate. Physiologically iron transport is carefully regulated via a highly specific set of transport proteins, for example, after internalization transferrin routes iron into the cytoplasm via the endosomal compartment minimizing lysosomal exposure. In contrast, exposure of cells to nanosized iron colloid leads to trafficking to lysosomes and exposure iron to acidic pH and potentially copper released during autophagy-mediated protein turnover—a highly reactive environment that will have significant implications for the toxicological profile seen.

- Altered body distribution of the active substance due to a nanocarrier: The profile of dose-limiting toxicity seen for a drug given in the form of a nanomedicine can be totally different from that of the parent drug. This is not unexpected as their pharmacokinetics, body distribution, and, as mentioned above, even subcellular distribution can be (and is often designed to be) very different. Altered body distribution is often desired for improved therapeutic efficacy (e.g., macrophage targeting for amphotericin B delivery, hepatocyte targeting for treatment of hepatocellular carcinoma; discussed in Ref. [1]). However, therapeutic agents may be unintentionally directed to sites that they would not normally localize to, causing toxicities not normally associated with their use. Good examples include the dose-limiting "hand and foot syndrome" (or Palmar–Plantar erythrodysesthesia [PPE]) reported for Doxil [33, 41] and in certain cases for Abraxane [38], which arises as a result of their long circulation and deposition in the peripheral tissues due to leaky capillary vasculature, exposing these sites to highly potent, cytotoxic drugs. The failure of an *N*-(2-hydroxypropyl)methacrylamide (HPMA)–copolymer conjugate camptothecin (PNU 166148) in early clinical trials due to dose-limiting cumulative bladder toxicity [42, 43] is another good example. These conjugates exhibit a very high degree of urinary excretion (clinically 50%–90% at 24 hours), and this, coupled with the fact that the polymer–drug linker was urine labile, led to bladder toxicity exacerbated by repeated dosing.

- Other toxicities: There is now significant clinical experience with the use of monoclonal antibody therapeutics. Even humanized antibodies can cause a variety of side effects, in addition to anaphylaxis, immunotoxcity, and serum sickness, the nature of which depends on the specific target for the antibody therapeutic. These include infections, cancer, autoimmune disease, and organ-specific adverse events such as cardiotoxicity (reviewed in Ref. [39]). Therapeutic proteins can exhibit other adverse effects, for example, growth factors. Platelet-derived growth factor (PDGF) has been applied topically for the treatment of neuropathic diabetic foot ulcers [44], and bone regeneration [45], and the bone morphogenetic proteins (BMPs) [46, 47] used to promote bone regrowth. However, the rhPDGF-containing product Regranex caused increased risk of death from cancer in patients treated with three or more tubes and has recently been withdrawn in Europe [48]. For an rhBMP, life-threatening complications associated with use in the cervical spine and the potential of unwanted bone growth, also referred to as ectopic bone growth, have been widely reported [46, 47].

What, you might ask, do the toxicological effects of antibodies and protein therapeutics have to do with nanomedicines? Many papers enthusiastically proposing next-generation nanomedicines show cartoons representing a multicomponent nanocarrier with targeting residues (maybe an antibody) and imaging agents decorating the surface, one or more "drug molecules" (maybe a protein) either attached to the surface or entrapped within the nanomatrix, and often the added benefit of some nanomaterial in the core to enable external activation or triggering of disassembly (e.g., [49]). Such reviews rarely acknowledge the known toxicological considerations relating to all these components, even if safety of a "nanocarrier" material is alluded to. Decorating a nanocarrier with drugs or targeting ligands followed by systemic administration is more than likely to increase the probability of infusion reactions and/or immunotoxcity—hapten effects are rarely discussed.

The examples selected above serve only to illustrate the need to consider a nanomedicine in its entirety. It is important to learn from past clinical observations relevant to all the components of a nanomedicine when assembling novel second-generation constructs. In addition, the above examples highlight the need to consider the proposed route of administration and clinical setting. Topical application of Regranex for foot ulcer treatment [44] required consideration as to whether the patient also had cancer due to systemic availability via the wound—the damaged skin barrier. It is foolish to declare that any particular nanomaterial/nanocarrier is "safe or unsafe" per se. As emphasized, before safety relates to route of administration (e.g., intravenous, intramuscular, topical, oral, pulmonary), dose, frequency of use, and clinical setting are very much influenced by the properties of the bioactive drug payload, construct stability, and all pharmacokinetic–pharmacodynamic relationships.

1.2 Emerging Nanotechnologies: New Medicines or Nice Publications?

There are two main categories of emerging nanotechnologies being proposed for or currently undergoing clinical development. There are those building on successes in the classes of first-generation nanomedicines (Table 1.1, Fig. 1.3), and they include liposomes designed for site-specific activation [50], containing combination chemotherapy [51] or using surface ligands for disease-specific targeting [52]). In addition there are block copolymer micelles [53], albumin nanocarriers given in combination therapy [54], synthetic or natural polymer-based nanoparticles [55], iron oxide nanoparticles [56], and an array of antibody–drug [57], polymer–protein [58], polymer–drug [59, 60], and polymer–aptamer conjugates [61] (all reviewed in Ref. [1]). Indeed there are ~70 nanomedicinal products in clinical development in cancer alone [62], and their clinical safety and efficacy status are already beginning to emerge.

The second category includes (also see Ref. [1] for a lengthy discussion) putative nanomedicines based on novel nanomaterials: fullerenes, carbon nanotubes, inorganic nanoparticles (e.g., gold [including gold nanorods], silver, quantum dots, silica, and core-shell silica-based nanoparticles), natural and synthetic polymer-based nanoparticles, nanocapsules, block copolymer micelles, and other self-assembling nanostructures, and complex nanoarchitectures manufactured via new routes by innovative methods for example, template manufacturing. Very few are near to realizing a clinical candidate. The current understanding of the potential toxicological properties of many of these new nanomaterials has been discussed (e.g., [63–65]). In reality, the majority of studies that claim a new "nanomedicine," simply describe interesting, albeit important, advances in nanoscience. Few of these technologies will ever reach clinical evaluation due to fundamental toxicological concerns, inability to achieve reproducible scale-up manufacture (this has to be practical and cost effective), and/or full characterization of the product using techniques that can be validated to the level of accuracy needed to ensure pharmaceutical reproducibility batch by batch.

Dendrimers provide a good illustration of the challenges. This chemistry provides an elegant approach to polymer synthesis with unique possibility of controlling 3D architecture while tuning surface features (via terminal group modification). Although dendrimers have been heavily promoted for biomedical applications since the late 1980s, it soon became evident that many of the proposed chemistries were unsuitable for the applications for which they were being proposed [66]. More than two decades on, only two dendrimers have entered clinical development, Gadomer as an magnetic resonance imaging (MRI) agent (no longer being pursued clinically [67]) and VivaGel, a lysine-based dendrimer being developed as a preventative of HIV infection when administered vaginally as a topical gel in phase II/III clinical trial [68]. It has been noted that only one in 500,000 proposed drugs ever make it to market [29] and that the vast majority of scientific articles we see never arrive to lead candidate for clinical testing. Moreover, approximately 95% of candidates that do enter clinical development actually fail (not only for reasons of safety [69, 70]). Advent of nanomedicines over the last decade has actually reduced the number of advanced drug delivery systems progressing to clinical evaluation; "over the last decade in Europe the translation of nanomedicines and drug delivery into useful therapeutics has been modest" [71]. No doubt all nanomaterials will find important commercial uses, but many will likely be applications outside the nanomedicines/nanopharmaceuticals arena, for example, applications in diagnostics, biosensors, surgical tools (use outside the patient), or alternatively improved laboratory reagents and tools to assist in definition of cellular mechanisms and pathophysiology in health and disease. Current efforts relating to nanotoxicology should be considered in this context.

It is important to note those innovative second-generation nanomedicines that have made it to early clinical development. For example, complex self-assembled nanoparticles based on cyclodextrin containing polymer conjugates for transferrin receptor targeted siRNA delivery [72], PEGylated gold nanoparticles carrying tumor necrosis factor (TNF) [73], and PEGylated poly(lactide-co-glycolide) nanoparticles [74.] in early clinical trials as anticancer therapeutics. Hafnium oxide nanoparticles are being developed as a new radiotherapy treatment for cancer—interestingly being proposed as a nanodevice in Europe but are viewed as a (nano) medicine by the FDA for regulatory purposes in the United States [75]. PEGylated silica-gold nanoshells designed for localized near-infrared-mediated laser tumor ablation [76] and multimodal silica nanoparticles for tumor imaging and potentially treatment [77].

1.3 Nanomedicine Safety-Nanotoxicology: Lessons to Share

1.3.1 General Areas of Overlapping Interest

The disciplines of toxicology; absorption; pharmacokinetics; adsorption, distribution, metabolism, and excretion (ADME) (see Chapter 7 for a further discussion); and pharmacovigilance/epidemiology have a long and well-found history. Globally many hundreds of thousands of scientists are working in these areas daily. Moreover, in Europe regulatory staff scientists work with the input of ~4,500 scientific experts that daily support the medicines evaluation process. Last year (2011) ~200 papers were published that related to "nanotoxicology" (PUBMED search using this keyword) compared to >6,000 papers on "toxicology." Some argue that this underlines the need for greater funding of nanotoxicology research per se, but this is oversimplistic. There is certainly also an urgent need to ensure the results obtained in all disciplines are properly cross-referenced. There are many historical and current studies in the areas of environmental toxicology, medical geology, and accidental human exposure (i.e., in the workplace) that are also pertinent to nanomedicine development and use. For example, the following:

i. Regulatory requirements for assessment of environmental exposure. It has been suggested that there is a need for a greater understanding of the environmental impact of nanomedicine products, with a call to review the justification for the exposure limits below which the exemption of a medicinal product from environmental assessment can be made [78].

ii. Potential for laboratory researcher and/or manufacturer exposure. The issue of "nanosafety" in the workplace (academic laboratory or industry) is widely discussed. It is widely agreed that sensitive tools are needed for workplace monitoring and that there may be a need to scientifically justify exposure limits for new nanomaterials [79]. This discussion includes workers involved in the research and development of nanomedicines. However, it should be noted that there are already in place very strict requirements for risk assessment/containment for all European academic and commercial premises that work with medicines per se (e.g., manufacturing can involve very potent cytotoxic drugs). Sharing new data relating to novel nanomaterial toxicity that could impact on either accidental (workplace/environmental) or purposeful (nanomedicine) exposure is essential.

1.3.2 Manufacture, Characterization, and Formulation: Quality by Design

The challenges for characterization of nanomaterials and nanomedicines are often discussed. For any medicine, the chemical nature and stability, including aggregation and degradation, must be rigorously determined following manufacture, following formulation, and when present in different containers (rarely discussed in academic labs, but the composition of container materials, container caps, infusion kits, etc., can have a major impact on the physicochemical stability of a medicinal product and potentially dangerous impurities present), and during storage at various temperatures. Stability, aggregation, and biodegradation are also important to define following addition to solutions, for example, water for injection, tissue culture medium, and in solutions chosen to mimic the biological environment to be encountered following administration.

Results gained from in vitro and in vivo biological studies involving nanomaterial/nanomedicine samples are of little value if the samples are not properly characterized, account is not made of sample heterogeneity (polydispersity) and/or rate of degradation, and the manufacturing process is not sufficiently well controlled to enable batch-to-batch reproducibility. It is also ethically unjustifiable to screen poorly characterized materials in animal experiments. Over the last decade much effort has been invested in development of new tools for characterization of identity, nature of impurities, and sample heterogeneity in terms of size distribution and surface heterogeneity and for polymer samples and polymer coatings the molecular weight distribution [80]. Much can be learned from the methodology developed for the validated characterization of the first- and second-generation nanomedicines in this respect, for example, [80, 81].

The "quality by design (QbD)" approach now being promoted to assist the development of new medicines can also provide important insights regarding the critical product characteristics governing biological properties, including safety (see Ref. [82, 83] for a full explanation of QbD). QbD begins with predefined objectives (target product profile for patient use) and emphasizes the need to optimize product design and the process for manufacture that is guided by sound science and quality risk management The goal is to achieve a full understanding of how the product's physicochemical attributes and the process of manufacture/final formulation relate to the ultimate clinical performance—safety and efficacy. Given the emergence of innovator and follow-on nanomedicinal products [27, 28] there has been increasing interest in a QbD approach to manufacture and formulation. It is particularly important to understand how any variability in raw materials (this might include certain nanomaterials) can impact the critical product attributes of often complex, multicomponent novel medicines. For example, recent studies have taken a QbD approach to investigate liposomal formulations of tenofovir as a model hydrophilic drug [84] and superoxide dismutase (SOD) as a model protein [85], to investigate the effect of formulation variables of cyclosporine nanosuspensions [86] and to help with the development of Dermavir as a commercially viable vaccine product [87].

1.3.3 Definition of the Toxicity of a Nanomaterial/Nanomedicine

Bearing in mind the regulatory tests routinely needed for medicine development (see Ref. [2] and Box 1.2), it is evident that there are many overlaps with the issues addressed elsewhere in this volume, for example, genotoxicity, carcinogenicity, reproductive and developmental toxicology, local toxicity (e.g., for topical administration), immunotoxcity, and hematoxicity. In addition, prior to clinical studies, the pharmacokinetics, toxicokinetics, and metabolic fate of a nanomedicine must be defined. GLP in vivo toxicological evaluation using a single dose and a repeated dose protocol optimized to suit the product and its route, frequency, and duration of use are undertaken in species selected to provide the best preclinical-clinical prediction in relation to safety and to enable estimation of an appropriate starting dose for first-in-man studies. All such studies are carried out with the nanomedicine that the patient will be exposed to, *not* the component parts. Nanotoxicology evaluation of particular nanomaterials may be totally irrelevant (overestimate or underestimate) to the safety of the specific nanomedicinal product that contains it. When designing nanomedicines it is typical to first evaluate the biological behavior (including all aspects of safety and physicochemical properties) of all the proposed components and then systematically assemble them using a reiterative approach to design in the hope of ultimately arriving at a nanomedicine suitable to become a lead candidate for clinical trial. Each experimental step can be considered a stop/go point for further development. This approach has been discussed elsewhere in the context of polymer therapeutics [25] and is shown schematically in Fig. 1.4. Certain types of experiments may be performed both with the nanomaterial and the nanomedicine during the optimization process, and at any stage should the toxicological consequences seem to preclude further development, an alternative would be sought.

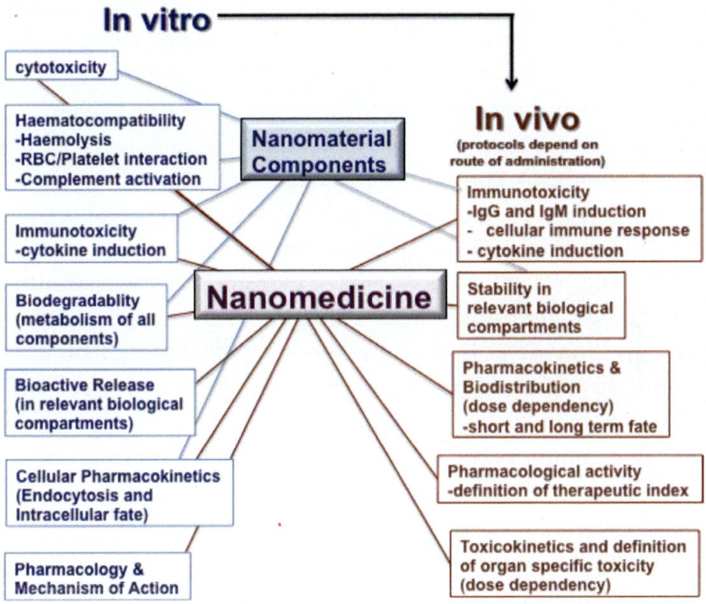

Figure 1.4 Typical panel of tests used to define the preliminary safety of a nanomaterial (blue) that is a candidate for incorporation into a nanomedicine, and the safety and efficacy of a nanomedicine (red). Toward a critical path for nanomedicine optimization (after [25]).

The results emerging from current nanotoxicology investigations have the potential to make important contributions at several stages in this process. They can help in the early identification of artifacts arising in routinely used assays due to the presence of a nanomaterial [88]; in the development of new, more sensitive tests; and in identification of species differences in pharmacokinetic/ toxicological response to nanomaterials. Documentation of nanomaterial toxicity can itself help to define whether a specific material may be more, or less, suitable for a particular nanomedicine application. Development of a suitable experimental protocol for the routine assessment of either a nanomedicine or a nanomaterial presents similar challenges, including definition of the dose range to be used, length of the incubation/test period, the culture medium (± serum) to be used for in vitro studies, the need to be aware of any sample interference with the assay readout, selection of appropriate controls, awareness as to the endotoxin content of samples, and also an appreciation of sample stability/biodegradation under the assay conditions used and over the time frame proposed. However, is worthy to emphasize that the most appropriate dose range and experimental time frame may be very different when trying to define the chronic toxicological effects of low-level exposure to a nanomaterial compared to that needed when assessing a nanomedicine to be administered at a specific dose, at a specific schedule of dosing, and via a defined route.

Given the toxicities observed for some of the parenterally administered first-generation nanomedicines (as discussed above), it is evident that there are particular fields of interest common to nanotoxicology and nanomedicine safety. These include hematotoxicity (including complement activation [89–91]), immunotoxicity, induction of inflammation, and the need to better understand mechanisms of cell damage, including triggering apoptosis and oxidative stress [92–94]. While the ability of nanomaterials and nanoparticles to induce oxidative stress after pulmonary administration is widely discussed [95], in parallel studies nanomedicines are being specifically developed to deliver nitroxide radicals with the goal of quenching reactive oxygen species to provide treatments for diseases such as cerebral and renal ischemia reperfusions, ulcerative colitis, and Alzheimer's disease [96, 97]. A better understanding of the basis of underlying interindividual variation in adverse responses to nanomaterials and nanomedicines is also a common goal in all settings [98].

1.3.4 Pharmacokinetics, Body Distribution, and Passage across Biological Barriers

The fact that "nanosized materials do behave differently to low-molecular-weight drugs, . . . that biological properties of nanomaterials are mainly dependent on relevant physiology and anatomy . . . and endocytosis, . . . and that this is likely to have an impact on potential toxicity" has already been discussed [99]. Almost all nanomedicines rely on a pharmacokinetically guided design to achieve their objectives of targeting the pharmacological target (organ, cell, subcellular compartment), prevention of drug access to sites of toxicity, improved transport across biological barriers, bioactive liberation at a desired rate to maximize therapeutic benefit, and ensuring of elimination of all components in an acceptable time frame to minimize the chance of cumulative toxicity after repeated administration. Thus in vivo pharmacokinetics, tissue distribution, and metabolism studies are an essential part of nanomedicine evaluation and design, and validated (preferably GLP) techniques are essential to monitor all components. Knowledge of residence time in specific compartments (e.g., blood, tissues) also helps to guide the time frame used for both safety and efficacy studies. Quantitation of both the percentage of the dose arriving in specific tissues/cells as well as the area under the curve for levels present over time is needed. When intravenous administration is intended, the rate of infusion (rate of dilution) is an important consideration when establishing protocols for in vitro hematoxicity studies.

Design of nanocarriers able to promote drug delivery (especially macromolecular delivery) across biological barriers including the gastrointestinal (GI) tract, the lung, the skin, and the BBB has long been a goal. However, the low rates of transfer achievable to date, for example, typically passage across the BBB is <1% of the dose [100], have prevented the realization of such dreams. However, concerns raised by nanotoxicologists as to the potential harm of chronic, low level exposure of tissues to nanosized materials traversing such biological barriers has emphasized the need to document carefully the potential fate of all nanocarriers. Dendrimers provide an interesting example. Certain chemistries have demonstrated the ability to traverse biological barriers at higher rates, for example, the GI tract [101], the vascular endothelium [102], and even the placenta [103], but absolute transfer in vivo is still very low. Nevertheless such observations may have important implications for acute and chronic toxicity in general, and regarding placental transfer there are obvious implications in relation to teratogenicity—an issue rarely discussed in the context of nanomaterial toxicology. Methods able to quantitate the passage of nanomaterials across biological barriers are vitally important both in relation to nanosafety and in relation to guiding the design of improved nanomedicines for barrier transfer. The possibility that nanomaterials may cause dermal effects and/or pass across the skin [104] is another topic of common interest. In this case transport rate will be related to the physicochemical properties of the nanomaterial under discussion as well as the barrier integrity. The latter may be compromised in acute and chronic wounds and/or by repeated topical administration of a pharmaceutical formulation. Interestingly it has recently been shown nanosized titanium dioxide present in sunscreen formulations does not appear to significantly penetrate the skin [105].

1.3.5 Endocytosis and Intracellular Trafficking

Last but not least, it is important to emphasize the importance of considering intracellular fate when defining the toxicological properties of nanomaterials and nanomedicines. Unlike many low-molecular-weight chemicals, which often rapidly pass across cell membranes and thus distribute randomly in the body, the cellular uptake of most nanomaterials/nanomedicines is limited to the mechanisms of endocytosis This includes both phagocytosis and pinocytosis depending on cell type and/or nanomaterial characteristics. This route of cellular internalization results in very different patterns of intracellular organelle localization (discussed in Ref. [106]). It has been suggested that certain nanomaterials, including carbon nanotubes, may be able to penetrate the cell membrane to access the cytosol [107]; however, other carbon nanotubes do not do this [108]. Only with robust analytical methodology and an understanding of concentration dependence (at not toxic concentrations), is it really possible to define mechanisms of cellular uptake. Those methods routinely used, and the challenges associated with gaining an accurate quantitation of "uptake," have recently been comprehensively described and discussed in Ref. [106]. The widespread use of fluorescence and other microscopy techniques can lead to artifacts and misguided conclusions. Others have also noted that you shouldn't simply believe all that you see [109]. The rates of uptake of nanosized materials differ depending on cell type (it also changes in disease), according to the physicochemical properties of the nanomaterial/nanomedicine, and according to size/shape, and they can, in addition, be influenced by the presence of specifically introduced targeting residues (nanomedicines) and/or the proteins that become absorbed during passage through extracellular fluids. It should also be remembered that endomembrane trafficking is a complex series of vesicle–vesicle fusion events with secretion (exocytosis) occurring in parallel with endocytosis and autophagy [110].

Whether or not a nanomaterial or nanomedicine is completely biodegradable and what its rate of degradation is are pivotal factors that, in concert with the whole body and subcellular distribution, determine which tissues, cells, and intracellular compartments are exposed to the nanomaterial/nanomedicine and for how long. (To note that the route of administration is important here, after oral administration the GI flora can play an important role in degradation of materials that mammalian enzymes cannot degrade.) Nanomedicines optimized for lysosomotropic drug delivery [106] are often designed to bypass mechanisms of resistance such plasma membrane–localized efflux pumps, to target intracellular pathogens (e.g., the leishmania parasite), or to localize to organelle-specific diseases (e.g., lysosomal storage diseases) (see Ref. [106] for a full discussion). Thus timely elimination of all components of the administered nanomedicine from intracellular vesicular compartments is crucially important. Nonbiodegradable, or slowly degrading, materials/polymers have the potential to create a "lysosomal storage" disease syndrome, as recently discussed in the context of PEG [106, 111], and they may interfere with normal lysosomal catabolic function and/or the endomembrane trafficking pathways, including secretion and autophagy [110, 112]. Any changes in normal cellular endomembrane homeostasis have the potential to induce acute or chronic toxicity. Moreover, the potential consequences of leakage of lysosomal enzymes and/or protons into the cytosol when using endosomolytic carriers are rarely discussed [106]. Advances in our understanding of endocytosis and intracellular trafficking in different cell types, changes in endocytic internalization and intracellular trafficking in disease, and also awareness of the potential impact of both clinically used and emerging nanomedicines on endomembrane trafficking are now opening the door to new methods for early evaluation of potential risks in this area.

1.4 Conclusions

Even though there are clearly overlapping interests in the areas of nanomedicine design, nanomedicine safety evaluation, and nanotoxicology, the notion that nanotoxicology and nanomedicine are "two sides of the same coin" [113] is oversimplistic due to the fact that many components make up a nanomedicine. The bioactive payload may be a primary determinant of toxicity and/or characteristics of the nanomedicine pharmaceutical formulation ultimately administered to a patient. For a nanomedicine the dose and route of administration are known. Ongoing research in nanotoxicology is more usually seeking to define the likely levels of human and/or environmental exposure to a specific nanomaterial and with this in mind determine the potential for subsequent adverse biological impact.

A recent statement that "interestingly pharmaceutical sciences are using nanoparticles to reduce toxicity and side effects of drugs and up to recently did not realise that carrier systems themselves may impose risks to the patient" [114] is unfair and actually wrong. The successful introduction of many first-generation nanomedicines (i.e., of acceptable risk versus benefit) is testament to the efforts of pharmaceutical scientists and those regulatory agencies appointed by society to ensure timely introduction of the emerging, innovative (safe) products that can bring improved healthcare. This is not, however, an argument for complacency. Many new, and extremely novel, advanced drug delivery systems and nanomaterials are emerging. During design and preclinical development the potential toxicity of all components must be carefully considered in respect to the drug payload (toxicological profile of any bioactive(s) being carried), the route of administration, and the dose and frequency of use. Continuing collaboration across the fields of nanomedicine safety and nanotoxicology are very important to aid the design and clinical introduction of second-generation nanomedicines that all hope will bring healthcare benefits (maximum therapeutic benefit and minimum risk) to all.

1.5 Update 2012–2014

This Chapter was written in 2012. The nanomedicines field is moving rapidly and since 2012 there have been new products coming to market and emerging technologies entering first clinical trials. Lessons are being learnt from products in clinical trials (successes and failures), and post marketing surveillance has led to the withdrawal of some products (e.g., examples that relate to polymer therapeutics are given in Ref. [115]). Moreover, there is continuing dialogue across Regulatory Agencies globally in order to ensure safe entry of new technologies into first in man studies (see Ref. [116]), and to ensure that "follow-on" nanomedicines enter the market in a safe and timely way for the benefit of public health [117]. There is continuing effort across disciplines to ensure that any perceived "regulatory gaps" are quickly addressed, and where needed, they are closed [118].

Acknowledgments and Disclaimer

Thanks to Prof. Rogerio Gaspar (Lisbon, Portugal) and Dr. Simon Richardson (Kent, UK) as this chapter draws heavily on the recent reviews relating to nanomedicines [1] and endocytosis [106] that we have coauthored. All comments relating to nanomedicine regulation are the opinions of the author.

References

1. Duncan, R., and Gaspar, R. (2011). Nanomedicine(s) under the microscope. *Mol. Pharm.*, **8**(6), pp. 2101–2141.
2. International Conference on Harmonisation. *International Conference on Harmonisation (ICH) of Technical Requirements for Registration of Pharmaceuticals for Human Use*, http://www.ich.org/.
3. Vogel, V. (2009). Nanomedicine: the next waves of medical innovations, in *Nanotechnology, Volume 5: Nanomedicine*, ed. Vogel, V. (WILEY-VCH Verlag GmbH, Weinheim), pp. 1–16.
4. Oberdörster, G., Oberdörster, E., and Oberdörster, J. (2005). Nanotoxicology: an emerging discipline evolving from studies of ultrafine particles. *Environ. Health Persp.*, **113**(7), pp. 823–839.
5. Donaldson, K., Stone, V., Tran, C. L., Kreyling, W., and Borm, P. J. (2004). Nanotoxicology. *Occup. Environ. Med.*, **61**(9), pp. 727–728.
6. Feliu, N., and Fadeel, B. (2010). Nanotoxicology: no small matter. *Nanoscale*, **2**(12), pp. 2514–2520.

7. European Union Scientific Committee on Emerging and Newly Identified Health Risks (2010). *Scientific Basis for the Definition of the Term "Nanomaterial"* ISSN 1831-4783 ISBN 978-92-79-12757-1; doi:10.2772/39703 ND-AS-09-004-EN-N.

8. Maynard, A. D. (2011). Don't define nanomaterials. *Nature,* **475**(7354), p. 31.

9. Maynard, A., Bowman, D., and Hodge, G. (2011). The problem of regulating sophisticated materials. *Nat. Mater.,* **10**(8), pp. 554–557.

10. ISO/TS 80004-7:2011 (2011). *Nanotechnologies Vocabulary Part 7: Diagnostics and Therapeutics for Healthcare,* http://www.iso.org/.

11. European Science Foundation (2005). *European Science Foundation's Forward Look Nanomedicine: An EMRC Consensus Opinion,* http://www.esf.org/.

12. *UK Nanotechnologies Strategy Small Technologies, Great Opportunities* (2010). March 2010 © Crown Copyright URN 10/825.

13. Munos, B. (2009). Lessons from 60 years of pharmaceutical innovation. *Nat. Rev. Drug Discovery,* **8**(12), pp. 959–968.

14. Hodel, C., and Bass, R. (1992). Are newer scientific concepts in regulatory toxicology used timely and appropriately? *Toxicol. Lett.* **64–65**, pp. 149–155.

15. Winship, K., Hepburn, D., and Lawson, D. H. (1992). The review of medicines in the United Kingdom. *Br. J. Clin. Pharmacol.,* **33**(6), pp. 583–587.

16. MHRA (2012). *How We Make Regulatory Decisions about Medicines and Medical Devices,* http://www.mhra.gov.uk/Howweregulate/Whatprinciplesinformthedecisions/index.htm.

17. Chowdhury, N. (2010). Regulation of nanomedicines in the EU: distilling lessons from the pediatric and the advanced therapy medicinal products approaches. *Nanomedicine (London),* **5**(1), pp. 135–142.

18. Bawa, R. (2011). Regulating nanomedicine: can the FDA handle it? *Curr. Drug Delivery,* **8**(3), pp. 227–234.

19. Dorbeck-Jung, B. R., and Chowdhury, N. (2011). Is the European medical products authorisation regulation equipped to cope with the challenges of nanomedicines? *Law Policy,* **33**, pp. 276–303.

20. Gaspar, R. (2007). Regulatory issues surrounding nanomedicines: setting the scene for the next generation of nanopharmaceuticals. *Nanomedicine (London),* **2**(2), pp. 143–147.

21. Gaspar, R. (2010). Therapeutic products: regulating drugs and medical devices, in *International Handbook on Regulating Nanotechnologies,* eds., Hodge, G. A., Bowman, D. M., and Maynard, A. D., (Edward Elgar, Cheltenham), pp. 291–320.
22. FDA Nanotechnology Task Force, http://www.fda.gov/nanotechnology/.
23. FDA fact sheet: nanotechnology April (2012), http://www.fda.gov/Food/GuidanceComplianceRegulatoryInformation/GuidanceDocuments/ucm300914.htm.
24. European Medicines Agency, Reflection paper on nanotechnology-based medicinal products for human use (EMEA/CHMP/79769/2006), http://www.ema.europa.eu/docs/en_GB/document_library/Regulatory_and_procedural_guideline/2010/01/WC500069728.pdf
25. Gaspar, R., and Duncan, R. (2009). Polymeric carriers: preclinical safety and the regulatory implications for design and development of polymer therapeutics. *Adv. Drug Delivery Rev.,* **61**, pp. 1220–1231.
26. Williams, D. F. (1989). A model of biocompatibility and its evaluation. *J. Biomed. Eng.,* **11**, pp. 185–191.
27. European Medicines Agency (2011). Reflection paper on non-clinical studies for generic nanoparticle iron medicinal product applications. EMA/CHMP/SWP/100094/2011.
28. European Medicines Agency (2011). Reflection paper on the data requirements for intravenous liposomal products developed with reference to an innovator liposomal product. EMA/CHMP/806058/2009.
29. Zamboni, W. C., Torchilin, V., Patri, A. K., Hrkach, J., Stern, S., Lee, R., Nel, A., Panaro, N. J., and Grodzinski, P. (2012). Best practices in cancer nanotechnology: perspective from NCI nanotechnology alliance. *Clin. Cancer Res.,* **18**(12), pp. 3229–3241.
30. Baker, J. R., Jr. (2011). The need to pursue and publish clinical trials in nanomedicine. *WIREs Nanomed. Nanobiotechnol.,* **3**, pp. 341–342.
31. US National Institutes of Health Clinical Trials Registry, http://clinicaltrials.gov.
32. US National Cancer Institute Clinical Trials Registry, http://www.cancer.gov/clinicaltrials.
33. Doxil product information, http://doxil.com/assets/DOXIL_PI_Booklet.pdf.

34. Qunibi, W. Y. (2010). The efficacy and safety of current intravenous iron preparations for the management of iron-deficiency anaemia: a review. *Arzneimittelforschung*, **60**(6a), pp. 399–412.

35. Kattan, J., Droz, J. P., Couvreur, P., Marino, J. P., Boutan-Laroze, A., Rougier, P., Brault, P., Vranckx, H., Grognet, J. M., Morge, X., and Sancho-Garnier, H. (1992). Phase I clinical trial and pharmacokinetic evaluation of doxorubicin carried by polyisohexylcyanoacrylate nanoparticles. *Invest. New Drugs,* **10**(3), pp. 191–199.

36. Sausville, E. A., Garbo, L. E., Weiss, G. J., Shkolny, D., Yurkovetskiy, A. V., Bethune, C., Ramanathan, R. K., and Fram, R. J. (2010). Phase I study of XMT-1001 given IV every 3 weeks to patients with advanced solid tumors. *J. Clin. Oncol.,* **28**(suppl), abstr. e13121.

37. Macdougall, I. C. (2009). Evolution of IV iron compounds over the last century. *J. Ren. Care.* **35**(Suppl 2), pp. 8–13.

38. Abraxane product information, http://abraxane.com/docs/Abraxane_PresecribingInformation.pdf.

39. Hansel, T. T., Kropshofer, H., Singer, T., Mitchell, J. A., and George, A. J. (2010). The safety and side effects of monoclonal antibodies. *Nat. Rev. Drug Discovery*, **9**(4), pp. 325–338.

40. Bishu, K., and Agarwal, R. (2006). Acute injury with intravenous iron and concerns regarding long-term safety. *Clin. J. Am. Soc. Nephrol.,* **1**(Suppl 1), pp. S19–S23.

41. Soloman, R., and Gabizon, A. (2008). Clinical pharmacology of liposomal anthracyclines: focus on pegylated liposomal doxorubicin. *Clin. Lymphoma Myeloma,* **8**(1), pp. 21–32.

42. Bissett, D., Cassidy, J., de Bono, J. S., Muirhead, F., Main, M., Robson, L., Fraier, D., Magne, M. L., Pellizzoni, C., Porro, M. G., Spinelli, R., Speed, W., and Twelves, C. (2004) Phase I and pharmacokinetic (PK) study of MAG-CPT (PNU 166148): a polymeric derivative of camptothecin (CPT). *Br. J. Cancer,* **91**, pp. 50–55.

43. Duncan, R. (2009). Development of HPMA copolymer anticancer conjugates: clinical experience and lessons learnt. *Adv. Drug. Delivery Rev.,* **61**, pp. 1131–1148.

44. Papanas, N., and Maltezos, E. (2010). Benefit-risk assessment of becaplermin in the treatment of diabetic foot ulcers. *Drug Saf.,* **33**(6), pp. 455–456

45. Solchaga, L. A., Hee, C. K., Roach, S., and Snel, L. B. (2012). Safety of recombinant human platelet-derived growth factor-BB in Augment® Bone Graft. *J. Tissue Eng.,* **3**(1), 2041731412442668. Epub April 4, 2012.

46. Carragee, E. J., Hurwitz, E. L., and Weiner, B. K. (2011). A critical review of recombinant human bone morphogenetic protein-2 trials in spinal surgery: emerging safety concerns and lessons learned. *Spine J.*, **11**(6), pp. 471–491.

47. Mroz, T. E., Wang, J. C., Hashimoto, R., and Norvell, D. C. (2010). Complications related to osteobiologics use in spine surgery: a systematic review. *Spine*, **35**(9 Suppl), pp. S86–S104.

48. European Commission decision to withdraw the marketing authorisation for Regranex. 16 July 2012, http://www.ema.europa.eu/docs/en_GB/document_library/Public_statement/2012/08/WC500130948.pdf.

49. Rivera, G. P., Hühn, D., del Mercato, L. L., Sasse, D., and Parak, W. J. (2010). Nanopharmacy: inorganic nanoscale devices as vectors and active compounds. *Pharmacol. Res.*, **62**(2), pp. 115–125.

50. Wood, B., Poon, R. T., Neeman, Z., Eugeni, M., Locklin, J., Dromi, S., Kachala, S., Prabhakar, R., Hahne, W., and Libutti, S. K. (2007). Phase I study of thermally sensitive liposomes containing doxorubicin (ThermoDox [TD]) given during radiofrequency ablation (RFA) in patients with unresectable hepatic malignancies. *ASCO Gastrointest. Cancer Symp.*, abstr. 188.

51. Batist, G., Gelmon, K. A., Chi, K. N., Miller, W. H. Jr., Chia, S. K. L., Mayer, L. D., Swenson, C. E., Janoff, A. S., and Louie, A. C. (2009). Safety, pharmacokinetics, and efficacy of CPX-1 liposome injection in patients with advanced solid tumors. *Clin. Cancer Res.*, **15**(2), pp. 692–700.

52. Munster, P. N., Miller, K., Krop, I. E., Dhindsa, N., Reynolds, J., Geretti, E., Niyikiza, C., Nielsen, U., Hendriks, B., Wickham, T. J., Moyo, V. M., and LoRusso, P. (2012). A phase I study of MM-302, a HER2-targeted liposomal doxorubicin, in patients with advanced, HER2-positive (HER2+) breast cancer. *J. Clin. Oncol.*, **30**(suppl), abstr. TPS663.

53. Matsumura, Y., and Kataoka, K. (2009). Preclinical and clinical studies of anticancer agent-incorporating polymer micelles. *Cancer Sci.*, **100**(4), pp. 572–579.

54. Sinclair, N., Abu-Khalaf, M. M., Sakr, B. J., Rizack, T., Lannin, D. R., Gass, J. S., Strenger, R., Bossuyt, V., Fenton, M. A., Harris, L., and Sikov, W. M. (2011). Carboplatin (Cb), weekly nanoparticle, albumin-bound paclitaxel (wAb), and bevacizumab (Av) neoadjuvant chemotherapy (NAC) in HER2-negative breast cancer (BrCA): a BrUOG study. *J. Clin. Oncol.*, **29**(suppl), abstr. e11573.

55. Initiation of ReLive, phase III clinical trial with Liva tag^{TM} in primary liver cancer May 14th 2012, http://www.bioalliancepharma.com/eng/Media2/Communiques-de-presse2/Press-releases/(offset)/8.

56. Johannsen, M., Gneveckow, U., Eckelt, L., Feussner, A., Waldöfner, N., Scholz, R., Deger, S., Wust, P., Loening, S. A., and Jordan, A. (2005). Clinical hyperthermia of prostate cancer using magnetic nanoparticles: presentation of a new interstitial technique. *Int. J. Hyperthermia*, **21**(7), pp. 637–647.

57. Krop, I. E., Beeram, M., Modi, S., Jones, S. F., Holden, S. N., Yu, W., Girish, S., Tibbitts, J., Yi, J. H., Sliwkowski, M. X., Jacobson, F., Lutzker, S. G., and Burris, H. A. (2010). Phase I study of trastuzumab-DM1, a HER2 antibody-drug conjugate, given every 3 weeks to patients with HER2-positive metastatic breast cancer. *J. Clin. Oncol.,* **28**, pp. 2698–2704.

58. Tran, S., Ready, N., Krug, L. M., Pietanza, M. C., Jungbluth, A. A., Pan, L. S., Venhaus, R. R., Hoffman, E. W., Peters, A.-M., Dukelow, K., Bomalaski, J. S., Wu, B.-W., and Old, L. J. (2012). Phase II study of ADI-PEG 20 in patients with relapsed sensitive or refractory small cell lung cancer. *J. Clin. Oncol.*, **30**(suppl), abstr. e17558.

59. Ng, T., Fontaine, J., Suntharalingam, M., Dipetrillo, T., Horiba, M. N., Oldenburg, N. B., Perez, K., Chen, W., Habr, F., and Safran, H. (2010). Neoadjuvant paclitaxel poliglumex (PPX), cisplatin, and radiation (RT) for esophageal cancer. *J. Clin. Oncol.,* **28**(Suppl 15), abstr. 4085.

60. Nowotnik, D. P., and Cvitkovic, E. (2009). ProLindac (AP5346): a review of the development of an HPMA DACH platinum polymer therapeutic. *Adv. Drug Delivery Rev.,* **61**(13), pp. 1214–1219.

61. Keefe, A. D., Pai, S., and Ellington, A. (2010). Aptamers as therapeutics. *Nat. Rev. Drug Discovery,* **9**, pp. 537–550.

62. Farrell, D., Ptak, K., Panaro, N. J., and Grodzinski, P. (2011). Nanotechnology-based cancer therapeutics-promise and challenge-lessons learned through the NCI Alliance for Nanotechnology in Cancer. *Pharm. Res.*, **28**(2), pp. 273–278.

63. Kunzmann, A., Andersson, B., Thurnherr, T., Krug, H., Scheynius, A., and Fadeel, B. (2011). Toxicology of engineered nanomaterials: focus on biocompatibility, biodistribution and biodegradation. *Biochim. Biophys. Acta*, **1810**(3), pp. 361–373.

64. Fadeel, B., and Garcia-Bennett, A. E. (2010). Better safe than sorry: understanding the toxicological properties of inorganic nanoparticles manufactured for biomedical applications. *Adv. Drug Delivery Rev.*, **62**, pp. 362–374.

65. Johnson-Lyles, D. N., Peifley, K., Lockett, S., Neun, B. W., Hansen, M., Clogston, J., Stern, S. T., and McNeil, S. E. (2010). Fullerenol cytotoxicity in kidney cells is associated with cytoskeleton disruption, autophagic

vacuole accumulation, and mitochondrial dysfunction. *Toxicol. Appl. Pharmacol.*, **248**(3), pp. 249–258.

66. Malik, N., Wiwattanapatapee, R., Klopsch, R., Lorenz, K., Frey, H., Weener, J. W., Meijer, E. W., Paulus, W., and Duncan, R. (2000). Dendrimers: relationship between structure and biocompatibility in vitro, and preliminary studies on the biodistribution of 125I-labelled polyamidoamine dendrimers in vivo. *J. Controlled Rel.*, **65**, pp. 133–148.
67. Herborn, C. U., Barkhausen, J., Paetsch, I., Hunold, P., Mahler, M., Shamsi, K., and Nagel, E. (2003). Coronary arteries: contrast-enhanced MR imaging with SH L 643A-experience in 12 volunteers. *Radiology*, **229**(1), pp. 217–223.
68. Tyssen, D., Henderson, S. A., Johnson, A., Sterjovski, J., Moore, K., La, M., Zanin, J., Sonza, S., Karellas, P., Giannis, M. P., Krippner, G., Wesselingh, S., McCarthy, T., Gorry, P. R., Ramsland, P. A., Cone, R., Paull, J. R., Lewis, G. R., and Tachedjian, G. (2010). Structure activity relationship of dendrimer microbicides with dual action antiviral activity. *PLOS ONE*, **5**(8), e12309.
69. Zamboni, W. C., Torchilin, V., Patri, A. K., Hrkach, J., Stern, S., Lee, R., Nel, A., Panaro, N. J., and Grodzinski, P. (2012). Best practices in cancer nanotechnology: perspective from NCI nanotechnology alliance. *Clin. Cancer Res.*, **18**(12), pp. 3229–3241.
70. Arrowsmith, J. (2011). Trial watch: phase III and submission failures: 2008–2010. *Nat. Rev. Drug Discovery*, **10**, pp. 328–329.
71. Eaton, M. A. (2012). Improving the translation in Europe of nanomedicines (a.k.a. drug delivery) from academia to industry. *J. Controlled Rel.*, **164**(3), pp. 370–371
72. Davis, M. E., Zuckerman, J. E., Choi, C. H. J., Seligson, D., Tolcher, A., Alabi, C. A., Yen, Y., Heidel, J. D., and Ribas, A. (2010). Evidence of RNAi in humans from systemically administered siRNA via targeted nanoparticles. *Nature*, **464**, pp. 1067–1071.
73. Libutti, S. K., Paciotti, G. F., Byrnes, A. A., Alexander, H. R., Jr., Gannon, W. E., Walker, M., Seidel, G. D., Yuldasheva, N., and Tamarkin, L. (2010). Phase, I and pharmacokinetic studies of CYT-6091, a novel PEGylated colloidal gold-rhTNF nanomedicine. *Clin. Cancer Res.*, **16**(24), pp. 6139–6149.
74. Summa, J., LoRusso, P. M., Eisenberg, P., Hrkach, J., Schnipper, E., and Von Hoff, D. (2012). A phase I, open label, safety, pharmacokinetic and pharmacodynamic dose escalation study of BIND-014 given by IV

infusion to patients with advanced or metastatic cancer. *Proc. AACR*, abstr. LB-452.

75. Nanobiotix starts clinical trial with lead product NBTXR3 (2011), http://www.nanobiotix.com/news/release, clinical trial registration number RCB 2011-A00342-39.

76. Maier-Hauff, K., Ulrich, F., Nestler, D., Niehoff, H., Wust, P., Thiesen, B., Orawa, H., Budach, V., and Jordan, A. (2011). Efficacy and safety of intratumoral thermotherapy using magnetic iron-oxide nanoparticles combined with external beam radiotherapy on patients with recurrent glioblastoma multiforme. *J. Neurooncol.*, **103**(2), pp. 317–324.

77. Benezra, M., Penate-Medina, O, Zanzonico, P. B., Schaer, D., Ow, H., Burns, A., DeStanchina, E., Longo, V., Herz, E., Iyer, S., Wolchok, J., Larson, S. M., Wiesner, U., and Bradbury, M. S. (2011). Multimodal silica nanoparticles are effective cancer-targeted probes in a model of human melanoma. *J. Clin. Invest.*, **121**(7), pp. 2768–2780.

78. Baun, A., and Hansen, S. F. (2008). Environmental challenges for nanomedicine. *Nanomedicine,* **3**(5), pp. 605–608.

79. Kuempel, E. D., Geraci, C. L., and Schulte, P. A. (2012). Risk assessment and risk management of nanomaterials in the workplace: translating research to practice. *Ann. Occup. Hyg.*, **56**(5), pp. 491–505.

80. McNeil, S. E. (2011). Challenges for nanoparticle characterization. *Methods Mol. Biol.*, **697**, pp. 9–15.

81. Wu, L., Zhang, J., and Watanabe, W. (2011). Physical and chemical stability of drug nanoparticles. *Adv. Drug Delivery Rev.*, **63**(6), pp. 456–469.

82. Yu, L. X. (2008). Pharmaceutical quality by design: product and process development, understanding, and control. *Pharm. Res.*, **25**(4), pp. 781–791.

83. Riley, B. S., and Li, X. (2011). Quality by design and process analytical technology for sterile products-where are we now? *AAPS Pharm. Sci. Tech.*, **12**(1), pp. 114–118.

84. Xu, X., Khan, M. A., and Burgess, D. J. (2012). A quality by design (QbD) case study on liposomes containing hydrophilic API: II. Screening of critical variables, and establishment of design space at laboratory scale. *Int. J. Pharm.*, **423**(2), pp. 543–553.

85. Xu, X., Costa, A. P., Khan, M. A., and Burgess, D. J. (2012). Application of quality by design to formulation and processing of protein liposomes. *Int. J. Pharm.*, **434**(1–2), pp. 349–359.

86. Zidan, A. S., Sammour, O. A., Hammad, M. A., Megrab, N. A., Habib, M. J., and Khan, M. A. (2007). Quality by design: understanding the formulation variables of a cyclosporine a self-nanoemulsified drug delivery systems by Box-Behnken design and desirability function. *Int. J. Pharm.*, **332**(1–2), pp. 55–63.

87. Toke, E. R., Lorincz, O., Somogyi, E., and Lisziewicz, J. (2010). Rational development of a stable liquid formulation for nanomedicine products. *Int. J. Pharm.*, **392**(1–2), pp. 261–267.

88. Crist, R. M., Grossman, J. H., Patri, A. K., Stern, S. T., Dobrovolskaia, M. A., Adiseshaiah, P. P., Clogston, J. D., and McNeil, S. E. (2013). Common pitfalls in nanotechnology: lessons learned from NCI's Nanotechnology Characterization Laboratory. *Integr. Biol.*, **5**(1), pp. 66–73.

89. Moghimi, S. M., Andersen, A. J., Ahmadvand, D., Wibroe, P. P., Andresen, T. L., and Hunter, A. C. (2011). Material properties in complement activation. *Adv. Drug Delivery Rev.*, **63**(12), pp. 1000–1007.

90. Hamad, I., Hunter, A. C., Szebeni, J., and Moghimi, S. M. (2008). Poly(ethylene glycol)s generate complement activation products in human serum through increased alternative pathway turnover and a MASP-2-dependent process. *Mol. Immunol.*, **46**(2), pp. 225–232.

91. Moghimi, S. M., Andersen, A. J., Hashemi, S. H., Lettiero, B., Ahmadvand, D., Hunter, A. C., Andresen, T. L., Hamad, I., and Szebeni, J. (2010). Complement activation cascade triggered by PEG-PL engineered nanomedicines and carbon nanotubes: the challenges ahead. *J. Controlled Rel.*, **146**(2), pp. 175–181.

92. Torres Andon, F., and Fadeel, B. (2013). Programmed cell death: molecular mechanisms and implications for safety assessment of nanomaterials. *Acc. Chem. Res.*, **46**(3), pp. 733–742.

93. Andersen, A. J., Wibroe, P. P., and Moghimi, S. M. (2012). Perspectives on carbon nanotube mediated adverse immune effects. *Adv. Drug Delivery Rev.*, **64**(15), pp. 1700–1705.

94. Fadeel, B., and Orrenius, S. (2005). Apoptosis: a basic biological phenomenon with wide-ranging implications in human disease. *J. Intern. Med.*, **258**(6), pp. 479–517.

95. Vanwinkle, B. A., de Mesy Bentley, K. L., Malecki, J. M., Gunter, K. K., Evans, I. M., Elder, A., Finkelstein, J. N., Oberdörster, G., and Gunter, T. E. (2009). Nanoparticle (NP) uptake by type I alveolar epithelial cells and their oxidant stress response. *Nanotoxicology*, **3**(4), pp. 307–318.

96. Nagasaki, Y. (2012). Nitroxide radicals and nanoparticles: a partnership for nanomedicine radical delivery. *Ther. Delivery*, **3**(2), pp. 165–179.

97. Hood, E., Simone, E., Wattamwar, P., Dziubla, T., and Muzykantov, V. (2011). Nanocarriers for vascular delivery of antioxidants. *Nanomedicine (London)*, **6**(7), pp. 1257–1272.

98. Moghimi, S. M., Wibroe, P. P. Helvig, Shen, Y. Z., Farhangrazi, S., and Hunter, A. C. (2012). Genomic perspectives in inter-individual adverse responses following nanomedicine administration: the way forward. *Adv. Drug Delivery Rev.*, **64**(13), pp. 1385–1393.

99. Garnett, M. C., and Kallinteri, P. (2006). Nanomedicines and nanotoxicology: some physiological principles. *Occup. Med.*, **56**, pp. 307–311.

100. Pardridge, W. M. (2012). Drug transport across the blood brain barrier. *J. Cereb. Blood Flow Metab.*, **32**(11), 1959–1972.

101. Wiwattanapatapee, R., Carreno-Gomez, B., Malik, N., and Duncan, R. (2000). Anionic PAMAM dendrimers rapidly cross adult rat intestine in vitro: a potential oral delivery system. *Pharm. Res.*, **17**(8), pp. 991–998.

102. Najlah, M., and Emanuele, A. (2006). Crossing cellular barriers using dendrimer nanotechnologies. *Curr. Opin. Pharmacol.*, **6**(5), pp. 522–527.

103. Menjoge, A. R., Rinderknecht, A., Navath, R. S., Faridnia, M., Kim, C. J., Romero, R., Miller, R. K., and Kannan, R. M. (2011). Transfer of PAMAM dendrimers across human placenta: prospects of its use as drug carrier during pregnancy. *J. Controlled Rel.*, **150**(3), pp. 326–338.

104. Choksi, A. N., Poonawalla, T., and Wilkerson, M. G. (2010). Nanoparticles: a closer look at their dermal effects. *J. Drugs Dermatol.*, **9**(5), pp. 475–481.

105. Sadrieh, N., Wokovich, A. M., Gopee, N. V., Zheng, J., Haines, D., Parmiter, D., Siitonen, P. H., Cozart, C. R., Patri, A. K., McNeil, S. E., Howard, P. C., Doub, W. H., and Buhse, L. F. (2010). Lack of significant dermal penetration of titanium dioxide from sunscreen formulations containing nano- and submicron-size TiO_2 particles. *Toxicol. Sci.*, **115**(1), pp. 156–166.

106. Duncan, R., and Richardson, S. C. W. (2012). Endocytosis and intracellular trafficking as gateways for nanomedicine delivery: opportunities and challenges. *Mol. Pharm.*, **9**, pp. 2380–2402.

107. Kostarelos, K., Lacerda, L., Pastorin, G., Wu, W., Wieckowski, S., Luangsivilay, J., Godefroy, S., Pantarotto, D., Briand, J.-P., Muller, S.,

Prato, M., and Bianco, A. (2007). Cellular uptake of functionalized carbon nanotubes is independent of functional group and cell type. *Nat. Nanotechnol.,* 2007, **2**, pp. 108–113.

108. Raffa, V., Ciofani, G., Vittorio, O., Riggio, C., and Cuschieri, A. (2010). Physicochemical properties affecting cellular uptake of carbon nanotubes. *Nanomedicine,* **5**, pp. 89–97.

109. Brandenberger, C., Clift, M. J. D., Vanhecke, D., Muhlfeld, C., Stone, V., Gehr, P., and Rothen-Rutishauser, B. (2010). Intracellular imaging of nanoparticles: is it an elemental mistake to believe what you see? *Part. Fibre Toxicol.,* **7**, p. **15**, doi: 10.1186/1743-8977-7-15.

110. Stern, S. T., Adiseshaiah, P. P., and Crist, R. M. (2012). Autophagy and lysosomal dysfunction as emerging mechanisms of nanomaterial toxicity. *Part. Fibre Toxicol.,* **9**, p. 20, doi: 10.1186/1743-8977-9-20.

111. Young, M. A., Malavalli, A., Winslow, N., Vandegriff, K. D., and Winslow, R. M. (2007). Toxicity and hemodynamic effects after single dose administration of MalPEG-hemoglobin (MP4) in rhesus monkeys. *Transl. Res.,* **149**, pp. 333–342.

112. Zabirnyk, O., Yezhelyev, M., and Seleverstov, O. (2007). Nanoparticles as a novel class of autophagy activators. *Autophagy,* **3**, pp. 278–281.

113. Nyström, A. M., and Fadeel, B. (2012). Safety assessment of nanomaterials: implications for nanomedicine. *J. Controlled Rel.,* **161**(2), pp. 403–408.

114. De Jong, W. H., and Borm, P. J. A. (2008). Drug delivery and nanoparticles: applications and hazards. *Int. J. Nanomed.,* **3**, pp. 133–149.

115. Duncan, R. (2014). Polymer therapeutics: Top 10 selling pharmaceuticals—What next? *J. Controllled Rel.,* in press, http://dx.doi.org/10.1016/j.jconrel.2014.05.001.

116. Joint MHLW/EMA reflection paper on the development of block copolymer micelle medicinal products (2013), http://www.ema.europa.eu/docs/en_GB/document_library/Scientific_guideline/2013/02/WC500138390.pdf 17th January 2013.

117. Ehmann, F., Sakai-Kato, K., Duncan, R., Hernan, D., Pita, R., Vidal, J.-M., Kohli, A., Tothfalusi, L., Sanh, A., Tinton, S., Robert, J.-L., Silva Lima, B., and Papaluca Amati, M. (2013). Next generation nanomedicines and nano-similars: EU Regulators' initiatives relating to the development and evaluation of nanomedicines. *Nanomedicine,* **8**(5), pp. 849–856.

118. Tinkle, S., McNeil, S.E., Mühlebach, S., Bawa, R., Borchard, G., Barenholz, Y. C., Tamarkin, L., and Desai, N. (2014). Nanomedicines: addressing the scientific and regulatory gap. *Ann. N. Y. Acad. Sci.,* **1313**, pp. 35–56.

Chapter 2

Nanotoxicology: Focus on Nanomedicine

Helinor Johnston, Ali Kermanizadeh, and Vicki Stone
NanoSafety Research Group, School of Life Sciences, Heriot Watt University, Edinburgh, UK
v.stone@hw.ac.uk

2.1 Introduction

Medicine is evolving to include new forms of therapeutics, drug delivery devices, and diagnostic tools that harness nanotechnology. Nanotechnology involves the development of substances that are intentionally produced to have dimensions in the nanoscale (1–100 nm). As a consequence of their small size the properties of nanomaterials (NMs) differ from larger forms of the same material. Although the definition of an NM is debatable, it is generally agreed that an NM has at least one dimension that is 1–100 nm in diameter (e.g., [1]). Nanotechnology has led to an expansion in NM production and use in a diverse array of applications, including medicine.

The advantages of developing nanomedicines includes the ability to use lower doses, leading to fewer side effects, improved targeting, and modified material properties. However, the modified properties

Handbook of Safety Assessment of Nanomaterials: From Toxicological Testing to Personalized Medicine
Edited by Bengt Fadeel

ISBN 978-981-4463-36-2 (Hardcover), 978-981-4463-37-9 (eBook)
www.panstanford.com

of materials at the nanoscale can also lead to a lack of understanding about how such NMs behave in and react with the human body. The uncertainties surrounding NM safety and the anticipated increase in exposure of humans and the environment to NMs gave rise to the discipline of nanotoxicology, which is now a well-established area of research.

This chapter reflects the progress made within nanotoxicology research over recent years and discusses how hazard investigations may be better designed to more accurately predict NM toxicity. A focus will be placed on the exploitation of NMs as nanomedicines and the implications of this for nanotoxicology investigations but a number of the issues raised are more widely applicable to other applications of NMs.

2.2 Nanomedicine and Nanotoxicology

One of the most prominent areas of research and interest in NM exploitation is the field of nanomedicine (see Chapter 1). NMs may be used as a means of accurate, early diagnosis and effective treatment of disease through the development of novel drug delivery systems, new therapeutics, and diagnostics. Importantly, a diverse array of NMs may be used as delivery agents as they are excellent carriers of therapeutic or diagnostic agents. For example, vesicles, micelles, and polymer particles, as well as metal or metal oxides of various types, have been developed for drug delivery (e.g., [2–4]). Alternatively, NMs themselves may have diagnostic or therapeutic properties that can be exploited in a clinical setting. Some applications include using the same NM for both purposes, such as paramagnetic iron oxide or gold NM, both of which are used as a contrast agent in imaging as well as for photothermal ablation (hyperthermia therapy) within the same patient [5]. The simultaneous diagnosis and treatment of disease is therefore possible through the use of NMs, and the term "theranostics" has been coined to describe this phenomenon. The use of NMs is anticipated to be extremely advantageous as a means of noninvasive, early diagnostic imaging, which combined with a therapeutic function could revolutionize current medical practice. The continued interest and optimism surrounding the utilization of NMs within medical applications is encouraged by the knowledge that a number of novel NM-based therapeutics are in transition from a research to a clinical phase [6] (and see Chapter 1 for a historical perspective).

Despite the diversity of nanomedicines under development, many of these new devices share the same principal structure. Thus, they are typically composed of three components: an NM core, a shell/coating (typically a polymer to improve biocompatibility and stability), and functional moieties on the surface (e.g., targeting ligand, drug) (Fig. 2.1). The intelligent design and precise control of the physicochemical properties of NMs can be used to ensure efficacy, whilst minimizing the likelihood of adverse effects developing. For example, the specific targeting of NMs to their required site of action is essential in order to minimize adverse effects associated with systemic administration, and this can be achieved through passive or active processes. For example, controlling NM size can enable their passive delivery to tumors. Specifically, the fenestrations in tumor capillaries provide a relatively leaky barrier compared to other tissues, and the size of these fenestrations is optimal for the capture of NMs. In addition, modification of NM surface properties using carefully selected targeting moieties can enable the specific delivery of NMs to the required target site. Surface modifications are also routinely used to increase the circulation time of NMs in blood (e.g., polyethylene glycol [PEG]).

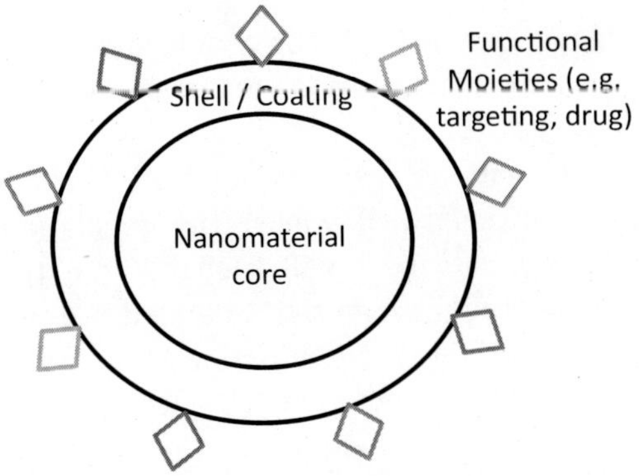

Figure 2.1 The structure of a nanomedicine typically has three main components: an NM core, a shell/coating, and functional moieties on the surface, which may include targeting ligands or drugs. Importantly, the structure and composition of a nanomedicine controls the efficacy and toxicity of the NM.

There is much optimism surrounding the benefits associated with the development of novel NM-based diagnostics and therapeutics. However, as for other NM applications, a rigorous assessment of their safety is essential. The successful exploitation of NMs within a clinical setting will require a robust risk evaluation for each nanomedicine to be performed prior to their widespread use. Without a full understanding of the implications of NM exposure the success of this rapidly expanding field could be severely impeded [7].

At this time there are no standard methodologies for assessing NM hazards. As such, there has been a significant amount of effort placed on the development of suitable testing approaches to assess NM safety. Due to the unique properties exhibited by NMs existing "traditional" testing approaches for assessing chemical safety may not be appropriate or may need to be adapted to allow for a more accurate prediction of NM safety. Much progress has been made within the field of nanotoxicology over recent years and in particular in the identification of aspects of the experimental design that should be considered when developing an appropriate testing strategy for addressing NM safety. However, nanotoxicology must constantly adapt to incorporate new evidence in order to ensure that the approach used to predict NM risk is suitable and based on the available evidence. Progress has been made in addressing the uncertainties surrounding NM toxicity, but there are still a number of knowledge gaps that need to be resolved, and a number of attempts have been made to prioritize research priorities in the area of nanotoxicology (e.g., [8, 9]).

The rigorous toxicity assessment required of pharmaceuticals should ensure their safety. However, the regulations that are applicable to nanomedicines will depend on the intended application of the NM and how an NM is defined (e.g., what size of a material constitutes an NM, 1–100 nm?), whether NMs are defined as a new type of substance, whether NMs are defined as a device or pharmaceutical, etc., as this will dictate the extent of testing that is required. The approach used to assess the safety of "traditional" chemical substances will need to be adapted. For example, special considerations for NMs that will need to be accounted for within the testing strategy adopted for their safety evaluation may include the need for physicochemical characterization of the material, procedures used to disperse NMs for hazard investigations should be carefully considered, and the potential for NM interference within toxicology assays should be evaluated, to name a few. These issues will be discussed in greater detail later. Current medical regulations do not take into account the need to assess the impact of subtle changes in particle size or how the particle is made, prepared, or dispersed—all factors that can impact NM toxicity.

2.3 Nanomaterial Physicochemical Properties

By definition all NMs share the common property that they are "small." However, there are many other physicochemical variables that have a significant influence on NM properties and toxicity. These include morphology, chemical composition, surface structure, charge, surface chemistry, solubility (dissolution), and agglomeration state of NMs [10, 11]. The contribution of each of these characteristics to NM toxicity is covered below.

2.3.1 Size

As stated previously, decreasing particle size has the advantage that novel properties are introduced into materials that can be harnessed within nanomedicines. There has been particular interest in the development of nanomedicines for the treatment of cancer, as the small size of NMs may enable their passive delivery to tumors. The size-dependent biomedical applications of gold and iron oxide nanoparticles continue to be investigated with much enthusiasm due to the novel properties that emerge within these materials when they are developed at the nanoscale.

Gold NMs can strongly absorb and scatter near-infrared (NIR) light, which allows their exploitation as novel imaging modalities [12]. This scattering of light in the NIR range allows gold NMs to be used as contrast agents and hence aids their visualization in the body. Furthermore, the conversion of the absorbed NIR light into heat and can also be used to generate localized photothermal effects to enable the destruction of tumor cells as a noninvasive hyperthermia-based treatment for cancer [12]. It is essential that the toxicity of gold NMs be assessed alongside their clinical efficacy. It is generally assumed that gold NMs are relatively nontoxic. However, existing studies have revealed that their size, shape, surface properties (e.g., coating), and charge are all important to the biological response (e.g., [13]). Cell–nanoparticle interactions have been investigated using primary human dermal microvascular endothelial cells and a panel of 15 gold NMs that had a different sized cores (18, 35, or 65 nm) and surface (polymer) coatings (including PEG, glucosamine, hydroxypropylamine, ethanediamine, and taurine) [14]. None of the NMs tested induced a cytotoxic response, but the size and surface coating used were able to impact the internalization of the NMs by cells. The uptake of 18 nm particles was greater than that of their larger counterparts (35 nm or 65 nm). In addition, PEGylated and negatively charged NMs were not effectively internalized by cells, whereas the uptake of ethandiamine-modified particles, which exhibit a positive surface charge, was extremely high. Modifying the size and surface properties of NMs is therefore able to influence their interaction with cells, with further studies required to investigate the toxicological consequences of this. The size-dependent toxicity of gold NMs (0.8 nm, 1.2 nm, 1.4 nm, and 15 nm diameter), which were all stabilized with triphenylphosphine monosulfonate (TPPMS), has been investigated in epithelial cells (HeLa cell line), phagocytes (macrophages; J774 cell line), and tissue stromal cells (connective tissue fibroblasts; L929 cell line) [15]. This study demonstrated that smaller particles were more toxic than their larger counterparts. The biodistribution pattern of gold NMs following exposure via intratracheal instillation [16], intravenous injection [16, 17], or ingestion [18] has revealed that approximately 1% of the inhaled or ingested dose of gold NMs is able to translocate into blood, they distribute within the body, depending on their size and charge, but they preferentially (>95% of translocated dose) accumulate within the liver. In general, there is an absence of studies that have assessed gold NM toxicity, with a greater emphasis placed on exploring their potential exploitation within biomedical applications, such as tumor targeting and capacity to kill tumor cells.

The magnetic properties of iron oxide can be used for magnetic resonance imaging (MRI). In addition it is possible to use iron oxide NMs as drug delivery vehicles. Although, "bulk" iron oxide is currently used in a clinical setting for imaging purposes, the development of NMs for this purpose may enhance their effectiveness compared to conventional approaches. However, the toxicity of iron oxide NMs has been demonstrated to be greater than that of their larger counterparts, following intratracheal instillation of rats [19] and exposure of lung epithelial cells [20]. Of particular concern is the ability of iron to induce oxidative stress due to its ability to stimulate reactive oxygen species (ROS) production via Fenton chemistry. A commonly used strategy to reduce particle toxicity and improve biocompatibility is the use of surface coatings (e.g., polymers). This approach is also used to improve colloidal stability of the particle suspensions (i.e., to reduce their agglomeration). The type of modification used is able to influence particle toxicity (e.g., [21, 22]).

The size of particles has been widely demonstrated to be critical to their potential toxicity, so a decrease in particle size often corresponds to enhanced toxicity for many different materials, including carbon black (e.g., [23]), polystyrene (e.g., [24]), titanium dioxide (e.g., [25]), and silver (e.g., [26]). Unraveling how and why the behavior of NMs is distinct to that of their larger counterparts is essential as their toxicity cannot simply be extrapolated from bulk forms of the same material. This requires that the contribution of size, and other physicochemical properties of NMs, to their toxicity be investigated to support their safe development as nanomedicines.

2.3.2 Morphology

The morphology of NMs (spheres, tubes, wires, rods, etc.) plays a crucial role in their classification, properties, and toxicity. Carbon nanotubes (CNTs) are fiber-like structures termed high-aspect-ratio nanomaterials (HARNs). There has been much discussion about the use of CNTs within the area of nanomedicine due to the potential to use their hollow interior for drug delivery and the ability to readily functionalize their surface [27].

Due to their structural similarity (shape and length) to pathogenic fibers there is concern that CNTs may elicit toxicity in a similar manner to asbestos. The relationship between asbestos toxicity and properties is well documented in a robust structure–activity relationship exists that relates fiber length (>10 mm), biopersistence, and aspect ratio (>3:1) to their toxicity [28]. It has been demonstrated that CNTs appear to adhere to the fiber pathogenicity paradigm, with existing studies demonstrating that their length is critical to their pathogenicity [29–31]. Careful consideration of the modifications (e.g., shape and surface properties) required to reduce CNT toxicity is therefore required for their development in biomedical applications. For example, the extent of CNT functionalization is able to control their tissue distribution and elimination from the body [32]. In addition, there is an urgent need to develop biodegradable HARN structures [33] that in future can ensure the safe design of HARNs within medical applications (see Chapter 10).

Gold (spherical) particles and gold nanorods have been developed for biomedical applications, and it is of interest to consider how particle shape impacts their toxicity. The toxicity of gold nanorods (~17 × 44 nm) has been compared to that of gold nanospheres (~20 nm) within a keratinocyte cell in vitro model [13]. The rods were less effectively internalized by cells, stimulated cytotoxicity, and ROS production and changes in gene expression relating to inflammation, apoptosis, and DNA damage; however the nanospheres were relatively nontoxic. This study could suggest that particle shape may be important to their toxicity; however, these NMs had different surface coatings, which importantly may influence the biological response. Specifically the spheres had a mercaptopropanesulfonate (MPS) coating and the rods a PEG coating, which would reduce uptake. Such findings warrant a more detailed investigation of the impact of NM shape on their toxicity in the future.

2.3.3 Composition

An NM can be composed of a single constituent material or be a composite of two or more materials. For the purposes of classification, NMs are often categorized according to their composition: carbon-based, organic, and inorganic polymers, as well as metal and metal oxide compounds or composites [34]. Existing evidence suggests that not all NMs are equally harmful. For example, silver NMs are consistently toxic across different models (e.g., [35]), while TiO_2 is often demonstrated to be of low toxicity (e.g., [36]).

There has been much interest regarding the exploitation of quantum dots (QDs) as imaging agents for the diagnosis of disease (e.g., cancer). QDs are generated from a heavy metal core (typically cadmium selenide or telluride), often with a surface coating (e.g., zinc sulfide shell). However, there are concerns regarding QD safety due to their composition (as cadmium is inherently toxic) and their nano dimensions. Surface treatments (coating and functionalization) are often introduced to enhance QD stability, preventing the leakage of cadmium from the QD core and for targeting purposes. It is generally accepted that core-shell particles are less toxic than core-only particles due to the protective role of the shell layer [37]. Indeed, it has been demonstrated that a zinc sulfide coating is able to protect against the release of cadmium from QDs to reduce their toxicity [38]. A greater awareness of the strategies that can be used to improve QD stability could be exploited within the safe design of QDs.

Modifying the biopersistence of NMs is critical to ensure their degradation or elimination from the body, as accumulation at primary and secondary target sites may lead to unwanted side effects [33]. For example, as previously discussed the persistent nature of CNTs and their (long) length can increase CNT residence time in the body to enhance their pathogenicity (according to the fiber pathogenicity paradigm), and the degradation of QDs to release cadmium can result in toxicity. An important aspect within the development of nanomedicines is therefore the ability to modify the biodegradability of NMs, which is required to control the duration of action of the nanomedicine but minimize the likelihood of side effects manifesting.

2.3.4 Surface Properties

The surface area of NMs has been demonstrated to be fundamental to their toxicity [24, 39] as it relates to what or how the cell "sees" and responds to. This has led to suggestions that surface area could be a more useful than mass as a way to express dose for some NMs [39]. This study also demonstrates that surface area is not the only driver of biological reactivity, with surface properties also being important. Modification to the surface of NMs has been used as a strategy to improve the stability of NMs, reduce their agglomeration, target NMs to specific sites/cells in the body, introduce "stealth" properties to increase circulation time, and improve NM biocompatibility. It is therefore prudent to reflect on the influence of such modifications on the toxicity of NMs.

Investigation of the surface coatings that can be applied to gold NMs have demonstrated the ability of the NM coating to drive toxicity. For example, cetyltrimethylammonium bromide (CTAB) is commonly used within the synthesis process as a capping agent to control the size and shape of gold NMs [13]. However, CTAB can elicit toxicity, while the removal of CTAB can promote the aggregation or agglomeration of gold NMs [13], which may compromise their effectiveness within clinical applications. The detoxification of CTAB-stabilized gold NMs has been attempted using a number of approaches to improve their biocompatibility and allow for their use in biomedical applications (e.g., [40, 41]). For example, the surface of gold NMs may be coated with a polymer or CTAB exchanged for a substance that is less toxic [13] to reduce the likelihood of CTAB inducing toxicity.

There is a desire to increase circulation time and avoid accumulation of nanomedicines in the reticuloendothelial system, including the liver. The modification of the NM surface with PEG is a commonly used strategy within the area of drug delivery, which increases the circulation time of drugs by preventing recognition by the reticuloendothelial system through the introduction of "stealth" properties by reducing protein absorption (i.e., opsonization). QDs with a PEG coating have reduced uptake by macrophages in vitro [42], which is likely to increase their circulation time in the body. Furthermore, PEG-modified gold nanorods that were injected into mice exhibited an increased circulation time compared to unmodified gold nanorods that accumulated in the liver [41].

Various strategies have been used to improve the colloidal stability (to avoid NM agglomeration), targeting, and biocompatibility of iron oxide NMs. A common approach is the use of polymer coatings [22]. However, it has been demonstrated that the choice of polymer is critical to achieve the desired outcome and to control toxicity. The cationic polymer poly(ethylene imine) (PEI) has been used to stabilize iron oxide NMs, but the polymer is able to enhance the toxicity of NMs [22]. A copolymer coating of PEI and PEG was effective at minimizing toxicity and enhancing colloidal stability. However, this study focused on assessment of the stability of the NMs and their ability to induce cytotoxicity within lung A549 epithelial cells evaluated using the trypan blue exclusion assay. Since this is just one way to assess viability, using an in vitro cell line a more thorough investigation of the impact of surface coatings on iron oxide NM–induced toxicity is required.

2.3.5 Dissolution

The contribution of NM solubility (dissolution) to the biological response also needs to be evaluated, where applicable, as it can be difficult to decipher if it is the NMs themselves or the release of ions or a combination of both factors that is responsible for any observed toxicity. The antibacterial properties of silver NMs are exploited in medical devices (e.g., catheters), wound dressings, and a number of consumer products such as textiles and food packaging. There is concern that the widespread use of silver NMs may lead to the development of bacterial resistance to silver, and this needs to be carefully monitored.

Silver NMs are highly toxic, and their toxicity has been investigated at a number of target sites, including, for example, the lung [43], skin (e.g., [44]), gut (e.g., [26]), and liver (e.g., [36]). There is debate surrounding whether the observed toxicity derives from particle or ion mediated effects. The antibacterial properties of silver particles are size dependent, with smaller particles exhibiting a greater effect. It has been suggested that the antibacterial properties of silver emanates from their ability to release silver ions, with smaller particles have a greater capacity for this mode of action due to their relatively large surface area [45]. In contrast, the solubility of Ag in hepatocyte cell culture medium could not account for the relatively high toxicity of NM to the C3A hepatocyte cell line [36] or primary human hepatocytes [46].

The contribution of particle dissolution to other NM types, such as zinc and copper oxides, may also be relevant and warrants further investigation. For example, the toxicity of ZnO NMs to monocytes in vitro, however, also could not be attributed entirely to the solubility [47]. The solubility of NMs is therefore required within the experimental design of nanotoxicology investigations, and "ionic controls" are often included for comparative purposes.

2.3.6 Agglomeration

NMs have a tendency to form larger agglomerates in biological media, and this instability of NM dispersions can influence their efficacy and toxicity. NMs are obviously prepared in biological media within hazard investigations; hence if NMs agglomerate/settle out of the aqueous phase over time this could impact their toxicity [48]. Furthermore, there is a desire to obtain well-dispersed suspensions of NMs for biomedical applications. For example, CNTs are insoluble in aqueous media, and to improve their dispersion and therefore biocompatability their surface has been functionalized to make their use within biomedical applications possible [33]. As discussed above, the use of surface coatings is a frequently used strategy to improve the colloidal stability (reduce agglomeration) of NM suspensions (e.g. [22]).

The rate of agglomeration of NMs in an aqueous medium will partly depend on particle–particle collision frequency and the attractive–repulsive properties of the materials involved (i.e., surface charges on two positively charged materials) [49]. After the initial collision, particles may remain as single particles or form particle–particle, particle–cluster, or cluster–cluster agglomerates. This phenomena has many implications, such as the attachment of materials to the walls of the experimental equipment (glassware, scientific instruments) [49], or can influence the toxicological response as NMs may behave as one large particle and thereby lose the chemical and toxicological characteristics associated with their "nano" form. Therefore it is necessary to consider the likely existence of free monodispersed and agglomerated NMs in real-life exposure scenarios and use this information within toxicity tests, as human exposure may not involve individual particulates but instead much larger agglomerates and this needs to be replicated within toxicity tests [50]. There is currently inadequate information available on the exposure of humans to NMs or the environment to guide relevant dispersion protocols for toxicological investigations, both in vitro and in vivo. However, the dispersion of NMs may be improved, including the use of dispersants (such as proteins, detergents, and solvents), mechanical and physical processes (such as sonication and stirring), and a manipulation of ionic strength and pH of the dispersion medium.

2.3.7 Charge

The surface charge of an NM is principally imposed by its composition and whether the surface of NMs has been functionalized. The charge of NMs has the capacity to dictate the extent and mechanism of toxicity. Recent studies have demonstrated that cationic NMs tend to be more toxic than anionic or neutral NMs (e.g., [47]). This may derive from electrostatic interactions between the positively charged particles and anionic cytosolic surface of cell membranes.

2.4 Assessment of Nanomaterial Toxicity

There is a wealth of information available on the pathogenicity of various types of materials, which can be used to inform the most relevant means to assess NM safety. A vast amount of knowledge exists on the toxicology of metal fumes, dust, silica, asbestos and other synthetic fibers, and air pollution particles such as PM_{10}. These early studies were integral in demonstrating that small particles (<100 nm) have unique toxicological properties. This information has proved to be extremely useful in guiding the design of nanotoxicology studies.

Critical to the assessment of NM safety is the development of robust methodologies that reliably predict the toxicity of NMs in humans. Currently there has been a concerted effort from the international scientific community to develop harmonized approaches to NM toxicity assessment. This is crucial as there is a paucity of information on the response of humans to NM exposure, and only a few controlled human exposure studies have been conducted to date (e.g., [51]). In the absence of information relating to the toxicity of NMs to humans both rodent and cell models have been routinely used to predict the response of humans. Due to the diverse array of NMs whose safety requires assessment and an increase in production and the use of NMs, emphasis has been placed on minimizing animal testing and promoting the development of in vitro alternatives to animal experimentation (preferably that can be adapted to screen NM safety within high-throughput systems). However, this will require that the findings from in vitro studies be validated in vivo. For a detailed discussion on in vitro versus in vivo tests, see Chapters 4 and 5. Due to the concern surrounding the ability of in vitro models to mimic realistic in vivo situations more sophisticated models that are able to incorporate cell–cell interactions and movement of cells in the body are being developed (e.g., [52]). The development of predictive in silico approaches that allow a prediction of NM toxicity to be made on the basis of the physicochemical properties of NMs is also encouraged as a longer-term goal for the assessment of NM safety.

It is inevitable that the intentional exposure to NMs will rise due to their increasing exploitation within nanomedicines, and thus it is essential that their safety be thoroughly evaluated. Hazard investigations should assess the acute and chronic toxicity of NMs at relevant exposure (e.g., lungs, skin, gut) and secondary target sites (e.g., liver, cardiovascular system, brain, kidneys). Furthermore it is crucial to conduct studies to understand the mechanism underlying any toxic responses (mode of action). In parallel to toxicology studies knowledge of the biodistribution (kinetics) of NMs following exposure is critical to understanding NM localization and capacity for translocation from the exposure site, biopersistence, and elimination. There has been a tremendous increase in the number of studies assessing NM hazard, and thus some advances have been made with regard to the approach used to evaluate NM safety. It is ultimately desirable to combine exposure and hazard information to conduct risk assessments for NMs.

Available evidence illustrates that different NMs will vary in their toxic potency and the mechanism by which they elicit adverse effects. It is therefore likely that NM toxicity is controlled by their physicochemical properties and sensitivity (susceptibility) of the target site or individual exposed. However, although it may not be possible to make generalizations regarding NM toxicity, investigations routinely assess the contribution of oxidative stress and inflammatory responses following exposure of cells or animals to NMs, as such responses are found to commonly contribute to the mechanism of toxicity or biological effect. The ability of NMs to stimulate an oxidative driven response is typically assessed by evaluating the production of ROS, changes in the expression, level, and activity of antioxidants (such as glutathione) and the ability of a pretreatment of antioxidants to ameliorate any observed NM-induced toxicity. Pro-inflammatory responses are evaluated by determining the cellular infiltration (e.g., neutrophils or macrophages) following exposure as well as changes in the expression or production of pro-inflammatory mediators (e.g., cytokines) or activation of intracellular signaling cascades and transcription factors (e.g., nfκβ). Assessment of the uptake of NMs is frequently investigated to assess the interaction that occurs between NMs and cells and whether NM uptake is a requirement for their toxicity. Uptake is typically assessed using electron microscopy, while elemental analysis can be used to confirm NM presence, which is required due to the similarity in particle appearance to cell structures. The ability of NMs to elicit cytotoxicity is routinely assessed within in vitro models. This allows a useful means of benchmarking NM toxicity, especially when a panel of NMs is being investigated and researchers are attempting to dissect out which attributes of the NMs are driving any observed toxicity. The ability of NMs to elicit DNA damage is of concern, and there are several means of investigating genotoxicity (e.g., Comet and micronucleus assays) (see Chapter 11 for a detailed discussion).

There are certain aspects of the experimental design that should be carefully considered prior to the start of a new study, which are summarized below [9].

- Toxicity studies are often criticized for using unrealistic doses of NMs that do not reflect "real life" exposures. Often, dose–response experiments are conducted, but it is currently difficult to decipher what really is a "high" or "low" dose. As such, it is necessary for investigators to justify the dose of NMs used as currently dose selection varies greatly between investigations. This issue is more relevant for environmental exposures compared to medical applications where dose ranges are driven by efficacy and pharmacokinetics. Medical hazard studies typically use very high exposure concentrations; this may not be suitable for NMs that may behave very different at high concentrations where agglomeration is more likely.
- It is also important to consider what dose metric is most appropriate to quantify NM exposure (e.g., mass, number, surface area). This will be especially important when comparing the findings of different investigations but is not a trivial task. In vitro doses may also be complicated by agglomeration and settling rate of the particles onto cells [53].

- In the main, the acute effects of NMs (typically 24 hours) are investigated, but there is a need for the chronic effects of NMs to be evaluated (following repeated exposure) to understand the longevity of an observed response and to determine if it changes with time.
- The preparation of NMs is critical, as investigators use various strategies to prevent against NM agglomeration. Researchers must provide information on how they have suspended NMs within toxicology studies.
- Critical to all aspects of toxicology is the inclusion of appropriate controls within the experiment. This is relevant to understand the suitability of the model being used to measure the biological response of interest but also to understand how the physicochemical properties (e.g., dissolution) of NMs contribute to their toxicity.
- It has been observed that NMs are able to interfere with toxicology assays [54] and the capacity of NMs to generate false positive or negative data should be assessed. This also promotes the use of two or more approaches to confirm the response observed is "real."

2.5 Nanomaterial Physicochemical Characterization

The extent of characterization of NMs that is required in parallel to hazard investigations is a hotly debated subject. Ideally, some characterization of the physical and chemical characteristics of NMs (e.g., size, shape, charge, solubility, surface area) in their "as produced" form and when dispersed in relevant biological media should be performed in parallel to hazard investigations. However, it may not be possible to achieve this due to methodological, time, and financial constraints. The pressure to conduct detailed physicochemical characterization of NMs continues to increase from reviewers, journals, and the scientific community. There are several important drivers within the characterization of NMs. The characterization of NMs as provided by the supplier may be required to confirm the properties of a material toxicity to be investigated. The characterization of NMs in an "as produced/supplied" form also allows identification of the properties of NMs that will ultimately dictate how the NM will behave, which may be important within the development of structure–activity relationships. The characterization of NMs within biological media enables evaluation of how NMs behave in conditions more relevant to real-life exposures, but importantly this will be dictated by the properties of the as-produced/supplied NMs. Characterization of NMs in more relevant media is challenging due to the often complex composition of biological media and the interaction of different components (e.g., protein, salts) with the NMs, which can impede their characterization by traditional approaches. It is also relevant to consider how the characteristics of an NM evolve over time once in contact with a biological system (in vitro or in vivo). It is unfeasible to conduct all types of characterizations described within all hazard studies, and instead it will be necessary to prioritize the characterization conducted in relation to the hypotheses of the study.

An area that has not yet received much attention in nanomedicine is how the method of particle production can influence the efficacy and toxicity of a particle/NM. A better understanding of how physicochemical characteristics influence efficacy and toxicity will be useful in this context. A number of projects (see, for instance, http://www.enpra.eu/) aim to use in silico approaches to address this issue.

2.6 Relationship between Exposure Route and Toxicity

Exposure to nanomedicines may occur via inhalation (lungs), ingestion (gut), dermal application (skin), and injection (blood). On entering the body NMs will interact and form complexes with an array of proteins and other biological molecules (e.g., lipids), which will be dictated by the route of exposure. The small size and large surface area of NMs allow for the adsorption of biological molecules such as proteins onto their surface. The term "protein corona" has been used to describe the coating of the NM surface with proteins [55]. This phenomenon is well established in the drug delivery community and termed "opsonization." The composition of the protein coating (or corona) is likely to be dictated by the route of exposure and physicochemical properties of the NM (including size, hydrophobicity, surface area, and charge) [56–60]. The composition of the corona will also evolve over time, whereby proteins associate and disassociate from the NM surface as proteins compete for binding to the NM surface [55]. The translocation of NMs from their exposure site to secondary target sites is also likely to influence NM behavior and toxicity due to the acquisition of different surface coatings during their transit in the body. The surface properties of NMs are highly influential to their interaction with cells in the body, and thus the composition of the protein corona can influence the biological and toxicological response to NMs. As such, it is suggested that the anticipated route of exposure and information on NM biodistribution should inform the preparation of the NM for the hazard investigation (i.e., how the NM is dispersed).

Traditionally, NMs are prepared in an aqueous suspension with a protein component to aid in their dispersion, as the coating of the NM surface with protein can help to prevent against NM agglomeration. For example, within in vitro studies, the cell culture medium is supplemented with serum, which improves the dispersion of the NMs (e.g., [61]). It is essential to compare the findings from in vitro cell models to those of in vivo responses in order to evaluate whether the simple serum dispersions that have been used widely to date to prepare NMs for the exposure of cells in vitro are appropriate or whether the NM suspensions generated should better reflect the exposure route and transit of NMs in the body. This issue is of importance as it may account for the lack of correlation often observed between in vivo and in vitro studies. To more realistically mimic real-life exposure conditions, the preparation of NMs for exposure via the lungs (using animal or cell models) could incorporate lung lining fluid (e.g., [62]) or surfactants (e.g., [63]). However, the exposure of secondary target sites for NM toxicity, such as liver cells (e.g., hepatocytes), may require a different approach. NMs are known to accumulate within the liver, following exposure via the lungs or blood [16]. With respect to the exposure of hepatocytes following injection the dispersion of NMs in a serum-containing medium is relevant. However, when investigating the response of the liver following the translocation of NMs from the lungs it may be necessary to adapt the dispersion protocol to better reflect exposure in vivo. For example, NMs could be precoated with a lung-lining fluid or surfactant and then suspended in a serum-containing cell culture medium to represent a more realistic exposure of the cells [62] (Fig. 2.2). This is important as the composition of NM–protein complexes have the capacity to influence NM toxicity.

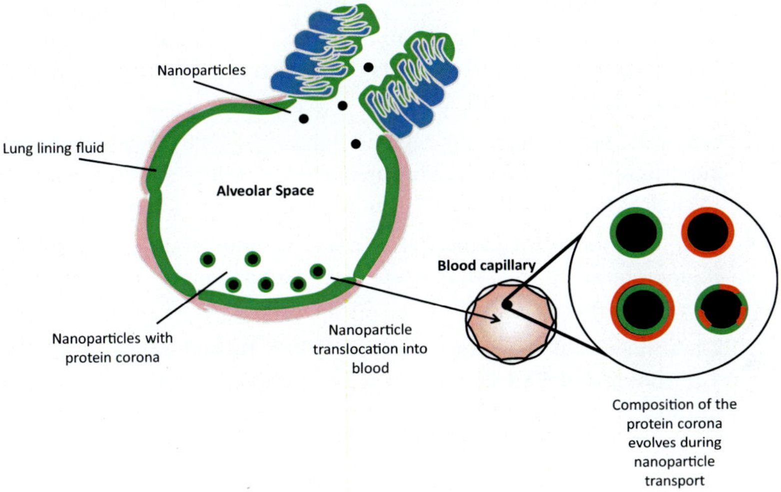

Figure 2.2 The deposition of NMs in the lung will mean that the NMs become coated in lung-lining fluid (green). Following the translocation of the NMs into blood the composition of the protein–NM complexes is likely to change due to the interaction of NMs with proteins in blood (red). The nature of the protein corona is likely to influence the toxicity of NMs at secondary target sites (e.g., the liver).

2.7 Conclusions

There is much optimism surrounding the exploitation of NMs as nanomedicines. However, it is essential that the safety of NMs be thoroughly investigated before their widespread use. The design of nanotoxicology investigations is critical to the usefulness of NMs within biomedical applications. Recommendations have been provided to better direct the design of hazard studies so that they more accurately reflect real-life exposure. It is integral to understand the physicochemical properties of NMs that are responsible for the efficacy and toxicity within medical applications, as this knowledge can be incorporated into their safe design in the future.

References

1. European Commission (2011). *Recommendation on a Definition of a Nanomaterial,* http://ec.europa.eu/environment/chemicals/nanotech/index.htm#definition.
2. Liang, Z., Liu, Y., Li, X., Wu, Q., Yu, J., Luo, S., Lai, L., and Liu, S. (2011). Surface-modified goldnanoshells for enhanced cellular uptake. *J. Biomed. Mater. Res. A,* **98**, 479–487.
3. Lin, M. M., Kim do, K., El Haj, A. J., and Dobson, J. (2008). Development of superparamagnetic iron oxide nanoparticles (SPIONS) for translation to clinical applications. *IEEE Trans. Nanobiosci.,* **7**, pp. 298–305.
4. Loo, C., Lowery, A., Halas, N., West, J., and Drezek, R. (2005). Immunotargeted nanoshells for integrated cancer imaging and therapy. *Nano Lett.,* **5**, 709–711.
5. Zhan, T., Li, P., Bi, S., Dong, B., Song, H., Ren, H., and Wang, L. (2012). 12P-conjugated PEG-modified gold nanorods combined with near-infrared laser for tumor targeting and photothermal therapy. *J. Nanosci. Nanotechnol.,* **12**, 7198–7205.
6. Kostarelos, K., Bianco, A., and Prato, M. (2009). Promises, facts and challenges for carbon nanotubes in imaging and therapeutics. *Nat. Nanotechnol.,* **10**, pp. 627–633.
7. Maynard, M. A. (2007). Nanotechnology: the next big thing, or much ado about nothing? *Ann. Occup. Hyg.,* **51**, pp. 1–12.
8. Royal Society and Royal Academy of Engineering (2004). *Nanotechnology Report, Nanoscience and Nanotechnologies: Opportunities and Uncertainties,* July 2004, http://www.nanotec.org.uk.
9. Johnston, H., Pojana, G., Zuin, S., Jacobsen, N. R., Møller, P., Loft, S., Semmler-Behnke, M., McGuiness, C., Balharry, D., Marcomini, A., Wallin, H., Kreyling, W., Donaldson, K., Tran, L., and Stone, V. (2013). Engineered nanomaterial risk. Lessons learnt from completed nanotoxicology studies: potential solutions to current and future challenges. *Crit. Rev. Toxicol.,* **43**, pp. 1–20.
10. Nel, A., Xia, T., Madler, L., and Li, N. (2006). Toxic potential of materials at the nanolevel. *Science,* **311**, pp. 622–627.
11. Tsuji, J. S., Maynard, A. D., Howard, P. C., James, J. T., Lam, C. W., Warheit, D. B., and Santamaria, A. B. (2006). Research strategies for safety evaluation of nanomaterials, part IV: risk assessment of nanoparticles. *Toxicol. Sci.,* **89**, pp. 42–50.

12. Tong, L., Wei, Q., Wei, A., and Cheng, J. X. (2009). Gold nanorods as contrast agents for biological imaging: optical properties, surface conjugation and photothermal effects. *Photochem. Photobiol.*, **85**, pp. 21–32.

13. Schaeublin, N. M., Braydich-Stolle, L. K., Maurer, E. I., Park, K., MacCuspie, R. I., Afrooz, A. R., Vaia, R. A., Saleh, N. B., and Hussain, S. M. (2012). Does shape matter? Bioeffects of gold nanomaterials in a human skin cell model. *Langmuir*, **28**, pp. 3248–3258.

14. Freese, C., Gibson, M. I., Klok, H. A., Unger, R. E., and Kirkpatrick, C. J. (2012). Size- and coating-dependent uptake of polymer-coated gold nanoparticles in primary human dermal microvascular endothelial cells. *Biomacromolecules*, **13**, pp. 1533–1543.

15. Pan, Y., Neuss, S., Leifert, A., Fischler, M., Wen, F., Simon, U., Schmid, G., Brandau, W., and Jahnen-Dechent, W. (2007). Size-dependent cytotoxicity of gold nanoparticles. *Small*, **3**, pp. 1941–1949.

16. Semmler-Behnke, M., Kreyling, W. G., Lipka, J., Fertsch, S., Wenk, A., Takenaka, S., Schmid, G., and Brandau, W. (2008). Biodistribution of 1.4- and 18-nm gold particles in rats. *Small*, **4**, pp. 2108–2111.

17. Hirn, S., Semmler-Behnke, M., Schleh, C., Wenk, A., Lipka, J., Schäffler, M., Takenaka, S., Möller, W., Schmid, G., Simon, U., and Kreyling, W. G. (2011). Particle size-dependent and surface charge-dependent biodistribution of gold nanoparticles after intravenous administration. *Eur. J. Pharm. Biopharm.*, **77**, pp. 407–416.

18. Schleh, C., Semmler-Behnke, M., Lipka, J., Wenk, A., Hirn, S., Sch Ffler, M., Schmid, G. N., Simon, U., and Kreyling, W. G. (2011). Size and surface charge of gold nanoparticles determine absorption across intestinal barriers and accumulation in secondary target organs after oral administration. *Nanotoxicology*, **6**, pp. 36–46.

19. Zhu, M. T., Feng, W. Y., Wang, B., Wang, T. C., Gu, Y. Q., Wang, M., Wang, Y., Ouyang, H., Zhao, Y. L., and Chai, Z. F. (2008). Comparative study of pulmonary responses to nano- and submicron-sized ferric oxide in rats. *Toxicology*, **247**, pp. 102–111.

20. Bhattacharya, K., Hoffmann, E., Schins, R. F., Boertz, J., Prantl, E. M., Alink, G. M., Byrne, H. J., Kuhlbusch, T. A., Rahman, Q., Wiggers, H., Schulz, C., and Dopp, E. (2012). Comparison of micro- and nanoscale Fe^{3+}-containing (hematite) particles for their toxicological properties in human lung cells in vitro. *Toxicol. Sci.*, **126**, pp. 173–182.

21. Hong, S. C., Lee, J. H., Lee, J., Kim, H. Y., Park, J. Y., Cho, J., Lee, J., and Han, D. W. (2011). Subtle cytotoxicity and genotoxicity differences

in superparamagnetic iron oxide nanoparticles coated with various functional groups. *Int. J. Nanomed.*, **6**, pp. 3219–31.

22. Schweiger, C., Pietzonka, C., Heverhagen, J., and Kissel, T. (2011). Novel magnetic iron oxide nanoparticles coated with poly(ethylene imine)-g-poly(ethylene glycol) for potential biomedical application: synthesis, stability, cytotoxicity and MR imaging. *Int. J. Pharm.*, **408**, pp. 130–137.
23. Li, X. Y., Brown, D., Smith, S., MacNee, W., and Donaldson, K. (1999). Short-term inflammatory responses following intratracheal instillation of fine and ultrafine carbon black in rats. *Inhal. Toxicol.*, **11**, pp. 709–731.
24. Brown, D. M., Wilson, M. R., MacNee, W., Stone, V., and Donaldson, K. (2001). Size-dependent proinflammatory effects of ultrafine polystyrene particles: a role for surface area and oxidative stress in the enhanced activity of ultrafines. *Toxicol. Appl. Pharmacol.*, **175**, pp. 191–199.
25. Ferin, J., Oberdorster, G., and Penney, D. P. (1992). Pulmonary retention of ultrafine and fine particles in rats. *Am. J. Respir. Cell. Mol. Biol.*, **6**, pp. 535–542.
26. Gaiser, B. K., Fernandes, T. F., Jepson, M., Lead, J. R., Tyler, C. R., and Stone, V. (2009). Assessing exposure, uptake and toxicity of silver and cerium dioxide nanoparticles from contaminated environments. *Environ. Health.*, **8**(Suppl 1), p. S2.
27. Prato, M., Kostarelos, K., and Bianco, A. (2008). Functionalized carbon nanotubes in drug design and discovery. *Acc. Chem. Res.*, **41**, pp. 60–68.
28. Donaldson, K., Murphy, F., Schinwald, A., Duffin, R., and Poland, C. A. (2011). Identifying the pulmonary hazard of high aspect ratio nanoparticles to enable their safety-by-design. *Nanomedicine (London)*, **6**(1), pp. 143–156.
29. Brown, D. M., Kinloch, I. A., Bangert, U., Windle, A. H., Walter, D. M., Walker, G. S., Scotchford, C. A., Donaldson, K., and Stone, V. (2007). An in vitro study of the potential of carbon nanotubes and nanofibres to induce inflammation mediators and frustrated phagocytosis. *Carbon*, **45**, pp. 1743–1756.
30. Murphy, F. A., Poland, C. A., Duffin, R., Al-Jamal, K. T., Ali-Boucetta, H., Nunes, A., Byrne, F., Prina-Mello, A., Volkov, Y., Li, S., Mather, S. J., Bianco, A., Prato, M., Macnee, W., Wallace, W. A., Kostarelos, K., and Donaldson, K. (2011). Length-dependent retention of carbon nanotubes in the pleural space of mice initiates sustained inflammation and progressive fibrosis on the parietal pleura. *Am. J. Pathol.*, **178**, pp. 2587–600.

31. Poland, C. A., Duffin, R., Kinloch, I., Maynard, A., Wallace, W. A., Seaton, A., Stone, V., Brown, S., Macnee, W., and Donaldson, K. (2008). Carbon nanotubes introduced into the abdominal cavity of mice show asbestos-like pathogenicity in a pilot study. *Nat. Nanotechnol.*, **3**, pp. 423–428.

32. Al-Jamal, K. T., Nunes, A., Methven, L., Ali-Boucetta, H., Li, S., Toma, F. M., Herrero, M. A., Al-Jamal, W. T., ten Eikelder, H. M., Foster, J., Mather, S., Prato, M., Bianco, A., and Kostarelos, K. (2012). Degree of chemical functionalization of carbon nanotubes determines tissue distribution and excretion profile. *Angew. Chem., Int. Ed. Engl.*, **51**, pp. 6389–6393.

33. Bianco, A., Kostarelos, K., and Prato, M. (2011). Making carbon nanotubes biocompatible and biodegradable. *Chem. Commun. (Camb.)*, **47**, pp. 10182–10188.

34. Grill, A. E., Johnston, N. W., Sadhukha, T., and Panyam, J. (2009). Reversible "irreversible" inhibition of chymotrypsin using nanoparticle receptors. *J. Am. Chem. Soc.: Recent Patents Drug Delivery Formulat.*, **3**, pp. 137–142.

35. Gaiser, B. K., Fernandes, T. F., Jepson, M. A., Lead, J. R., Tyler, C. R., Baalousha, M., Biswas, A., Britton, G. J., Cole, P. A., Johnston, B. D., Ju-Nam, Y., Rosenkranz, P., Scown, T. M., and Stone, V. (2012). Interspecies comparisons on the uptake and toxicity of silver and cerium dioxide nanoparticles. *Environ. Toxicol. Chem.*, **31**, pp. 144–154.

36. Kermanizadeh, A., Pojana, G., Gaiser, B. K., Birkedal, R., Bilaničová, D., Wallin, H., Jensen, K. A., Sellergren, B., Hutchison, G. R., Marcomini, A., and Stone, V. (2013). In vitro assessment of engineered nanomaterials using a hepatocyte cell line: cytotoxicity, pro-inflammatory cytokines and functional markers. *Nanotoxicology*, **7**, pp. 301–313.

37. Soenen, S. J., Demeester, J., De Smedt, S. C., and Braeckmans, K. (2012). The cytotoxic effects of polymer-coated quantum dots and restrictions for live cell applications. *Biomaterials*, **33**(19), pp. 4882–4888.

38. Derfus, A. M., Chan, W. C. W., and Bhatia, S. N. (2004). Probing the cytotoxicity of semiconductor quantum dots. *Nano Lett.*, **4**, pp. 11–18.

39. Duffin, R., Tran, L., Brown, D., Stone, V., and Donaldson, K. (2007). Proinflammogenic effects of low-toxicity and metal nanoparticles in vivo and in vitro: highlighting the role of particle surface area and surface reactivity. *Inhal. Toxicol.*, **19**, pp. 849–856.

40. Leonov, A. P., Zheng, J., Clogston, J. D., Stern, S. T., Patri, A. K., and Wei, A. (2008). Detoxification of gold nanorods by treatment with polystyrenesulfonate. *ACS Nano*, **2**, pp. 2481–2488.

41. Niidome, T., Yamagata, M., Okamoto, Y., Akiyama, Y., Takahashi, H., Kawano, T., Katayama, Y., and Niidome, Y. (2006). PEG-modified gold nanorods with a stealth character for in vivo applications. *J. Controlled Rel.*, **114**, pp. 343–347.

42. Clift, M. J., Rothen-Rutishauser, B., Brown, D. M., Duffin, R., Donaldson, K., Proudfoot, L., Guy, K., and Stone, V. (2008). The impact of different nanoparticle surface chemistry and size on uptake and toxicity in a murine macrophage cell line. *Toxicol. Appl. Pharmacol.*, **232**, pp. 418–427.

43. Foldbjerg, R., Dang, D. A., and Autrup, H. (2011). Cytotoxicity and genotoxicity of silver nanoparticles in the human lung cancer cell line, A549. *Arch. Toxicol.*, **85**, pp. 743–750.

44. Arora, S., Jain, J., Rajwade, J. M., and Paknikar, K. M. (2008). Cellular responses induced by silver nanoparticles: in vitro studies. *Toxicol. Lett.*, **179**, pp. 93–100.

45. Lok, C. N., Ho, C. M., Chen, R., He, Q. Y., Yu, W. Y., Sun, H., Tam, P. K., Chiu, J. F., and Che, C. M. (2007). Silver nanoparticles: partial oxidation and antibacterial activities. *J. Biol. Inorg. Chem.*, **12**, pp. 527–534.

46. Kermanizadeh, A., Gaiser, B. K., Ward, M. B., and Stone, V. (2013). Primary human hepatocytes vs. hepatic cell line: assessing their suitability for in vitro nanotoxicology. *Nanotoxicology*, **7**, pp. 1255–1271.

47. Prach, M., Stone, V., and Proudfoot, L. (2013). Zinc oxide nanoparticles and monocytes: impact of size, charge, and solubility on activation status. *Toxicol. Appl. Pharmacol.*, **266**, pp. 19–26.

48. Buzea, C., Pacheco, I. I., and Robbie, K. (2007). Nanomaterials and nanoparticles: sources and toxicity. *Biointerphases*, **2**, pp. 18–67.

49. Handy, R. D., Owen, R., and Valsami-Jones, E. (2008). The ecotoxicology of nanoparticles and nanomaterials: current status, knowledge gaps, challenges, and future needs. *Ecotoxicology*, **17**, 315–325.

50. Hirano, S. (2009). A current overview of health effect research on nanoparticles. *Environ. Health Prevent. Med.*, **14**, 223–225.

51. Mills, N. L., Amin, N., Robinson, S. D., Anand, A., Davies, J., Patel, D., de la Fuente, J. M., Cassee, F. R., Boon, N. A., MacNee, W., Millar, A. M., Donaldson, K., and Newby, D. E. (2005). Do inhaled carbon nanoparticles translocate directly into the circulation in man? *Am. J. Respir. Crit. Care Med.*, **173**, pp. 426–431.

52. Rothen-Rutishauser, B., Kiama, S. G., and Gehr, P. (2005). A three-dimensional cellular model of the human respiratory tract to study the interaction with particles. *Am. J. Respir. Cell Mol. Biol.*, **32**, pp. 281–289.

53. Teeguarden, J. G., Hinderliter, P. M., Orr, G., Thrall, B. D., and Pounds, J. G. (2007). Particokinetics in vitro: dosimetry considerations for in vitro nanoparticle toxicity assessments. *Toxicol. Sci.,* **95**, pp. 300–312.

54. Wörle-Knirsch, J. M., Pulskamp, K., and Krug, H. F. (2006). Oops they did it again! Carbon nanotubes hoax scientists in viability assays. *Nano Lett.,* **6**, 1261–1268.

55. Lundqvist, M., Stigler, J., Cedervall, T., Berggård, T., Flanagan, M. B., Lynch, I., Elia, G., and Dawson, K. (2011). The evolution of the protein corona around nanoparticles: a test study. *ACS Nano,* **5**, pp. 7503–7509.

56. Brown, D. M., Dickson, C., Duncan, P., Al-Attili, F., and Stone, V. (2010). Interaction between nanoparticles and cytokine proteins: impact on protein and particle functionality. *Nanotechnology,* **21**, p. 215104.

57. Deng, Z. J., Liang, M., Toth, I., Monteiro, M., and Minchin, R. F. (2013). Plasma protein binding of positively and negatively charged polymer-coated gold nanoparticles elicits different biological responses. *Nanotoxicology,* **7**, pp. 314–322.

58. Dobrovolskaia, M. A., Patri, A. K., Zheng, J., Clogston, J. D., Ayub, N., Aggarwal, P., Neun, B. W., Hall, J. B., and McNeil, S. E. (2009). Interaction of colloidal gold nanoparticles with human blood: effects on particle size and analysis of plasma protein binding profiles. *Nanomedicine,* **5**, pp. 106–117.

59. Lundqvist, M., Stigler, J., Elia, G., Lynch, I., Cedervall, T., and Dawson, K. A. (2008). Nanoparticle size and surface properties determine the protein corona with possible implications for biological impacts. *Proc. Natl. Acad. Sci. U S A,* **105**, pp. 14265–14270.

60. Tenzer, S., Docter, D., Rosfa, S., Wlodarski, A., Kuharev, J., Rekik, A., Knauer, S. K., Bantz, C, Nawroth, T., Bier, C., Sirirattanapan, J., Mann, W., Treuel, L., Zellner, R., Maskos, M., Schild, H., and Stauber, R. H. (2012). Nanoparticle size is a critical physicochemical determinant of the human blood plasma corona: a comprehensive quantitative proteomic analysis. *ACS Nano,* **5**, pp. 7155–7167.

61. Johnston, H. J., Semmler-Behnke, M., Brown, D. M., Kreyling, W., Tran, L., and Stone, V. (2010). Evaluating the uptake and intracellular fate of polystyrene nanoparticles by primary and hepatocyte cell lines in vitro. *Toxicol. Appl. Pharmacol.,* **242**, pp. 66–78.

62. Johnston, H., Brown, D., Kermanizadeh, A., Gubbins, E., and Stone, V. (2012). Investigating the relationship between nanomaterial hazard and physicochemical properties: informing the exploitation of nanomaterials within therapeutic and diagnostic applications. *J. Controlled Rel.,* **164**, pp. 307–313.

63. Gasser, M., Wick, P., Clift, M. J., Blank, F., Diener, L., Yan, B., Gehr, P., Krug, H. F., and Rothen-Rutishauser, B. (2012). Pulmonary surfactant coating of multi-walled carbon nanotubes (MWCNTs) influences their oxidative and pro-inflammatory potential in vitro. *Part. Fibre Toxicol.*, **9**, p. 17.

64. Woodrow Wilson database, http://www.nanotechproject.org/inventories/consumer/.

Chapter 3

Nanomaterial Characterization for Toxicity Testing

David B. Warheit and Christie M. Sayes
[a]*DuPont Haskell Global Centers for Health and Environmental Sciences, Newark, DE, USA*
[b]*Center for Aerosol & Nanomaterials Engineering, RTI International, Research Triangle Park, NC, USA*
david.b.warheit@usa.dupont.com, csayes@rti.org

3.1 Introduction

Over the past 15 years, the field of nanotoxicology has developed at a rapid pace. Interdisciplinary research, ranging from chemistry and physics to biochemistry and toxicology, has intrigued scientists in a variety of sectors, including academics and industry, and enabled collaborations in unprecedented numbers. This chapter highlights some of the most interesting collaborative methods and results of a nanochemist and a pulmonary toxicologist. Techniques from both fields are needed to be able to accurately and precisely measure the characteristics of engineered nanomaterials exposed to the lung or to cells that compose the lung. The discussion in the present chapter is limited to three popular nanoparticle types: metal oxide powders,

Handbook of Safety Assessment of Nanomaterials: From Toxicological Testing to Personalized Medicine
Edited by Bengt Fadeel

ISBN 978-981-4463-36-2 (Hardcover), 978-981-4463-37-9 (eBook)
www.panstanford.com

metal colloids, and polymeric particles. The characterization methods described include measurement techniques of particles both in pre-exposure and postexposure modes to the biological test system. The tools presented may also be useful for nanobiomedicinal research.

3.1.1 Nanoparticles Used in the Industry

There are a variety of particles used in industrial applications. Two of the most common particles are in the metal oxide class of engineered nanomaterials and include titanium dioxide (aka titania, TiO_2) and silicon dioxide (aka silica, SiO_2). Titanium dioxide (TiO_2) is a naturally occurring oxide of the element titanium, whose physical form is a fine white powder. The crystal structure of this substance is identified by three mineral compounds known as anatase, rutile, and brookite, with the first two structures being the most prevalent. Numerous industrial applications are associated with this mineral; however, the most common uses are in (1) pigments or opacifiers, (2) cosmetics, and (3) catalysts.

Synthetic amorphous silica (AMO) (defined as noncrystalline forms—with a detection limit of 0.01% by weight relative to quartz) can be divided into two general categories, according to whether the manufacturing process occurs by synthesis using the "wet route" (i.e., precipitated silica, silica gel) or by formulation of the "thermal route" (i.e., pyrogenic silica). Colloidal silica is known to represent stable dispersions of synthetic AMO types in a liquid, frequently in the form of water. Synthetic AMO generally is hydrophilic but may be transformed to hydrophobic properties via surface treatments. In general, AMO particulates are formed as highly pure, white fluffy powers or milky-white dispersions. AMO particle types are used in a wide variety of industrial applications, including reinforcement and thickening agents in elastomers, resins, or inks. Additional applications include flow aids for powdery coatings and dry toners, while consumer product applications are represented in components of cosmetic pharmaceuticals, food and feed additives, and beer and wine clarification.

3.1.2 Nanoparticles Used in Medicine

The use of nanomaterials in biology and medicine is one of the most intensely researched areas in nanotechnology. Mass production and ubiquitous use of this novel area are not as common. Because of this, there is ample opportunity for researchers proposing the use of engineered nanomaterials in medicinal application, such as drug formulations and implantable devices, to incorporate tools used routinely in the nanotoxicological sciences. Comprehensive particle characterization in a variety of complex physiologically relevant matrices over time is a necessary component of designing both safe and effective drugs and devices. Analytical tools such as electron microscopy, light scattering, mass spectroscopy, and liquid chromatography can be used in the prediction of the particles' eventual fate and transformation, while dose responses, histopathology, and behavioral assessments of animals can be used in the prediction of the particles' toxicity.

In the last few years, polymer-based particles and metal colloids have become some of the most widely used nanomaterials for medical applications. Polymer-based particles are usually no smaller than 200 nm in diameter, come in a variety of shapes and sizes, and are often encapsulating a drug/molecule or surface-functionalized by a drug/molecule. Polyethylene glycol (PEG), poly(lactic-co-glycolic acid) (PLGA), poly(ethyleneimine) (PEI), poly(styrene sulfonate) (PSS), and poly(allylamine hydrochloride) (PAH) are just some of the polymers used in biomedical applications. Metal colloids, however, are often engineered to be very small particles (i.e., less than 20 nm in diameter), are most commonly spherical or spheroidal, and are surface-functionalized by a drug or biomolecule with an additional linker molecule. Depending on the metal, and the size of the metal particle, these particles could also be used in a variety of rigid shapes such as prisms, tetrahedrons, ellipses, or rods. Gold, silver, iron, copper, and nickel—all transition metals—are the most studied particle types in the literature.

3.2 Characterization of Particles Used in the Industry

3.2.1 Titanium Dioxide

Perhaps the most prevalent industrial applications for TiO_2 are in the use of pigments and paints. Titanium dioxide has a high refractive index property and is characterized by the properties of brightness and natural white coloring, which serve as an effective pigment and optimizes the color of paints, coatings, plastics, papers, food items, and toothpaste products. Light-scattering properties by TiO_2 particles are optimized in the particle size range of 0.2–0.3 μm [1]. In addition, TiO_2 particulates also provide important applications in cosmetic-related functions such as sunscreens. This is because the particle type functions as an ultraviolet (UV) ray scavenger (i.e., reflecting UV rays away from the skin), concomitant with high-refractive-index properties, and resistance to discoloration under UV light conditions. A third major commercial application for TiO_2 particulates is as a catalyst or a photocatalyst function. The crystal structure of anatase is the most effective form and functions best under both UV conditions and visible light. Titanium dioxide can also oxidize organic materials directly. Titanium dioxide is also added to paints, cements, tiles, or other products for its sterilizing, deodorizing, and antifouling properties.

Titanium dioxide is produced in varying particle sizes, as oil and water dispersible, and with numerous coatings for the cosmetic industry. The pigment is used extensively in plastics and other applications for its UV-resistant properties, where it acts as a UV absorber, efficiently transforming destructive UV light energy into heat. Titanium dioxide or zinc oxide particles are integral ingredients in almost every sunscreen formulation containing a physical blocker because of their high refractive index, potent UV light–absorbing capabilities, and resistance to discoloration under UV light. This inherent property serves to augment TiO_2's stability and proclivity to protect the skin from UV light.

With regard to production and synthesis, TiO_2 pigments are manufactured from a number of naturally occurring ores that contain ilmenite ($FeTiO_3$), rutile, anatase, and others sources that are mined from deposits derived from igneous rocks and various sands [2]. The production and processing of titanium dioxide occurs via several stages. The titanium dioxide–containing ore is converted to either titanium tetrachloride or an aqueous titanium-containing sulfate solution. The process then transforms the intermediate substances to crystalline, size-specific rutile or anatase particle types. In the final stages, the pigment is coated, ground, filtered, washed, and dried [1].

Titanium dioxide pigment is produced from titanium-containing ores, either via a chloride or via a sulfate-type process. In the sulfate process, ilmenite or other mined ore is reacted with sulfuric acid and the resulting titanium-containing hydroxide product is precipitated by hydrolysis, filtered, and calcined. In the chloride process, starting materials are converted to titanium tetrachloride, which is oxidized and calcined to facilitate removal of any residual or hydrochloric acid, which may have resulted from the reaction process [3]. There are a number of industrial applications for this mineral. For one thing, it has very high refraction properties. In fact, titanium dioxide is one of the whitest materials known to exist on earth, which has earned it the nickname "titanium white." For this reason, it is often included in many cosmetic preparations to reflect light away from the skin. It is also forms a major component of sunblock formulations to deter the absorption of UV rays from the sun, the concentration of which determines the product's sun protection factor, or SPF. As a pigment, titanium dioxide is used to enhance the white color of certain foods. It also contributes brightness to toothpaste and some medications.

3.2.2 Amorphous Silica

Silicon ranks next to oxygen in abundance in the earth's crust in the form of silicate materials. It is also present as the oxide (of silica in soils/sediments, in some organisms such as diatoms, and in some plants). Since dissolved silica is the bioavailable form for living organisms, the total flux of dissolved silica into the aquatic environment can be compared with anthropogenic sources. AMO is widespread in nature as biogenic silica and nonbiogenic silica glass. Analytical data on the kinetics of AMO particle deposition in the lungs of experimental animals during and after prolonged exposure are largely consistent. The initial uptake phase is followed by dissolution in lung fluids. AMO particulates are rapidly eliminated from lung tissue. In contrast, crystalline forms of silica are durable or biopersistent in the respiratory tract, often producing pathological effects in the lung following inhalation exposures. In addition, following oral ingestion, there is limited accumulation of AMO in body tissues and rapid elimination occurs.

There are a number of different forms or polymorphs of AMO particulates. Some of the most prevalent forms are described below:

- *Fumed silica.* Fumed silica refers to the fume released during the manufacturing process of silicon alloys and elemental silicon. Fumed silica has applications as an anticaking and thickening agent, primarily in powders. Other commercial applications include functioning as a desiccant and in cosmetics for its light-diffusing properties and as an abrasive in toothpaste formulations. The primary particle size is in the range of 5–50 nm, corresponding to specific surface area measures of 50–600 m^2/g [4].
- *Precipitated silica and silica gels.* Precipitated AMO particulates have both internal and external surface areas, which facilitates adsorption of liquids. Precipitated AMO particulates are composed primarily of aggregates of particles of colloidal size that have not been linked to a gel network during the synthetic process. Precipitated silica particle types are formed from a sodium silicate solution and have a more open structure with a higher pore volume than dried silica gels [5]. Precipitated AMO and silica gels have applications as fillers for paint, rubber, and paper, as well as other functions such as grease thickening and carrying agents for catalysts [6].
- *Colloidal silica.* Colloidal silicas are suspensions of fine amorphous-type silica particles in a liquid or aqueous phase. Most colloidal silica suspensions are in a monodisperse form with particle size distributions ranging from approximately 20 nm to 100 nm in diameter. The commercial applications are numerous but include some of the following: absorbent properties, binders, catalysts, lubricants, and abrasion-resistant coatings. An electron micrograph of colloidal silica particles is provided in Fig. 3.1.

3.2.2.1 Production of synthetic amorphous silica

Commercial/synthetic AMO has been classified as "wet process" silica (including precipitated silica and silica gels), pyrogenic (fumed) silica, and surface-modified silica. Surfaces of the modified silica have been rendered hydrophobic, for example, by silyation with dimethyl dichlorosilane.

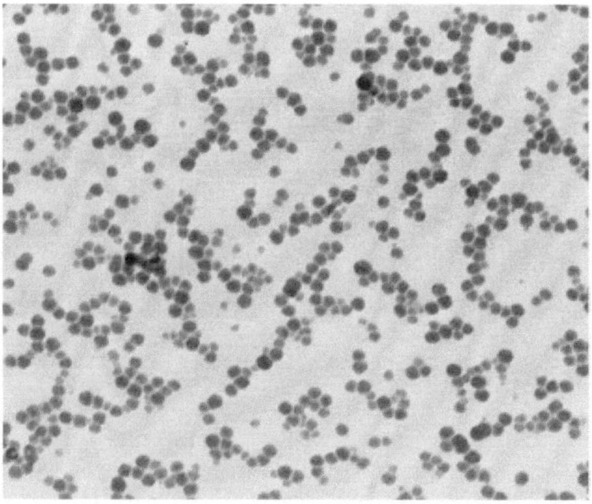

Figure 3.1 Electron micrograph of colloidal silica particles. In terms of size scale, each particle is 20 nm in diameter.

3.2.2.2 Silica production based on the "wet process"

This manufacturing process is based mainly on the precipitation of amorphous silicon dioxide particles from an aqueous alkali metal silicate solution by acid neutralization. Usually, sulfuric acid is used, although carbon dioxide and hydrochloric acid can be used. Depending on the final pH of the solution, the following two different classes of synthetic AMO can be obtained: precipitated silica (obtained in neutral or alkaline conditions) and silica gels (obtained under acidic conditions). The main manufacturing steps include precipitation, filtration, washing, drying, and grinding.

3.2.2.3 Production of pyrogenic silica

The manufacturing process for pyrogenic silica is based mainly on the combustion of volatile silanes, especially silicon tetrachloride, in an oxygen–hydrogen burner. Primary particles (7–50 nm particle size) of AMO fuse together in the high-temperature flame to yield stable aggregates of between 100 nm and 500 nm in diameter. These aggregates form micron-sized agglomerates. The finely divided silica is separated from the hydrochloric acid–containing off-gas stream in filter stations. The hydrochloric acid content of the product is commonly reduced to less than 100 ppm by desorbing the hydrochloric acid with air in a fluid-bed reactor. Pyrogenic silica appears as a fluffy white power.

3.2.2.4 Surface-modified synthetic amorphous silica

All forms of synthetic AMO can be surface-modified either physically or chemically. Methods for chemical modification of the silica particle surface (e.g., silyation) are many and various. Most common treating agents are organosilicon compounds. Less than 10% of the total production volume of synthetic AMO is surface-modified.

3.2.3 Health Risks

The major health risks associated with titanium dioxide and AMO particles are related to the inhalation (i.e., pulmonary) routes of exposure. When considering the extensive pulmonary toxicity database for these particle types, it is generally concluded that both pigment-grade titanium and rutile-type ultrafine forms of titanium dioxide particle types and the various polymorph forms of AMO particulate exposures produce low hazards at most workplace and ambient exposure concentrations. Other forms of ultrafine TiO_2 particle types containing different crystal structures and different particle surface reactivity parameters are potentially more hazardous to the lungs than some forms of rutile TiO_2 particles. However, the anatase:rutile ultrafine forms are much less hazardous when compared to cytotoxic forms of crystalline silica particulates. Similarly, amorphous forms of silica are much less toxic than crystalline polymorphs such as quartz and cristobalite silica. Although amorphous forms are less hazardous than crystalline forms of silica, there is a differential hazard impact among some of the AMO polymorphs. Some forms of AMO such as precipitated and colloidal polymorphs are generally considered to be less active when compared to fumed AMO particulates.

3.3 Characterization of Particles Used in Medicine

One of the most intensely studied areas in nanotechnology is the potential placation of engineered nanomaterials in medicine. The area of research brings scientists and engineers from material sciences and biological sciences together. The result is often collaboration, where both types of scientists have access to a valuable set of research tools and medical devices. Nanomaterials used in medicine hold the promise to not only provide *therapy* to conditions or *diagnosis* to identify conditions but also provide *theranostics* to identify and provide treatment within the same "nanomedicine."

Recently, polymers and metal colloids have become the most commonly used nanomaterials for medical applications. Their utility has ranged from the prevention and treatment of cancer to antimicrobial use to inhibit bacterial growth. While polymer particles usually remain in the interstitial spaces between cells within tissues, metal colloids with a diameter of less than 10 nm have been used as imaging agents or carriers of drugs and have been observed to penetrate the cells and aggregate around the nuclear membrane [7–11]. The majority of published data agrees that the cellular uptake mechanism occurs via endocytosis, an energy-independent cell uptake process [8, 10]. Understanding these differential uptake mechanisms of the cellular entry of various nanoparticles is largely dependent on the physicochemical properties of individual nanoparticles. Properties such as size, shape, and surface charge dictate the localization of particles within tissues and within cellular components. A nanoparticle's intracellular translocation, subcellular locations, and excretion [8] are of utmost importance in order to comprehend possible nanoparticle hazards and their potential for drug delivery or cytotoxic-induced effect.

3.3.1 Polymeric Materials

Polymeric materials on the submicrometer-size scale offer novel nanotechnology platforms that have the potential to impact medicine and biomedical research. Some of the most common properties that these materials are theorized to possess include, but are not limited to:

- Disease diagnosis and imaging: the detection of changes at the molecular level that are indicative of the onset of disease
- Drug delivery and therapy: multifunctional polymeric platforms well suited for combined diagnostics and therapeutics

- Research and development: nanometer-size-scale-enabling technologies accelerating basic research to translational studies

Controlled-release polymeric materials represent an effective carrier platform at the nanometer-size scale [12, 13]. Polymeric materials can be exploited to encapsulate drugs and then subsequently release the drug in a controlled way. The release is controlled through physicochemical properties of the polymer particles, specifically through surface erosion, pore size, and/or polymer swelling. In addition, drug release could also be promoted via the surrounding matrix, environmental conditions (such as temperature or pH), or the presence or absence of active analytes. Polymer platforms for nanomedicine that have the potential to release drugs in a controlled manner are beneficial to doctors and their patients because these systems can decrease a drug's dosage over extended periods while increasing the efficacy of the drug.

3.3.2 Metal Colloids

Colloids are a suspension of small particles dispersed in another substance, such as solvents or buffers. Some common examples of metal-based colloidal suspensions are fluorescent quantum dot nanocrystals and gold nanoshells. A micrograph of quantum dot nanocrystals in human skin cells is shown in Fig. 3.2 below.

These suspensions, like polymeric-based materials, are promising platforms for molecular imaging and in some cases drug delivery applications. Studies have shown that quantum dots may be effective for lymph node mapping [14], and nanoshells have been shown to be effective as a *theranostic* tool in animal models of cancer [15]. Research and development in this field of nanomedicine will likely lead to enhanced surface characteristics for improved biocompatibility and targeting moieties for enhanced sensitivity and specificity [16]. The successful translation of any research and development activity on an engineered nanomaterial platform for biomedical use requires optimization of many physicochemical properties, including (but not limited to):

- Nontoxic
- Efficacious
- Hydrophilic
- Charges' surface

- Particle size on the nanometer-size scale
- Ability to functionalize with ligands for targeting

Figure 3.2 Quantum dot nanocrystals (~16 nm in diameter) are shown distributed through HDFs stained for the nuclear region (blue) the and cytosolic region (red). *Abbreviation*: HDF, human dermal fibroblast.

3.4 Nanomaterial Characterization Methods

3.4.1 Transmission Electron Microscopy and Energy-Dispersive Spectroscopy

Transmission electron microscopy (TEM) has become a necessary characterization tool in nanobiomedicinal research. TEM utilizes wavelengths of electrons to increase resolution and magnification of nanometer-sized inorganic, organic, and biological components. TEM can usually provide information such as the size, shape, agglomeration state, and crystalline structure of particles. It is also used to image cells and tissues to determine the intracellular localization of nanoparticles as well as cellular structural changes.

Preparation for TEM analyses is a multistep process consisting of fixation, dehydration, embedding, and sectioning. Poststaining usually enhances contrast and allows for better visualization of structures. If samples are poststained, however, one must be assured that visualized nanoparticles are not artifacts from the protocol itself.

While TEM is a useful characterization tool, combinations of TEM with alternative methodologies such as energy-dispersive spectroscopy (EDS), electron diffraction (ED), or high-resolution TEM may be useful for determining additional characterization parameters such as chemical composition and speciation. ED is a tool used by materials scientists to determine the crystal structure of a solid particle. High-resolution TEM is used to study the crystallographic structure at the atomic scale. EDS as a microanalytical tool can yield both quantitative and qualitative results on chemical composition.

A common alternative technique to TEM is scanning transmission electron microscopy, STEM. The use of STEM as a characterization tool provides a method to determine the composition of individual nanoparticles in a complex matrix.

3.4.2 Emission and Absorption Spectroscopy

Spectroscopy is one of the most common analytical techniques in nanoscience. This characterization technique measures and interprets various areas of the electromagnetic spectra from either the emission or the absorption of energy by different materials. Information about the chemical composition, structure, surface functionality, and optical/electronic properties of the sample, for example, can be obtained.

Different types of spectroscopic techniques can be used to characterize engineered nanomaterials used for biomedical purposes. For example, mass spectroscopy is used to determine the masses of small electrically charged particles and can be utilized to measure properties of a particle surface, such as bound (absorbed or adsorbed) proteins. Raman spectroscopy is used to study vibrational, rotational, and other low-frequency modes in a system and can be utilized to determine the type and degree of functionalization on the sidewall of a carbon nanotube. Absorption spectroscopy is used to quantify the amount of photons a substance absorbs and can be utilized to measure the size of gold nanoshells. Fluorescence spectroscopy is used to analyze the different frequencies of light emitted by a substance, which is then used to determine the structure of the vibrational levels of that substance.

3.4.3 Dynamic Light Scattering and Zeta Potential

Dynamic light scattering (DLS) (also known as photon correlation spectroscopy, PCS) is used for measuring the size and size distribution of a nanoparticle sample when suspended in a medium. This is accomplished by measuring the Brownian motion of the particle sample. Brownian motion is the random movement of suspended particles due to bombardment of the particulate by solvent molecules. As solvent molecules collide with the nanoparticles, they exert a force that is capable of moving the particles in a random direction. This random movement will be much more exaggerated in a small particle than it will be in a large particle of the same composition over the same time period.

DLS is also used to measure biological samples such as proteins when in a biological buffer. It is difficult to measure single particles using light scattering when aggregates or agglomerates are present in the focal suspension, such as the case for many biological samples. A polydisperse sample mixture may read as if it were void of the smaller fraction of particles. Many nanoparticles tend to agglomerate once in solution, especially when surrounded by complex biological matrices such as a serum or a buffer. This agglomeration state may yield particulates in the micrometer (10^{-6}) range, primarily due to protein absorption onto the surface of a nanomaterial. This protein adsorption makes the nanoparticle size larger due to the increased hydrodynamic radius.

The zeta potential of a nanoparticle represents the charge on the surface, in addition to the adsorbed layer of counter ions. The *adsorbed layer* is generated from oriented solute molecules and ions surrounding the nanoparticle surface. The electric double layer is sensitive to various changes in pH and ionic strength. The zeta potential is also used as a measurement of colloidal particle stability. A zeta potential value above ±30 mV is considered a stable system. Zeta potential values with an absolute value less than 30 mV are considered an unstable system and are prone to agglomeration. This agglomeration is due to the lack of charge–charge repulsion between individual nanoparticles.

3.5 Conclusions

When considering the most accurate and relevant characterization of engineered nanomaterials in industrial and biomedical science, it is import to understand that (1) each nanomaterial is unique, (2) each characterization technique is nontrivial and requires specific preparation methodologies, and (3) each application of a nanomaterial calls for a differing set of characterized properties. In this chapter, we briefly summarized some of the most important collaborative principles currently employed for characterizing nanoparticles in both industrial and biomedical research. In industrial applications, metal oxide nanomaterials, such as titanium dioxide (TiO_2) and silicon dioxide (SiO_2) are commonly used, while in biomedical research, polymer-based materials and metal colloids, such as gold and iron, are often studied. The methods to characterize all of these nanomaterial types range from microscopy to spectroscopy, yet the complex matrix that surrounds each of these materials changes over time. Examples of complex matrices for industrially relevant nanomaterials include cellulose or polymers, and examples of complex matrices for medicinally relevant nanomaterials include salt buffers and blood serum. No matter what the characterization technique or surrounding matrix, comprehensive characterization for toxicity testing is needed. The findings from a variety of recent nanotoxicology studies have correlated hazard effects with some aspects of nanomaterial physicochemical characteristics. Therefore, as a prerequisite, it is critical to rigorously characterize the nanomaterial test substance prior to toxicity testing (i.e., pre-exposure) and then carefully monitor the particle–cellular interactions in the biological system at several time points postexposure. The robust characterization of nanomaterials for toxicity testing should provide vital, required information for better comprehension of the physiological and toxicological effects following exposure in the exposed populations of workers, consumers, and patients undergoing diagnostic or therapeutic procedures.

References

1. Braun, J. H. (1997). Titanium dioxide: a review. *J. Coatings Technol.*, **69**, pp. 59–72.

2. Chang, L. Y. (2002). *Industrial Minerals: Materials, Processes and Uses* (Prentice Hall, Upper Saddle River, NJ).
3. Kischkewitz, J., Woditsch, P., Westerhaus, A., et al. (2002). Pigments, inorganic, in *Ullmann's Encyclopedia of Industrial Chemistry* (Wiley-VCH Verlag GmbH & Co.) [online version].
4. *ACGIH TLV Documentation for Silica, Amorphous-Fume* (1992).
5. Kirk-Othmer (1978). *Encyclopedia of Chemical Technology, Volume 20* (Wiley-Interscience, New York), p. 776.
6. Schepers, G. W. H., et al. (1957). IARC monographs on the evaluation of carcinogenic risks to humans. *Arch. Ind. Health*, **16**, pp. 125–146.
7. Mahmood, M., Kurmakar, A., Fejleh, A., Mocan, T., Iancu, T., Mocan, L., Todea Iancu, D., Xu, Y., Dervishi, E., Li, Z., Biris, A. R., Agarwal, R., Ali, N., Galanzha, E. I., Biris, A. S., and Zharov, V. (2009). Synergistic enhancement of cancer therapy using a combination of carbon nanotubes and antitumor drug. *Nanomedicine,* **4**(8), pp. 1–11.
8. Mu, Q. X., Broughton, D. L., and Yan, B. (2009). Endosomal leakage and nuclear translocation of multiwalled carbon nanotubes: developing a model for cell uptake. *Nano Lett.*, **9**(12), pp. 4370–4375.
9. Xu, Y., Mahmood, M., Fejleh, A., Li, Z., Watanabe, F., Trigwell, S., Little, R. B., Kunets, V. P., Dervishi, E., Biris, A. R., Salamo, G. J., and Biris, A. S. (2010). Carbon-covered magnetic nanomaterials and their application for the thermolysis of cancer cells. *Int. J. Nanomed.,* **5**, pp. 167–176.
10. Kam, N. W. S., Liu, S., and Dai, H. (2006). Carbon nanotubes as intracellular transporters for proteins and DNA: an investigation of the uptake mechanism and pathway. *Angew. Chem., Int. Ed.,* **45**, pp. 577–581.
11. Pantarotto, D., Briand, J., Prato, M., and Bianco, A. (2004). Translocation of bioactive peptides across cell membranes by carbon nanotubes. *Chem. Commun.,* **1**, p. 16.
12. Langer, R. (1998). Drug delivery and targeting. *Nature,* **392**, pp. 5–10.
13. L. Brannon-Peppas, J. O. (2004). Blanchette, nanoparticle and targeted systems for cancer therapy. *Adv. Drug Delivery Rev.*, **56**, pp. 1649–1659.
14. Kim, S., Lim, Y. T., Soltesz, E. G., De Grand, A. M., Lee, J., Nakayama, A., Parker, J. A., Mihaljevic, T., Laurence, R. G., and Dor, D. M. (2004). Near-infrared fluorescent type II quantum dots for sentinel lymph node mapping. *Nat. Biotechnol.*, **22**, pp. 93–97.
15. Loo, C., Lowery, A., Halas, N., West, J., and Drezek, R. (2005). Immunotargeted nanoshells for integrated cancer imaging and therapy. *Nano Lett.*, **5**, pp. 709–711.

16. Farokhzad, O. C., and Langer, R. (2006). *Ad. Drug Delivery Rev.*, **58**, pp. 1456–1459.

17. IARC (2010). *IARC Monographs on the Evaluation of Carcinogenic Risks to Humans: Volume 93: Carbon Black, Titanium Dioxide, and Talc* (Lyon, France), pp. 193–210.

18. IARC (1997). *IARC Monographs on the Evaluation of Carcinogenic Risks to Humans: Volume 68: Silica, Some Silicates, Coal Dust and Para-Aramid Fibrils* (Lyon, France), pp. 41–62.

19. Iavicoli, I., Leso, V., and Bergamaschi, A. (2012). Toxicological effects of titanium dioxide nanoparticles: a review of in vivo studies. *J. Nanomater.*, doi:10.1155/2012/964381.

20. Warheit, D. B., Webb, T. R., Reed, K. L., Frerichs, S., and Sayes, C. M. (2007). Pulmonary toxicity study in rats with three forms of ultrafine-TiO2 particles: differential responses related to surface properties. *Toxicology*, **230**, pp. 90–104.

21. Warheit, D. B. (2001). Inhaled amorphous silica particulates: what do we know about their toxicological profiles? *J. Environ. Pathol. Toxicol. Oncol.*, **20**(Suppl 1), pp. 133–141.

22. Warheit, D. B., McHugh, T. A., and Hartsky, M. A. (1995). Differential pulmonary responses in rats inhaling crystalline, colloidal or amorphous silica dusts. *Scand. J. Work Environ. Health*, **21**(Suppl 2), pp. 19–21.

23. European Centre for Ecotoxicology and Toxicology of Chemicals (2006). *ECETOC JACC Report No. 51. Synthetic Amorphous Silica* (CAS No. 7631-86-9, September, 2006.

Chapter 4

The Synthetic and Biological Identities of Nanomaterials*

Bengt Fadeel[a] and Wolfgang J. Parak[b]
[a]*Nanosafety & Nanomedicine Laboratory, Division of Molecular Toxicology, Institute of Environmental Medicine, Karolinska Institutet, Stockholm, Sweden*
[b]*Fachbereich Physik, Philipps Universität Marburg, Marburg, Germany, and CIC Biomagune, San Sebastian, Spain*
bengt.fadeel@ki.se

4.1 Safety Assessment of Nanomaterials

Engineered nanomaterials are in the same size range as biomolecules and cellular structures, which makes them highly useful for various nanomedicine applications. However, this very same fact also raises concerns regarding the safety of nanomaterials: the interference of man-made nanomaterials with biological systems could lead to hazardous effects on human health and the environment [1]. While the interest in nanoscale materials has increased tremendously in recent years, important observations on their interactions with biological systems were reported already much earlier. For instance,

*This chapter is based, in part, on Ref. [46].

Handbook of Safety Assessment of Nanomaterials: From Toxicological Testing to Personalized Medicine
Edited by Bengt Fadeel

ISBN 978-981-4463-36-2 (Hardcover), 978-981-4463-37-9 (eBook)
www.panstanford.com

the fact that nanoparticles are typically incorporated by cells via endocytosis was known for decades [2]. In addition, colloidal nanoparticles were shown to induce alterations in the blood–air barrier in the mouse lung more than half a century ago [3].

Numerous studies have been published in the past decade on the toxicity potential of engineered nanomaterials [reviewed in 4–6]. However, understanding which physicochemical properties of nanomaterials are driving toxicity remains a challenge. If one could connect material properties (size, shape, surface charge, porosity, colloidal stability, purity/degree of contamination, etc.) with toxicity, then this would enable the prediction of potential hazards and could also lead to the design of nanomaterials with minimal toxicity [7]. In addition, a thorough understanding of the properties of nanomaterials that determine biological responses to these materials would also facilitate the design of novel nanomedicines for the treatment of human disease.

In this chapter, we discuss the synthetic and biological "identities" of nanomaterials. We suggest that a careful understanding of nanomaterial physicochemical properties, that is, the "synthetic identity" as well as the "biological identity" of nanomaterials, determined by the adsorption of biomolecules (proteins, lipids, nucleic acids, etc.) onto the nanomaterial surface upon introduction into a living system, is required. In fact, as pointed out by Walkey and Chan [8] in their recent review, "Once fully mapped, the relationships between synthetic identity, biological identity, and physiological response will enable researchers to predict the physiological response of a nanomaterial by characterizing its synthetic identity." While such conceptual dichotomization into different "identities" may not be fully applicable in the real world, such models are, nevertheless, very useful for our understanding of the behavior of complex systems and may thus aid in the development of *predictive nanotoxicology*, the ultimate goal of which is to decode nanomaterial properties to enable the redesign of materials that are both useful and safe.

4.2 Understanding Nanomaterial Properties

To systematically investigate toxic effects of nanoparticles it would be strongly desirable to correlate their toxic effects with their physicochemical properties [9]. However, unfortunately, this approach is not straightforward, as many physicochemical properties are strongly entangled and are difficult to control independently [7]. Moreover, as pointed out by Schrurs and Lison [10], more coherence in the way that studies are conducted is needed in nanotoxicological research. Nevertheless, in the following section, we discuss studies showing how material properties may be linked to toxicity.

4.2.1 Linking Physicochemical Properties to Toxicity

Size matters, in particular for cellular uptake of nanoparticles. Moreover, in general, the greater the intracellular dose of nanoparticles, the more toxic effects they generate. Chan et al. demonstrated in a series of experiments that there can be an optimal size for nanoparticle uptake [11, 12]. Similar claims have been made by several other authors, but the latter work stands out, as size was controlled in an exclusive way. Indeed, the nanoparticles were colloidally stable and thus were not agglomerated, that is, they did not have the effectively larger diameter of an agglomerate of nanoparticles, and surface chemistry was the same for all sizes. In this way, the size dependence of nanoparticle uptake and cytotoxicity could be investigated. However, it is noted that cellular uptake is not mandatory for cytotoxicity to occur. For example, cobalt–chromium nanoparticles can damage fibroblasts across an intact cellular barrier without having to cross the barrier. The outcome, characterized by DNA damage without significant cell death, was found to be different from that observed in cells directly exposed to nanoparticles [13]. Size may also influence the behavior of nanoparticles in vivo. Choi et al. [14] studied the fate of intratracheally instilled nanoparticles that were varied in size, surface modification, and core composition in a rat model after lung instillation. They showed that nanoparticle distribution strongly depends on the hydrodynamic radius of the particles. Moreover, nanoparticle behavior also depended on surface coating, which affected protein adsorption, which, in turn, affected the hydrodynamic size of the particles. Furthermore, using a panel of well-characterized, radio-labeled gold nanoparticles, Kreyling et al. [15] showed in a rat model that translocation across the air–blood barrier and accumulation and retention in secondary organs are two distinct processes, both depending specifically on particle physicochemical properties, namely, specific surface area (size) and surface charge.

Also shape can be important. The most frequently cited example is carbon nanotubes, which are thought to exert toxicity by virtue of their fiber-like shape. This may be true for multiwalled carbon nanotubes (MWCNTs) with high width and, therefore, high rigidity [16]. Interestingly, in the latter study, thin and thick nanotubes similarly affected macrophages, while the deleterious effects on mesothelial cells were shown to be diameter dependent. However, it is important to ask when a fiber is a fiber and when it is, effectively, a particle. Murray et al. [17] have shown that agglomeration is an important factor when assessing the in vivo toxicity of single-walled carbon nanotubes (SWCNTs). Shape can influence the mode of cellular uptake. In the case of a rod-shaped nanoparticle and a spherical nanoparticle of the same volume the leading edge of the rod-shaped nanoparticle has a much smaller cross section and may therefore penetrate cell membranes more effectively. Schaeublin et al. [18] investigated two gold nanoparticles with different aspect ratios using a keratinocyte cell line and found that gold nanospheres were nontoxic, whereas the gold nanorods induced apoptosis. Notably, both nanoparticles formed agglomerates in the cell culture medium but the spherical particles had a large fractal dimension (i.e., were tightly bound and densely packed), while the nanorod agglomerates had a smaller fractal dimension (i.e., were loosely bound). However, in many studies, in particular in theoretical simulations, aspect ratios are calculated for the nanoparticle cores, neglecting the surface coating and the adsorbed protein corona (see below), which reduces the effective aspect ratio and thus may nullify potential shape effects. Furthermore, agglomeration in physiological media may rule out effects of the shape of individual nanoparticles [19]. An agglomerate of "sharp" (high aspect ratio) particles is no longer "sharp."

Surface charge strongly influences uptake of nanoparticles. In general, positively charged nanoparticles are incorporated more rapidly by cells than negatively charged ones, which is typically explained by the overall net negative charge of cellular surfaces. Though studies exist that demonstrate that in some cases positively charged nanoparticles interact with cells differently when compared to negatively charged ones, resulting in different mechanisms of cytotoxicity, the higher toxicity of positively charged nanoparticles is generally correlated to their enhanced cellular uptake [20]. To elucidate surface charge-dependent toxicity, nanoparticles with different surface charges but with other physicochemical parameters constant are required, which often is experimentally complicated to achieve [21, 22]. It has been noted that the protein corona tends to give nanomaterials a zeta potential of about −10 mV to −20 mV irrespective of the nanomaterial chemistry. This "normalization" of zeta potentials is related to the fact that most plasma proteins carry a net negative charge at physiological pH [8]. However, a recent study has shown that positively charged nanoparticles were incorporated by cells to a larger extent than negatively charged ones, both in serum-free and serum-containing media, that is, with and without a corona [23]. Thus, internalization depends directly on the sign of the net charge. Cytotoxicity assays revealed a higher cytotoxicity for the positively charged nanoparticles. Notably, due to the high colloidal stability of the nanoparticles, the effects of charge could be examined virtually without interference by changes in size or colloidal stability [23].

In fact, perhaps the most important physicochemical parameter that interferes with other parameters is colloidal stability. Obviously, agglomerated nanoparticles do not have the size of the individual nanoparticles but the size of the agglomerate. This means that unless nanoparticles are very well dispersed, any statement about size- or shape-dependent uptake or cytotoxicity is not sound, as the cell would interact with the agglomerates and not with the individual nanoparticles. Besides the fact that agglomeration masks effects of other physicochemical parameters, it also can directly affect interaction with cells. "Sticky" agglomerates of nanoparticles tend to precipitate on top of cells and thus can cause cytotoxic effects [24].

Many metal and metal oxide nanoparticles can undergo dissolution within acidic compartments (lysosomes) in the cell, which could drive toxicity. This phenomenon, sometimes referred to as a Trojan horse–type uptake mechanism because it circumvents the plasma membrane barrier and allows toxic ions to "sneak" into cells, has been shown, for instance, for oxides of zinc, iron, manganese, and cobalt [25]. Gliga et al. [26] reported that silver nanoparticles exert cytotoxicity in the human lung cell line BEAS-2B as a function of the intracellular Ag release, as determined using atomic absorption spectroscopy, while the fraction of Ag that was released extracellularly did not induce any cytotoxicity. Cho et al. [27] evaluated the pulmonary inflammogenicity of 15 different metal/metal oxide nanoparticles and showed that toxicity of the nanomaterials displayed a significant correlation with one of two physicochemical parameters, zeta potential under acid conditions for low-solubility nanoparticles and solubility (degree of dissolution) for high-solubility nanoparticles. The authors suggested that in the case of high-solubility nanoparticles, inflammogenicity depends on the ions that are produced during dissolution of nanoparticles inside the acidic phagolysosomes of the cells.

Catalytic effects at the nanoparticle surface play an important role in the generation of reactive oxygen species (ROS) (see the review by Nel et al. in Ref. [28]). Sayes et al. [29] studied the effects of titanium dioxide nanoparticles in cell culture and found that the extent to which nanoscale titania affected cellular behavior was not dependent on surface area. However, the phase composition of the nanoscale titania did correlate strongly to cytotoxicity, insofar as anatase TiO_2 was 100 times more toxic than rutile TiO_2. The most cytotoxic nanoparticle samples were also the most effective in generating ROS.

In summary, it may seem disappointing that in most of the cases one cannot directly pinpoint how a certain physicochemical parameter influences the toxicity of nanomaterials. This is due, in part, to the fact that many studies published to date are based on poorly defined nanoparticles, in which many physicochemical parameters are entangled. In addition, not all nanomaterials are created equal, and it remains possible that classes of nanomaterials may behave differently: size or shape or surface charge may drive toxicity for certain categories of nanomaterials, while the rate of dissolution may be a primary determinant of toxicity for others. To be more conclusive, toxicity studies should be performed with well-defined model nanoparticles in which specific particle physicochemical properties can be independently varied. In addition, attention should be given to the potential role of the biocorona, discussed below. However, some general trends can be extracted from the available literature. Thus, generally speaking, small, sharp, and positively charged nanoparticles are internalized in vitro to a higher extent by cells than big, flat, and negatively charged ones, and consequently such nanoparticles are also often found to be more toxic. Again, such trends should not be confused with statements of direct correlation, which would be of a much stronger nature.

4.2.2 Predictive Modeling of Nanomaterial Toxicity

The so-called 21st-century toxicology paradigm aims to employ high-throughput, systems biology, and computational toxicology approaches to yield data predictive of results from animal toxicity studies, which will allow prioritization of chemicals for further testing and assist in prediction of risk to humans [30]. This paradigm may also be applied to engineered nanomaterials. Nel et al. [31] provided an excellent summary of potentially useful mechanistic injury pathways that could be employed to develop *predictive* toxicological paradigms for nanomaterials. Indeed, high-throughput screening (HTS), based on assays that capture specific mechanistic injury pathways, would enable rapid testing of vast numbers of new nanomaterials [32].

In an illustrative example of high-content screening, Shaw et al. [33] reported on the assessment of nanoparticle effects using multiple cell types and multiple in vitro assays. The cell types were selected to reflect a range of tissues relevant for evaluation of intravascularly administered (nanomedicine) agents: vascular cells (endothelial and smooth muscle cells), monocytes, and hepatocytes. Hierarchical clustering of the data identified nanomaterials with similar patterns of biologic activity across a broad sampling of cellular contexts, yielding robust structure–activity relationships for the nanomaterials tested. Furthermore, a subset of nanoparticles was tested in mice, and nanoparticles with similar activity profiles in vitro exerted similar effects in vivo, using the monocyte number as the end point. This data suggested a strategy of multipronged characterization of nanomaterials in vitro that can inform the design of novel nanomaterials and guide studies of in vivo activity [33]. George et al. [34] applied high-content screening methods for rapid cytotoxicity screening of metal oxide nanoparticles. The authors developed a fluorescence assay that simultaneously measured four different responses in cells to oxidative stress caused by ZnO, CeO_2, and TiO_2 nanoparticles. Using the information, the authors were able to reduce the cytotoxicity of ZnO nanoparticles by decreasing

ZnO dissolution through Fe doping. In another very illustrative example, Zhang et al. [35] were able to demonstrate for 24 metal oxide nanoparticles that it is possible to use conduction band energy levels to delineate their toxicological potential under illumination at the cellular and whole-animal level. The latter study is important because it points toward a predictive toxicological paradigm, at least for one class (metal oxides) of nanomaterials. These examples suggest that high-content and high-throughput screening is a powerful tool for the prediction of nanomaterial hazards. However, the fact that nanomaterials may interfere with commonly used in vitro assays needs to be taken into account [36]. Standardized and validated toxicity and ecotoxicity tests for nanomaterials are therefore needed [37].

In addition to experimental approaches for hazard assessment of nanomaterials, there is a need for in silico methods with which one can develop structure–activity relationships. Indeed, this paradigm allows for predicting of toxicological effects induced by chemicals on the basis of their structural similarity with other chemicals for which toxicological end points have been previously measured. These structure–toxicity relationships can be quantitative or qualitative in nature, and they can predict toxicological effects directly from the physicochemical properties of the entities/chemicals of interest [38]. Therefore, this approach can aid in prioritizing resources in toxicological investigations, while reducing the ethical and monetary costs that are related to animal testing. Quantitative structure–activity relationship (QSAR) models have been successfully applied in the pharmaceutical field and in regulatory toxicology [38] (see Chapter 7 for a further discussion). While QSAR approaches have only recently been used to predict biological effects of nanomaterials, some encouraging initial results have already shown. Puzyn et al. [39] reported on the cytotoxicity of 17 different types of metal oxide nanoparticles to *Escherichia coli*, and Epa et al. [40] generated quantitative, predictive, and informative models that describe QSAR relationships for cellular uptake and apoptosis induced by metal oxide nanoparticles in several types of cells. Fourches et al. [41] studied the cellular uptake of 109 different nanoparticles with similar cores but diverse surface modifiers. Interestingly, the authors noted that the cellular uptake of nanoparticles possessing the same metal core but different organic molecules on their surface can be predicted by taking into account the chemical structure of the coating molecules. Therefore, the structural determinants of the biological behavior of nanoparticles can be found both at the core of nanoparticles and at their surface. Indeed, the biocorona is a key determinant of the "identity" of a nanomaterial and should be taken into account when modeling the behavior of nanomaterials in biological systems. To this end, Xia et al. [42] recently developed a *biological surface adsorption index* to predict the molecular interactions of nanoparticles with proteins on the basis of the competitive adsorbtion of a set of small-molecule probes onto a set of nanoparticles. The measured adsorption coefficients were used to develop descriptors, which, in turn, could be used for in silico modeling to predict the adsorption of small molecules to other nanomaterials. Finally, in a recent study, serum protein corona "fingerprinting" of a library of 105 surface-modified gold nanoparticles was performed, implicating a set of hyaluronan-binding proteins as mediators of nanoparticle–cell interactions [43]. Notably, the authors developed a multivariate model that used the protein corona "fingerprint" to predict cell association 50% more accurately than a model that uses parameters describing nanoparticle size, aggregation state, and surface charge.

4.3 The Nanomaterial Biocorona

When nanomaterials are presented to living systems they interact with biomolecules, including proteins, lipids, and polysaccharides, conferring a "biological identity" to the engineered nanostructures (see Refs. [44, 45] for recent reviews). In the following sections, we discuss the biocorona concept and its implications for the toxicological outcomes of nanomaterial exposure, as well as for clinical applications of nanomaterials, not least for targeted drug delivery.

4.3.1 The Biocorona Concept

The biocorona is the dynamic layer of biomolecules that adsorbs to nanoparticle surfaces immediately upon their contact with living systems and is presumed to be what organisms or cells "see" and interact with [45]. Thus, as we have suggested in a recent review

[46], and as exemplified by several studies cited in the present chapter, it is the combination of material-intrinsic properties (the "synthetic identity") and context-dependent properties determined, in part, by the biocorona of a given biological compartment (the "biological identity") that will determine the interactions of engineered nanomaterials with cells and tissues and subsequent (toxicological) outcomes. Nanomedicines are designed to interact with biological systems at the nanoscale. Thus, to improve biosafety of novel nanomedicines, one must understand the bio–nano interface between these man-made nanostructures and cellular nanostructures or "nanomachines" [47]. Of course, the observation that proteins interact with, and opsonize, particles is not new. Saba and Di Luzio [48] noted, in a study published almost half a century ago on phagocytosis of a variety of particles as a function of opsonization, that "it is now obvious that the removal of colloidal and particulate materials, either inert or viable, is dependent, in part, on such variables as the physico-chemical nature of the particle, particle number, particle size, surface charge, blood flow, opsonic capacity, and species, as well as the functional state of the reticuloendothelial cell." Indeed, recent studies on nanoparticles have shown that the adsorption of biomolecules onto the surface of nanomaterials is linked to nanomaterial-intrinsic properties, such as size/surface curvature, hydrophobicity, and surface charge [49–51]. In a recent study, the surface topography of colloidal mesoporous silica nanoparticles was shown to play a key role in driving the interactions with proteins [52]. Walkey and Chan [8] compiled data on the plasma-derived protein corona on nanoparticles from 26 published studies and concluded that "the protein corona is complex, that there is no one 'universal' plasma protein corona for all nanomaterials, and that the relative densities of the adsorbed proteins do not, in general, correlate with their relative abundances in plasma" (in other words, there is a degree of specificity). In this context, it needs to be pointed out that there are hundreds of abundant plasma-derived proteins, while depending on their size only a few tens of proteins can adsorb on the surface of one nanoparticle. Thus, there will be a huge variability in the distinct protein coronas around individual nanoparticles, and mass spectrometry–derived compositions of coronas around nanoparticle ensembles describe only the average coverage but not what is

present on each nanoparticle. It was suggested that the protein corona depends on the "synthetic identity" of each nanomaterial [8]. Furthermore, the formation of a biocorona on nanoparticles is a bilateral phenomenon, as proteins that adsorb to nanoparticle surfaces may also alter their behavior as a result of unfolding [53] or fibrillation [54]. Indeed, Mortimer et al. [55] demonstrated that the exposure of cryptic epitopes of albumin upon binding of albumin to layered silicate nanoparticles or "platelets" determines macrophage clearance of these particles via scavenger receptors. Thus, the synthetic and biological identities of nanomaterials are interrelated.

Dawson et al. have suggested differentiating between a "hard" (long-lived) corona of slowly exchanging proteins and an outer "soft" corona collection of (weakly interacting and rapidly exchanging) proteins [56, 57]. It has been suggested recently that the difference between the "hard" and "soft" corona might be in first order determined by the dissociation constant of the proteins on the nanoparticle surface [58]. In fact, overall, the literature on the protein corona on nanoparticles is largely based on qualitative data, while more quantitative data on, for instance, dissociation constants, on- and off-rate coefficients, etc., is needed and would further our understanding of the bio–nano interface. Notably, the experimental design, with separation of unbound proteins from protein–nanoparticle conjugates, needs to be considered when decoding the protein corona [58]. Nonetheless, it is generally believed that while the hard corona is formed because of the direct interaction of proteins with the surface of the nanoparticles, instead protein–protein interactions dominate the interactions of the soft corona with the hard corona. It should also be noted that changes in the hard corona may occur when nanoparticles are transferred to a new compartment, for example, upon translocation of nanoparticles into cells, which may influence their toxicity [59].

4.3.2 Pathophysiological Impact of the Biocorona

The protein corona has been shown to play an important role in modulating uptake and toxicity of SWCNTs [60, 61]. However, it remains to be firmly established whether the biological identity of nanoparticles, and the attendant toxicity, is dependent on a specific protein(s) in the nanoparticle corona or is a nonspecific effect related to the fact that proteins may alter the agglomeration behavior of nanoparticles, leading, for instance, to a difference in cellular uptake. Ehrenberg et al. [62] reported that the capacity of polystyrene nanoparticle surfaces to adsorb proteins is indicative of their tendency to associate with cells. However, removal of the most abundant proteins from cell culture media did not affect the level of cell association, and the authors therefore concluded that cellular association is not dependent on the identity of adsorbed proteins. Lartigue et al. [63] studied the adsorption of proteins on iron oxide nanoparticles and assessed the effect of plasma at different concentrations ranging from 1% to 100% on nanoparticle behavior. It was noted that at low plasma concentrations (representative of most in vitro conditions), the nanoparticles tended to form clusters triggered by proteins like fibrinogen, whereas at high plasma concentration (closer to the physiological situation) other proteins such as apolipoproteins tended to coat and subsequently to stabilize individual nanoparticles. This, in turn, affected in vitro uptake by macrophages. Lesniak et al. [64] reported that silica nanoparticles incubated with lung carcinoma cells in the absence of serum have a stronger adhesion to the cell membrane and higher internalization efficiency when compared to nanoparticles with a preformed surface corona. This finding may be a peculiarity of the cell model used as most phagocytic cells (macrophages) are more prone to engulf opsonized particles and microorganisms. Saha et al. [65] reported that the formation of a plasma protein corona on nanoparticle surfaces protects from both hydrophilic and hydrophobic nanoparticle-mediated hemolysis, while Dobrovolskaia et al. [66] found that the composition of the plasma protein corona did not correlate with hematocompatibility of colloidal gold nanoparticles. Finally, in a recent study using silica and polystyrene nanoparticles of various size and surface functionalization, Tenzer et al. [67] found that plasma protein adsorption was rapid and affected hemolysis, thrombocyte activation, cellular uptake, and endothelial cell death. The latter study—and the recent study by Walkey et al. [43]—serves as an excellent example of the comprehensive approach that is needed to derive structure–activity relationships for nanomaterials.

While most of the examples cited above failed to pinpoint a role for any specific corona protein(s), or where not designed to do so, Deng et al. [53] demonstrated that negatively charged poly(acrylic acid)-conjugated gold nanoparticles bound to and induced unfolding of fibrinogen, which promoted interaction with integrin receptors on THP.1 cells, resulting in the release of inflammatory cytokines. In a follow-up study, the authors showed that fibrinogen bound with high affinity to positively and negatively charged gold nanoparticles [68]. However, only the negatively charged nanoparticles triggered cytokine release in THP.1 cells, perhaps due to a different orientation of the protein on the different particles. Thus, while common proteins may bind to different nanoparticles, the physiological response may differ.

The complement system constitutes an important barrier to infection or other foreign intrusion. Nanoparticles may also activate the complement system. This may be viewed as a special case of biocorona formation and one that is of particular relevance in nanomedicine (see Chapter 10 for a detailed discussion). Nanomaterial interaction with the complement system is complex and regulated by inter-related physicochemical parameters such as size, morphology, and surface properties [69]. Hamad et al. [70] investigated polystyrene nanoparticles with surface-projected polyethyleneoxide (PEO) chains in "mushroom-brush" and "brush" configurations and found that distinct polymer architectures mediate the switching of complement activation pathways.

The majority of biocorona studies have been performed with plasma proteins [71], which is certainly relevant in cases when nanoparticles are administered into the bloodstream. Nevertheless, it is important to also consider other portals of entry of nanomaterials into the body, for example, via inhalation, through the skin, or via the gastrointestinal tract, as the corona composition is likely to change as a function of the anatomical site and the specific biofluids encountered at each of these sites. Kapralov et al. [72] reported on the in vivo formation of a lipid–protein corona on the surface of SWCNTs following administration by pharyngeal aspiration in mice. The biocorona was identical to the lung surfactant, and subsequent in vitro studies demonstrated a role for the surfactant corona of lipids and proteins in RAW264.7 macrophage uptake of SWCNTs.

4.3.3 Implications of the Biocorona for Targeting

Nanomedicine offers tantalizing opportunities for targeted drug delivery, potentially combined with simultaneous imaging modalities to monitor disease processes and/or drug delivery to the desired tissues [73]. Indeed, nanoparticles functionalized with either transferrin or prostate-specific antigen for delivery of short interfering RNA (siRNA) or drugs have been applied in human cancer patients [74, 75]. The biocorona should be taken into consideration in the design of targeted nanomedicines: the interplay between the engineered "corona" of targeting ligands (such as transferrin) and the adsorbed biocorona that is formed in situ upon administration to patients must be understood. Salvati et al. [76] recently claimed in an in vitro study using transferrin-functionalized nanoparticles that the adsorption of serum proteins obscures the targeting ligand grafted to the nanoparticles, thereby preventing their targeted uptake. While it remains debatable if this is a generally applicable quantitative effect, this study is important as it underscores that nanoparticles (nanomedicines) need to be evaluated under relevant conditions and cannot be viewed as pristine entities. In contrast, however, recent data from our laboratory (BF) suggests that specific, receptor-mediated uptake of folic acid–functionalized nanoparticles by ovarian cancer cells occurs only in the presence, but not in the absence, of serum [77]. We speculated that although excessive protein adsorption could perhaps partially mask targeting ligands, a degree of protein binding may serve to stabilize nanoparticles, thereby facilitating the presentation of targeting ligands and subsequent receptor-mediated uptake. It may be pertinent in this context to ask whether the biocorona covers the nanoparticle surfaces completely or whether (some) targeting ligands remain accessible. Using superparamagnetic iron oxide nanoparticles (SPIONs), Simberg et al. [78] found that both the dextran coat and the iron oxide core of dextran-coated SPIONs remained accessible to specific probes after their incubation in plasma, suggesting that the nanoparticle surface could be available for recognition by cells, regardless of protein coating. Importantly, in a recent preclinical study, Pridgen et al. [79] demonstrated targeting of nanoparticles to the neonatal Fc receptor (FcRn) following oral administration. The clinical effect of the targeted, insulin-containing nanoparticles was abolished in FcRn-deficient mice, thus providing evidence that the enhanced nanoparticle transport across the intestinal epithelium was due to FcRn.

In nanomedicine, grafting poly(ethylene glycol) (PEG) onto nanoparticle surfaces is often undertaken in order to avoid nonspecific protein adsorption and clearance by cells of the reticuloendothelial system, thereby extending the half-life of nanoparticles in circulation. However, PEGylation does not prevent protein adsorption altogether [80]. Sacchetti et al. [81] showed that PEG conformation on SWCNTs may influence the protein corona. In fact, PEG conformation had a greater influence on the repertoire of corona proteins than the SWCNTs' surface charge, and changing PEG conformation from "mushroom" to "mushroom brush" affected the pharmacokinetic profile of the SWCNTs in mice.

On the other hand, one may consider exploiting the biocorona phenomenon for loading and delivery of drugs [82, 83]. Prapainop et al. [84] attempted cell-specific uptake of nanomaterials by "reprogramming" the protein corona. Specifically, the surface of CdSe/ZnS quantum dots possessing an amino-functionalized, PEGylated hydrophilic surface was decorated with the inflammatory metabolite, cholesterol 5,6-secosterol atheronal-B, and the resulting nanoparticles were shown to bind to and induce the misfolding of apolipoprotein B, leading to uptake by RAW264.7 murine macrophages. Caracciolo et al. [85] provided evidence that nanoparticles can acquire a selective targeting capability through the adsorption of a protein corona. The authors demonstrated that lipid/DNA nanoparticles, upon interaction with human plasma, spontaneously became coated with vitronectin, which, in turn, promoted efficient uptake by cancer cells expressing high levels of the vitronectin $\alpha v \beta 3$ integrin receptor. Furthermore, Pozzi et al. [86] exploited the fact that cationic liposomes adsorbed an abundance of apolipoproteins to deliver the liposomes to human prostate cancer cells that express a high level of scavenger receptor class B type 1.

4.3.4 Nanoparticles vs. Molecules: The Case of Dendrimers

Stark [25] has pointed out that nanoparticles differ from molecules in several respects. Nevertheless, he concluded that "from a functional point of view, chemically well-defined nanoparticles are an extension of the classical concept of the molecule," as they combine the properties of solids with mobility (a property of molecules).

Dendrimers are polymeric nanoparticles with a perfectly defined structure and molecular weight [87]. Hayder et al. [88] reported that azabisphosphonate (ABP)-capped dendrimers selectively target monocytes and directs them toward anti-inflammatory activation. The dendrimers also exhibited antiosteoclastic activity, thus preventing bone erosion. Intravenous injections of ABP-capped dendrimers inhibited the development of inflammatory arthritis in two animal models. This exciting study suggests that dendrimers could function as novel therapeutics for rheumatoid arthritis. Moreover, dendrimers conjugated to glucosamine and glucosamine 6-sulfate have been shown to possess immunomodulatory and antiangiogenic properties, respectively, and when administered together, the nanoparticles increased the long-term success of glaucoma surgery in an animal model by preventing scar tissue formation [89]. Thus, the synthetic and biological identities of a nanomaterial may, in a manner of speaking, "blend" into one: dendrimers may function as drugs per se by virtue of their unique physicochemical properties, that is, size and multivalent surface functionalities, which allow these nanoparticles to directly engage biological receptors and modulate cell function. Do dendrimers, which are comparable in size and density to most blood proteins, adopt a biocorona of proteins on their surface? Some studies suggest that this is the case (and that complement proteins, in particular, are prone to bind to dendrimers in a generation-dependent manner) [90]. However, conceptually, this is not so much a case of proteins coating a particle as a form of heteroaggregation, as the particles and the proteins are of similar size. On the other hand, the biocorona is not necessarily composed only of proteins. In fact, polyamidoamine (PAMAM) dendrimers are essentially nucleic acid scavengers, and cationic PAMAM G-3 dendrimers were shown in a preclinical study to prevent thrombosis following carotid artery injury and pulmonary thromboembolism through the binding of prothrombotic nucleic acids without significantly increasing bleeding [91].

4.4 Future Perspectives

The fundamental challenge in nanosafety (nanotoxicology) research—and one that is equally relevant for medical applications

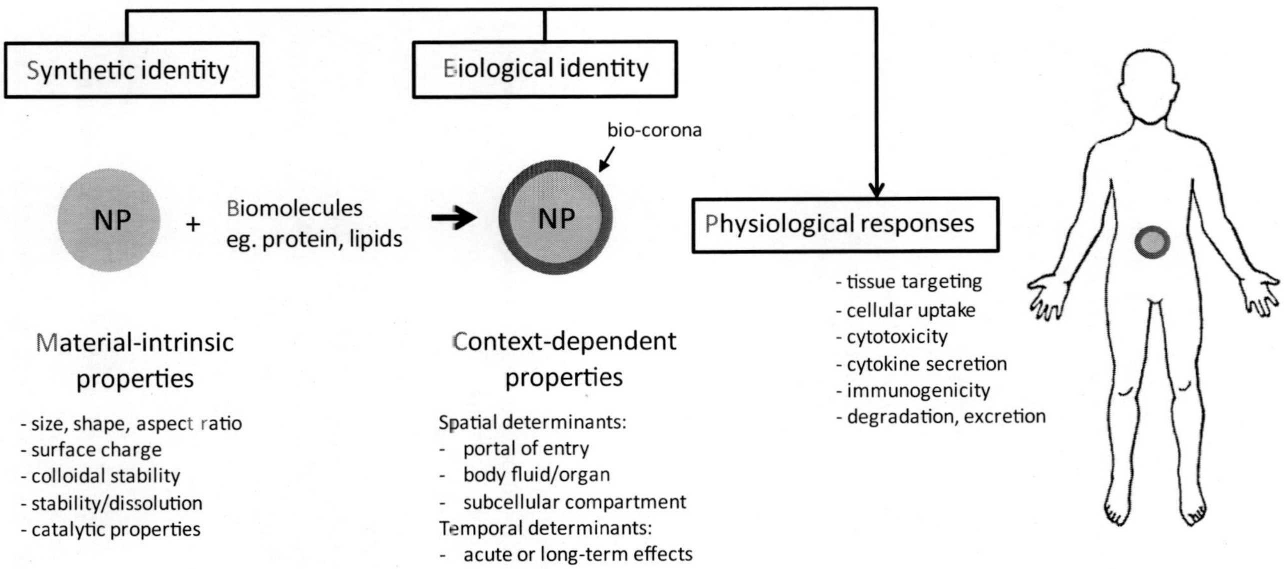

Figure 4.1 Synthetic and biological identities of nanomaterials. This schematic depicts the "synthetic" identity of nanomaterials, which is determined by material-intrinsic properties, and the "biological" identity, which is manifested in a living system and can be viewed as the sum of the context-dependent properties of the nanomaterial. As discussed in this chapter, the biological identity is shaped, in part, by the adsorption of biomolecules (proteins, lipids, etc.) that form a "corona" on the surface of nanoparticles; the composition of the biocorona depends on the particular biofluid (e.g., blood, lung fluid, gastrointestinal fluid, etc.) and may exhibit dynamic changes as the nanoparticle passes from one biological compartment to another. The physiological responses to nanomaterials are dictated by the synthetic and biological identities. A partial list of possible biological/toxicological outcomes is shown in this figure. Adapted from Ref. [46], with permission from Wiley.

of nanotechnologies—is how to correlate nanomaterial physicochemical properties with biological responses at the organismal, cellular, and subcellular levels [92]. Many studies in the past were hampered by a lack of characterization of the nanoparticles used, making it difficult to ascribe their biological behavior to a specific property [7]. In addition, insufficient attention was given to the "biological identity" of nanomaterials in living systems. More recently, studies have emerged on the biocorona on nanoparticles (Fig. 4.1), but there is still a paucity of quantitative information to guide our understanding of the biocorona [58].

Careful assessment of material-intrinsic properties and how these properties are linked to physiological responses is thus essential both in nanotoxicology and in nanomedicine [93]. Notably, the very same property may be highly desirable for certain clinical applications (e.g., the use of small particles to exploit the enhanced permeability and retention [EPR] effect) but could also yield unwanted hazardous effects. Taking into consideration not only the synthetic identity but also the biological identity of nanomaterials, and how these identities may evolve over time and as a function of different biological compartments in the body or at the subcellular level, may enable a better understanding of nanomaterial-induced physiological responses [94]. New approaches, including HTS for rapid screening and ranking of the hazard potential of vast numbers of engineered nanomaterials and mathematical modeling of structure–activity relationships of nanomaterials, may also facilitate the development of safe and useful nanomaterials for in vivo imaging, drug delivery, and other clinical applications. Overall, we subscribe to the view that nanoparticles in the context of in vitro and in vivo studies may be viewed as *biological entities* and that they should be studied (and regulated) as such [95]. However, material-intrinsic properties also play a key role in their behavior, as outlined in this chapter.

References

1. Shvedova, A. A., Kagan, V. E., and Fadeel, B. (2010). Close encounters of the small kind: adverse effects of man-made materials interfacing with the nano-cosmos of biological systems. *Annu. Rev. Pharmacol. Toxicol.*, **50**, pp. 63–88.

2. Harford, C. G., Hamlin, A., and Parker, E. (1957). Electron microscopy of HeLa cells after the ingestion of colloidal gold. *J. Biophys. Biochem. Cytol.*, **3**, pp. 749–756.

3. De Groodt, N., Lagasse, A., and Sebruyns, M. (1958). Alterations in the ultrastructure of the blood-air barrier in the mouse lung after inhalation of colloidal gold particles. *Nature*, **181**, pp. 1418–1419.

4. Lewinski, N., Colvin, V., and Drezek, R. (2008). Cytotoxicity of nanoparticles. *Small*, **4**, pp. 26–49.

5. Stern, S. T., and McNeil, S. E. (2008). Nanotechnology safety concerns revisited. *Toxicol. Sci.*, **101**, pp. 4–21.

6. Feliu, N., and Fadeel, B. (2010). Nanotoxicology: no small matter. *Nanoscale*, **2**, pp. 2514–2520.

7. Rivera-Gil, P., Jimenez de Aberasturi, D., Wulf, V., Pelaz, B., del Pino, P., Zhao, Y., de la Fuente, J. M., Ruiz de Larramendi, I., Rojo, T., Liang, X. J., and Parak, W. J. (2013). The challenge to relate physicochemical properties of colloidal nanoparticles to their cytotoxicity. *Acc. Chem. Res.*, **46**, pp. 743–749.

8. Walkey, C. D., and Chan, W. C. (2012). Understanding and controlling the interaction of nanomaterials with proteins in a physiological environment. *Chem. Soc. Rev.*, **41**, pp. 2780–2799.

9. Rivera-Gil, P., Oberdörster, G., Elder, A., Puntes, V., and Parak, W. J. (2010). Correlating physico-chemical with toxicological properties of nanoparticles: the present and the future. *ACS Nano*, **4**, pp. 5527–5531.

10. Schrurs, F., and Lison, D. (2012). Focusing the research efforts. *Nat. Nanotechnol.*, **7**, pp. 546–548.

11. Chithrani, B. D., Ghazan, A. A., and Chan, C. W. (2006). Determining the size and the shape dependence of gold nanoparticle uptake into mammalian cells. *Nano Lett.*, **6**, pp. 662–668.

12. Chithrani, B. D., and Chan, W. C.W. (2007). Elucidating the mechanism of cellular uptake and removal of protein-coated gold nanoparticles of different sizes and shapes. *Nano Lett.*, **7**, pp. 1542–1550.

13. Bhabra, G., Sood, A., Fisher, B., Cartwright, L., Saunders, M., Evans, W. H., Surprenant, A., Lopez-Castejon, G., Mann, S., Davis, S. A., Hails, L. A., Ingham, E., Verkade, P., Lane, J., Heesom, K., Newson, R., and Case, C. P. (2009). Nanoparticles can cause DNA damage across a cellular barrier. *Nat. Nanotechnol.*, **4**, pp. 876–883.

14. Choi, H. S., Ashitate, Y., Lee, J. H., Kim, S. H., Matsui, A., Insin, N., Bawendi, M. G., Semmler-Behnke, M., Frangioni, J. V., and Tsuda, A. (2010). Rapid translocation of nanoparticles from the lung airspaces to the body. *Nat. Biotechnol.*, **28**, pp. 1300–1303.

15. Kreyling, W. G., Hirn, S., Möller, W., Schleh, C., Wenk, A., Celik, G., Lipka, J., Schäffler, M., Haberl, N., Johnston, B. D., Sperling, R., Schmid, G., Simon, U., Parak, W. J., and Semmler-Behnke, M. (2014). Air-blood barrier translocation of tracheally instilled gold nanoparticles inversely depends on particle size. *ACS Nano*, **8**, pp. 222–233.

16. Nagai, H., Okazaki, Y., Chew, S. H., Misawa, N., Yamashita, Y., Akatsuka, S., Ishihara, T., Yamashita, K., Yoshikawa, Y., Yasui, H., Jiang, L., Ohara, H., Takahashi, T., Ichihara, G., Kostarelos, K., Miyata, Y., Shinohara, H., and Toyokuni, S. (2011). Diameter and rigidity of multiwalled carbon nanotubes are critical factors in mesothelial injury and carcinogenesis. *Proc. Natl. Acad. Sci. U S A*, **108**, pp. E1330–1338.

17. Murray, A. R., Kisin, E. R., Tkach, A. V., Yanamala, N., Mercer, R., Young, S. H., Fadeel, B., Kagan. V. E., and Shvedova, A. A. (2012). Factoring-in agglomeration of carbon nanotubes and nanofibers for better prediction of their toxicity versus asbestos. *Part. Fibre Toxicol.*, **9**, p. 10.

18. Schaeublin, N. M., Braydich-Stolle, L. K., Maurer, E. I., Park, K., MacCuspie, R. I., Nabiul Afrooz, A. R. M., Vaia, R. A., Saleh, N. B., and Hussain, S. M. (2012). Does shape matter? Bioeffects of gold nanomaterials in a human skin cell model. *Langmuir*, **28**, pp. 3248–3258.

19. Liu, S., Wei, L., Hao, L., Fang, N., Chang, M. W., Xu, R., Yang, Y., and Chen, Y. (2009). Sharper and faster "nano darts" kill more bacteria: a study of antibacterial activity of individually dispersed pristine single-walled carbon nanotube. *ACS Nano*, **3**, pp. 3891–3902.

20. Oh, W.-K., Kim, S., Choi, M., Kim, C., Jeong, Y. S., Cho, B.-R., Hahn, J.-S., and Jang, J. (2010). Cellular uptake, cytotoxicity, and innate immune response of silica-titania hollow nanoparticles based on size and surface functionality. *ACS Nano*, **4**, pp. 5301–5313.

21. Sayes, C. M, Fortner, J. D., Guo, W., Lyon, D., Boyd, A. M., Ausman, K. D., Tao, Y. J., Sitharaman, B., Wilson, L. J., Hughes, J. B., West, J. L., and Colvin, V. L. (2004). The differential cytotoxicity of water-soluble fullerenes. *Nano Lett.*, **4**, pp. 1881–1887.

22. Geidel, C., Schmachtel, S., Riedinger, A., Pfeiffer, C., Müllen, K., Klapper, M., and Parak, W. J. (2011). A general synthetic approach for obtaining cationic and anionic inorganic nanoparticles via encapsulation in amphiphilic copolymers. *Small*, **7**, pp. 2929–2934.

23. Hühn, D., Kantner, K., Geidel, C., Brandholt, S., Cock, I. D., Soenen, S. J. H., Rivera-Gil, P., Montenegro, J.-M., Braeckmans, K., Müllen, K., Nienhaus, G. U., Klapper, M., and Parak, W. J. (2013). Polymer-coated nanoparticles interacting with proteins and cells: focusing on the sign of the net charge. *ACS Nano*, **7**, pp. 3253–3263.

24. Kirchner, C., Liedl, T., Kudera, S., Pellegrino, T., Muñoz Javier, A., Gaub, H. E., Stölzle, S., Fertig, N., and Parak, W. J. (2005). Cytotoxicity of colloidal CdSe and CdSe/ZnS nanoparticles. *Nano Lett.*, **5**, pp. 331–338.

25. Stark, W. J. (2011). Nanoparticles in biological systems. *Angew. Chem., Int. Ed. Engl.*, **50**, pp. 1242–1258.

26. Gliga, A. R., Skoglund, S., Wallinder, I. O., Fadeel, B., and Karlsson, H. L. (2014). Size-dependent cytotoxicity of silver nanoparticles in human lung cells: the role of cellular uptake, agglomeration and Ag release. *Part. Fibre Toxicol.*, **11**, p. 11.

27. Cho, W. S., Duffin, R., Thielbeer, F., Bradley, M., Megson, I. L., Macnee, W., Poland, C. A., Tran, C. L., and Donaldson, K. (2012). Zeta potential and solubility to toxic ions as mechanisms of lung inflammation caused by metal/metal oxide nanoparticles. *Toxicol. Sci.*, **126**, pp. 469–477.

28. Nel, A., Xia, T., Madler, L., and Li, N. (2006). Toxic potential of materials at the nanolevel. *Science*, **311**, pp. 622–627.

29. Sayes, C.M., Wahi, R., Kurian, P. A., Liu, Y., West, J. L., Ausman, K. D., Warheit, D. B., and Colvin, V. L. (2006). Correlating nanoscale titania structure with toxicity: a cytotoxicity and inflammatory response study with human dermal fibroblasts and human lung epithelial cells. *Toxicol. Sci.*, **92**, pp. 174–185.

30. Collins, F. S., Gray, G. M., and Bucher, J. R. (2008). Toxicology. Transforming environmental health protection. *Science*, **319**, pp. 906–907.

31. Nel, A., Xia, T., Meng, H., Wang, X., Lin, S., Ji, Z., and Zhang, H. (2013). Nanomaterial toxicity testing in the 21st century: use of a predictive toxicological approach and high-throughput screening. *Acc. Chem. Res.*, **46**, pp. 607–621.

32. Damoiseaux, R., George, S., Li, M., Pokhrel, S., Ji, Z., France, B., Xia, T., Suarez, E., Rallo, R., Mädler, L., Cohen, Y., Hoek, E. M., and Nel, A. (2011). No time to lose: high throughput screening to assess nanomaterial safety. *Nanoscale*, **3**, pp. 1345–1360.

33. Shaw, S. Y., Westly, E. C., Pittet, M. J., Subramanian, A., Schreiber, S. L., and Weissleder, R. (2008). Perturbational profiling of nanomaterial biologic activity. *Proc. Natl. Acad. Sci. U S A*, **105**, pp. 7387–7392.

34. George, S., Pokhrel, S., Xia, T., Gilbert, B., Ji, Z., Schowalter, M., Rosenauer, A., Damoiseaux, R., Bradley, K. A., Mädler, L., and Nel, A. E. (2010). Use of a rapid cytotoxicity screening approach to engineer a safer zinc oxide nanoparticle through iron doping. *ACS Nano*, **4**, pp. 15–29.

35. Zhang, H., Ji, Z., Xia, T., Meng, H., Low-Kam, C., Liu, R., Pokhrel, S., Lin, S., Wang, X., Liao, Y. P., Wang, M., Li, L., Rallo, R., Damoiseaux, R., Telesca,

D., Mädler, L., Cohen, Y., Zink. J. I., and Nel, A. E. (2012). Use of metal oxide nanoparticle band gap to develop a predictive paradigm for oxidative stress and acute pulmonary inflammation. *ACS Nano*, **6**, pp. 4349–4368.

36. Guadagnini, R., Halamoda Kenzaoui, B., Cartwright, L., Pojana, G., Magdolenova, Z., Bilanicova, D., Saunders, M., Juillerat, L., Marcomini, A., Huk, A., Dusinska, M., Fjellsbø, L. M., Marano, F., and Boland, S. (2013). Toxicity screenings of nanomaterials: challenges due to interference with assay processes and components of classic in vitro tests. *Nanotoxicology*, July 27 [Epub ahead of print].
37. Savolainen, K., Backman, U., Brouwer, D., Fadeel, B., Fernandes, T., Kuhlbusch, T., Landsiedel, R., Lynch, I., and Pylkkänen, L. (2013). *Nanosafety in Europe 2015-2025: Towards Safe and Sustainable Nanomaterials and Nanotechnology Innovations* (Työterveyslaitos).
38. Winkler, D. A., Mombelli, E., Pietroiusti, A., Tran, L., Worth, A., Fadeel, B., and McCall, M. J. (2013). Applying quantitative structure-activity relationship approaches to nanotoxicology: current status and future potential. *Toxicology*, **313**, pp. 15–23.
39. Puzyn, T., Rasulev, B., Gajewicz, A., Hu, X., Dasari, T. P., Michalkova, A., Hwang, H.-M., Toropov, A., Leszczynska, D., and Leszczynski, J. (2011). Using nano-QSAR to predict the cytotoxicity of metal oxide nanoparticles. *Nat. Nanotechnol.*, **6**, pp. 175–178.
40. Epa, V. C., Burden, F. R., Tassa, C., Weissleder, R., Shaw, S., and Winkler, D. A. (2012). Modeling biological activities of nanoparticles. *Nano Lett.*, **12**, pp. 5808–5812.
41. Fourches, D., Pu, D., Tassa, C., Weissleder, R., Shaw, S. Y., Mumper, R. J., and Tropsha, A. (2010). Quantitative nanostructure-activity relationship modeling. *ACS Nano*, **4**, pp. 5703–5712.
42. Xia, X. R., Monteiro-Riviere, N. A., and Riviere, J. E. (2010). An index for characterization of nanomaterials in biological systems. *Nat. Nanotechnol.*, **5**, pp. 671–675.
43. Walkey, C. D., Olsen, J. B., Song, F., Liu, R., Guo, H., Olsen, D. W., Cohen, Y., Emili, A., and Chan, W. C. (2014). Protein corona fingerprinting predicts the cellular interaction of gold and silver nanoparticles. *ACS Nano*, **8**, pp. 2439–2455.
44. Nel, A. E., Mädler, L., Velegol, D., Xia, T., Hoek, E. M. V., Somasundaran, P., Klaessig, F., Castranova, V., and Thompson, M. (2009). Understanding biophysicochemical interactions at the nano-bio interface. *Nat. Mater.*, **8**, pp. 543–557.

45. Monopoli, M. P., Åberg, C., Salvati, A., and Dawson, K. A. (2012). Biomolecular coronas provide the biological identity of nanosized materials. *Nat. Nanotechnol.*, **7**, pp. 779–786.

46. Fadeel, B., Feliu, N., Vogt, C., Abdelmonem, A. M., and Parak, W. J. (2013). Bridge over troubled waters: understanding the synthetic and biological identities of engineered nanomaterials. *WIRE Nanomed. Nanobiotechnol.*, **5**, pp. 111–129.

47. Nyström, A. M., and Fadeel, B. (2012). Safety assessment of nanomaterials: implications for nanomedicine. *J. Controlled Rel.*, **161**, pp. 403–408.

48. Saba, T. M., and Di Luzio, N. R. (1965). Kupffer cell phagocytosis and metabolism of a variety of particles as a function of opsonization. *J. Reticuloendothel. Soc.*, **2**, pp. 437–453.

49. Lundqvist, M., Stigler, J., Elia, G., Lynch, I., Cedervall, T., and Dawson, K. A. (2008). Nanoparticle size and surface properties determine the protein corona with possible implications for biological impacts. *Proc. Natl. Acad. Sci. U S A*, **105**, pp. 14265–14270.

50. Tenzer, S., Docter, D., Rosfa, S., Wlodarski, A., Kuharev, J., Rekik, A., Knauer, S. K., Bantz, C., Nawroth, T., Bier, C., Sirirattanapan, J., Mann, W., Treuel, L., Zellner, R., Maskos, M., Schild, H., and Stauber, R. H. (2011). Nanoparticle size is a critical physicochemical determinant of the human blood plasma corona: a comprehensive quantitative proteomic analysis. *ACS Nano*, **5**, pp. 7155–7167.

51. Casals, E., Pfaller, T., Duschl, A., Oostingh, G. J., and Puntes, V. F. (2011). Hardening of the nanoparticle–protein corona in metal (Au, Ag) and oxide (Fe3O4, CoO, and CeO2) nanoparticles. *Small*, **7**, pp. 3479–3486.

52. Paula, A. J., Silveira, C. P., Martinez, D. S., Souza Filho, A. G., Romero, F. V., Fonseca, L. C., Tasic, L., Alves, O. L., and Durán, N. (2014). Topography-driven bionano-interactions on colloidal silica nanoparticles. *ACS Appl. Mater. Interfaces*, **6**, pp. 3437–3447.

53. Deng, Z. J., Liang, M., Monteiro, M., Toth, I., and Minchin, R. F. (2011). Nanoparticle-induced unfolding of fibrinogen promotes Mac-1 receptor activation and inflammation. *Nat. Nanotechnol.*, **6**, pp. 39–44.

54. Linse, S., Cabaleiro-Lago, C., Xue, W. F., Lynch, I., Lindman, S., Thulin, E., Radford, S. E., and Dawson, K. A. (2007). Nucleation of protein fibrillation by nanoparticles. *Proc. Natl. Acad. Sci. U S A*, **104**, pp. 8691–8696.

55. Mortimer, G. M., Butcher, N. J., Musumeci, A. W., Deng, Z. J., Martin, D. J., and Minchin, R. F. (2014). Cryptic epitopes of albumin determine

mononuclear phagocyte system clearance of nanomaterials. *ACS Nano*, **8**, pp. 3357–3366.

56. Lundqvist, M., Stigler, J., Cedervall, T., Berggård, T., Flanagan, M. B., Lynch, I., Elia, G., and Dawson, K. (2011). The evolution of the protein corona around nanoparticles: a test study. *ACS Nano*, **5**, pp. 7503–7509.
57. Milani, S., Bombelli, F. B., Pitek, A. S., Dawson, K. A., and Rädler, J. (2012). Reversible versus irreversible binding of transferrin to polystyrene nanoparticles: soft and hard corona. *ACS Nano*, **6**, pp. 2532–2541.
58. del Pino, P., Pelaz, B., Zhang, Q., Maffre, P., Nienhaus, G. U., and Parak, W. J. (2014). Protein corona formation around nanoparticles: from the past to the future. *Mater. Horiz.*, **1**, pp. 301–313.
59. Pietroiusti, A., Campagnolo, L., and Fadeel, B. (2013). Interactions of engineered nanoparticles with organs protected by internal biological barriers. *Small*, **9**, pp. 1557–1572.
60. Dutta, D., Sundaram, S. K., Teeguarden, J. G., Riley, B. J., Fifield, L. S., Jacobs, J. M., Addleman, S. R., Kaysen, G. A., Moudgil, B. M., and Weber, T. J. (2007). Adsorbed proteins influence the biological activity and molecular targeting of nanomaterials. *Toxicol. Sci.*, **100**, pp. 303–315.
61. Ge, C., Du, J., Zhao, L., Wang, L., Liu, Y., Li, D., Yang, Y., Zhou, R., Zhao, Y., Chai, Z., and Chen, C. (2011). Binding of blood proteins to carbon nanotubes reduces cytotoxicity. *Proc. Natl. Acad. Sci. U S A*, **108**, pp. 16968–16973.
62. Ehrenberg, M. S., Friedman, A. E., Finkelstein, J. N., Oberdörster, G., and McGrath, J. L. (2009). The influence of protein adsorption on nanoparticle association with cultured endothelial cells. *Biomaterials*, **30**, pp. 603–610.
63. Lartigue, L., Wilhelm, C., Servais, J., Factor, C., Dencausse, A., Bacri, J.-C., Luciani, N., and Gazeau, F. (2012). Nanomagnetic sensing of blood plasma protein interactions with iron oxide nanoparticles: impact on macrophage uptake. *ACS Nano*, **6**, pp. 2665–2678.
64. Lesniak, A., Fenaroli, F., Monopoli, M. P., Åberg, C., Dawson, K. A., and Salvati, A. (2012). Effects of the presence or absence of a protein corona on silica nanoparticle uptake and impact on cells. *ACS Nano*, **6**, pp. 5845–5857.
65. Saha K, Moyano, D. F., and Rotello, V. M.(2014). Protein coronas suppress the hemolytic activity of hydrophilic and hydrophobic nanoparticles. *Mater. Horiz.*, **2014**, pp. 102–105.

66. Dobrovolskaia, M. A., Neun, B. W., Man, S., Ye, X., Hansen, M., Patri, A. K., Crist, R. M., and McNeil, S. E. (2014). Protein corona composition does not accurately predict hematocompatibility of colloidal gold nanoparticles. *Nanomedicine*, pii: S1549-9634(14)00033-1. doi: 10.1016/j.nano.2014.01.009.

67. Tenzer, S., Docter, D., Kuharev, J., Musyanovych, A., Fetz, V., Hecht, R., Schlenk, F., Fischer, D., Kiouptsi, K., Reinhardt, C., Landfester, K., Schild, H., Maskos, M., Knauer, S. K., and Stauber, R. H. (2013). Rapid formation of plasma protein corona critically affects nanoparticle pathophysiology. *Nat. Nanotechnol.*, **8**, pp. 772–781.

68. Deng, Z. J., Liang, M., Toth, I., Monteiro, M., and Minchin, R. F. (2013). Plasma protein binding of positively and negatively charged polymer-coated gold nanoparticles elicits different biological responses. *Nanotoxicology*, **7**, pp. 314–322.

69. Moghimi, S. M., Andersen, A. J., Ahmadvand, D., Wibroe, P. P., Andresen, T. L., and Hunter, A. C. (2011). Material properties in complement activation. *Adv. Drug Delivery Rev.*, **63**, pp. 1000–1007.

70. Hamad, I., Al-Hanbali, O., Hunter, A. C., Rutt, K. J., Andresen, T. L., and Moghimi, S. M. (2010). Distinct polymer architecture mediates switching of complement activation pathways at the nanosphere–serum interface: implications for stealth nanoparticle engineering. *ACS Nano*, **4**, pp. 6629–6638.

71. Dufort, S., Sancey, L., and Coll, J. L. (2012). Physico-chemical parameters that govern nanoparticles fate also dictate rules for their molecular evolution. *Adv. Drug Delivery Rev.*, **64**, pp. 179–189.

72. Kapralov, A. A., Feng, W. H., Amoscato, A. A., Yanamala, N., Balasubramanian, K., Winnica, D. E., Kisin, E. R., Kotchey, G. P., Gou, P., Sparvero, L. J., Ray, P., Mallampalli, R. K., Klein-Seetharaman, J., Fadeel, B., Star, A., Shvedova, A. A., and Kagan, V. E. (2012). Adsorption of surfactant lipids by single-walled carbon nanotubes in mouse lung upon pharyngeal aspiration. *ACS Nano*, **6**, pp. 4147–4156.

73. Riehemann, K., Schneider, S. W., Luger, T. A., Godin, B., Ferrari, M., and Fuchs, H. (2009). Nanomedicine-challenge and perspectives. *Angew. Chem., Int. Ed. Engl.*, **48**, pp. 872–897.

74. Davis, M. E., Zuckerman, J. E., Choi, C. H. J., Seligson, D., Tolcher, A., Alabi, C. A., Yen, Y., Heidel, J. D., and Ribas, A. (2010). Evidence of RNAi in humans from systemically administered siRNA via targeted nanoparticles. *Nature*, **464**, pp. 1067–1070.

75. Hrkach, J., Von Hoff, D., Mukkaram Ali, M., Andrianova, E., Auer, J., Campbell, T., De Witt, D., Figa, M., Figueiredo, M., Horhota, A., Low, S.,

McDonnell, K., Peeke, E., Retnarajan, B., Sabnis, A., Schnipper, E., Song, J. J., Song, Y. H., Summa, J., Tompsett, D., Troiano, G., Van Geen Hoven, T., Wright, J., LoRusso, P., Kantoff, P. W., Bander, N. H., Sweeney, C., Farokhzad, O. C., Langer, R., and Zale, S. (2012). Preclinical development and clinical translation of a PSMA-targeted docetaxel nanoparticle with a differentiated pharmacological profile. *Sci. Transl. Med.*, **4**, p. 128ra139.

76. Salvati, A., Pitek, A. S., Monopoli, M. P., Prapainop, K., Bombelli, F. B., Hristov, D. R., Kelly, P. M., Åberg, C., Mahon, E., and Dawson, K. A. (2013). Transferrin-functionalized nanoparticles lose their targeting capabilities when a biomolecule corona adsorbs on the surface. *Nat. Nanotechnol.*, **8**, pp. 137–143.

77. Krais, A., Wortmann, L., Hermanns, L., Feliu, N., Vahter, M., Stucky, S., Mathur, S., and Fadeel, B. (2014). Targeted uptake of folic acid-functionalized iron oxide nanoparticles by ovarian cancer cells in the presence but not in the absence of serum. *Nanomedicine*, pii: S1549-9634(14)00030-6. doi: 10.1016/j.nano.2014.01.006.

78. Simberg, D, Park, J. H., Karmali, P. P., Zhang, W. M., Merkulov, S., McCrae, K., Bhatia, S. N., Sailor, M., and Ruoslahti, E. (2009). Differential proteomics analysis of the surface heterogeneity of dextran iron oxide nanoparticles and the implications for their in vivo clearance. *Biomaterials*, **30**, pp. 3926–3933.

79. Pridgen, E. M., Alexis, F., Kuo, T. T., Levy-Nissenbaum, E., Karnik, R., Blumberg, R. S., Langer, R., and Farokhzad, O. C. (2013). Transepithelial transport of Fc-targeted nanoparticles by the neonatal fc receptor for oral delivery. *Sci. Transl. Med.*, **5**, p. 213ra167.

80. Mahon, E., Salvati, A., Baldelli Bombelli, F., Lynch, I., and Dawson, K. A. (2012). Designing the nanoparticle-biomolecule interface for "targeting and therapeutic delivery." *J. Controlled Rel.*, **161**, pp. 164–174.

81. Sacchetti, C., Motamedchaboki, K., Magrini, A., Palmieri, G., Mattei, M., Bernardini, S., Rosato, N., Bottini, N., and Bottini, M. (2013). Surface polyethylene glycol conformation influences the protein corona of polyethylene glycol-modified single-walled carbon nanotubes: potential implications on biological performance. *ACS Nano*, **7**, pp. 1974–1989.

82. Kah, J. C., Chen, J., Zubieta, A., and Hamad-Schifferli, K. (2012). Exploiting the protein corona around gold nanorods for loading and triggered release. *ACS Nano*, **6**, pp. 6730–6740.

83. Cifuentes-Rius, A., de Puig, H., Kah, J. C., Borros, S., and Hamad-Schifferli, K. (2013). Optimizing the properties of the protein corona

surrounding nanoparticles for tuning payload release. *ACS Nano*, **7**, pp. 10066–10074.

84. Prapainop, K., Witter, D. P., and Wentworth, P. A (2012). Chemical approach for cell-specific targeting of nanomaterials: small-molecule-initiated misfolding of nanoparticle corona proteins. *J. Am. Chem. Soc.*, **134**, pp. 4100–4103.

85. Caracciolo, G., Cardarelli, F., Pozzi, D., Salomone, F., Maccari, G., Bardi, G., Capriotti, A. L., Cavaliere, C., Papi, M., and Laganà, A. (2013). Selective targeting capability acquired with a protein corona adsorbed on the surface of 1,2-dioleoyl-3-trimethylammonium propane/DNA nanoparticles. *ACS Appl. Mater. Interfaces*, **5**, pp. 13171–13179.

86. Pozzi, D., Colapicchioni, V., Caracciolo, G., Piovesana, S., Capriotti, A. L., Palchetti, S., De Grossi, S., Riccioli, A., Amenitsch, H., and Laganà, A. (2014). Effect of polyethyleneglycol (PEG) chain length on the bio-nano-interactions between PEGylated lipid nanoparticles and biological fluids: from nanostructure to uptake in cancer cells. *Nanoscale*, **6**, pp. 2782–2792.

87. Menjoge, A. R., Kannan, R. M., and Tomalia, D. A. (2010). Dendrimer-based drug and imaging conjugates: design considerations for nanomedical applications. *Drug Discovery Today*, **15**, pp. 171–185.

88. Hayder, M., Poupot, M., Baron, M., Nigon, D., Turrin, C. O., Caminade, A. M., Majoral, J. P., Eisenberg, R. A., Fournié, J. J., Cantagrel, A., Poupot, R., Davignon, J. L. (2011). A phosphorus-based dendrimer targets inflammation and osteoclastogenesis in experimental arthritis. *Sci. Transl. Med.*, **3**, p. 81ra35.

89. Shaunak, S., Thomas, S., Gianasi, E., Godwin, A., Jones, E., Teo, I., Mireskandari, K., Luthert, P., Duncan, R., Patterson, S., Khaw, P., and Brocchini, S. (2004). Polyvalent dendrimer glucosamine conjugates prevent scar tissue formation. *Nat. Biotechnol.*, **22**, pp. 977–984.

90. Åkesson, A., Cárdenas, M., Elia, G., Monopoli, M. P., and Dawson, K. A. (2012). The protein corona of dendrimers: PAMAM binds and activates complement proteins in human plasma in a generation dependent manner. *RSC Adv.*, **2**, pp. 11245–11248.

91. Jain, S., Pitoc, G. A., Holl, E. K., Zhang, Y., Borst, L., Leong, K. W., Lee, J., and Sullenger, B. A. (2012). Nucleic acid scavengers inhibit thrombosis without increasing bleeding. *Proc. Natl. Acad. Sci. U S A*, **109**, pp. 12938–12943.

92. Fadeel, B. (2013). Nanosafety: towards safer design of nanomedicines. *J. Intern. Med.*, **274**, pp. 578–580.

93. Andon, F. T., and Fadeel, B. (2014). Nanotoxicology: towards safety-by-design, in *Nano-Oncologicals: Targeting and Delivery Approaches*, eds. Jose Alonso, M., and Garcia-Fuentes, M. (Controlled Release Society; Springer) [in press].

94. Albanese, A., Tang, P. S., and Chan, W. C. (2012). The effect of nanoparticle size, shape, and surface chemistry on biological systems. *Annu. Rev. Biomed. Eng.*, **14**, pp. 1–16.

95. Byrne, H. J., Ahluwalia, A., Boraschi, D., Fadeel, B., Gehr, P., Gutleb, A. C., Kendall, M., Papadopoulos, M., and Lynch, I. (2013). The bio-nano-interface in predicting nanoparticle fate and behaviour in living organisms: towards grouping and categorising nanomaterials and ensuring nanosafety by design. *BioNanoMaterials*, **14**, pp. 195–216.

Chapter 5

Nanotoxicology: The Case for in vitro Tests

Thomas Hartung
Center for Alternatives to Animal Testing (CAAT), Bloomberg School of Public Health, Johns Hopkins University, Baltimore, MD, USA
CAAT Europe, University of Konstanz, Konstanz, Germany
thartung@jhsph.edu

5.1 Introduction

Nanomaterials are acclaimed for their novel properties, for which broad, new uses are being discovered with increasing frequency. It is obvious that, as the properties change, unwanted properties (toxicity) are to be expected as well. Today's toxicology, however, is already overwhelmed with the challenge of addressing new chemicals, not to mention the enormous number of old chemicals never properly assessed. Limitations of traditional approaches range from animal welfare issues, which were a strong driving force for alternative approaches (the 3Rs concept) over the last two decades, to aspects of throughput and accuracy of the predicted toxicities. The latter has prompted discussion about a revolutionary change [1] in

Handbook of Safety Assessment of Nanomaterials: From Toxicological Testing to Personalized Medicine
Edited by Bengt Fadeel

ISBN 978-981-4463-36-2 (Hardcover), 978-981-4463-37-9 (eBook)
www.panstanford.com

chemical safety assessment, now known as "toxicology for the 21st century" (Toc-21c) [2, 3]. The multitude of possible formulations of nanomaterials to be assessed for novel toxic properties makes these alternative approaches especially attractive, given the well-recognized limitations of traditional animal-based approaches [4]—limitations that might be even more pronounced for nanomaterials, which have notably altered biokinetics (refer to Chapter 7 for a discussion).

Nano is a "big thing" in toxicology, as summarized in some recent reviews [5–12]. Articles, journals, and conferences are mushrooming, paralleling the rise of nanotechnologies but also showing the hunger of toxicology for new objects to study. Perhaps it would be better to focus on new approaches first? Nanomedicine promises new solutions for old problems, but what about the old problems of toxicology? It is a fallacy to assume that one can gain beneficial effects in the human organism without unwanted collateral effects. First, any biologically active agent perturbs physiology, hopefully as a corrective for the patient but at least requiring compensatory reactions of the healthy. Second, few agents are specific enough to have only one effect, but it is rare that we want all the effects and in the given mix of strengths. Third, many agents show excess toxicity—even desired effects often become negative if excessively stimulated. This increase in negative effects is directly linked to more sensitive subpopulations (children, the elderly, the diseased, those with genetic polymorphisms, etc.). Thus, the promise of nanoparticles (NPs) may be paid for in possible side effects, that is, toxicities [13]. For most manufactured NPs, toxicity data is sparse, with some exceptions for carbon black, titanium dioxide, iron oxides, and amorphous silica [14].

So far, NPs and other nanomaterials (the abbreviation NP is used in the present chapter for both but primarily with reference to particles or fibers and not, for example, nano-thick films) are, for the most part, treated by regulatory toxicology as chemicals [15]. The most common definitions for NPs include materials with dimensions from 1 nm (size of a sugar molecule) to 100 nm (size of a virus). Regulatory frameworks are on the way, opening up possibilities for alternative approaches [16]. Whether the differences between NPs and their parent or bulk compounds are actually small or big problems remains to be seen. Notwithstanding, to quote Albert

Einstein, "Anyone who doesn't take truth seriously in small matters cannot be trusted in large ones either."

5.2 Alternative or Advanced Methods in Toxicology

The term "alternative methods" is most commonly understood as "alternative to animal experiments" or at least using refinement and reduction alternatives to traditional animal experiments. Recently, the catch phrase "toxicology for the 21st century" (Tox-21c) generated an atmosphere of departure, more on the west side of the Atlantic, emphasizing the technological needs and opportunities for change. Fortunately, the two paths normally converge, and we can see them as two sides of the same coin—the most humane science is also the best science. There is a broad base of literature highlighting the ethical concerns, costs [17, 18], limited predictivity [19–21], and limited throughput [22, 23] of current approaches that, for the most part, were developed some decades ago for drug safety testing and subsequently were adapted to pesticides, chemicals, cosmetics, and foodstuffs. These limitations serve as the driving forces for change on both sides of the Atlantic [24]. While there is progress, particularly in some areas of topical and acute toxicity [25], progress for the more demanding systemic and chronic toxicities has been limited [26].

Is there any reason to assume that nanotoxicology would not benefit from alternative methods, for example, that they are less applicable to particles than to dissolved substances? Indeed, some theoretical considerations apply: the in vitro kinetics of particles might differ, that is, their behavior in cell culture. This might include particle clumping (aggregation), binding to plastic, or floating on the cell culture media surface, all of which would alter cellular exposure and, thus, the concentration–response curve. Similarly, exposure to air and nonphysiological culture conditions might affect the experiments. Also, specific artifacts have interfered with cytotoxicity measures (MTT) 3-(4,5-dimethylthiazol-2-yl)-2,5-diphenyltetrazolium bromide as typically applied in alternative methods [27]. Later, we will discuss some general problems of alternative methods use for NPs. However, altogether nanotoxicology is likely a driving force and not a stumbling block toward the use of modern approaches in toxicology [28]. These modern approaches offer the necessary level of standardization [29] that other experimental approaches often lack. The call for in vitro approaches is loud [6, 30, 31].

5.2.1 Do We Need Special Methods for Nanotoxicology?

The first major question for nanotoxicology is, does it even exist as a scientific discipline? Is it any different from the current risk paradigm, that is, hazard, kinetics, exposure measurement, and overall risk assessment? First, completely new modes of action for NPs have been found—if we think of asbestos as a natural nanofiber, where a key mechanism is macrophage activation after ingestion of needles of asbestos, it also applies to nanofibers in general. However, the hazards are still classical ones, that is, fibrosis and cancer. We can argue that this is only an additional mode of action, which can either be anticipated by the size and shape of particles or be simply added to the assessment and be found by traditional approaches. From this point of view, it is rather unlikely that a really new hazard that could not be seen in repeated dose studies or cancer bioassays would be attributed to particles, even at the nanoscale.

However, lung toxicology, for example, lags far behind other areas of concern, while NPs are especially likely to reach the alveoli of the lung and exert toxicity there [32, 33] (refer to Chapter 12). Airborne exposure testing is experimentally cumbersome and is avoided, when possible, due not only to the effort (costs) involved but also to the poor reflection rodents give of human exposure. Testing for respiratory irritation and sensitization is not standard for industrial chemicals, and guideline tests have yet to be developed. We also should be clear that the particular health effects of industrial chemicals—endocrine disruption, immunotoxicity, and developmental neurotoxicity—are among the more recent additions and are not yet reflected in testing programs. It would just take one scandal, however, to change this.

It may be unlikely that a completely new adverse health effect is induced by NPs, but there are many human diseases where we do not suspect any chemical involvement and where, in fact, one might exist. In particular, the chronic effects of chemicals are so poorly understood that we have no idea whether we would get any relevant alert from routine animal tests, which are often inadequate even with regard to well-known hazards. Possible examples of continuously increasing health problems include atherosclerosis, male infertility, autism, and diabetes. It is worth noting that air pollution involving natural NPs led mainly to excess deaths associated with cardiovascular illness [34], a hazard not generally addressed in toxicology. Arteriosclerosis, in fact, is very difficult to induce in animals. Determination of the pulmonary and systemic inflammatory hazards typically seen with NPs [35] is not among the strengths of the toxicological toolbox.

There might well be hazards not present for a parent compound due to kinetics, as the adsorption, distribution, metabolism, and excretion (ADME) of NPs can differ greatly from that of larger particles or soluble substances. Changes in kinetics (bioavailability) alone [36] are sufficient to create additional hazards not seen with the parent compound, since whenever higher plasma or tissue levels of the substance are obtained, thresholds of toxicity might be exceeded. We know very well from formulations of drugs that solubility after oral administration depends on particle size influencing peak plasma levels—a crucial determinant for toxicity. Similarly, higher concentrations can be achieved if transports through barriers are accelerated. However, the faster elimination—for example, by cellular uptake or chemical reactivity—acts against this.

For toxic effects, size matters, as a number of studies show size-dependent effects [37]. Good examples are gold and silver, which normally are minimally reactive but become much more so at NP sizes. Silver NPs, for this reason, are used as bactericidal coatings for clothes (e.g., "Wash your socks less often, thanks to a silver NP coating"). At least for the antimicrobial properties of silver NPs, shape dependence has been shown [38]. How this translates to toxic effects on human cells is not known. Cellular uptake of gold NPs has been found to be shape dependent [39], while others reported no differences in a number of cell systems for silica NPs [40].

Similarly, reactive chemistry, a key feature of many toxicants, is strongly influenced by the particle surface. One milliliter of 10 nm NPs has the surface area of a soccer field. Thus, we might see hazards with NPs at lower concentrations than the maximally tested or testable doses currently applied to the parent compound. As one consequence, the exposure might require different measurements, for example, instead of dose measures in milligrams per kilogram, particle numbers or the particle surface might be more meaningful both in vivo and in vitro.

We might also see some effects only in vitro. We should not underestimate the hazards that are masked by current tests, because the animals defend themselves successfully. More than 90% of substances that can exert genotoxicity in cells are not mutagenic in animal tests. This is not to say that the cell result was wrong; rather, it usually means we did not achieve the concentrations in vivo that we can apply in vitro or that some defenses were not reflected in vitro. The substance may still present a hazard, which might become relevant for humans or subpopulations. We do not know whether the defense mechanisms against some hazards are as effective when administered as NPs. The novel properties of NPs also can lead to new biological interactions [41], which could result in toxicities not shown by the parent compound. This extends to refinement methods, where, for instance, the exposure to airborne particles for the obligatory nose-breathing rat and mouse require attention to avoid overloading the respiratory tract.

There might be opportunities for reduction alternatives, too. Though it is rather unlikely that lower variability of responses to NPs would allow a reduction in animal group numbers, they should still be considered, as should designs with one control group for multiple tests or longitudinal studies following the same group of animals rather than sacrificing a group per time point. Here, the imaging opportunities for NPs might make a difference for kinetics experiments. Opportunities to combine studies, for example, mutagenicity and repeat-dose studies, enhanced chronic studies, including carcinogenicity, or the inclusion of developmental neurotoxicity in reproductive toxicity studies should be considered as reduction alternatives. NPs do not really differ from other test materials in this regard.

5.2.2 Do We Need a Traditional or an Alternative Toxicology for NPs?

The number of different NPs we might need to address is potentially extremely high, with various shapes, size distributions, and coatings for each material. This alone suggests the importance of using alternative methods, which often allow higher throughput, replicates, and parallel tests. Thus, the limit of test throughput is more relevant for NPs. First, the main health concerns in particle toxicity hint at risks for the most complex endpoints (cancer, lung toxicity), which require most test capacities. Rodent inhalation models are especially prohibitive in terms of time and expense [42]. Second, since any given substance, at least theoretically, can be formulated to particles of very different sizes, size distributions, shapes, and modifications, an almost unlimited testing demand could be considered. Choi, Ramachandrran, and Kandlikar [43] calculated the costs for traditional testing of NPs already on the market to be between $250 million and $1.2 billion and the time required at 34–53 years.

In addition, we should be aware that current regulatory toxicology was established for drugs under development. A very precautionary approach was taken to avoid putting volunteers and patients at risk [21]. While this might be appropriate for nanomedicine products, we have to ask ourselves what development opportunities we sacrifice if we apply the same precautionary methods to products with lesser exposure to humans or those that are not intended to be biologically active. The problem becomes most pronounced when precautionary methods (many false positives) are used for substance groups with rare side effects [19, 21].

This reasoning makes evident the need to explore the toxic profiles of a broad variety of NPs to learn what we must control for. Furthermore, novel approaches need to be developed, since traditional approaches might have even greater limitations for NPs than for other industrial chemicals and may not offer the throughput and velocity to cope with the dynamic developments in nanotechnologies.

What are the specific opportunities to use alternative methods? First, one major concern in cosmetics is dermal penetration of NPs. Rodent and rabbit skin has little to do with human skin, and both artificial human skin and explants offer opportunities with available, accepted methods. Some of these also allow testing for mechanical stress or inflammation, as well as penetration via hair follicles, as specific concerns for NPs. Other barrier models for gut uptake, the blood–brain barrier, or the placental barrier are prevalidated but not yet validated. Still, they might be useful to characterize NPs and identify or rule out specific concerns.

Kinetics will be affected, which can shift concentration–response curves, limits, and threshold concentrations, as well as no-effect levels, thus altering critical components of the risk assessment paradigm. We will need to explore whether this can be handled with, for example, safety or assessment factors. This means requiring, for example, an additional factor as safety margin for the use of NPs in humans. For current estimations of safe dose uses from animal no-effect levels, we often require 1 of 10 for interspecies differences and 1 of 10 for sensitive subpopulations. The numbers appear to correspond more with the decimal system than with sound science—if we had 12 fingers instead of 10, we would likely be 1.44 times better protected by regulation. Completely new ports of entry of substances have already been described, when NP sizes fall below barrier cutoffs [44]. Note that the ease of cellular uptake may also result in bioaccumulation of NPs [45]. Furthermore, the larger surface area per unit weight often makes NPs more reactive. Since many forms of toxicity are mediated by chemical reactivity, such as mutagenicity by formation of DNA adducts or sensitization by hapten binding, this raises the possibility of increased toxic potentials.

A number of alternative methods validated for chemicals and drugs might be useful for NPs (Table 5.1), but none has been validated for this purpose. Direct comparisons of in vitro and in vivo findings are still rare [46]. The modular approach to validation [47], it should be noted, allows expanding applicability domains for validated methods, a possible fast track to obtain regulatory acceptance for NP evaluation. The potential carcinogenicity of NPs is of concern to toxicologists because of several specific properties—the potential to activate inflammatory mechanisms as promoters of cancer, the ability to reach alveolar compartments when inhaled, altered cellular uptake allowing NPs to reach DNA, and the fact that mutagenicity is linked with reactive chemistry, which is often amplified by the large surfaces of NPs. A number of in vitro models for mutagenicity are available, but they also are known to have many false positive results [48, 49]. Many of them also can be integrated in enhanced animal studies, allowing a reduction in animal use. A specific opportunity is offered by the cell transformation assays for cancer, which currently are peer-reviewed after validation.

Table 5.1 Validated alternative methods for chemicals and their likely relevance for NPs

No.	Method	Date of ESAC statement	Suitability for NPs
1	3T3 NRU phototoxicity test	03/11/1997	Likely applicable but rare testing demand
2	EpiSkin skin corrosivity test	03/04/1998	Likely applicable
3	Rat TER skin corrosivity test	03/04/1998	
4	Application of the 3T3 NRU phototoxicity test to UV filter chemicals	20/05/1998	See 1
6	Local lymph node assay for skin sensitization	21/03/2000	Reduction / Refinement method likely applicable
7	EpiDerm skin corrosivity test	21/03/2000	See 2–3
8	CORROSITEX skin corrosivity test	06/12/2000	
11	Micromass embryotoxicity assay	01/05/2002	Relevance unclear; might need to be combined with a placenta barrier model
12	Whole rat embryotoxicity assay	01/05/2002	
13	Embryonic stem cell test for embryotoxicity	01/05/2002	
17	UTC step-down approach for acute aquatic toxicity testing	21/03/2006	Reduction method likely applicable
18	CFU-GM assay for predicting acute neutropenia in humans	21/03/2006	Relevance unclear; possible value for acute toxicity testing strategy

(*Continued*)

Table 5.1 (*Continued*)

No.	Method	Date of ESAC statement	Suitability for NPs
19	Human whole blood IL-1 for in vitro pyrogenicity testing	21/03/2006	Likely applicable; especially important because of large surface area and thereby questionable applicability of the Limulus assay
20	Human whole blood IL-6 for in vitro pyrogenicity testing	21/03/2006	
21	PBMC IL-6 for in vitro pyrogenicity testing	21/03/2006	
22	MM6 IL-6 for in vitro pyrogenicity testing	21/03/2006	
23	Human cryopreserved whole blood IL-1 for in vitro pyrogenicity testing	21/03/2006	
24	In vitro micronucleus test as an alternative to the in vitro chromosome aberration assay for genotoxicity testing	17/11/2006	Likely applicable; high relevance
25	Application of the SkinEthic human skin model for skin corrosivity testing	17/11/2006	See 2–3 and 7–8
27	Bovine corneal opacity and permeability (BCOP) test method	27/04/2007	Likely applicable
28	Isolated chicken eye (ICE) test method	27/04/2007	
29	rLLNA	27/04/2007	See 6
30	EpiDerm (with MTT reduction) for skin irritation	27/04/2007	Likely applicable, highly relevant
31	EPISKIN (with MTT reduction) for skin irritation	27/04/2007	

No.	Method	Date of ESAC statement	Suitability for NPs
32	FDP	31/10/2007	Reduction methods likely applicable
33	ATC method	31/10/2007	
34	UDP	31/10/2007	
35	EpiDerm SIT (with MTT reduction) for skin irritation	5/11/2008	See 30–31
36	SkinEthic (with MTT reduction) for skin irritation	5/11/2008	
37	CellSystems human skin model EST-1000 for skin corrosivity testing	12/06/2009	See 2–3 and 7–8
38	Cellsensor microphysiometer for eye irritation	10/07/2009	See 27–28
39	Fluorescence leakage assay for severe eye irritation	10/07/2009	
40	Cell transformation assays	18/02/2011	Likely applicable, highly relevant

Noteworthy, methods 18–23 were not developed for the purpose of chemical testing, but current validation activities explore their use for acute toxicity testing, which might lead to an extension of the applicability domain. Validity statements not listed are not relevant for chemicals/NPs. Noteworthy, several methods for endocrine disruptor testing have been validated and accepted by US agencies and the OECD.
Abbreviations: UTC, upper threshold concentration; IL-1, interleukin 1; rLLNA, reduced local lymph node assay; FDP, fixed-dose procedure; ATC, acute toxic class; UDP, up-and-down procedure.

Inflammation can be studied in monocyte activation tests, such as the validated alternative pyrogen tests [50, 51]. Human whole-blood assays offer specific opportunities, as a cell suspension is used [52]. Experimental setups for airborne exposure of particles to air/liquid interface cultures of cells are available but have not yet been validated.

5.2.3 Special Problems for in vitro Nanotoxicology

A number of articles addressing problems of in vitro nanotoxicology are available [53–56]. Below a number of common themes are summarized (and see Chapters 2 and 3 for a related discussion on physicochemical properties).

5.2.3.1 Agglomeration

Not everything called nano is actually nano. Aggregation or agglomeration of NPs is very common and difficult to prevent [57]. NPs can have complex aggregation behaviors in aqueous solutions [36] with substantial impact on their toxicity. Many of the studies published so far did not exclude aggregation, but even as non-mono-dispersed particles, the smaller particles are more potent in many respects [58]. Aggregation effects also have been recognized in the ecotoxicity of silica, titanium dioxide, and zinc oxide NPs [59]. Some systematic approaches to dispersed NPs have been proposed [60], but the problem still needs to be addressed on a case-by-case basis.

5.2.3.2 Stability

The stability of NPs is not often discussed, but the sheer surface area represents a problem, as it not only attracts substances offering binding sites (for pyrogens, for example) but also lends itself to chemical reactions such as oxidation. Many NPs might actually be coated. We know as little about the modification and degradation occurring over time as we do about the metabolic fate of NPs.

5.2.3.3 Dosimetry

In toxicology, we have seen a move from primarily weight-based doses (mg/kg or ppm) to (molar) concentrations, especially when kinetic measures (plasma concentrations, for example) could be assessed or when experiments were done in vitro. For NPs, weight, particle number, and surface area are typical dose measures, but shape, coating, electrophysical properties, etc., can have a further impact. The chemical characterization [61] and dosimetry clearly require closer attention than for traditional chemicals [41]. In some cases, toxicity correlated best with NP surface area [62], but it is still to be established whether this is a more general rule. It makes sense for reactive chemistry, which is a leading mechanism of toxicological damage, and for oxidative processes; in fact, generation of reactive oxygen species has been a key mode of action associated with NP toxicity.

5.2.3.4 In vitro biokinetics

This term has been coined to indicate that test substances in vitro also exhibit kinetics: they are adsorbed (e.g., by plastic or the albumin of fetal calf serum), stay soluble or precipitate, are taken up by cells, or are oxidized by air or metabolized by the cells, and we interfere with their presence when changing cell culture media. The situation is not as complex as in vivo kinetics, but certainly the actual effective concentration reaching the cells is not the one we added. Just as in vivo work has been augmented by introducing kinetics, we might likewise give consideration to these factors as we move the field of in vitro toxicology forward [63]. The situation is not less complex for NPs, where aggregation and particle coating must be considered. Cell membranes, mitochondria, and nuclei are considered major compartments for NP toxicity [62]. Thus, uptake and intracellular trafficking must also be addressed. There is increasing literature on cellular uptake of NPs [64], in part stimulated by drug targeting, that is, the targeted delivery of drugs to their desired site of action, using NPs [65]. Uptake of NPs into cells appears to be an effect specific for materials in the range of 50–200 nm. Surface modifications (positively or negatively charged side groups of the polymers, amino acids, or peptides/proteins) enhance this uptake. It is not clear, to date, which types of particles are taken up by pinocytosis, receptor-mediated endoytosis, or phagocytosis, respectively, to mention the most prominent uptake routes, which might cooperate at least in some instances. Research about factors influencing cellular uptake, like size, surface properties, cell type, and endocytotic pathways will be critical to interpret in vitro toxicity data and the suitability of test systems.

5.2.3.5 Cell contact of NPs

Actual exposure of cells to NPs needs to be assured, as NPs might swim on the culture media. Also, NPs are known to pass from cell to cell. Cell monolayers resemble pan-fried eggs lying next to each other, giving only minimal cell-to-cell contact areas. Furthermore, cell density in a typical culture is only 1% of normal tissue [66], which changes dosimetry and the likelihood of NP-to-cell contact.

5.2.3.6 Artifacts

Many artifacts in cell culture are because of the enormous surface of NPs, which not only allows absorption of many components but also augments chemical reactivity. The abundant serum albumin from fetal calf serum of most cell cultures is likely coating many NPs; similarly, surfactant proteins in the lung or mucus in the gastrointestinal tract would cover NPs but might not be able to in the presence of albumin. Anecdotal observations have shown that NPs can absorb many dyes, enzymes, or inflammatory mediators, thereby influencing test results. They can carry pyrogenic contaminations, difficult to detect but biologically strongly active [67]. Single-walled carbon nanotubes have been shown to interact with MTT, commonly used to test cell viability [27]. More such artifacts are likely to be discovered, which may be prompted by a large surface area, electrostatic properties, or increased reactivity.

5.3 Existing Alternative Methods and Their Suitability for Nanotoxicology

It is disturbing that nanotoxicology is reinventing alternative approaches, often without bridging the two decades of development and validation done for chemicals and cosmetic ingredients. Cytotoxicity and mutagenicity assays are broadly used [54, 56] without necessarily bridging the validated methodologies. Others have highlighted the need to optimize validated toxicity and ecotoxicity tests for NPs [58]. A variety of approaches lend themselves to adaptation for NPs (Table 5.1), but none has been formally validated for NPs. Here an overview on the currently validated tests with potentials for NP toxicity is given; for a more extensive review see Ref. [68].

- *Skin corrosion.* The skin corrosion endpoint for chemicals has been replaced, following European Centre for the Validation of Alternative Methods (ECVAM) validation, by seven validated and accepted methods (transepithelial electrical resistance test with rat skin, Corrositex, and five skin models). It is certainly critical that the US accepts only positive results and requires animal testing for the negative ones. A long planned study to address false negative skin corrosives in the novel alternative skin irritant tests is essential to demonstrate the safety of the combined use for US acceptance. The methods primarily in use for skin corrosion require commercial, artificial human skin. To safeguard availability and free market, new products (similar methods, sometimes also referred to as "me-too" methods) require assessment of equivalence. The endpoint seems to be of lesser importance for NPs, but given the augmented reactivity of NPs due to the large surface area, NPs from irritant substances might actually become corrosive.

- *Skin irritation*. The endpoint has been replaced for chemicals by a successful validation in 2007 with inclusion in Registration, Evaluation, Authorisation and Restriction of Chemicals (REACH) test strategies. Test guideline acceptance succeeded in the EU and has been completed at the Organisation for Economic Co-operation and Development (OECD) level. A number of me-too developments are ongoing. This is important to maintain competition in the market and to gain acceptance. Research activities using omics technologies are being considered to reduce the false positive rate in future test strategies. The methods appear to be suitable for NPs but have not been formally evaluated.
- *Skin absorption*. The method for this endpoint has been accepted by the OECD, as a political compromise, in parallel to the creation of an in vivo method. The methods are certainly key for adaptation to NPs.
- *Eye irritation*. This endpoint might, in fact, be considered the "holy grail" of alternative methods because of its emotional impact on many animal lovers. Various assays are being evaluated and most recently accepted as OECD test guidelines, which, combined in a test strategy, might have the potential to replace the animal test for this endpoint. The strategy aims at utilizing the strengths of specific in vitro assays to address required ranges of irritation and/or chemical classes (applicability domains) to classify test substances [69].

Around 20 methods have been in industrial use for more than two decades, and 6 large validation trials have evaluated them. However, no single method was found to fully replace the complexity of the animal responses. Two most promising methods based on reconstituted human corneal epithelial cells are currently under validation with the support of COLIPA. These models appear promising for the identification of nonirritant substances like the in vivo application with a large applicability domain. It appears that the various methods can be highly relevant for NPs, but they have not been evaluated to larger extent.

- *Acute toxicity*. This test could probably be considered obsolete—in the pharmaceutical arena, the International Conference on Harmonization has abandoned it because other preliminary single-dose tolerability studies and/or dose range–finding data are sufficient. A validation study between the Interagency Coordinating Committee on the Validation of Alternative Methods (ICCVAM) and ECVAM to predict acute toxicity by cytotoxicity assays was completed in 2005. The study confirmed previous findings suggesting that the in vitro tests could identify nontoxic chemicals. A number of activities follow up on this but have not yet achieved a breakthrough. Cytotoxicity assays will represent a key tool for screening NPs and identifying novel toxic properties.
- *Genotoxicity*. Several accepted in vitro tests exist (also as OECD test guidelines), but they often identify substances as false positives, which then need to be sorted out using animal tests. Various in vitro assays are available; the problem, however, is the low quality of these nonvalidated tests, particularly their oversensitivity. When three such tests are combined, as is typical, only 3% of the positive results are correct, while 97% are negative in the subsequent animal test [48]. ECVAM held a workshop on the reduction of false positives in in vitro genotoxicity tests [49], and the recommendations are being followed with funding from COLIPA, NC3Rs, and ECVAM [70, 71]. Validation of skin models for genotoxicity testing is cosponsored by COLIPA and ECVAM. The in vitro micronucleus test (MNT) was validated in 2006 [72]. This is a remarkable test case for various reasons: it was the first

completely retrospective validation ever, as suggested by the modular approach to validation [47], and it made use of a compilation of existing data from various sources instead of designing a new validation study. In fact, the validation was completed in roughly half the time of a prospective study and with minimal costs. The MNT validation also represents the new close collaboration with the OECD, as test guideline development was carried out in parallel to validation. Thus, consensus at the OECD level could already be reached in 2008. It is, however, only an in vitro test replacing another in vitro test of lower quality. The Japanese Center for the Validation of Alternative Methods (JaCVAM) is leading the validation of the COMET assay with EU and US involvement. Both the MNT and the COMET assay promise to be more predictive. More importantly, however, they can be incorporated into the repeated-dose animal studies, making a further mutagenicity test unnecessary. Other options not formally validated include reducing the number of animals in in vivo genotoxicity testing, for example, the use of only one gender, the omission of positive controls. New DNA repair-based assays were submitted or identified and are currently under discussion. For NPs, mutagenicity tests are certainly first-line tests and need proper adaptation. The putative carcinogenic potential of nanofibers is mainly attributed to inflammatory effects and is less likely to be covered by these genotoxicity tests.

- *Phototoxicity*. The endpoint has been replaced by a validated method for which minor follow-up activities are required. However, the validated method is considered overly sensitive. Work using skin models instead or together with cell cultures in a tiered testing strategy is on the way from companies and the Centre for Documentation and Evaluation of Alternatives to Animal Experiments (ZEBET). It appears that the tests can be adapted to NPs but have lesser priority for the risk assessment process.
- *Skin sensitization*. The local lymph node assay, an ICCVAM-validated refinement method, was scientifically endorsed by the ECVAM Scientific Advisory Committee (ESAC) and is now an OECD adopted test guideline (TG 429). Nonradioactive variants have been adopted. A number of relevant workshops

have been organized to identify the most promising methods and to give advice on how to progress toward their validation (peptide-binding assay, skin permeability, and dendritic cell–based methods). Three promising methods, the peptide-binding assay, the U937 test, and the hCLAT test, were validated with help from COLIPA and are currently analyzed. The Integrated Project Sens-it-iv was set up (http://www.sens-it-iv.eu/) following an ECVAM workshop. Four different cell systems and four different markers have shown promising results. These are to be standardized and assessed in an interlaboratory trial. Beside Sens-it-iv, the general interest in developing relevant alternative methods for skin sensitization testing is supported by the industry's large investment in this area. The area has been reviewed as part of the European Commission's stock taking [73, 74] and the road map activity of the Center for Alternatives to Animal Testing (CAAT) [75]. The primary goal at this stage is to combine the various approaches to an integrated testing strategy. For NPs, the sensitization endpoint is of high interest as reactive chemistry is the first step in the sensitization process and surface expansion of NPs might have an impact here.

- *Toxicokinetics.* Toxicokinetics is best defined as an adjunct technique to support the interpretation of toxicological results in risk assessment. At present, animal-free techniques of prediction of pharmaco- or toxicokinetics are used as a screen in pharmaceutical research. They rely essentially upon the use of physiologically based pharmacokinetic (PBPK) modeling techniques, with input of quantitative parameter values generated by various techniques in silico, in vitro, ex vivo, and in vivo. This field is still under development, and the use and acceptance of PBPK in risk assessment is increasing. For a review and the way forward see Ref. [75]. For NPs, it appears that kinetics represent a key concern, as unusual behavior of NPs has been documented. It is not clear yet how to address this, as modeling might be as misleading as animal models as a result of species differences.

  The most promising approach for the moment are barrier models [76], for example, for the gastrointestinal barrier, skin penetration, the blood–brain barrier, and the placenta barrier. The exclusion of transfer through these barriers might help to reduce the concern about toxic effects in the respective compartment. This could mean simply no bioavailability, no systemic toxicity testing, or no blood–brain barrier transfer for central neurotoxicity. For NPs these should have highest priority.

- *Carcinogenicity*. The area was analyzed in the road map activity of CAAT [75]. ECVAM carried out the validation of three cell transformation assays parallel to OECD test guideline development. The peer review has been recently completed with a positive outcome. The tests have the potential to partially replace the two-year bioassay in the rat. At up to $1 million per substance, the assay is so expensive that it is rarely carried out (from almost 5,000 new chemicals notified in Europe over the last 25 years, 14 had a cancer bioassay in the new chemical database). Thus, any in vitro test would first enable a larger number of substances to be tested. However, the general concerns about the high false positive rate of the cancer bioassay (some aspects summarized in Hoffmann and Hartung [20]) prompt a need for new approaches. For NPs, the cell transformation assays represent a key opportunity to assess some aspects of the carcinogenic potential.
- *Reproductive toxicity*. The area was analyzed in the road map activity of CAAT [75]. Because of the complexity of reproductive toxicity and the variety of toxicological mechanisms involved, only (partial) replacement can be achieved in the meantime by focusing on toxicological targets with high prevalence. This requires setting priority on the relevant target cells and/or biological mechanisms, as the complete mammalian reproductive cycle cannot currently be mimicked in vitro. To not compromise consumer safety, the use of in vitro tests for assessing reproductive toxicity therefore depends on a detailed analysis of historical in vivo data of reproductive toxicants belonging to the various chemical groups. In this context, the histopathological data of reproductive organs in repeated dose experiments is of high interest, as this endpoint seems to be very sensitive. Within FP6 the integrated project ReProTect [77, 78] has been established to develop a toolbox of in vitro tests assessing various cell functions relevant for mammalian reproduction, such as hormone production,

germ cell maturation, and uterine function. However, as not all possible target cells/mechanisms are covered within this project, further research activity is necessary. For NPs, this area is of concern, and placenta barrier models might help a lot here. However, as this is an area of massive costs and animal use, it is most important to define strategies for key endpoints such as embryonic stem cell tests for malformations or endocrine disruptor assays to cover key concerns in a screenable manner.

- *Repeated-dose toxicity*. The area was analyzed in the road map activity of CAAT [75]. The integrated project Predict-IV (http://www.predict-iv.toxi.uni-wuerzburg.de/en/) addresses strategies to improve the assessment of drug safety in the early stage of development and in the late discovery phase by an intelligent combination of nonanimal-based test systems, cell biology, mechanistic toxicology, and in silico modeling in a rapid and cost-effective manner. As the assessment does not differ for NPs, the project is also fully relevant here. The project will integrate new developments to improve and optimize cell culture models for toxicity testing and to characterize the dynamics and kinetics of cellular responses to toxic effects in vitro. It was based on two ECVAM workshops on chronic toxicity and PBPK modeling. Repeated-dose systemic toxicity requires a new systematic approach in order to move from a traditional, animal-based assessment to a mechanism-based approach. In collaboration between the cosmetic industry and the European Commission, a 50 million euro project SEURAT-1 has been launched (http://www.seurat-1.eu). It will pioneer the field exploring various new technologies. These five-year projects started in early 2011. NPs are not considered, but translation of results could be most helpful. Repeat-dose toxicity is the cornerstone of risk assessment. It is the animal test most likely to be used for NP assessments. For NPs as well as for general chemicals, however, alternative approaches are only emerging.
- *Ecotoxicology*. Activities in ecotoxicology/environmental toxicity are currently focused on short-term (acute) aquatic toxicity and bioconcentration. Endpoints such as long-term (chronic) aquatic toxicity testing or toxicity to birds have

not yet been tackled [79]. Changes in legislation, such as environmental assessment of pharmaceuticals and REACH, have increased and will increase the numbers of fish used. The threshold approach (= upper threshold concentration [UTC] step-down approach) [80] for acute aquatic toxicity testing is based on a retrospective analysis of data in the New Chemicals Database carried out by ECVAM and the European Chemical Bureau (ECB). Its validity for the reduction of the numbers of fish was endorsed by the ECVAM Scientific Advisory Committee (ESAC) in 2006. It is part of the intelligent testing strategy in the requirements for REACH and the OECD guidance. The most promising replacement method is the fish embryo test (FET), for which a lot of data is available and which is already used for effluent testing in Germany. Studies using fish cell lines showed that fish gill cell lines might be useful for acute aquatic toxicity testing. The endpoint bioconcentration assesses whether a substance accumulates in fish. In a first step, bioconcentration models based on physicochemical properties and physiological parameters (uptake, metabolism, and excretion) are used. If the outcome is above a certain threshold a test in fish becomes mandatory. The current strategy followed by the industry and also proposed in the intelligent testing strategy for data requirements under REACH is to improve the existing in silico methods by including in vitro data on metabolism. Substances that are metabolized do not accumulate in the organism. For NPs, to the best of our knowledge, the adaptation of the respective in vitro methods has not yet taken place.

- *Immunotoxicity*. A number of in vitro tests have been identified [81, 82] and have undergone some prevalidation [81, 83–85]. Quite a few studies have been done in vitro on NPs [86], but their predictive value is not clear.
- *Pyrogenicity*. This might be considered a special form of immunotoxicity; however, it is typically not produced by the test material but by contaminations, mostly bacterial endotoxins. NPs represent a most interesting test material because of their large surface area, putatively binding these endotoxins. The Limulus assay, the most prominent alternative pyrogen test, has problems with the testing of solid materials.

It remains to be explored whether NPs qualify for testing. However, a series of cellular methods based on the activation for cytokine release of blood monocytes or derived cells has been validated [50, 51] and accepted for regulatory use. The author has been particularly involved in the development of a whole-blood pyrogen test [52, 87]. This test is unique, as it works with a cell suspension, which might be especially suited to test NPs.

5.3.1 Alternative Methods Based on Nanotechnologies

The question whether nanotechnologies offer specific opportunities to create new alternative methods deserves some consideration. Nanostructuring the surfaces of cell culture dishes is one example to induce or maintain the differentiation of cells. Coating techniques also are often required for approaches such as "cells on chips." Imaging technologies using quantum dots might also enhance (non) invasive imaging technologies for laboratory animals as they are developed in humans. NPs already are used to deliver genes or other materials into cells, enhancing and broadening opportunities for in vitro approaches. The opportunities offered by nanotechnologies, however, are only starting to be exploited.

5.3.2 Opportunities for in silico Alternatives in Nanotoxicology

Computational approaches to nanotoxicology so far are rather limited [88] (see Chapter 6). With increasing datasets, however, modeling some aspects of interest might become feasible. Data mining of large datasets and the interspecies extrapolation of kinetics are most promising. Size and shape variations add dimensions of complexity to the correlative approaches, however, which will require enormous datasets. The tremendous opportunities and challenges to in silico toxicity approaches have been discussed recently [89]. In the meantime, modeling of kinetics, starting with airway disposition, might hold the most promise. However, nanotoxicology could be very stringent from the beginning, making the best use of biometry and avoiding the common pitfalls (multiple testing, lack of power analysis, significance vs. relevance, lack of meta-analysis, etc.)

5.3.3 Are There Reasons to Make Current Alternative Tests Less Applicable to NPs?

The answer to the above question is yes, unfortunately, since the biokinetics of NPs will affect in vivo and in vitro results very differently. Consequently, many prediction models developed for general chemicals will not work. NP aggregation and the difficulty of application to cells and animals affect the execution of routine tests. Databases allowing computational approaches are so rare that, for the immediate future, no major contribution can be expected.

There might also be reasons to question the extrapolation of the NP parent compound to humans, but it is difficult to say whether animal results for NPs are better or worse when extrapolated to humans. NPs differ in exposure/kinetics, and with regard to hazards, even higher potentials for interspecies differences exist. What is required is the systematic evaluation of validated alternatives for their applicability to NPs and, if necessary, a modification of the prediction model.

5.4 Towards a Human Toxome Project

The US National Academies/National Research Council (NRC) report from 2007 "Toxicity Testing in the 21st Century: A Vision and a Strategy" [90] has created an atmosphere of departure in the US [91]. It suggests moving away from traditional (animal) testing to modern technologies based on pathways of toxicity. The concept is based on the hypothesis that there might be only a couple of hundred ways to harm a cell. These pathways of toxicity could be modeled in relatively simple cell tests, which can be run by robots. The goal is to develop a public database for such pathways, the Human Toxome (http://humantoxome.com), to enable scientific collaboration and exchange.

There is a continuously growing awareness—not to say excitement—about Tox-21c in all stakeholder groups. It was first embraced by scientists and in the US—it was difficult to find any US toxicology meeting over the last few years that did not have it as a topic. The US Society of Toxicology instantly started a series of comments in its journal. Most importantly, the US agencies followed fast on the 2007 NAS/NRC report: the Tox-21 alliance in 2008 [92], the Environmental Protection Agency (EPA) made it its chemical testing paradigm in 2009, and the Food and Drug Administration (FDA) followed [93]. The chemical and consumer product industries got engaged, for example, with the Human Toxicology Project Consortium, the pharmaceutical industry somewhat more reluctantly. In Europe, all this is rather delayed, with some adaptation of the vocabulary but without necessarily grasping the new approach. This is not alternative methods under a new name. However, interest is lately increasing strongly in Europe.

Tox-21c suggests to move to a new resolution, that is, pathways of toxicity. The problem is that the respective science is only emerging. Everybody feels that after 40 years without major change in regulatory sciences, there is something coming from the biotech and bioinformatics revolution. The call for mechanism-based approaches is around for a while. The new concept is a change in resolution to molecularly defined pathways. The new technologies (especially omics) allow this. What will be needed is using the Human Toxome as a comprehensive pathway list [94]; an annotation of cell types, species, toxicant classes and hazards to these pathways; an integration of information in systems toxicology approaches; and in-vitro-to-in-vivo extrapolation by reversed dosimetry and finally making sense of the data, most probably in a probabilistic way.

However, there is also the perception that the system is most resistant to change. Proof of concepts for risk assessment is still rare; the best example so far was the 2010 evaluation of dispersants for the oil spill disaster. Here, the test battery was used to choose between eight dispersants for the Gulf oil spill disaster: Results were available within four weeks, where a traditional assessment would have lasted more than a year at, at least, 10 times higher costs. For comparison, an assessment of one drug or pesticide totals around $5–$20 million. The ToxCast battery of more than 600 tests is carried out at less than $20.000 per substance.

Most notable activities:

- ToxCast of the US EPA based on available high-throughput tests; the EPA made Tox-21c officially its toxicity-testing paradigm for chemicals in 2009 [95, 96].
- The Tox-21 alliance of the EPA, the National Institute of Environmental Health Sciences (NIEHS), the NIH Chemical

Genomics Center (NCGC), and the FDA extending this to more chemicals [92, 97].
- Case study approaches at the Hamner institute, which was originally created by the chemical industry [98].
- The Human Toxicology Project Consortium (seven global companies and three stakeholders, including CAAT) [99].
- The Human Toxome project [100–102] led by CAAT and financed by an National Institutes of Health (NIH) Transformative Research Grant.
- The OECD embracing this in its new Adverse Outcome Pathway concept [103].

There is no European branch of the Human Toxome project yet, though the EU project AXLR8 aims to reach out to the US activities and the 50 million euro project SEURAT-1, financed by the cosmetic industry and the European Commission, goes into this direction. A recent event in the European Parliament was alerting of the need for a European branch. The European Commission has given more than 200 million euros for alternative approaches so far (plus about the same by member states). However, these are mainly one-by-one replacement methods, where a cell system shall reflect key aspects of an animal test. The new vision calls for a larger international project, with a different resolution of analyzing the problems, that is, molecular pathways.

The most advanced regulatory application is certainly the proposal to use some of the high-throughput assays from ToxCast to prioritize endocrine disruptor screening (EDSP21 program). However, this proposal is only on the way and seems to create also quite some resistance. It is very important that the OECD has embraced some of the concepts under the label of adverse outcome pathways. Advances in in vitro testing, measurement technologies (omics, image analysis), high throughput, etc., over the last decade were crucial. Some technology providers (ThermoFisher, Agilent, Life Technologies) are in the $10 billion turnover range now, which brings in standardization and market forces not seen before.

Early on, the need for quality assurance for the new approaches as a sparring partner for their development and implementation has been noted. Formal validation as developed for the first generation of alternative methods can only partially serve this purpose [104].

For this reason, the Evidence-based Toxicology Collaboration (EBTC) was created in the US and Europe in 2011 and 2012, respectively (www.ebtox.com). This collaboration of representatives from agencies, industry, academia, and stakeholder groups aims to develop tools of evidence-based medicine for toxicology [20, 105, 106]. The secretariat is run by CAAT, the first conference was held in early 2012 hosted by the US EPA, and working groups have started to address pertinent issues and methodologies. Taken together, Tox-21c and its implementation activities, including the Human Toxome and the EBTC, promise a credible approach to revamp regulatory toxicology.

Obviously, the approach of Tox-21c, the Human Toxome, and the EBTC lends itself to extension to nanotoxicology. It has also been prioritized by the road map activity of 2011–2012 as most promising to improve systemic toxicity testing [75], though there is still quite a road to go [107]. A first CAAT workshop on nanotoxicology embraced this concept [108]. However, no major practical attempts for NP evaluation have been done so far.

5.5 Conclusions

The toxicology of NPs is a rapidly emerging concern. It is driven by the dramatic increase in industrial uses of NPs and by public debate. Increasing funding and studies inevitably will result in reports of toxicological effects of NPs—both publication bias for positive findings and the multiple testing fallacy (if 20 experiments or endpoints are analyzed with $p = 0.05$ for significance, one should be false positive) will come to bear here. They will spur further research, and it will require decades to sort out what is true and what is relevant. We will need validated tools that offer the throughput, reliability, and relevance to address these key features of alternative approaches. The additional testing demand for NPs adds to the urgency of developing new approaches in toxicology. Whether this will only add some tools for NPs to the traditional approach or help to create an entire new paradigm for toxicology awaits an answer. So far, it appears that the problem of nanotoxicology is mainly a kinetic one—some safety factors could help to account for differences in ADME, but we need to keep in mind that the enhanced bioavailability of NPs at the body, organ, and cell levels might result in thresholds of toxicity being overstepped, in which case a change in hazards suddenly does become relevant. The fact that higher exposure levels in target cells can be more easily modeled in cell systems than in vivo, and thus such hazards identified, argues again for the use of alternative methods.

Taken together, it appears that nanotoxicology, to a large extent, is dependent on the use and further development of alternative approaches. The more we know what we are looking for, the better we can target our testing. If we have no hypothesis, screening in many models and black-box type of animal tests might be the only way forward. NPs are different, but they are not so different that we should expect completely new hazards. Hazards not necessarily shown by the parent compound may be seen, however, due to higher concentrations achieved at target tissues. In vitro approaches represent a reasonable compromise between effort and information gain, allowing direct comparison of various NPs and their parent compounds. A broad, animal-based screening approach is not feasible with regard to laboratory capacities and costs, and it certainly is not desirable from an animal welfare viewpoint.

A number of alternative approaches have undergone the optimization and validation process to make them suitable for regulatory purposes. It appears to be most promising to adapt these to NPs in order to have a testing platform for broader characterization. When combined with a somewhat more extensive physicochemical characterization than normally applied to industrial chemicals, this will help us derive some more general rules about the hazards of NPs. The field of alternative approaches has paid for its lessons on the importance of good practices and standardization for the success of validation and regulatory acceptance of methods. It is strongly advised that the respective guidance on good laboratory practice for in vitro toxicity [109] and good cell culture practice [110] be followed from the beginning. It is promising that some good practices for how to test NPs have emerged from expert workshops [111–114]. In the near future, the respective quality assurance for the execution of such tests will be integrated.

Due to the high number and heterogeneity of particle samples and experimental systems, it is still difficult to find common principles of NP toxicity [115]. We have been rightly warned [116], however, that we are witnessing only the first generation of NPs; more sophisticated NPs (active nanostructures, coated NPs, integrated nanosystems, etc.) will make this even more complicated. Thus, it might well be that each and every NP formulation of a substance will have to be considered an individual entity requiring at least some risk assessment. Elliott [117] characterized nanotoxicology as a prenormal science, in which researchers have no widely accepted paradigm to guide their investigations.

We must not forget that not only NPs themselves but also contaminations may have adverse effects. Carbon nanotubes, for instance, were shown to include metals, amorphous carbon, and other compounds [116, 118]. A special case of high relevance is the contamination with pyrogens due to the large surface area and the high lipophilicity of these compounds [119]. It remains to be seen whether current pyrogenicity tests can retrieve such contamination before applying nanomedicines by injection.

The major problem for NP risk assessment is kinetics. Though we expect differences from the parent compound due to size and shape, we do not really know how to test for them. Species differences are not really well established. A key problem is that we still do not know how NPs are metabolically processed [120]. The field of alternatives mainly has to offer some barrier models, which certainly represent a key priority.

Last but not least, toxicity is not always bad news, since sometimes it can be exploited for therapeutic purposes [58]. The main difference between toxicology and pharmacology is whether an effect is desired. NPs offer fascinating opportunities to interfere with the organism in new ways. We must take care to find the right balance between opportunities and safety concerns. In vitro approaches promise to provide an affordable database on the biological activities to help understand the risks and opportunities.

Acknowledgments

This article is based on Refs. [68, 121].

The discussion was updated to reflect the literature until the end of 2012. The valuable discussions with Enrico Sabbioni and the nanotoxicology group at ECVAM and, more recently, with Ellen

Silbergeld at Johns Hopkins are gratefully appreciated—they have shaped my understanding of NP toxicology. The whole blood monocyte activation test mentioned in this report is licensed to Merck KGaA. I invented this test and receive royalties from Merck KGaA from sales of this test.

References

1. Hartung, T. (2008a). Towards a new toxicology: evolution or revolution? *ATLA: Altern. Lab. Anim.*, **36**, pp. 635–639.
2. Hartung, T., and Leist, M. (2008). Food for thought ... on the evolution of toxicology and phasing out of animal testing. *ALTEX*, **25**, pp. 91–96.
3. Leist, M., Hartung, T., and Nicotera, P. (2008). The dawning of a new age of toxicology. *ALTEX*, **25**, pp. 103–114.
4. Hartung, T. (2008b). Food for thought ... on animal tests. *ALTEX*, **25**, pp. 3–9.
5. Krug, H. F. H., and Wick, P. P. (2011). Nanotoxicology: an interdisciplinary challenge. *Angew. Chem., Int. Ed. Engl.*, **50**, pp. 1260–1278.
6. Clark, K. A., White, R. H., and Silbergeld, E. K. (2011). Predictive models for nanotoxicology: current challenges and future opportunities. *Regul. Toxicol. Pharmacol.*, **59**, p. 3.
7. Maynard, A. D., Warheit, D. B., and Philbert, M. A. (2011). The new toxicology of sophisticated materials: nanotoxicology and beyond. *Toxicol. Sci. Soc. Toxicol.*, **120**, pp. S109–S129.
8. Elsaesser, A., and Howard, C. V. (2012). Toxicology of nanoparticles. *Adv. Drug Deliv. Rev.*, **64**(2), pp. 129–137.
9. Bolt, H. M., Marchan, R., and Hengstler, J. G. (2012). Recent developments in nanotoxicology. *Arch. Toxicol.*, **87**(6), pp. 927–928.
10. Santamaria, A. (2012). Historical overview of nanotechnology and nanotoxicology. *Methods Mol. Biol.*, **926**, pp. 1–12.
11. Greish, K., Thiagarajan, G., and Ghandehari, H. (2012). In vivo methods of nanotoxicology. *Methods Mol. Biol.*, **926**, pp. 235–253.
12. Johnston, H., Pojana, G., Zuin, S., Jacobsen, N. R., Møller, P., Loft, S., Semmler-Behnke, M., McGuiness, C., Balharry, D., Marcomini, A., Wallin, H., Kreyling, W., Donaldson, K., Tran, L., and Stone, V. (2013). Engineered nanomaterial risk. Lessons learnt from completed nanotoxicology studies: potential solutions to current and future challenges. *Crit. Rev. Toxicol.*, **43**(1), pp. 1–20.

13. Garnett, M. C., and Kallinterni, P. (2006). Nanomedicine and nanotoxicology: some physiological principles. *Occup. Med.*, **56**, pp. 307–311.

14. Di Gioacchino, M., Verna, N., Gornati, R., Sabbioni, E., and Bernadini, G. (2009). Metal nanoparticle health risk assessment, in *Nanotoxicity: From in vitro Models to Health Risks*, eds. Sahu, S. C., and Casciano, D. A. (John Wiley & Sons, Chichester, UK), pp. 519–541.

15. Hartung, T. (2010b). Food for thought... on alternative methods for chemical safety testing. *ALTEX*, **27**, pp. 3–14.

16. Sauer, U. G. (2009). Political incentives towards replacing animal testing in nanotechnology? *ALTEX*, **26**, pp. 285–294.

17. Bottini, A. A., Amcoff, P., and Hartung, T. (2007). Food for thought... on globalization of alternative methods. *ALTEX*, **24**, pp. 255–261.

18. Bottini, A. A., and Hartung, T. (2009). Food for thought... on economics of animal testing. *ALTEX*, **26**, pp. 3–16.

19. Hoffmann, S., and Hartung, T. (2005). Diagnosis: toxic! Trying to apply approaches of clinical diagnostics and prevalence in toxicology considerations. *Tox. Sci.*, **85**, pp. 422–428.

20. Hoffmann, S., and Hartung, T. (2006). Towards an evidence-based toxicology. *Human Exp. Toxicol.*, **25**, pp. 497–513.

21. Hartung, T. (2009). Toxicology for the twenty-first century. *Nature*, **460**, pp. 208–212.

22. Rovida, C., and Hartung, T. (2009). Re-evaluation of animal numbers and costs for in vivo tests to accomplish REACH legislation requirements. *ALTEX*, **26**, pp. 187–208.

23. Hartung, T., and Rovida, C. (2009). Chemical regulators have overreached. *Nature*, 2009, **460**, pp. 1080–1081.

24. Hartung, T. (2010). Lessons learned from alternative methods and their validation for a new toxicology in the 21st century. *J. Toxicol. Environ. Health*, **13**, pp. 277–290.

25. Hartung, T. (2008c). Food for thought ... on alternative methods for cosmetics safety testing. *ALTEX*, **25**, pp. 147–162.

26. Hartung, T., and Daston, G. (2009). Are in vitro tests suitable for regulatory use? *Tox. Sci.*, **111**, pp. 233–237.

27. Worle-Knirsch, J. M., Pulskamp, K., and Krug, H. F. (2006). Oops they did it again! Carbon nanotubes hoax scientists in viability assays. *Nano Lett.*, **6**, pp. 1261–1268.

28. Nyland, J. F., and Silbergeld, E. K. (2009). A nanobiological approach to Nanotoxicology. *Human Exp. Toxicol.*, **28**, pp. 393–400.

29. Valant, J., Iavicoli, I., and Drobne, D. (2012). The importance of a validated standard methodology to define in vitro toxicity of nano-TiO2. *Protoplasma*, **249**, pp. 493–502.

30. Clift, M. J. D. M., Gehr, P. P., and Rothen-Rutishauser, B. B. (2011). Nanotoxicology: a perspective and discussion of whether or not in vitro testing is a valid alternative. *Arch. Toxicol.*, **85**, pp. 723–731.

31. Arora, S., Rajwade, J. M., and Paknikar, K. M. (2012). Nanotoxicology and in vitro studies: the need of the hour. *Toxicol. Appl. Pharmacol.*, **258**, pp. 151–165.

32. Donaldson, K., Stone, V., Tran, C. L., Kreyling, W., and Borm, P. J. A. (2004). Nanotoxicology. *Occup. Environ. Med.*, **61**, pp. 727–728.

33. Kagan, V. E., Bayir, H., and Shvedova, A. A. (2005). Nanomedicine and nanotoxicology: two sides of the same coin. *Nanomed. Nanotech. Biol. Med.*, **1**, pp. 313–316.

34. Seaton, A., and Donaldson, K. (2005). Nanoscience, nanotoxicology, and the need to think small. *Lancet*, **365**, pp. 923–924.

35. Kipen, H. M., and Laskin, D. L. (2005). Smaller is not always better: nanotechnology yields nanotoxicology. *Am. J. Physiol. Lung. Cell. Mol. Physiol.*, **289**, pp. L696–L697.

36. Holl, M. M. B. (2009). Nanotoxicology: a personal perspective. *WIREs Nanomed. Nanobiotechnol.*, **1**, pp. 353–359.

37. Gornati, R., Papis, E., Di Gioacchino, M., Sabbioni, E., Dalle Donne, I., Milzani, A., and Bernadini, G. (2009). In vivo and in vitro models for nanotoxciology testing, in *Nanotoxicity: From in vitro Models to Health Risks*, eds. Sahu, S. C., and Casciano, D. A. (John Wiley & Sons, Chichester UK), pp. 279–302.

38. Pal, S., Tak, Y. K., and Song, J. M. (2007). Does the antibacterial activity of silver nanoparticles depend on the shape of the nanoparticle? A study of the gram-negative bacterium Escherichia coli. *Appl. Environ. Microbiol.*, **73**, pp. 1712–1720.

39. Chen, M., and Goodman, D. W. (2006). Catalytically active gold: from nanoparticles to ultrathin films. *Acc. Chem. Res.*, **39**, pp. 739–746.

40. Cha, E. K., and Myung, H. (2007). Cytotoxic effects of nanoparticles assessed in vitro and in vivo. *J. Microbiol. Biotechnol.*, **17**, pp. 1573–1578.

41. Walker, N. J., and Bucher, J. R. (2009). A 21st century paradigm for evaluating the health hazards of nanoscale materials? *Toxicol. Sci.*, **110**, pp. 251–254.

42. Hillegass, J. M., Shukla, A., Lathrop, S. A., MacPherson, M. B., Fukagawa, N. F., and Mossman, B. T. (2009). Assessing nanotoxicity in cells in vitro. *WIREs Nanomed. Nanobiotechnol.*, doi: 10.1002/wnan.054.

43. Choi, J.-Y., Ramachandrran, G., and Kandlikar, M. (2009). The impact of toxicity testing costs on nanomaterial regulation. *Environ. Sci. Technol.*, **43**, pp. 3030–3034.

44. Hillyer, J. F., and Albrecht, R. M. (2001). Gastrointestinal persorption and tissue distribution of differently sized colloidal gold nanoparticles. *J. Pharm. Sci.*, **90**, pp. 1927–1936.

45. Oberdörster, G., Stone, V., and Donaldson, K. (2007). Toxicology of nanoparticles: a historical perspective. *Nanotoxicology*, **1**, pp. 2–25.

46. Han, X. X., Corson, N. N., Wade-Mercer, P. P., Gelein, R., Jiang, J., Sahu, M., Biswas, P., Finkelstein, J. N., Elder, A., and Oberdörster, G. (2012). Assessing the relevance of in vitro studies in nanotoxicology by examining correlations between in vitro and in vivo data. *Toxicology*, **297**, pp. 1–9.

47. Hartung, T., Bremer, S., Casati, S., Coecke, S., Corvi, R., Fortaner, S., Gribaldo, L., Halder, M., Hoffmann, S., Roi, A. J., Prieto, P., Sabbioni, E., Scott, L., Worth, A., and Zuang V. (2004). A modular approach to the ECVAM principles on test validity. *ATLA: Altern. Lab. Anim.*, **32**, pp. 467–472.

48. Kirkland, D., Aardema, M., Henderson, L., and Muller, L. (2005). Evaluation of the ability of a battery of three in vitro genotoxicity tests to discriminate rodent carcinogens and noncarcinogens I. Sensitivity, specificity and relative predictivity. *Mutat. Res.*, **584**, pp. 1–256.

49. Kirkland, D., Pfuhler, S., Tweats, D., Aardema, M., Corvi, R., Darroudi, F., Elhajouji, A., Glatt, H., Hastwell, P., Hayashi, M., Kasper, P., Kirchner, S., Lynch, A., Marzin, D., Maurici, D., Meunier, J. R., Müller, L., Nohynek, G., Parry, J., Parry, E., Thybaud, V., Tice, R., van Benthem, J., Vanparys, P., and White, P. (2007). How to reduce false positive results when undertaking in vitro genotoxicity testing and thus avoid unnecessary follow-up animal tests: report of an ECVAM workshop. *Mutat. Res.*, **628**, pp. 31–55.

50. Hoffmann, S., Peterbauer, A., Schindler, S., Fennrich, S., Poole, S., Mistry, Y., Montag-Lessing, T., Spreitzer, I., Loschner, B., van Aalderen, M., Bos, R., Gommer, M., Nibbeling, R., Werner-Felmayer, G., Loitzl, P., Jungi, T., Brcic, M., Brugger, P., Frey, E., Bowe, G., Casado, J., Coecke, S., de Lange, J., Mogster, B., Naess, L. M., Aaberge, I. S., Wendel, A., and Hartung, T. (2005). International validation of novel pyrogen tests based on the human fever reaction. *J. Immunol. Methods*, **298**, pp. 161–173.

51. Schindler, S., Spreitzer, I., Loschner, B., Hoffmann, S., Hennes, K., Halder, M., Brügger, P., Frey, E., Hartung, T., and Montag, T. (2006). International validation of pyrogen tests based on cryopreserved human primary blood cells. *J. Immunol. Methods*, **316**, 42–51.

52. Schindler, S., von Aulock, S., Daneshian, M., and Hartung, T. (2009). Development, validation and applications of the monocyte activation test for pyrogens based on human whole blood. *ALTEX*, **26**, 293–305.

53. Monteiro-Riviere, N. A., Inman, A. O., and Zhang, L. W. (2009). Limitations and relative utility of screening assays to assess engineered nanoparticle toxicity in a human cell line. *Toxicol. Appl. Pharmacol.*, **234**, pp. 222–235.

54. Kroll, A., Pillukat, M. H., Hahn, D., and Schnekenburger, J. (2009). Current in vitro methods in nanoparticle risk assessment: limitations and challenges. *Eur. J. Pharm. Biophys.*, **72**, 370–377.

55. Stone, V., Johnston, H., and Schins, R. P. F. (2009). Development of in vitro systems for nanotoxicology: methodological considerations. *Crit. Rev. Toxicol.*, **39**(7), 613–626.

56. Hillegass, J. M., Shukla, A., Lathrop, S. A., MacPherson, M. B., Fukagawa, N. K., Mossman, B. T. (2010). Assessing nanotoxicity in cells in vitro. *Wiley Interdiscip. Rev. Nanomed. Nanobiotechnol.*, **2**, pp. 219–231.

57. Misra, S. K., Dybowska, A., Berhanu, D., Luoma, S. N., and Valsami-Jones, E. (2012). The complexity of nanoparticle dissolution and its importance in nanotoxicological studies. *Sci. Total Environ.*, **438C**, pp. 225–232.

58. Oberdörster, G., Stone, V., and Donaldson, K. (2007). Toxicology of nanoparticles: a historical perspective. *Nanotoxicology*, **1**, pp. 1–25.

59. Adams, L. K., Lyon, D. Y., and Alvarez, P. J. J. (2006). Comparative ecotoxicity of nanoscale TiO2, SiO2 and ZnO water suspensions. *Water Res.*, **40**, 3527–3532.

60. Sager, T. M., Porter, D. W., Robinson, V. A., Lindsley, W. G., Schwegler-Berry, D. E., and Castranova, V. (2007). Improved method to disperse nanoparticles for in vitro and in vivo investigation of toxicity. *Nanotoxicology*, **1**, pp. 118–129.

61. Powers, K. W., Palazuelos, M., Moudgil, B. M., and Roberts, S. M. (2007). Characterization of the size, shape, and state of dispersion of nanoparticles for toxicological studies. *Nanotoxicology*, **1**, pp. 42–51.

62. Unfried, K., Albrecht, C., Klotz, L.-O., von Mikecz, A., Grether-Beck, S., and Schins, R. F. F. (2007). Cellular responses to nanoparticles: target structure and mechanisms. *Nanotoxicology*, **1**, pp. 52–71.

63. Bouvier d'Yvoire, M., Prieto, P., Blaauboer, B. J., Bois, F. Y., Boobis, A., Brochot, C., Coecke, S., Freidig, A., Gundert-Remy, U., Hartung, T., Jacobs, M. N., Lavé, T., Leahy, D. E., Lennernäs, H., Loizou, G. D., Meek, B., Pease, C., Rowland, M., Spendiff, M., Yang, J., and Zeilmaker, M. (2007). Physiologically-based Kinetic Modelling (PBK Modelling): meeting the 3Rs agenda. The Report and Recommendations of ECVAM Workshop 63. *ATLA: Altern. Lab. Anim.*, **35**, pp. 661–671.

64. Mailänder, V., and Landfester, K. (2009). Interaction of nanoparticles with cells. *Biomacromolecules*, **10**, pp. 2379–2400.

65. Caldorera-Moore, M., Guimard, N., Shi, L., and Roy, K. (2010). Designer nanoparticles: incorporating size, shape and triggered release into nanoscale drug carriers. *Expert Opin. Drug Delivery*, **7**, pp. 479–495.

66. Hartung, T. (2007). Food for thought ... on cell culture. *ALTEX*, **24**, pp. 143–147.

67. Jones, C. F., and Grainger, D. W. (2009). In vitro assessments of nanomaterial toxicity. *Adv. Drug Delivery Rev.*, **61**, pp. 438–456.

68. Hartung, T., and Sabbioni, E. (2011). Alternative in vitro assays in nanomaterial toxicology. Wiley interdisciplinary reviews. Nanomedicine and nanobiotechnologyn/a–n/a.

69. Scott, L., Eskes, C., Hoffmann, S., Adriaens, E., Alepée, N., Bufo, M., Clothier, R., Facchini, D., Faller, C., Guest, R., Harbell, J., Hartung, T., Kamp, H., Le Varlet, B., Meloni, M., McNamee, P., Osborne, R., Pape, W., Pfannenbecker, U., Prinsen, M., Seaman, C., Spielmann, H., Stokes, W., Trouba, K., Van den Berghe, C., van Goethem, F., Vassallo, M., Vinardell, P., and Zuang, V. (2009). A proposed eye irritation testing strategy to reduce and replace in vivo studies using bottom-up and top-down approaches. *Toxicol. in vitro*, **24**, pp. 1–9.

70. Pfuhler, S., Kirkland, D., Kasper, P., Hayashi, M., Vanparys, P., Carmichael, P., Dertinger, S., Eastmond, D., Elhajouji, A., Krul, C., Rothfuss, A., Schoening, G., Smith, A., Speit, G., Thomas, C., van Benthem, J., and Corvi, R. (2009). Reduction of use of animals in regulatory genotoxicity testing: identification and implementation opportunities—report from an ECVAM workshop. *Mutat. Res.: Genet. Toxicol. Environ. Mutagen.*, **680**, pp. 31–42.

71. Kirkland, D. (2010). Evaluation of different cytotoxic and cytostatic measures for the in vitro micronucleus test (MNVit): introduction to the collaborative trial. *Mutat. Res.: Genet. Toxicol. Environ. Mutagen.*, pp. 1–4.

72. Corvi, R., Albertini, S., Hartung, T., Hoffmann, S., Maurici, D., Pfuhler, S., van Benthem, J., and Vanparys, P. (2008). ECVAM retrospective

validation of in vitro micronucleus test (MNT). *Mutagenesis*, **23**, pp. 271–283.

73. Adler, S., Basketter, D, Creton, S., Pelkonen, O., van Benthem, J., Zuang, V., Andersen, K. E., Angers-Loustau, A., Aptula, A., Bal-Price, A., Benfenati, E., Bernauer, U., Bessems, J., Bois, F. Y., Boobis, A., Brandon, E., Bremer, S., Broschard, T., Casati, S., Coecke, S., Corvi, R., Cronin, M., Daston, G., Dekant, W., Felter, S., Grignard, E., Gundert-Remy, U., Heinonen, T., Kimber, I., Kleinjans, J., Komulainen, H., Kreiling, R., Kreysa, J., Leite, S. B., Loizou, G., Maxwell, G., Mazzatorta, P., Munn, S., Pfuhler, S., Phrakonkham, P., Piersma, A., Poth, A., Prieto, P., Repetto, G., Rogiers, V., Schoeters, G., Schwarz, M., Serafimova, R., Tähti, H., Testai, E., van Delft, J., van Loveren, H., Vinken, M., Worth, A., and Zaldivar, J. M. (2011). Alternative (non-animal) methods for cosmetics testing: current status and future prospects: 2010. *Arch. Toxicol.*, **85**, pp. 367–485.

74. Hartung, T., Blaauboer, G. J., Bosgra, S., Carney, E., Coenen, J., Conolly, R. B., Corsini, E., Green, S., Faustman, E. M., Gaspari, A., Hayashi, M., Hayes, A. W., Hengstler, J. G., Knudsen, L. E., Knudsen, T. B., McKim, J. M., Pfaller, W., and Roggen, E. L. (2011). An expert consortium review of the EC-commissioned report "alternative (non-animal) methods for cosmetics testing: current status and future prospects: 2010." *ALTEX*, **28**, pp. 183–209.

75. Basketter, D. A., Clewell, H., Kimber, I., Rossi, A., Blaauboer, B., Burrier, R., Daneshian, M., Eskes, C., Goldberg, A., Hasiwa, N., Hoffmann, S., Jaworska, J., Knudsen, T. B., Landsiedel, R., Leist, M., Locke, P., Maxwell, G., McKim, J., McVey, E. A., Ouédraogo, G., Patlewicz, G., Pelkonen, O., Roggen, E., Rovida, C., Ruhdel, I., Schwarz, M., Schepky, A., Schoeters, G., Skinner, N., Trentz, K., Turner, M., Vanparys, P., Yager, J., Zurlo, J., and Hartung, T. (2012). A roadmap for the development of alternative (non-animal) methods for systemic toxicity testing: t4 report*. *ALTEX*, **29**, pp. 3–91.

76. Lehr, C.-M., Daum, N., Schneider, M., Schäfer, U. F. (2011). Biological barriers: a need for novel tools in nanotoxicology and nanomedicine. Preface. *Eur. J. Pharm. Biopharm.*, **77**, p. 337.

77. Hareng, L., Pellizzer, C., Bremer, S., Schwarz, M., and Hartung, T. (2005). The Integrated Project ReProTect: a novel approach in reproductive toxicity hazard assessment. *Reprod. Toxicol.*, **20**, pp. 441–452.

78. Schenk, B., Weimer, M., Bremer, S., van der Burg, B., Cortvrindt, R., Freyberger, A., Lazzari, G., Pellizzer, C., Piersma, A., Schäfer, W. R., Seiler, A., Witters, H., and Schwarz, M. (2010). The ReProTect Feasibility Study, a novel comprehensive in vitro approach to detect reproductive toxicants. *Reprod. Toxicol.*, **30**, pp. 200–218.

79. Behra, R., and Krug, H. (2008). Nanoecotoxicology: nanoparticles at large. *Nat. Nanotechnol.*, **3**, pp. 253–254.

80. Jeram, S., Sintes, J. M. R., Halder, M., Fentanes, J. B., Sokull-Klüttgen, B., and Hutchinson, T. H. (2005). A strategy to reduce the use of fish in acute ecotoxicity testing of new chemical substances notified in the European Union. *Regul. Toxicol. Pharmacol.*, **42**, pp. 218–224.

81. Gennari, A., Ban, M., Braun, A., Casati, S., Corsini, E., Dastych, J., Descotes, J., Hartung, T., Hooghe-Peters, R., House, R., Pallardy, M., Pieters, R., Reid, L., Tryphonas, H., Tschirhart, E., Tuschl, H., Vandebriel, R., and Gribaldo, L. (2005). The use of in vitro systems for evaluating immunotoxicity: the report and recommendations of an ECVAM workshop. *J. Immunotoxicol.*, **2**, pp. 61–83.

82. Hartung, T., and Corsini, E. (2013). Immunotoxicology: challenges in the 21st century and in vitro opportunities. *ALTEX*, **30**, pp. 411–426.

83. Langezaal, I., Hoffmann, S., Hartung, T., and Coecke, S. (2002). Evaluation and prevalidation of an immunotoxicity test based on human whole-blood cytokine release. *ATLA: Altern. Lab. Anim.*, **30**, pp. 581–595.

84. Carfi, M., Gennari, A., Malerba, I., Corsini, E., Pallardy, M., Pieters, R., Van Loveren, H., Vohr, H. W., Hartung, T., and Gribaldo, L. (2007). In vitro tests to evaluate immunotoxicity: a preliminary study. *Toxicology*, **229**, pp. 11–22.

85. Lankveld, D., Van Loveren, H., and Baken, K. (2009). In vitro testing for direct immunotoxicity: state of the art. *Methods Mol. Biol.*, **598**, 401–423.

86. Pfaller, T., Colognato, R., Nelissen, I., Favilli, F., Casals, E., Ooms, D., Leppens, H., Ponti, J., Stritzinger, R., Puntes, V., Boraschi, D., Duschl, A., and Oostingh, G. J. (2010). The suitability of different cellular in vitro immunotoxicity and genotoxicity methods for the analysis of nanoparticle-induced events. *Nanotoxicology*, **4**, pp. 52–72.

87. Hartung, T., and Wendel, A. (1995). Detection of Pyrogens using human whole blood. *ALTEX*, **12**, pp. 70–75.

88. Burello, E. E., and Worth, A. A. (2011). Computational nanotoxicology: predicting toxicity of nanoparticles. *Nat. Nanotechnol.*, **6**, pp. 138–139.

89. Hartung, T., and Hoffmann, S. (2009). Food for thought on ... in silico methods in toxicology. *ALTEX*, **26**, pp. 155–166.

90. National Research Council. (2007). *Toxicity Testing in the 21st Century: A Vision and a Strategy* (National Academies of Sciences, USA).

91. Hartung, T. (2009c). Toxicology for the twenty-first century. *Nature*, **460**, pp. 208–212.

92. Collins, F. S., Gray, G. M., and Bucher, J. R. (2008). Toxicology. transforming environmental health protection. *Science*, **319**, pp. 906–907.

93. Hamburg, M. A. (2011). Advancing regulatory science. *Science*, **331**, pp. 987–987.

94. Kleensang, A., Maertens, A., Rosenberg, M., Fitzpatrick, S., Lamb, J., Auerbach, S., Brennan, R., Crofton, K. M., Gordon, B., Fornace, A. J. Jr., Gaido, K., Gerhold, D., Haw, R., Henney, A., Ma'ayan, A., McBride, M., Monti, S., Ochs, M. F., Pandey, A., Sharan, R., Stierum, R., Tugendreich, S., Willett, C., Wittwehr, C., Xia, J., Patton, G. W., Arvidson. K., Bouhifd, M., Hogberg, H. T., Luechtefeld, T., Smirnova, L., Zhao, L., Adeleye, Y., Kanehisa, M., Carmichael, P., Andersen, E. M., and Hartung, T. (2014). Pathways of toxicity. *ALTEX*, **31**, pp. 53–61.

95. Kavlock, R., and Dix, D. (2010). Computational toxicology as implemented by the US EPA: providing high throughput decision support tools for screening and assessing chemical exposure, hazard and risk. *J. Toxicol. Environ. Health, Part B, Crit. Rev.*, **13**, pp. 197–217.

96. Firestone, M., Kavlock, R., Zenick, H., Kramer, M., and the US EPA Working Group on the Future of Toxicity Testing (2010). The US Environmental Protection Agency strategic plan for evaluating the toxicity of chemicals. *J. Toxicol. Environ. Health, Part B, Crit. Rev.*, **13**, pp. 139–162.

97. Schmidt, C. W. (2009). TOX 21: new dimensions of toxicity testing. *Environ. Health Perspect.*, **117**, pp. A348 A353.

98. Andersen, M. E., Clewell, H. J., Carmichael, P. L., and Boekelheide, K. (2011). Can case study approaches speed implementation of the NRC report: "toxicity testing in the 21st century: a vision and a strategy?" *ALTEX*, **28**, pp. 175–182.

99. Seidle, T., and Stephens, M. L. (2009). Bringing toxicology into the 21st century: a global call to action. *Toxicol. in vitro*, pp. 1–4.

100. Hartung, T., and Mcbride, M. (2011). Food for thought... on mapping the Human Toxome. *ALTEX*, **28**, pp. 83–93.

101. Hartung, T., van Vliet, E., Jaworska, J., Bonilla, L., Skinner, N., and Thomas, R. (2012). Food for thought... systems toxicology. *ALTEX*, **29**, pp. 119–128.

102. Bouhifd, M., Hogberg, H. T., Kleensang, A., Maertens, A., Zhao, L., and Hartung, T. (2014). Mapping the Human Toxome by systems toxicology. *Basic Clin. Pharmacol. Toxicol.*, 1–8, doi: 10.1111/bcpt.12198.

103. Ankley, G. T., Bennett, R. S., Erickson, R. J., Hoff, D. J., Hornung, M. W., Johnson, R. D., Mount, D. R., Nichols, J. W., Russom, C. L., Schmieder, P. K., Serrrano, J. A., Tietge, J. E., and Villeneuve, D. L. (2010). Adverse outcome pathways: a conceptual framework to support ecotoxicology research and risk assessment. *Environ. Toxicol. Chem.*, **29**, pp. 730–741.

104. Leist, M., Hasiwa, N., Daneshian, M., and Hartung, T. (2012). Validation and quality control of replacement alternatives: current status and future challenges. *Toxicol. Res.*, **1**, p. 8.

105. Hartung, T. (2009b). Food for thought... on evidence-based toxicology. *ALTEX*, **26**, pp. 75–82.

106. Hartung, T. (2010a). Evidence-based toxicology: the toolbox of validation for the 21st century? *ALTEX*, **27**, pp. 253–263.

107. Hartung, T. (2009a). A toxicology for the 21st century: mapping the road ahead. *Toxicol. Sci.*: an official *J. Soc. Toxicol.*, **109**, pp. 18–23.

108. Silbergeld, E. K., Contreras, E. Q., Hartung, T., Hirsch, C., Hogberg, H., Jachak, A. C., Jordan, W., Landsiedel, R., Morris, J., Patri, A., Pounds, J. G., de Vizcaya Ruiz, A., Shvedova, A., Tanguay, R., Tatarazako, N., van Vliet, E., Walker, N. J., Wiesner, M., Wilcox, N., and Zurlo, J. (2011). t^4 workshop report. Nanotoxicology: "the end of the beginning": signs on the roadmap to a strategy for assuring the safe application and use of nanomaterials. *ALTEX*, **28**, pp. 236–241.

109. Organisation for Economic Co-operation and Development (2004). Series on principles of good laboratory practice and compliance monitoring, number 14, advisory document of the Working Group on Good Laboratory Practice, *The Application of the Principles of GLP to in vitro Studies*, http://appli1.oecd.org/olis/2004doc.nsf/linkto/env-jmmono(2004)26.

110. Coecke, S., Balls, M., Bowe, G., Davis, J., Gstraunthaler, G., Hartung, T., Hay, R., Merten, O.-W., Price, A., Schechtman, L., Stacey, G., and Stokes, W. (2005). Guidance on good cell culture practice. *ATLA: Altern. Lab. Anim.*, **33**, pp. 261–287.

111. Maynard, A. D., Aitken, R. J., Butz, T., Colvin, V., Donaldson, K., Oberdörster, G., Philbert, M. A., Ryan, J., Seaton, A., Stone, V., Tinkle, S., Tran, L., Walker, N. J., and Warheit, D. B. (2006). Safe handling of nanotechnology. *Nature*, **444**, pp. 267–269.

112. Balbus, J. M., Maynard, A. D., Colvin, V. L., Castranova, V., Daston, G. P., Denison, R. A., Dreher, K. L., Goering, P. L., Goldberg, A. M., Kulinowski, K. M., Monteiro-Riviere, N. A., Oberdörster, G., Omenn, G. S., Pinkerton, K. E., Ramos, K. S., Rest, K. M., Sass, J. B., Silbergeld, E. K., and Wong, B. A.

(2007). Meeting report: hazard assessment for nanoparticles: report from an interdisciplinary workshop. *Environ. Health Perspect.*, **115**, pp. 1654–1659.

113. Warheit, D. B., Borm, P. J. A., Hennes, C., and Lademann, J. (2007). Testing strategies to establish the safety of nanomaterials: conclusions of an ECETOC workshop. *Inhal. Toxicol.*, **19**(8), 631–643.
114. Hoet, P., and Boczkowski, J. (2008). What's new in nanotoxicology? Brief review of the 2007 literature. *Nanotoxicology*, **2**, pp. 171–182.
115. Hoet, P., and Boczkowski, J. (2008). What's new in Nanotoxicology? Brief review of the 2007 literature. *Nanotoxicology*, **2**, 171–182.
116. Fadeel, B., Kagan, V., Krug, H., Shvedova, A., Svartengren, M., Tran, L., and Wiklund, L. (2007). There's plenty of room at the forum: potential risks and safety assessments of engineered nanomaterials. *Nanotoxicology*, **1**, pp. 73–84.
117. Elliott, K. C. (2007). Varieties of exploratory experimentation in nanotoxicology. *Hist. Phil. Life Sci.*, **29**, pp. 313–336.
118. Pulskamp, K., Diabate, S., and Krug, H. F. (2007). Carbon nanotubes show no sign of acute toxicity but induce intracellular reactive oxygen species in dependence on contaminants. *Toxicol. Lett.*, **168**, pp. 58–74.
119. Ashwood, P., Thompson, R. P., and Powell, J. J. (2007). Fine particles that adsorb lipopolysaccharide via bridging calcium cations may mimic bacterial pathogenicity towards cells. *Exp. Biol. Med.*, **232**, pp. 107–117.
120. Fisher, H. C., and Chan, W. C. W. (2007). Nanotoxicity: the growing need for in vivo study. *Curr. Opin. Biotech.*, **18**, pp. 565–571.
121. Hartung, T. (2010b). Food for thought... on alternative methods for nanoparticle safety testing. *ALTEX*, **27**, pp. 87–95.

Chapter 6

Nanotoxicology: The Case for in vivo Studies

David Y. Lai[a] and David B. Warheit[b]
[a]*US Environmental Protection Agency, Washington DC, USA*
[b]*DuPont Haskell Global Centers for Health and Environmental Sciences, Newark, DE, USA*
lai.david@epamail.epa.gov, david.b.warheit@usa.dupont.com

6.1 Introduction

With the increasing use and development of engineered nanomaterials in electronics, pesticides, consumer products, and chemical and pharmaceutical industries, there is a growing concern about their potential risks to humans and the environment. The toxic effects of most nanomaterials have not been characterized, but it is generally believed that nanoparticles can have toxicological properties that differ from their bulk materials. A challenge facing hazard identification and safety assessment of engineered nanomaterials is the diversity and complexity of the types of materials with varying physicochemical properties, many of which can affect their toxicity by different mechanisms [1, 2]. In 2005, a group of international experts

Handbook of Safety Assessment of Nanomaterials: From Toxicological Testing to Personalized Medicine
Edited by Bengt Fadeel

ISBN 978-981-4463-36-2 (Hardcover), 978-981-4463-37-9 (eBook)
www.panstanford.com

in fiber and particle toxicology convened by the International Life Sciences Institute Research Foundation/Risk Science Institute (ILSI) developed the first toxicity-testing strategy for nanoparticles. It is a broad data-gathering strategy for the initial hazard identification process of nanomaterial risk assessment consisting of a comprehensive array of in vitro and in vivo assays to investigate the toxicity of diverse nanoparticles in routes of entry and target tissues [3].

Short-term in vitro toxicity tests can provide a rapid and relatively inexpensive way to assess the potential toxicity of large numbers of untested nanoparticles, and in recent years, the number of in vitro studies of nanomaterials in the literature has skyrocketed. However, most in vitro test systems appear to have limited usefulness for hazard identification of nanoparticles due to a number of issues resulting in false positives and false negatives. In addition to some generic issues associated with the in vitro approach such as (i) high-dose effects, (ii) time course effects, and (iii) cell line effects, there are a number of issues specifically related to in vitro toxicity testing of nanoparticles [reviewed in 1, 2]. Furthermore, standard protocols for toxicity testing of nanoparticles are usually unavailable. Depending on the type of cells, the duration of exposure, the concentration of nanoparticles, and the composition of the culture media, testing of the same nanomaterial by different laboratories often resulted in different outcomes. Recent studies have shown little correlations between in vitro and in vivo toxicity of some nanomaterials [4]. Nonetheless, in vitro systems may be useful for studying the mechanisms and identifying specific biological and mechanistic pathways of nanotoxicity under controlled conditions.

Due to their small size, aerosolization of respirable nanomaterials is likely, either as singlet or as aggregated particles, and the inhalation route of human exposures is a major concern. The skin of workers is a potential route for exposure to nanoparticles during their manufacture. The general public may also be exposed to nanoparticles through the dermal route, as many textiles and cosmetics such as sunscreens contain nanoparticles [5]. The gastrointestinal (GI) tract is a possible portal of entry for nanoparticles since there are increasing uses of nanotechnology in food products [6]. Furthermore, it is possible that during the life of a nanomaterial (production, application, disposal, etc.) it may appear in the water supply or be inadvertently ingested. Depending on the nanoparticle size and surface chemistry, exposure to nanoparticles via the portal of entry also includes a probability of translocation to other organs and tissues. Their small size helps them to enter the cells and reach the circulatory system, eventually reaching various potential target sites [7, 8]. Therefore, in vivo animal studies are essential for evaluation of the toxicokinetics and potential pulmonary and extrapulmonary toxicity of nanomaterials. For hazard identification and dose–response analysis in quantitative risk assessment, animal studies in accordance with established test guidelines are required by many government agencies such as the US Environmental Protection Agency (EPA) and the Organisation for Economic Co-operation and Development (OECD).

Conducting chronic toxicity studies in rodents for every new nanomaterial introduced is impractical. These studies are very costly, use a large number of animals, take a long time to complete, require sophisticated exposure facilities, and are technically difficult. As a result, very few studies are available for engineered nanomaterials that report chronic or subchronic toxicity data. This chapter will focus on discussions of short-term in vivo study methods and designs, as well as the hazard potential of some commercially important nanomaterials based on available in vivo study data and mechanistic considerations.

6.2 In vivo Study Design and Methods

6.2.1 Inhalation Exposure

As airborne nanoparticles can easily enter the respiratory tract and some respirable, poorly soluble particulates are known to cause lung effects when inhaled, it is necessary to assess their potential toxic effects associated with inhalation exposure. Inhalation is the physiological process during which nanoparticles are deposited in the respiratory tract, including the nose and lungs, allowing for a slow buildup of the dose and for normal clearance processes to occur. So far, only a small number of inhalation studies have been conducted on a few nanomaterials. No standardized toxicity test guidelines have been developed for nanoparticles, although efforts have been initiated in the OECD to update the current inhalation toxicity test guidelines to take into account nanomaterial testing. As methods currently used for testing and assessing chemicals are generally accepted for safety tests of nanomaterials, many in vivo toxicological studies on nanomaterials have used the toxicity test guidelines developed by the US EPA or the OECD for testing chemicals. For acute, subchronic, and chronic inhalation studies, these guidelines include the Office of Chemical Safety and Pollution Prevention (OCSPP) Harmonized Test Guidelines 870.1300, 870.3465, and 870.8355 [9] and OECD test guidelines 403, 412, 413, and 436 and the associated *Guidance Document GD 39* [10].

In inhalation studies, particles should be characterized (primary particle size, particle size distribution, mass concentration, number concentration, surface area, etc.) so that the results can be useful for hazard and risk assessment. For hazard assessment, aerosol generation of nanoparticles should produce particle size distributions that allow the maximum lung burden (mass median aerodynamic diameter [MMAD] <3μm, including agglomerates). Effort should be made to aerosolize a powder and further deagglomeration of the particles in the aerosol. Dry powder generation methods are to be preferred. Using a pristine aerosol without alteration of the particle surface can best mimic the occupational situation. However, impurities and artifacts that may alter particle surfaces, leading to a modified toxic response in the biological test, should be avoided. During inhalation studies, constancy and homogeneity of atmospheric concentrations of the tested particles should be ensured. The flow of filtered air through the exposure chamber/system should be carefully controlled and continuously monitored.

Since bronchoalveolar lavage fluid (BALF) analysis can provide sensitive biochemical markers for toxic responses and information on early signs of cytotoxicity, inflammation, and fibrosis [11], it is recommended to include BALF analysis in the inhalation study of nanomaterials. Specific investigations of certain organs/tissues (e.g., cardiovascular system, brain with the olfactory bulb) for potential translocation should be conducted, in addition to a complete histopathology of the lungs and other organs/tissues specified in the US EPA or OECD test guidelines. It is desirable that inhalation studies on nanomaterials be designed to obtain quantitative data on organ burden and distribution and clearance half-life of the test material, which are useful for interpreting the toxicity data.

A review of the toxicity studies of fibers by an expert panel has concluded that no single short-term in vitro test or a battery of tests can be used to predict the carcinogenicity potential of fibrous particles, and a 90-day subchronic inhalation test was recommended [12]. A two-tier approach was recommended by the ILSI panel experts for in vivo assays [3]. Tier 1 in vivo assays are proposed for pulmonary, oral, and skin exposures. Tier 1 evaluations include markers of inflammation, oxidant stress, and cell proliferation in the portal of entry and selected remote organs and tissues. Tier 2 evaluations are proposed only for pulmonary exposures. A four-week inhalation study by nose-only or whole-body exposure in male or female rats with up to three months' observation is recommended for evaluating the toxicity and carcinogenic potential of nanomaterials. When inhalation exposure is not feasible, the panel concluded that other pulmonary exposure methods may be acceptable. Evaluation includes markers of damage, oxidant stress, cell proliferation, the degree/intensity and duration of pulmonary inflammation, and cytotoxic effects and histopathology of pulmonary and extrapulmonary organs/tissues. The tier 2 studies are aimed at obtaining additional information with respect to the deposition, translocation, biokinetics, and biopersistence of the nanomaterials to aid in interpreting the results for risk assessment. A similar testing strategy to assess the safety of nanomaterials was recommended by an expert panel convened at a European Centre for Ecotoxicology and Toxicology of Chemicals (ECETOC) workshop [13]. Recently, a short-term rat inhalation study (STIS) for nanomaterials has been developed within the European FP6 project *NanoSafe2* and the German BMBF project *nanoCare* [14]. The recommended STIS is a 28-day study with 5-day inhalation exposure (nose only) in male Wistar rats and a 3 weeks' postexposure period. It comprises a comprehensive scheme of biological effects and marker determination in order to generate appropriate information on early key elements of pathogenesis, such as inflammatory reactions in the lung and indications of effects in other organs. BALF analysis, gross examination, and histological examination are conducted either immediately after the last exposure or three days after exposure and at termination of the study. Information on the persistence, progression, and/or regression of effects is obtained, and organ burden in the lung and potential translocation to other tissues are addressed. The STIS uses less animals and resources, and the rankings of different nanomaterials relative to their potential to induce adverse effects by the STIS have been shown to be comparable to those of the corresponding subchronic and chronic studies [14, 15]. While the STIS cannot replace traditional chronic and subchronic inhalation studies, it may provide a tool for comparing toxicity of different nanomaterials and prioritizing them for further toxicity testing. This approach has been proposed to the OECD working party on manufactured nanomaterials.

6.2.2 Other Inhalation Exposure Methods

Inhalation is the "gold standard" in regard to the method of exposure of the respiratory tract for hazard identification of fibers and particles and to obtain dose–response data in quantitative risk assessment. However, inhalation studies are costly, use a large number of animals, take a long time to complete, and require sophisticated exposure facilities. Furthermore, a number of barriers exist for conducting inhalation studies of nanomaterials. For example, preparation of adequate aerosol dispersions suitable for inhalation studies is immensely difficult because of the high propensity of hydrophobic nanomaterials to agglomerate. Conducting chronic inhalation toxicity/carcinogenicity studies in rodents for every nanomaterial with varying physicochemical properties is impractical if not impossible. As a result, alternative exposure methods (e.g., intratracheal instillation, intratracheal inhalation, pharyngeal/laryngeal aspiration) have been used by various investigators for delivery of particles to the lungs.

6.2.2.1 Intratracheal instillation

Intratracheal instillation is a method whereby particles can be administered directly into the lower respiratory tract of rats [16–18]. It is essentially a gavage-like "injection" of the test particles suspended in a vehicle that gently delivers the dose to the bronchi and lung. Methodologies for intratracheal instillation/in vivo pulmonary toxicity measurements have previously been described [19]. In short, groups of rats (five rats/group/dose/time point) are intratracheally instilled with nanoparticle dose of 1 mg/kg or 5 mg/kg in phosphate-buffered saline (PBS). Groups of PBS-instilled rats serve as vehicle controls, and Min-U-Sil quartz is often utilized as a benchmark positive control. Following exposure, BALF analysis and histopathological examination are recommended at 24 hours, 1 week, 1 month, and 3 months postexposure.

Efforts should be made to disaggregate the nanomaterial suspended in vehicle. A number of chemical surfactants have been used to disperse nanoparticles prior to instilling into rat lungs, including Pluronic-F68 [20] and Tween [17], and a nanoparticle dispersion medium (DM) mimics components of the lung alveolar lining fluid, BALF, that is, Ca^{2+}- and Mg^{2+}-free PBS containing serum albumin and 1,2-dipalmitoyl-sn-glycero-3-phosphocholine (DPPC), a lung surfactant [21].

This exposure method is "nonphysiological," which delivers nanoparticles as a bolus to the lung airways, and therefore, the pulmonary response could potentially be caused by the high single dose of nanomaterials—artificial effects of large mass and agglomeration formed during exposure procedures may not be relevant to chronic lower exposure in occupational and environmental settings. Asphyxia has been reported in rats following intratracheal instillation of single-walled carbon nanotubes (SWCNTs) as a result of formation of large agglomerations in suspension [17]. Furthermore, this noninhalation method bypasses the nasal passages and upper respiratory tract, which can be a potentially important target site for inhaled nanoparticles. An advantage of instillation exposure is that it provides a known particulate dose to the relevant target tissue (i.e., lower respiratory tract), which is consistent and useful for hazard ranking and comparing the hazard potential. Moreover, it can be conducted with several particle types in a single experiment—provided that the experimental design is appropriate—with the inclusion of benchmark controls, dose response, and time–course studies.

Studies have been conducted to compare responses of the rat lung to inhaled and intractracheally instilled quartz and titanium dioxide (TiO_2). In an earlier study [22], intratracheal instillation resulted in greater response than inhalation at the same lung burden of quartz or titanium dioxide. The greater responses after instillation were believed to be due to differences in the dose rate and/or distribution of materials in the lungs. A comparison of the inflammation responses of the lungs following instillation and inhalation exposures to quartz also showed different patterns of injury in rats, with interstitial granulomas only observed in animals exposed by inhalation [23]. A study by Warheit et al. [24], however, found that the results generated from the instilled TiO_2 particles were consistent with findings from an inhalation assay of virtually identical particles. When the pulmonary toxicity of multiwalled carbon nanotubes (MWCNTs) following intratracheal instillation and short-term inhalation exposure to rats were compared, less pulmonary responses were observed by inhalation exposure than by intratracheal instillation. The difference was attributed to less amounts of MWCNTs delivered into the lungs by the inhalation exposures [25]. There were also large differences in pulmonary response in mice exposed to purified MWCNTs by instillation and inhalation routes of exposure; the differences were due to differences in aggregation and distribution of the test materials [26].

Pulmonary bridging studies probably can serve to provide relevant and accurate screening hazard data when assessing the safety of compounds in development or when making modifications to existing products, such as particle size alterations on particulates. The materials for which there is inhalation data can be used as reference particle types for comparison to the pulmonary bioassay (i.e., intratracheal instillation route of exposure) results. Thus, in describing the basic bridging concept, the effects of the instilled materials serve as a reference (known) particle type (e.g., pigment grade TiO_2 particle types) and then are "bridged" or benchmarked to the inhalation toxicity data for that particle type, concomitant with the new materials being tested (e.g., ultrafine TiO_2 particle types). The results of bridging studies in rats are useful as preliminary pulmonary toxicity screening (i.e., hazard) data, because consistency in the response of the inhaled and instilled control material serves to validate the responses with the newly tested particle type. The strength of the bridging strategy is dependent upon having good inhalation toxicity data for comparisons to data developed from other pulmonary studies (e.g., instillation).

6.2.2.2 Pharyngeal/laryngeal aspiration

The pharyngeal/laryngeal aspiration procedure represents a noninvasive alternative to the intratracheal instillation method for toxicity testing of respirable nanoparticles in vivo. Unlike intratracheal instillation, this technique does not require insertion of a needle or cannula through the laryngeal opening and thus eliminates the major source of trauma and technical difficulty associated with intratracheal instillation. The technique of pharyngeal aspiration involves anesthetization of animals, after which they are placed on a slant board and their tongues are pulled and maintained in full extension by small forceps. A 50 μL suspension of nanoparticles is pipetted at the base of the tongue, and the tongue restraint is continued until at least two deep breaths are completed. The placing of a suspension of the nanoparticles on the back of the tongue and the pulling of the tongue result in a reflex gasp and aspiration of the droplets. As food particles from the oral cavity may be unintentionally aspirated during the procedure, food should be withheld the night before the procedure. As an alternative to pharyngeal aspiration exposure and to avoid contamination with materials from the oral cavity, laryngeal aspiration may be used, particularly in rats since the pharyngeal aspiration technique has been only confined in mice.

The pharyngeal aspiration technique is also reputed to give better dispersion throughout the lung than intratracheal instillation. Studies in mice using fluorescent beads have demonstrated that following aspiration of pharyngeal deposited particles, exposures to the alveolar region of the lung are repeatable, technically simple, and highly correlated to the administered dose. There were no significant differences in particle distribution between the left side and the right side of the lung, and particles were widely disseminated within the alveolar region [27]. Studies using MWCNTs of similar structure sizes and at nearly equivalent lung burdens, the inflammatory responses of the lung in mice after pharyngeal aspiration do not significantly differ from those after inhalation exposure. Morphometric analysis of MWCNT distribution showed that one day after inhalation, 24% of MWCNTs were in the conducting airways, while 76% were in the alveolar region [28]. Morphometric data from the aspiration study reported that one day after aspiration, 18% of MWCNTs were in the conducting airways, while 84% were in the alveolar region [29]. Furthermore, MWCNTs were detected in the subpleural tissue in both inhalation and aspiration studies.

Comparision of pharyngeal aspiration with results on the inhalation protocols showed that pharyngeal aspiration to SWCNTs resulted in pulmonary responses very similar to those following inhalation exposure. However, inhalation exposure to SWCNTs was found to be more effective than aspiration of a bolus dose of SWCNTs in causing inflammation, oxidative stress, collagen deposition, and fibrosis, as well as mutations of the K-*ras* gene locus in the lungs of C57BL/6 mice [30]. In this inhalation study, the SWCNT aerosol was prepared by a special aerosolization device and was more dispersed than the SWCNT suspension used in the aspiration study. The more severe pulmonary responses induced by inhalation exposure than aspiration of SWCNTs is likely due to the use of stable SWCNT aerosols in this inhalation study. Mercer et al. [31] have demonstrated a four-fold greater fibrotic potency after pharyngeal aspiration of a well-dispersed SWCNT compared to a less dispersed suspension. When efforts were made to improve the dispersion of aspirated SWCNT suspensions, fibrotic potency was similar between aspiration and short-term inhalation exposure. Therefore, it appears that when nanomaterials of similar dispersion are used, aspiration studies can produce pulmonary responses comparable to those of inhalation studies.

6.2.2.3 Intratracheal inhalation

Intratracheal inhalation has the advantages of intratracheal administration but is an inhalation method that can deposit particles in a pattern similar to inhalation exposures. Using this method, rats can be exposed to high aerosol concentrations, resulting in high lung particle burdens in a short time period, with deposition occurring directly in the lower respiratory tract.

During intratracheal inhalation exposure, rats are anesthetized and mechanically ventilated at a constant rate with an aerosol of the test particle. To avoid the large-scale deposition that occurs in the nose, the animals breathe the aerosol but the trachea is cannulated transorally. The end of the endotracheal cannula is connected to the exposure port of the intratracheal inhalation system, which consisted of a manifold tubing system allowing for the simultaneous exposure of up to 16 rats. This method has the advantages of using small amounts of particles and is a closed system [32, 33]. However, like the intratracheal instillation method, it deposits particles in the lower respiratory tract and bypasses the upper respiratory tract, which can be a potentially important target site for inhaled nanoparticles. The response of rats exposed to "fine" (about 250 nm) and "ultrafine" (about 21 nm) TiO_2 particles by intratracheal inhalation and intratracheal instillation has been compared. The results showed a decreased pulmonary response in rats receiving particles through inhalation when compared to those receiving particles through instillation [34]. The difference in responses is likely due to differences in dose rate, particle distribution, and clearance between the two methods.

6.2.3 Dermal Exposure

For skin absorption and toxicity studies of nanomaterials, the rat and the pig have been the animal species of choice. The rats are used because of the established toxicology database by other routes of exposure. Because its skin is anatomically, physiologically, and biochemically similar to that of humans, the pig is also often utilized. In skin absorption or irritation studies, a small area on the back is clipped to remove hairs 24 hours prior to dosing of nanomaterials. Three doses at log intervals (mg/cm^2) plus vehicle and positive control materials are applied to normal skin and abraded skin to mimic human exposure scenarios. An occlusion device to retain the nanomaterial is attached to the surface of the skin by nonirritating tapes. About four to six animals per group/dose are used, and the treatment sites are scored for erythema and edema using the Draize test scores at 0.5, 1, 2, 4, 8, and 24 hours postexposure. Skin biopsies are taken for pathological examination and detection of penetrated particles within the skin by transmission electron microscopy. Nanomaterials can be applied to the skin for 5, 7, or 28 days for repeated-dose studies. At termination, hematology, clinical chemistry, evaluation of local lymph nodes, an immunotoxicology battery of tests, and standard full necropsy examination of all major organs/tissues should be performed [3]. Methods for performing skin absorption studies are given in US EPA test guidelines 870.1200, 2500, 3200, and 3250 [9] and in OECD guidelines 427 (in vivo) and *Guidance Document 28* as well as OECD Guidance Notes on Dermal Absorption 156 [10].

6.2.4 Oral Exposure

In addition to the increasing uses of nanotechnology in food products, release of nanomaterials in drinking water supplies is a health concern. As methods currently used for testing and assessing chemicals are generally accepted for safety tests of nanomaterials, toxicological studies of chemicals in experimental animals by the oral route of exposure in accordance to the EPA health effects test guidelines (e.g., OSCPP 870.1100, 3050, and 3100) [9] or OECD guidelines for testing of chemicals (e.g., no. 423) [10] can be used for evaluating the potential of nanomaterials to cause adverse effects following oral ingestion.

6.2.5 Parenteral Exposure

A number of injection methods have been used in studies of the absorption, distribution, metabolism, and excretion (ADME) of nanoparticles. It is usually done with formulations dispersed in saline or corn oil and administered into body cavities or circulation by intraperitonial (i.p.) injection and intravenous (i.v.) injection. Rats or mice of either sex are the recommended model systems. The distribution to various organs/tissues and elimination in the feces and urine are measured. Histopathological examinations of various organs/tissues at different time periods postexposure are conducted as following pulmonary exposure.

The needle-like fiber shape of some nanoparticles has been compared to asbestos, which caused mesothelioma in exposed workers [35]. Intraperitoneal (i.p.) and intrapleural injection methods have also been applied to evaluate the potential of carbon nanotubes (CNTs) and nanoparticles with a high aspect ratio to induce mesothelioma. In addition, the biocompatibility and potential toxicity of nanomaterials used for drug delivery after injection can be evaluated in animals by parenteral methods.

6.3 In vivo Toxicity Studies of Nanomaterials

There are only a small number of in vivo toxicity studies on nanomaterials, and few toxicological studies have addressed repeated dose exposures, reproductive toxicity, and carcinogenic effects. In most of these studies, nanomaterials were administered via the pulmonary route in rats or mice. Because of expected applications and toxicity at sites of exposure, dermal, oral, and injection routes of administration have also been used in some of the studies. Physicochemical properties, the agglomeration state, test sample preparation methods, impurities, concentration of materials administered, duration of exposure, and observation period are of paramount importance to identify hazards and assess the contribution of the experimental design and physicochemical characteristics to the observed toxicity. However, such information is not always reported by the investigators. In vivo studies of some commercially important nanomaterials are reviewed (for further discussion, see also Chapter 12 on pulmonary and cardiovascular effects, Chapter 14 on dermal effects, and Chapter 15 on effects on reproductive organs).

6.3.1 Nanotubes and Nanofibers

CNTs and carbon nanofibers (CNFs) are considered one of the most important classes of technological materials, with expectations for a wide variety of applications because of their numerous novel properties. SWCNTs consist of a single rolled graphene sheet with diameters from about 0.7 nm to 20 nm. MWCNTs consist of two or more concentric graphitic layers with diameters in the range of 2–100 nm. The length of SWCNTs and MWCNTs can reach up to tens of micrometers [36]. CNFs have diameters of about 40 nm to 200 nm and lengths ranging from tens of micrometers to several centimeters [37]. The long and thin structure of some CNTs and CNFs has features of both pathological fibers and nanoparticles and therefore has led to concern that they may exhibit some of the effects of asbestos fibers. From the mechanistic point of view, CNTs and CNFs that are long, thin, and straight may be expected to behave like pathological fibers. However, most of the long and thin CNTs/CNFs have the propensity to agglomerate/aggregate into intertwined, coiled, and particle-like materials, which may behave like low-density particles and be toxic via the "lung overload" mechanism. Therefore, depending on their agglomeration/aggregation state and physicochemical properties, CNTs/CNFs may elicit different toxic effects by different mechanisms.

During the manufacturing process, all CNTs/CNFs produced typically contain significant amounts of impurities, including residue metal catalysts such as iron and nickel. These metal catalysts may enhance their potential for toxicity [38]. Removing metallic impurities by strong acid treatments—one of the most commonly used methods—on the other hand can lead to structural defects and surface modifications and thus can alter the toxic properties of these nanomaterials. Furthermore, biopersistence is another important factor for the pathologenesis of fibers and particles. Pristine CNTs/CNFs are likely to be biopersistent since graphene is practically insoluble. However, functionalized CNTs/CNFs with various surface chemistries not only will have altered surface reactivity but may also become more soluble and less biopersistent. Differences in results from animal studies have been attributed to differences in physicochemical properties, the degree of agglomeration of the test material, and differences in study designs [39].

6.3.1.1 Pulmonary exposure

- *SWCNTs.* Two 90-day subchronic toxicity studies have been conducted to investigate the pulmonary toxicity of SWCNTs in the lungs of mice and rats by intratracheal instillation. In the mouse study [40], three types of sonicated SWCNT suspensions were administered to groups of male B6C3F1 mice at 0.1 mg or 0.5 mg per mouse. The three test samples (two produced by the high-pressure CO disproportionation [HiPCO] process) contained different types and amounts of residue catalytic metals; one contained primarily iron (27% by weight), another one contained almost 26% (by weight) nickel, and the third one was purified and contained less than 3 % (by weight) iron. Interestingly, regardless of metal impurities, dose-dependent lung lesions characterized by epithelioid granulomas, interstitial inflammation, and perichronchial inflammation were observed. Many of these lesions were first noted 7 days after exposure and persisted in the 90-day treatment groups. On an equal-weight basis, these SWCNTs were more toxic than quartz, the positive control [40]. In the rat study [17], the lungs of male Crl:CD rats were instilled 1 mg/kg or 5 mg/kg of SWCNTs synthesized by the laser ablation method. The nominal size of the SWCNTs was a 1.4 nm diameter and >1 μm length. However, the test material existed primarily as agglomerated "ropes" of about 30 nm in diameter and contained about 30%–40% amorphous carbon and 5% (by weight) each of nickel and cobalt. While exposures to quartz (positive control) produced significant increases in pulmonary inflammation, cytotoxicity, and cell proliferation indices, exposure to SWCNTs produced only transient inflammatory and cell injury responses with a non-dose-dependent series of multifocal granulomas that were not progressive beyond one month postexposure and evidence of a nonspecific foreign tissue body reaction. The differences between the findings of these two studies may be due to differences in test models, test sample characteristics, and experimental designs.

SWCNT-induced pulmonary toxicity in mice by intratracheal instillation was also observed in a pharyngeal aspiration study [41] in which suspensions of SWCNTs synthesized by the HiPCO method and then purified by acid treatment to remove the iron content to 0.23% (by weight) were administered to groups of female C57BL/6 mice at a dose of 0 μg, 10 μg, 20 μg, or 40 μg per mouse. The mean diameter and surface area of SWCNTs were reported to be 1–4 nm and 1,040 m^2/g, respectively. The test materials in the suspension droplets exhibited a continuous mat of intertwined carbon nanoropes with varying diameters. Two distinct particle morphologies were observed under transmission electron microscopy (TEM), compact SWCNT aggregates and disperse SWCNT structures. The results showed two distinct pulmonary responses associated with the two morphologically different SWCNT forms. Foci of granulomatous inflammation and discrete granulomas were associated with dense SWCNT aggregates. Diffuse interstitial fibrosis and alveolar wall thickening were associated with dispersed SWCNTs. The unique findings of this study were that SWCNTs elicited acute inflammation with early onset of granulomas and progressive fibrosis in mice and that the interstitial fibrotic response was not driven by chronic inflammation and chronic activation of alveolar macrophages. Mercer et al. [31] also observed that when a well-dispersed delivery of SWCNTs was given by aspiration to mice, a more rapid alveolar interstitial distribution and an accelerated increase in collagen production in the alveolar interstitium that progressed in the absence of persistent inflammation (with no development of granulomatous lesions) occurred. Similarly, in another study in which rats were exposed to 2 mg/kg SWCNTs via pharyngeal aspiration, focal interstitial fibrotic lesions in the alveolar region of the lungs were induced at 21 days postexposure. Furthermore, SWCNTs caused cell proliferation at sites of fibrotic lesions and increased platelet-derived growth factors (PDGFs) and their mRNA, suggesting that PDGFs, which mediate fibroblast proliferation during fibrogenesis, may play a role in the induction of fibrotic lesions by SWCNTs [20]. Intratracheal instillation of 0.5 mg SWCNTs into male mice not only induced pulmonary inflammatory responses and granuloma formation at 14 days and but also activated various transcription factors and cytokines, which may lead to oxidative stress [42].

Because of the high propensity of hydrophobic CNTs to aggregate, it is difficult to prepare an adequate aerosol for inhalation studies. So far, inhalation studies on SWCNTs are scarce. Using a special device and technique developed by Baron et al. [43] to prepare adequate aerosolization of SWCNTs, Shvedova et al. [30] conducted a four-day inhalation study of the more dispersed, nonpurified SWCNTs (iron content of 17.7% by weight) at 5 mg/kg and compared the responses resulting from exposure of the same SWCNTs via pharyngeal aspiration in an earlier study [41]. The SWCNTs used in both studies had dimensions of 0.8–1.2 nm diameters and 100–1000 nm lengths with a measured surface area (Brunauer–Emmett–Teller [BET] method) of 508 m^2/g. The results showed that all unusual features of pulmonary responses previously reported in the aspiration study were retained or enhanced in the inhalation study. It was concluded that inhalation was more effective than aspiration in causing inflammatory response, oxidative stress, collagen deposition, and fibrosis, as well as oncogene mutations in the lungs of C57BL/6 mice by SWCNTs. However, it is uncertain if the higher iron content of the SWCNTs used in the inhalation study may have contributed to some of the enhanced pulmonary responses.

- *MWCNTs.* A number of in vivo animal studies have shown that MWCNTs caused similar lung lesions observed in studies with SWCNTs. For instance, dose-dependent pulmonary lesions characterized by the formation of collagen-rich granulomas and inflammation were observed when purified (98%) intact MWCNTs (5.2 nm × 5.9 μm, an average of 15 carbon layers) and ground MWCNTs (5.1 nm × 0.7 μm) were administered intratracheally (0.5, 2, or 5 mg) to Sprague–Dawley rats. The more dispersed ground MWCNTs also induced fibrotic responses 60 days after treatment. The levels of tumor necrosis factor alpha (TNF-α) (a key mediator in inflammation and fibrosis) in the BALF were significantly increased three days after instillation of both intact and ground MWCNTs [18]. When groups of male Sprague–Dawley rats were intratracheally instilled with 0.04 mg/kg, 0.2 mg/kg, or 1 mg/kg of highly purified (99.8%) and dispersed MWCNTs (mean diameter 60 nm; mean length 1.5 μm), transient pulmonary inflammatory and granulomatous responses were observed only in the high-dosed rats (1 mg/kg) but not in rats exposed to low doses (0.04 mg/kg or 0.2 mg/kg) [44].

  The pathological lesions induced by MWCNTs in lungs of mice have been compared by intratracheal instillation and inhalation exposures [26]. There were large differences of the pulmonary response to purified (95%) MWCNTs (mean diameter 50 nm; mean length 10 μm) by instillation and inhalation routes of exposure due to differences in aggregation and distribution of the test materials. In the intratracheal instillation groups, the MWCNTs were aggregated into clumps in bronchi and alveoli, which led to inflammation to the lining wall of the bronchi and severe destruction to the alveolar structures. In the inhalation groups, the MWCNTs were more dispersed in the aerosol, which entered the alveolar walls and induced moderate proliferation and thickening of the alveolar walls. These studies exemplify the importance of the size, dose level, and aggregation/agglomerate state of test materials on the toxic responses of CNTs.

The pulmonary toxicity of MWCNTs following intratracheal instillation and short-term inhalation exposure has also been compared in rats [25]. In the intratracheal instillation study, groups of male Wistar rats received a single dose of 0.2 mg/rat (0.66 mg/kg) or 1 mg/rat (3.3 mg/kg) of MWCNTs and were sacrificed from three days to six months. In the inhalation study, rats were exposed (whole body) to 0.37 mg/m^3 aerosol of the same materials for four weeks and were sacrificed at three days, one month, and three months postexposure. The mean diameter of the bulk MWCNTs was 44 nm, and the mean BET surface area was reported to be 69$\pm$37 m^2/g. The mean diameter and length of the test materials in solution were determined to be 48 nm and 0.94 μm (0.2–8.9 μm), and the respective mean diameter and length in the aerosol were 63 nm and 1.1 μm. In both intratracheal instillation and inhalation studies, the test samples were prepared by the same dispersion procedure and were both well dispersed (>70% were individual particles). These well-dispersed MWCNTs induced pulmonary inflammation by intratracheal instillation and inhalation, as evidenced by neutrophil infiltration and cytokine measurement in BALF and lung tissues. As the inhalation exposures delivered less amounts of MWCNTs into the lungs (than by intratracheal instillation), less pulmonary responses were observed.

The dose and time course responses in the lungs induced by MWCNTs have been investigated in mice [29]. Groups of male C57BL/6J mice were exposed to purified and dispersed MWCNTs (median diameter about 49 nm and length about 3.9 μm; 20–50 walls) at 0 μg, 10 μg, 20 μg, 40 μg, or 80 μg per mouse by pharyngeal aspiration and pulmonary toxicity was evaluated at 1, 7, 28, and 56 days postexposure. Dose-dependent pulmonary inflammation and toxicity, as determined by BALF analysis, peaked at seven days postexposure. Like SWCNT exposure reported earlier by the same group [41], MWCNT exposure caused rapid development of pulmonary fibrosis as early as seven days postexposure, and granulomous inflammation persisted throughout the 56-day postexposure period. In addition, pleura penetration of MWCNTs and a dose- and time course–dependent MWCNT-induced inflammation in the pleura occurred [29, 45]. Asbestos-like pathogenicity was reported seven days after injection of certain length-dependent MWCNTs into the peritoneum of mice [46]. The MWCNTs associated with these effects had diameters of about 85 nm or 165 nm and lengths of at least 24% >15 μm and showed a stiff and straight structure. Shorter MWCNTs of a smaller diameter and with more tangled structures were not associated with the effects. Experiments in which nanofibers of different lengths were intrapleurally injected in female C57Bl/6 strain mice have recently led Schinwald et al. [47] to propose that the threshold for respirable fiber length in inducing pleural inflammation is 5 μm in the mouse, and probably in humans as well. It is believed that long fibers can cause pleural effects, including mesothelioma, because they cannot negotiate the stomata in the parietal pleura and are retained in the pleural space. Small particles, short nanotubes, or tangled nanotubes, on the other hand, can rapidly move through stomata in the parietal pleura and drain easily into the lymphatic system [35, 48].

Indeed, although lung inflammation, granulomas, fibrotic lesions, and mutations were induced in rats by ground MWCNTs (11 nm in diameter; 0.7 μm in length) injected intratracheally [18]; the same MWCNTs, with or without defects, did not induce mesotheliomas or other tumors when Wistar rats were injected intraperitoneally with a single dose of the test material at 20 mg/rat in a two-year bioassay [49]. The absence of a carcinogenic response in this study is likely due to insufficient number of MWCNTs longer than 5 μm. On the other hand, a high incidence of mesotheliomas (14/16, 87.5%) was observed at day 84 in p53+/− mice given a single dose (1×10^9 particles, 3 mg/mouse) of MWCNTs by i.p. injection [50]. The test materials had an average diameter of about 100 nm and 27.5% were >5 μm in length, containing aggregates among dispersed rod-shaped or fibrous particles. When the same MWCNTs were later administered to intact, nongenetically modified male Fisher 344 rats at a dose (0.85×10^8 particles, or 0.24 mg/rat) by a single intrascrotal injection, a similar incidence of mesotheliomas (6/7, 86%) was observed in the exposed rats [51]. However, the p53 gene-deficient mice

toxicity studies on MWCNTs (Baytubes®) probably because of the similarity in their morphological structures in the aerosols. Like MWCNTs (Baytubes®), electron microscopic examination of MWCNTs (Nanocyl NC 7000) in the aerosols showed particle clumps or agglomerates of a few hundred nanometers to several micrometers in diameter, suggesting that MWCNTs (Nanocyl NC 7000) may also behave more like poorly soluble particulates than asbestos fibers.

- *Functionalized CNTs.* The effects of the functional groups on the surface of CNTs are not clearly understood. Exposure of carboxylated MWCNTs by i.p. injection to mice has shown that functionalized MWCNTs have the potential to induce hepatotoxicity in mice through induction of reactive oxygen species (ROS). When a carboxylated MWCNTs (diameter 11.5 nm; length up to 12 μm; 2–7 wt.% COOH) suspension was administered intraperitoneally to Swiss–Webster mice at doses of 0.25 mg/kg/day, 0.5 mg/kg/day, and 0.75 mg/kg/day for five days, decreased body weight gain, increased activities of liver enzymes (e.g., ALT/AST, ALP), lipid peroxides, ROS, and histopathological changes in the liver were observed [58]. The carboxylated MWCNTs were also shown to have a higher clastogenic effect (using mitotic index, chromosomal aberrations, micronucleus formation, and DNA damage as the endpoint) than unfunctionalized MWCNTs when studied in bone marrow cells of Swiss–Webster mice [59]. In vitro studies in human T-lymphocytes have shown that oxidized CNTs were more cytotoxic than the nonfunctionalized form [60]. The functionalized CNTs are more toxic in these short-term tests probably because they are more readily dispersed than the nonfunctionalized, pristine form, which tends to agglomerate or aggregate into spheres of coiled, tangled, and intertwined structures. On the other hand, functionalized CNTs with carboxylic groups on the surface may be more susceptible to metabolism and thus less biopersistent and have lower potential for chronic toxicity and carcinogenicity.
- *CNFs.* A recently published study evaluated the subchronic toxicity of inhaled CNFs in rats [61]. Four groups of rats per sex were exposed, nose only, six hours/day, five days/week, to target concentrations of 0 mg/m^3, 0.50 mg/m^3, 2.5 mg/m^3,

or 25 mg/m^3 VGCF™-H nanofibers (surface area 13.8 m^2/g; mean lengths and diameters 5.8 μm and 158 nm, respectively) over a 90-day period (65 exposures). In addition, groups of high-concentration and control rats were also assessed at three months postexposure recovery. Exposure of rats to concentrations of 0.54 mg/m^3 (4.9 f/cc), 2.5 mg/m^3 (56 f/cc), and 25 mg/m^3 (252 f/cc) of VGCF™H resulted in a concentration-related small, detectable accumulation of extrapulmonary fibers, which produced no adverse tissue effects. At the two highest exposure concentrations, cellular inflammation of the terminal bronchiole and lung parenchymal areas of the lungs was noted in areas wherein fiber-laden alveolar macrophages had accumulated. This finding consisted of infiltrates of inflammatory cells with some thickening of interstitial walls and hypertrophy/hyperplasia of type II pneumocytes and was graded as slight and minimal for the 25 mg/m^3 and 2.5 mg/m^3 concentrations, respectively. In addition, a nonspecific reaction to exposure of the nasal passages resulted in minimal to mild inflammatory responses in the submucosa. BALF assessments demonstrated that exposures to 25 mg/m^3 VGCF™-H produced increases in BAL-recovered neutrophils (PMNs), lactate dehydrogenase (LDH), alkaline phosphatase (ALKP), and micrototal protein (MTP). Animals exposed to 0.54 mg/m^3 and 2.5 mg/m^3 demonstrated no differences from air-exposed control animals. Greater than 90% of VGCF™H-exposed, BAL-recovered pulmonary macrophages from the 25 mg/m^3 and 2.5 mg/m^3 exposure group contained nanofibers (>60% for 0.5 mg/m^3). Cell proliferation studies (using BrdU labeling) demonstrated increases versus controls in cell proliferation of the terminal bronchioles, lung parenchyma, and subpleural/mesothelial areas of both male and female rats exposed to 25 mg/m^3 VGCF™-H but no increases in cell proliferation indices in 0.54 mg/m^3 or 2.5 mg/m^3 exposure groups.

6.3.1.2 Effects on the cardiovascular system

Respiratory exposure to SWCNTs may cause not only pulmonary toxicity but also cardiovascular effects. An i.p. injection of purified

SWCNTs (diameter <2 nm; length 1–5 μm) into mice has been reported to exert prothrombotic effects in the microcirculation [62]. Pulmonary instillation of SWCNTs has also been shown to produce vascoconstrictive and prothrombotic effects in rats [63]. Similarly, studies in rats by Radomski et al. [64] showed that purified SWCNTs and MWCNTs, but not C_{60} particles, have the ability to enhance vascular thrombosis in carotid arteries.

Circular mitochondrial DNA (mtDNA) is highly susceptible to oxidative damage, and mtDNA damage has been suggested as an initiating event in atherogenesis [65]. Significant decreases in mtDNA amplification, reflecting an increase in mtDNA damage, was observed in the aortic tissue of C57BL/6 mice exposed to SWCNTs (10 μg or 40 μg per mouse) at 7, 28, and 60 days after a single intrapharyngeal instillation compared to vehicle-treated control mice [66]. Changes in mitochondria glutathione (GSH) levels are also manifestations of oxidative stress [67]. The aortic mitochondrial GSH/oxidized glutathione (GSSG) ratios were significantly reduced seven days after similar pulmonary exposure to a single dose (10 μg or 40 μg per mouse) of SWCNTs in mice [66]. To evaluate the effects of SWCNTs on atherosclerosis progression, ApoE–/– mice were exposed by pharyngeal aspiration to SWCNTs (20 μg/mouse) once every other week for eight weeks. SWCNT exposure was associated with a significantly larger atherosclerosis lesion area examined [66]. One mechanism of respiratory exposures to particles causing cardiovascular effects involved indirect effects mediated by inflammatory mediators, produced during pulmonary inflammation that is induced by particles deposited in the lung. These mediators are proposed to affect the autonomic control of the heart and alter the activity of the coagulator system [68].

In a 90-day inhalation study, exposure of VGCF™-H nanofibers in rats at 0 mg/m^3, 0.50 mg/m^3, 2.5 mg/m^3, or 25 mg/m^3, no significant increases versus sham controls were measured in cardiac muscle cell proliferation indices among any VGCF™H-exposed groups at any time point. Moreover, hematological/coagulation endpoints were not different among the groups, indicating that the effects measured in the respiratory tract did not impact systemic cardiovascular biomarkers [61].

6.3.1.3 Effects on the immune system

The adjuvant capacity of particles on allergic immune responses in mice has been shown to increase with decreasing particle size [69], even within the ultrafine range [70]. The allergic adjuvant effect of CNTs was investigated in two mouse models, first in a footpad injection model [71], followed by studies in an intranasal mouse model. SWCNTs and MWCNTs were given together with the allergen ovalbumin (OVA) during sensitization. After an allergen booster, allergen-specific antibody levels in serum were determined, and in the intranasal model, additional parameters associated with an allergic airway response were examined. Both SWCNTs and MWCNTs together with OVA increased serum levels of allergen-specific IgE, the number of eosinophils in BALF, and the secretion of Th2-associated cytokines in the MLN, all being important mediators in allergic airway responses. CNTs therefore have the capacity to promote allergic immune responses, as demonstrated by the adjuvant effect on allergen-specific antibody responses after subcutaneous injection into the footpad of mice. After intranasal exposure to MWCNTs alone, a pronounced increase in the number of neutrophils, accompanied by increased levels of MCP-1 and the proinflammatory cytokine TNF-α, was also observed [72].

While this study [72] demonstrates that CNTs promote allergic responses in mice, no information on the physcochemical properties and impurities was reported. In addition to size and shape, other properties such as surface area, degree of agglomeration, charge, coating, structural defect sites, metal contaminants, and protein- and thus allergen-binding properties may influence the allergy-promoting potential of CNTs.

In a short-term inhalation study [73], male C57BL/6 mice were exposed by whole-body inhalation to purified MWCNTs at 0 mg/m^3, 0.3 mg/m^3, 1 mg/m^3, or 5 mg/m^3 for 7 or 14 days. The MWCNTs were well characterized. The dimensions of the bulk materials are about 10–20 nm in diameter and 5–15 μm in length. The BET surface area was determined to be 100 m^2/g. Electron microscopy showed that the MWCNT aerosols were a mixture of materials in varying states of agglomeration. Particle median size distribution was 350–400 nm at a 1 mg/m^3 exposure level. Although no significant pulmonary toxicity was noted, systemic immunosuppression, as characterized by reduced T-cell-dependent antibody response, decreased proliferation of T-cells following mitogen stimulation, and altered natural killer (NK) cell killing was observed in all exposed mice [73]. The mechanism for the suppressed systemic immune function was found to involve activation of cyclo-oxygenase enzymes in the spleen in response to a signal from the lungs of the exposed mice [74].

6.3.1.4 Oral exposure

Toxicological studies on oral exposures to CNTs are scarce. Available studies indicate few or no adverse effects in rodents following acute ingestion of CNTs.

Three different types of SWCNTs with different dimensions and iron contents have been tested for acute oral toxicity in mice [75]. A single dose of 1000 mg/kg of body weight (b.w.) of raw SWCNTs (1 nm × 1–2 μm, 25% Fe), purified SWCNTs (1 nm × 1–2 μm, <4% Fe), or ultrashort tubes (1 × 20–80 nm, <1.5% Fe) was administered orally to Swiss mice. No death, body weight changes, or pathological abnormalities were observed 14 days after administration. The NOAEL was >1000 mg/kg.

SWCNTs (0.9–1.7 nm in diameter and <1 μm in length) were reported to generate oxidative DNA damages in the liver and lung of rats by oral administration. About 20% increases of 8-oxodG, a biomarker of oxidative DNA damage, were observed in the liver and lung of rats 24 hours after receiving a single dose (0.064 or 0.64 mg/kg b.w.) of the particle suspensions in saline or corn oil by gavage [76].

Murthy and coworkers [77] conducted an acute oral toxicity study in Wistar rats with two different sizes of MWCNTs, one with a diameter of 10–15 nm and a length of 1–10 μm and the other with a diameter of 140±30 nm and a length of 5–8 μm. A single dose of 2000 mg/kg b.w. was administered to 10 rats. There were no deaths, body weight gains, or adverse effects 10 days after exposure. The NOAEL is >2,000mg/kg.

In a 14-day repeated-oral-dosing study to evaluate the developmental effects of MWCNTs (10–15 nm in diameter and 20 μm in length), pregnant rats on gestation days 6–19 were administered the test material at 0 mg/kg/day, 40 mg/kg/day, 200 mg/kg/day, and 1000 mg/kg/day [78]. No developmental or other adverse effects were observed at gestation day 20, except for a decrease in thymus weight in the dams at the highest dose as compared to controls. The NOAEL was considered to be 1000 mg/kg for developmental toxicity and 200 mg/kg/day for the dams.

6.3.2 Fullerenes

Since their discovery in 1985, fullerenes have attracted considerable interest due to their potential applications in material and biomedical sciences. The C_{60} fullerene is a cage-like structure formed from 60 carbon atoms. The diameter is about 0.7, and it is remarkably stable and practically insoluble in water [79]. To explore biological activities for medical applications, various fullerene derivatives with greater water solubility have been prepared by chemical modification of the "pristine" hydrophobic molecule. In general, available in vivo toxicity data show that pristine C_{60} and most fullerene derivatives have low acute toxicity when tested by the inhalation, dermal, or oral route of exposure. However, some of its derivatives can be quite toxic by parenteral administration. Fullerenes may cause biological effects due exclusively to their adsorption to biomolecules, and the ability of fullerenes to adsorb biomolecules is dependent on the size and surface area of aggregates, water solubility, lipophilicity, and surface chemistry and reactivity. Therefore, it is difficult to generalize the in vivo behavior and toxicity of modified $C_{60.}$ The in vivo biology and toxicology of fullerenes and their derivatives have also been reviewed [80, 81].

6.3.2.1 Pulmonary effects

In contrast to results of in vitro toxicity reported for fullerenes, there was little or no lung toxicity in rats intratracheally instilled up to 3 mg/kg of pristine C_{60} or polyhydroxylated $C_{60}(OH)_{24}$. Groups of rats were exposed to a single intratracheal instillation of C_{60} (mean particle size about 160 nm) and an aqueous solution of $C_{60}(OH)_{24}$ at 0 mg/kg, 0.2 mg/kg, 0.4 mg/kg, 1.5 mg/kg, or 3 mg/kg, and lung toxicity was evaluated by BALF analysis, cell proliferation, and histopathological examination at one day, one week, one month, and three months postexposure. Exposure to these materials produced only transient inflammatory effects at one day postexposure. No adverse lung effects were observed at three months postexposure in the highest-dosed rats [82]. Mild lung inflammation, as indicated by an increased neutrophil influx in the lungs, occurred only in mice exposed to polyhydroxylated $C_{60}(OH)_{20\pm2}$ (fullerenol) by intratracheal instillation at a high dose (200 μg/mouse, or 10 mg/kg b.w.) but not at a low dose (20 μg/mouse, or 1 mg/kg b.w.). In fact, a low dose of fullerenol was shown to have anti-inflammatory effects as it attenuated the quartz-induced neutrophilic lung inflammation in mice [83]. Some fullerenes have been demonstrated to be effective radical scavengers and antioxidants [84].

In a short-term inhalation study in which groups of male Fischer 344 rats were exposed nose-only to C_{60} nanoparticles (55 nm in diameter, 2.22 mg/m^3, or microsized C_{60} particles 0.93 μm in diameter, 2.35 mg/m^3) three hours/day for 10 consecutive days, only minimal changes were noted in blood chemistry and BALF parameters at necropsy. These particles were cleared rapidly as their lung half-lives were 26 and 29 days for the nanosized and microsized C_{60} particles, respectively [85]. Similarly, after whole-body inhalation exposure to C_{60} (96 nm in diameter) at 0.12 mg/m^3 (4.1 × 104 particles/cm^3), 6 hours/day, 5 days/week for 28 days, no severe pulmonary toxicity was observed in Wistar rats up to 1 month postexposure. Only some genes associated with the immune system, and a few genes involved in the inflammatory response, oxidative stress or apoptosis were up-regulated [86].

The data of a 90-day subchronic inhalation study is consistent with the reported lack of toxicity of C_{60} after short-term inhalation exposures or administered by intratracheal instillation in rodents. Groups of Wistar–Han rats and B6C3F1 mice were exposed by inhalation (nose only) to two particle sizes of C_{60}(50 nm or 1000 nm) 3 hours/day, 5 days/week for 90 days. Exposure concentrations for the 50 nm C_{60} were 0 mg/m^3, 0.5 mg/m^3, and 2.5 mg/m^3, and those for the 1000 nm C_{60} were 0 mg/m^3, 5 mg/m^3, and 15 mg/m^3. There were no significant changes in organ weight, clinical chemistry, and hematology or significant adverse effects on the pulmonary tract, reproductive tissues, and immune system [87]. However, exposures of 5 mg or 10 mg per rat (but not 1 mg/rat) $C_{60}(OH)x$, (x = 22, 24; mean diameter = 68 nm) to Sprague–Dawley rats by intratracheal instillation for three days induced oxidative/nitrosative stress, inflammation, and other pathological lesions in the lungs in a dose-dependent manner [88].

6.3.2.2 Dermal and eye effects

Dermal absorption/penetration and toxicity of fullerenes are probably low. Studies of a fullerene-substituted phenylalanine (Baa) derivative of a nuclear localization peptide sequence (Baa-Lys(FITC)-NLS) showed that moderate dermal penetration of the nanoparticles occurred only in porcine skin flexed for 60 and 90 minutes, whereas no dermal penetration was seen in unflexed skin examined 24 hours after exposure [89].

In a skin-painting study in mice, a fullerene mixture (C_{60} and C_{70}) did not show a tumor-promoting effect. No induction of ornithine decarboxylse activity or DNA synthesis occurred after a single application of 200 μg on the clipped skin of mice. In a repeated-dose study, the C_{60} and C_{70} mixture did not promote the tumorigenicity of 7,12-dimethylbenzanthracene (DMBA) on mouse skin [90].

There is one human study of the dermal and eye effects of fullerenes. Thirty volunteers were exposed to a water suspension of fullerene soot by patch testing and were observed for 96 hours. No skin irritation was observed [91]. No eye effect was found in a Draize test in rabbits instilled with 0.2 mL of the water suspension of fullerene soot in one eye, while the other eye served as a control after 24, 48, and 72 hours [91].

6.3.2.3 Systemic effects

The absorption of a C^{14}-labeled carboxylic acid derivative of fullerene has been studied in Fischer rats by oral administration. Virtually all radioactivity (97%) was excreted in the feces within 48 hours after administration, and only trace amounts of radioactivity were detected in the liver and other tissues [92]. Because of their poor absorption from the GI tract, the acute toxicity of fullerenes has been demonstrated to be low in rats if administered orally. Male and female Sprague–Dawley rats were administrated a mixture of C_{60} and C_{70} orally at a single dose of 2000 mg/kg b.w. and observed for 15 days. No deaths and changes in body weight gain or other signs of toxicity were observed at necropsy [93]. All animals survived after a highly water-soluble poly-alkylsufonated fullerene with 4-6 sodium butyl sulfonate moieties, C_{60} $[(CH_2)_4SO_3Na]_{4\text{-}6}$, was administered a single oral dose of 2500 mg/kg b.w. to female Sprague–Dawley rats [94].

No acute toxicity was observed when a single dose of a pristine C_{60} solution (particle size range 60–1650 nm) was administered to rats by i.p. injection at 2 g/kg. Although the liver was colored brown, no enzyme activity increase, inflammation, or fibrosis of the liver was observed in the treated animals [95]. Most of a C^{14}-labeled carboxylic acid derivative of fullerene was found to retain in the liver of Fischer rats for only about a week after i.v. injection. The LD_{50} was estimated to be >500 mg/kg b.w. when the acute toxicity of the carboxylic acid derivative of fullerene was investigated in female ddY mice by a single i.p. injection of 200–500 mg/kg b.w. [92]. The water-soluble MSAD-C_{60} showed no apparent toxic effect after i.p. injection of 50 mg/kg per day for six days in mice [96].

However, some fullerenes appeared to be acutely toxic by parenteral administration. A single i.v. injection of the water-soluble MSAD-C_{60}, a bis(monosuccinimide) derivative of p,p'-bis(2-aminoethyl)-diphenyl-C_{60}, to rats at 25 mg/kg b.w. resulted in death within five minutes of dosing [96]. A single i.p. dose of the water-soluble derivative $C_{60}O_5(OH)_{18}$ increased the relative liver weight at 10 mg/kg. Death occurred at 100 mg/kg (10%) within the first three days after the administration [97].

The acute toxicity of C_{60} $[(CH_2)_4SO_3Na]_{4\text{-}6}$ was evaluated in rats given by a single i.p. injection at dose levels of 0 mg/kg, 500 mg/kg, 750 mg/kg, or 1000 mg/kg b.w. and by i.v. injections at 0 mg/kg, 10 mg/kg, or 100 mg/kg b.w [94]. After i.p. injection, the majority of rats died within 30 hours and the LD_{50} was determined to be about 600 mg/kg b.w. Gross and microscopic examinations revealed that the kidney appeared to be the target organ, although the test materials also distributed to the liver, spleen, and thymus. Phagolysosomal nephropathy and kidney lesions similar to those observed in the acute i.p. study also occurred in rats sacrificed two weeks after i.v. injection of the material at 100 mg/kg b.w.; no adverse effects were noted at 10 mg/kg b.w. In a subacute toxicity study of C_{60} $[(CH_2)_4SO_3Na]_{4\text{-}6,}$ rats were given i.p. injections at dose levels of 0 mg/kg, 0.6 mg/kg, 6 mg/kg, or 60 mg/kg b.w. for 12 consecutive days [94]. As in the acute i.p. or i.v. injection studies, histopathologic lesions of phagolysosomal nephropathy were observed in rats in the mid- and high-dosed (6 mg/kg and 60 mg/kg b.w.) groups. In addition, altered activities of a number of cytochrome P-450 enzymes were found in the liver and kidney.

6.3.2.4 Reproductive and developmental effects

Adequate studies on reproductive and development toxicity of fullerenes are not available. The distribution of C^{14}-labeled C_{60} in the pregnant rat, fetuses, lactating rats, and offspring has been studied [98]. Pregnant rats were dosed on gestation day 15, and lactating rats were dosed on postnatal day 8 via tail vein injection with a suspension of 0.3 mg/kg b.w. $[C^{14}]C_{60}$. At 24 and 48 days after dosing, large portions of the dosed material were found in the liver of the pregnant (43%) and lactating (35%) dams and the lungs of the lactating dams (25%). Low levels of radioactivity (6%) were detected in the reproductive tracts, placentas, and fetuses of the pregnant dams and in the mammary tissue and milk of the lactating rats. Furthermore, radioactivity was found in the GI tract and liver of the offspring. The study therefore showed that C_{60} can cross the placenta and be transmitted to offspring via the dam's milk and subsequently absorbed in the GI tract and distributed to the liver.

To investigate the effects of C_{60} on mouse embryos, pregnant mic were administered C_{60} solubilized in polyvinylpyrrolidone (PVP) at 25–137 mg/kg b.w. by i.p. injection. Embryos (day 11) were examined at 18 hours after administration. At 137 mg/kg b.w., all treated embryos died with severe abnormalities. At 50 mg/kg b.w., C_{60} caused abnormalities around the head and the tail and the yolk sac. The LOAEL was determined to be around 25 mg/kg b.w. based on abnormal enlargement of the head [99].

The developmental toxicity of C_{60} aggregates (about 100 nm in diameter) and a hydroxylated derivative, C_{60} $(OH)_{16\text{-}18}$, have been studied in the zebrafish model [100]. No toxicity was observed in zebrafish embryos exposed to C_{60} $(OH)_{16\text{-}18}$ at 50 mg/L. However, C_{60} aggregates (1.5 mg/L, about 100 nm in diameter) caused delayed embryo and larval development, decreased survival and hatching rates, and pericardial edema. Toxicity was less severe after coexposure to GSH, suggesting that free radical induction may play a role in developmental toxicity of C_{60} in zebrafish.

6.3.2.5 Genotoxic effects

Oral exposure to C_{60} (1.7 nm in diameter) was reported to induce oxidative DNA damages in the liver and lung of rats. Groups of female Fischer 344 rats were receiving a single dose (0.064 mg/kg or 0.64 mg/kg b.w.) of the particles (in saline or corn oil) by gavage. A dose-dependent increase of 8-oxodG, a biomarker of oxidative DNA damage, was observed in the livers of the rats 24 hours after treatment. In the high-dosed (but not the low-dosed) rats, a significantly elevated level (19%) of 8-oxodG was also observed in the lungs of the exposed rats as compared to the controls [76].

In vitro genotoxicity studies showed that fullerenes have no direct DNA-reactive effects. Mutagenicity of fullerite, a mixture of C_{60} and C_{70}, was studied in *Salmonella typhimurium* strains TA100, TA1535, TA98, and TA1537 and *Escherichia coli* WP2*uvr*A/pKM101 strains in the presence and absence of S9 bioactivation. No mutagenic effects were noted in any of the strains of bacteria at concentrations up to 5 mg/plate [93]. The material was also negative for clastogenic effects when tested in a Chinese hamster lung cell line with and without the S9. No structural and numerical chromosomal aberrations were observed in metaphase cells at concentrations up to 5 mg/mL [93].

The mutagenic effect of the C_{60} fullerene was studied in *Salmonella typhimurium* strains TA102, TA104, and YG3003 with and without S9 bioactivation. If the plates were prepared in the dark, no or few revertants were found. However, a dose-dependent mutagenic effect was observed in all strains if irradiated by visible light [101].

6.3.3 Titanium Dioxide

The greater surface area of nano-TiO_2 as compared to conventional TiO_2 affords greater potential for consumer and industrial applications. In recent years, nano-TiO_2 of different crystalline forms, particle sizes, surface areas, surface properties, photocatalytic activity, and chemical reactivity have been developed for a wide variety of applications, including sporting goods, water treatment agents, disinfectant sprays, cosmetics, and sunscreens (see Chapter 3). Most toxicological studies of have been conducted by respiratory exposure. However, dermal and oral routes of human exposures are likely due to their uses in sunscreens and water treatment.

6.3.3.1 Pulmonary effects

Pulmonary toxicology studies in rats have demonstrated that some nanoparticles administered to the lung are more toxic than larger, fine-sized particles of similar chemistry at identical mass concentrations.

Like other nanomaterials, nano-TiO_2 toxicity is strongly influenced by the size, crystal structure, and surface chemistry of the particles. In a 12-week inhalation study in rats, nano-TiO_2 particles (20 nm) have also been shown to elicit a persistently higher inflammatory reaction in the lungs compared to larger sizes (250 nm) of TiO_2 of the same crystalline structure (anatase) [102]. A correlation between particle surface area and effects was observed.

However, some studies of nano-TiO_2 appear to show that particle size may not be the most important physicochemical parameter for their toxicity. Warheit et al. [103] evaluated the acute lung toxicity in rats of intratracheally instilled pigment-grade TiO_2 particles (rutile-type particle size = ~300 nm) versus nanoscale TiO_2 rods (anatase = 200 × 35 nm) or nanoscale TiO_2 dots (anatase = ~10 nm) and compared with a positive control particle, quartz. Groups of rats were instilled with doses of 1 mg/kg or 5 mg/kg of the various particle types in PBS. The lungs of PBS- and particle-exposed rats were assessed using BALF biomarkers,cell proliferation methods and by the histopathological evaluation of lung tissue at 24 hours, 1 week, 1 month, and 3 months postinstillation exposure. The results showed that nanosized TiO_2 particles were not highly toxic and were not more cytotoxic or inflammatory to the lung compared to larger-sized particles of similar composition. Exposures to nanoscale TiO_2 rods or nanoscale TiO_2 dots produced only transient inflammatory and cell injury effects at 24 hours postexposure and were not different from the pulmonary effects of larger-sized TiO_2 particle exposures. In contrast, pulmonary exposures to quartz particles (not nanoscale size) in rats produced a sustained lung inflammatory response characterized by neutrophils and foamy lipid-containing alveolar macrophage accumulation, as well as evidence of early lung tissue thickening consistent with the development of pulmonary fibrosis.

Li et al. [104] compared the acute pulmonary toxicity induced by 3 nm and 20 nm TiO_2 primary particles after they were intratracheally instilled at doses of 0.4 mg/kg, 4 mg/kg, and 40 mg/kg into the lungs of mice. The biochemical parameters in BALF and pathological examination were used as endpoints to assess their pulmonary toxicity. At three-day postexposure, BALF analysis and pathological examination showed that the 3 nm TiO_2 primary particles induced no pulmonary toxicity at a dose of 0.4 mg/kg, moderate toxicity at

4 mg/kg, and lung overload at 40 mg/kg; the 3 nm TiO_2 particles did not produce higher pulmonary toxicity than the 20 nm ones at any instilled doses.

Similar results were obtained when the size effects of inflammatory response in instillation and acute inhalation exposures of TiO_2 nanoparticles with manufacturers' average particles sizes of 5 nm and 21 nm were investigated in mice [105]. C57Bl/6 mice were acutely exposed by inhalation (whole-body exposure) for four hours of 0.6–0.8 mg/m^3 or 7.20 mg/m^3 (about 1 μg or 12 μg per mouse) TiO_2 nanoparticles (5 nm or 21 nm) or by nasal instillation of 0.1 mg/mL, 0.4 mg/mL, and 0.6 mg/mL (about 5 μg, 20 μg, and 30 μg per mouse) TiO_2 nanoparticles (5 nm) or 0.5 mg/mL, 2.0 mg/mL, or 3.0 mg/mL (25 μg, 100 μg, and 150 μg per mouse) TiO_2 nanoparticles (21 nm). It was found that the smaller 5 nm particles did not create a larger inflammatory response than the 21 nm particles with either inhalation or instillation exposure methods based on BALF analysis and histopathological evaluation of the lungs 24 hours postexposure.

The pulmonary toxicities of two newly developed, nanoscale TiO_2 particle types in rats were compared to other TiO_2 samples in two different size ranges or crystal structures [19]. Groups of rats were exposed to particles via intratracheal instillation with doses of 1 mg/kg or 5 mg/kg b.w. of either two nano rutile TiO_2 (nano-1 or nano-2), fine-sized rutile TiO_2, or nano-TiO_2 (80%:20% anatase/rutile, nano-3). Min-U-Sil quartz particles were included in the experimental design as positive particulate control samples. The physicochemical characteristics of the nano and fine-sized test materials were rigorously characterized (see Fig. 6.1). Following exposures to the lungs, PBS- (vehicle controls) and particle-exposed rats were sampled for lung inflammatory and cytotoxicity endpoints and cell proliferation activityand assessed for lung tissue effects at postinstillation exposure time points of 24 hours, 1 week, 1 month, and 3 months. Evaluation of the BALF, cell proliferation data concomitant with histopathology observations demonstrated that the toxicity range (in descending order of potency) of lung inflammation/cytotoxicity/cell proliferation and histopathological responses was quartz > nano-3 > fine TiO_2 particles = nano-1 = nano-2. The findings demonstrated that lung exposures to the quartz particles and to a lesser degree nano-3 anatase/rutile TiO_2 particles produced sustained lung inflammation, cytotoxicity, and corresponding adverse lung tissue effects. In contrast, exposures to fine TiO_2 particles or to nano-1/nano-2 rutile nano-TiO_2 particles produced only transient inflammatory responses and no corresponding adverse lung tissue effects. The findings indicate that factors such as crystal structure, surface properties, or surface chemical reactivity but not necessarily particle size could be responsible for the differential effects measured in rats exposed to TiO_2 particles. Thus, the lung toxicity of anatase/rutile nano-3 should not be viewed as representative for all nano-TiO_2 particle types.

Characterization of fine and ultrafine titanium dioxide (TiO_2) particles used in the pulmonary bioassay studies

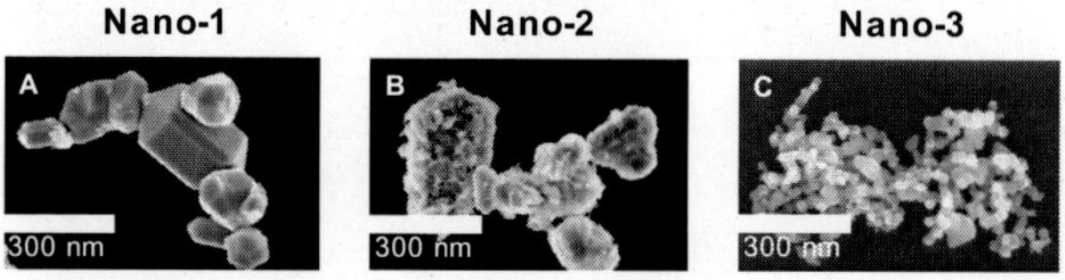

D

Sample	Crystalline phase	Median size and width distribution (nm)		Surface area (m^2/g)	pH		Chemical reactivity
		In water*	In PBS		Deionized water	In PBS	Δb^*
Fine TiO_2	Rutile	382.0 ± 36%	2667.2 ± 35%	5.8	7.49	6.75	0.4
Nano-1	Rutile	136.0 ± 35%	2144.3 ± 45%	18.2	5.64	6.78	10.1
Nano-2	Rutile	149.4 ± 50%	2890.7 ± 31%	35.7	7.14	6.78	1.2
Nano-3	20/80 rutile/anatase	129.4 ± 44%	2691.7 ± 31%	53.0	3.28	6.70	23.8

Figure 6.1 The characterization of the fine (fine rutile-type TiO_2 particles), nano-1, nano-2, and nano-3 samples used in pulmonary bioassay studies [13]. (A–C) HRSEMs of nano-1, nano-2, and nano-3 samples. Micrographs depict variations in surface treatment and particle size within each ultrafine TiO_2 particle type. (D) Table of physical properties of the samples used in the study, including size and size distribution, surface area, pH, and chemical reactivity. Particle size and pH measurements were taken in both water (*aqueous solution of 0.1% tetrasodium pyrophosphate) and PBS solution. *Abbreviation*: HRSEM, high-resolution scanning electron micrograph.

Coating and surface treatment affect surface properties and toxicity of nano-TiO_2. Rossi et al. [106] showed that the pulmonary effects induced in mice after acute and subacute inhalation exposures to several different types of coated or uncoated nano-TiO_2 particles could not be explained by the particle size or surface area but is rather explained by the surface coating of the particles. Female BALB/c/Sca mice were exposed to (i) nanosized (diameter 25 nm; surface area 222 m^2/g) anatase TiO_2, (ii) nanosized (diameter 30–40 nm; surface area 23 m^2/g) rutile/anatase TiO_2, (iii) nanosized (diameter 21 nm; surface area 61 m^2/g) anatase/brookite TiO_2, or (iv) silica-coated nanosized (diameter 10–40 nm; surface area 132 m^2/g) rutile TiO_2. Groups of eight mice were exposed by whole-body inhalation of the aerosols containing different types of nano-TiO_2 particles, either once for two hours on four consecutive days or for two hours on four days/week for four weeks at 10 mg/m^3. Pulmonary inflammation based on neutrophil increases was observed in mice exposed to nanosized rutile TiO_2 coated with silica for two hours for four consecutive days and after repeated exposure for four weeks. No signs of pulmonary inflammation were seen after exposure to any of the other three uncoated TiO_2 nanoparticles. Apparently, surface modifications can alter not only photoreactivity but also toxicity of nano-TiO_2. In rats intratracheally instilled with two types of nano-TiO_2 having the same core material, the nano-TiO_2 with a hydrophilic surface (Aeroxide® P25/Degussa P25) but not the hydrophobic (Aeroxide® P805/Degussa P805, coated with octylsilane) was carcinogenic [107].

A number of studies indicate that the short-term pulmonary effects of TiO_2 nanoparticles are only transient responses and are reversible. When C57Bl/6 mice were exposed by acute inhalation (whole-body exposure) for four hours or subacutely four hours/day for 10 days to 0.77 mg/m^3 or 7.22 mg/m^3 TiO_2 nanoparticles (2–5 nm), a moderate inflammatory response was observed in the high-dosed animals at one or two weeks after exposure. However, mice exposed subacutely recovered at week 3 postexposure [108]. Two intratracheal instillation experiments with TiO_2 nanoparticles of different sizes and agglomerations were conducted in rats to compare the biological responses induced by the different particles. In one experiment, 5 mg/kg of nano-TiO_2 particles of different primary sizes was intratracheally instilled in rats. In another experiment, a similar procedure was followed with 5 mg/kg of nano-TiO_2 particles of the same primary sizes but different agglomerations in liquid. At one week postinstillation, smaller particles induced greater pulmonary inflammation than larger particles. However, pulmonary inflammation remarkably recovered in all the TiO_2 particle-exposed group after one week postinstillation, with no differences between the groups regardless of particle size. The findings indicate that the short-term pulmonary effects observed induced by nano-TiO_2 are reversible. No clear relationship was observed between the TiO_2 particle-exposed groups with different agglomerations but with the same primary size [109].

The fate of nano-TiO_2 particles in the body after inhalation has been investigated [110]. Male Wistar rats were exposed to nano-TiO_2 (100 mg/m^3) by inhalation (head/nose only) six hours/day for five days. Deposition in the lung and tissue distribution was evaluated, and histological examination of the respiratory tract was performed upon termination of exposure and two weeks after the last exposure. BAL was carried out 3 and 14 days after the last exposure. The majority of the inhaled nano-TiO_2 was deposited in the lung, and translocation to the mediastinal lymph nodes, liver, and spleen was also noted. Nano-TiO_2 caused mild neutrophilic inflammation and activation of macrophages in the lung on the basis of changes in BALF composition and histological examination. However, the effects were reversible.

A multispecies, subchronic inhalation study comparing pulmonary responses to nano-TiO_2 was performed. Female rats, mice, and hamsters were exposed to aerosol concentrations of 0.5 mg/m^3, 2.0 mg/m^3, or 10 mg/m^3 nano-TiO_2 particles (primary size about 21 nm) for 6 hours/day, 5 days/week, for 13 weeks. Following the exposure period, animals were held for recovery periods of 4, 13, 26, or 52 weeks (49 weeks for the nano-TiO_2-exposed hamsters). Dose-dependent increases in lung burdens occurred in all three species. Lung burdens in all three species decreased with time after exposure, and at the end of the recovery period, the percentage of the lung particle burden remaining in the 10 mg/m^3 group was 57%, 45%, and 3% for the rat, the mouse, and the hamster, respectively. Pulmonary inflammation in rats and mice exposed to 10 mg/m^3 was evidenced by increased numbers of macrophages and neutrophils and increased concentrations of soluble protein markers in BALF, but the effects diminished with time postexposure. Increased cell proliferation was also induced in rats and mice exposed to 2 mg/m^3 or 10 mg/m^3 nano-TiO_2. There were no significant changes in cellular response or with BALF markers indicating toxicity in hamsters, reflecting the low lung burden of nano-TiO_2. The retardation of particle clearance from the lungs in rats and mice in the high-dosed (10 mg/m^3) group suggested that pulmonary particle overload may have been achieved in these animals. The pulmonary responses of rats and mice induced by nano-TiO_2 are likely due to lung overload of nano-TiO_2 [111].

To examine to what extent a short-term exposure study would be predictive for toxicity outcomes of long-term exposures, Wistar rats were exposed to aerosols of 0 mg/m^3 (control), 2 mg/m^3, 10 mg/m^3, and 50 mg/m^3 nano-TiO_2 particles (primary size about 25 nm) six hours/day for five days. BALF analysis and gross and histological examination of the lung were conducted either immediately after the last exposure or 3 days and 16 days after recovery. In agreement with a 90-day study [111] of a similar nano-TiO_2 of about the same size, the exposure of rats to 10 mg/m^3 nano-TiO_2 for five days also resulted in increases in BALF neutrophiles and total protein content after three days' recovery. As in the 90-day study, 2 mg/m^3 or 10 mg/m^3 nano-TiO_2 also induced significant increases in cell replication in the lungs. The results suggested that this short-term test may provide comparable prediction of respiratory tract toxicity to 90-day studies on nanomaterials [112].

6.3.3.2 Carcinogenic effects

Carcinogenicity of nano-TiO_2 was observed in three animal studies using photocatalytic nano-TiO_2 in rodents [107, 113, 114]. Increased lung tumor incidences were observed in rats [107, 113, 114], but not in mice [114], exposed to P25 through inhalation or intratracheal instillation.

In the first study, female Wister CRP/WU rats were administered photocatalytic nano-TiO_2 via intratracheal instillations (21 nm, 80% anatase, 20% rutile, uncoated, P25) five times at 6 mg each [113]. At these total doses (30 mg), the lung clearances appeared to be severely compromised. There was evidence of alveolar and interstitial inflammation and tumor induction 2.5 years after instillation. The histologically confirmed lung tumor incidence was 66%, while the macroscopic tumor incidence was only 27%.

In another intratracheal instillation study, female Wistar (HsdCpb:WU) rats received weekly doses of two types of nano-TiO_2, hydrophilic nano-TiO_2 (Aeroxide® P25/Degussa P25, average diameter about 25 nm; surface area 52 m^2/g) five times at 3 mg or 6 mg each and hydrophobic nano-TiO_2 (Aeroxide® P805/Degussa P805, coated with octylsilane; average diameter about 21 nm; surface area 32.5 m^2/g) 15 or 30 times at 0.5 mg each. Statistically significant increases in benign or malignant lung tumor incidences (52% and 70% at the low- and high-dose groups, respectively) were observed in rats about two years after instillation of hydrophobic nano-TiO_2. No increased tumor incidences were found in rats treated with hydrophilic nano-TiO_2, which is believed to be photostable [107].

Heinrich et al. [114] reported increased lung cancer in rats (but not in mice) that inhaled photocatalytic nano-TiO_2. Groups of female Wistar rats or female NMRI mice were exposed to P25 TiO_2 (80% anatase, 20% rutile) aerosols (18 hours/day, 5 days/week) in whole-body exposure chambers. Generated by a dry dispersion technique, the nano-TiO_2 aerosol had a mass median aerodynamic diameter of 0.80 μm. In the rat study, the nano-TiO_2 exposure was stopped after 24 months and followed by clean air exposure for 6 months. The 24-month nano-TiO_2 aerosol exposure consisted of 4 months at 7.2 mg/m^3, 4 months at 14.8 mg/m^3, and 16 months at 9.4 mg/m^3. At the end of the 30-month study, 32 of 100 nano-TiO_2-exposed rats (32%) had benign or malignant lung tumors (20 benign squamous cell tumors, 13 adenocarcinomas, 4 adenomas, and 2 squamous cell carcinomas), while only 1 of 217 control rats had lung adenocarcinoma. This study showed that inhaled photocatalytic nano-TiO_2 is a lung carcinogen in rats, but no dose–response relationship can be calculated due to the dosing design. In the mouse study, the nano-TiO_2 exposure was stopped after 13.5 months and followed by clean air exposure for 9.5 months. The 13.5-month nano-TiO_2 aerosol exposure was 4 months at 7.2 mg/ m^3, 4 months at 14.8 mg/ m^3, and 5.5 months at 9.4 mg/ m^3. Lung tumor incidence was not increased in exposed mice at the end of the study as compared to that of controls.

6.3.3.3 Dermal exposure

Because TiO_2 nanoparticles are colorless and reflect ultraviolet (UV) light more efficiently than larger TiO_2 particles, they are components of many cosmetic or sunscreen products. For healthy unflexed skin, adverse health effects are not expected from dermal exposure to photostable nano-TiO_2 in sunscreens. Photocatalytic nano-TiO_2, however, can generate ROS when exposed to UV light and has been reported to cause oxidative stress and cytotoxicity in cultured human fibroblasts [115].

There are relatively few in vivo dermal studies on the absorption/penetration and toxicity of TiO_2 nanoparticles. No skin irritation or sensitization was noted three days after application of a single topical application of 0.5 g photocatalytic nano-TiO_2 (P25, average particle size 129 nm in water) for four hours to the skin of rabbits [116]. Most available experimental evidence suggests that insoluble nanoparticles such as nano-TiO_2 do not penetrate into or through healthy human skin. Electron microscopy examinations of three different application forms of TiO_2 showed that no surface characteristics, particle size, or shape of the micronized TiO_2 particles results in any dermal absorption. Micronized TiO_2 was solely deposited on the outermost surface of the stratum corneum and was not detected in the deeper stratum corneum layers, epidermis, and dermis of human skin [117].

The penetration of nano-TiO_2 (rutile coated with Al_2O_3 steric acid) in a sunscreen formulation was investigated by repeated application to the human forearm for four days. The primary particle size was 17 nm, and the size of the aggregates in an oil-in-water emulsion was 150–170 nm. Most particles were found on and in the upper layers of the stratum corneum. A small amount (less than 1%) of TiO_2 nanoparticles were noted in the openings or channels of the hair follicles and sebaceous glands. No penetration into the skin was detected [118].

The penetration of sunscreen formulations containing TiO_2 nanoparticles (about 20 nm primary particle size) into hair follicles of both human and porcine skin was also reported. Particles were found as deep as approx. 400 μm in the follicles, obviously introduced mechanically rather than by a diffusive process. No particles were observed in vital tissue nor in sebaceous glands [119].

In a study exposing healthy and psoriatic human skin to a sunscreen formulation containing TiO_2 nanoparticles (particle size and type not specified) for two hours, TiO_2 nanoparticles were detected in deeper layers of the psoriatic skin than the healthy skin, but no penetration beyond the stratum corneum was found in either psoriatic or healthy skin [120]. The absorption/penertration studies of chemicals in sunscreens have been reviewed [121].

6.3.3.4 Oral exposure

Toxicological studies on TiO_2 nanoparticles by the oral route of exposure are scanty. Available oral toxicity data suggests that TiO_2 has low systemic toxicity. Due to the low toxicity, a fixed large dose of 5 g/kg b.w. of photocatalytic nano-TiO_2 (P25) suspensions was administrated by a single oral gavage to female rats. No body weight loss and no gross lesions at necropsy 14 days after the gavage were noted [116]. In another acute toxicity study, groups of 10 male and female mice received a single oral gavage of two types of TiO_2 at 5 g/kg b.w.: 25 nm rutile spindles and 80 nm rutile spindles. In two weeks, TiO_2 nanoparticles showed no obvious acute toxicity in male mice. However, elevated concentrations of nanosized (25 nm and 80 nm) TiO_2 were found in the liver, spleen, kidney, lung, and brain of female mice. Liver and kidney toxicity were indicated by biochemical parameters in the serum. Myocardial damage was also implied by increases in serum LDH and alpha-hydroxybutyrate dehydrogenase (α-HBDH). No major pathological changes were observed [122].

6.3.3.5 Systemic effects

The effects of respiratory exposure to nano-TiO_2 are not limited to the respiratory system. In rats exposed to 5 mg nano-TiO_2/kg b.w. of rutile nano-TiO_2 rods through a single intratracheal instillation, observed effects included increases in the numbers of monocytes and granulocytes in the blood (signs of systemic inflammation), decreases in the number of platelets in the blood (platelet aggregation), and cardiac edema [123]. Various cardiovascular dysfunctions have been reported in rats exposed to nano-TiO_2 aerosols by inhalation [124, 125]. Groups of rats were exposed by inhalation to fine (1000 nm) and nano-TiO_2 (21 nm P25 particles, at concentrations ranging from 0.5 $mg/m^3$to 20 mg/m^3) or filtered air for up to 12 hours. Both fine or nano-TiO_2 particles impaired arteriolar dilation 24 hours after the end of the exposure, and the effect was dose dependent. The study demonstrated that the degree of systemic microvascular dysfunction is related to the size and lung burden of the particles and that nano-TiO_2 is more potent than fine TiO_2 at an equivalent lung burden. However, evidence of lung inflammation was weak and a role of systemic inflammatory mediators was not clearly established in this study. A health review committee commented that the use of general anesthesia and surgical manipulations used to obtain the measurements in this study may affect microvascular function [125].

In mice exposed to rutile and anatase nano-TiO_2 through intranasal instillation, pathological changes were observed in the kidney and temporary liver injury was suggested by changes in serum biomarkers [126]. A few recent studies showed that anatase and rutile nano-TiO_2 translocate into the brain following intranasal instillations [122, 126, 127]. When mice were given 25 nm rutile, 80 nm rutile, or 155 nm anatase nano-TiO_2 (50 mg/kg b.w. every 2 days for 2, 10, 20, or 30 days) though intranasal instillation, inflammatory responses with changes in neurotransmitter levels and activity of cholinesterase in the brain were observed in mice exposed to 80 nm and 155 nm nano-TiO_2 [122]. The highest titanium concentrations in the brain were in the hippocampus and olfactory bulb, the two regions where most pathological changes were seen [126, 127].

The distribution and toxicity of nano-TiO_2 were investigated in mice by i.p. injection [128]. Groups of ICR mice were injected nano-TiO_2 (mean primary size 3.6 nm; size in suspension 80–110 nm) at dose levels of 0 mg/kg, 324 mg/kg, 648 mg/kg, 972 mg/kg, 1296 mg/kg, 1944 mg/kg, or 2592 mg/kg b.w. and the toxic effects of particles were evaluated 1, 2, 7, and 14 days postexposure. The highest concentration of TiO_2 nanoparticles was detected in the spleen, although they were also found in the liver, kidney, and lung. Pathological lesions in the spleen, thrombosis in the pulmonary vascular system, heptocellular necrosis and fibrosis, renal glomerulus swelling, and interstitial pneumonia were observed in the high-dosed mice.

The tissue distribution and toxicity of intravenously administered nanoparticles of TiO_2 (<100 nm size) were investigated in rats. Male Wistar rats were treated with single i.v. injections of a suspension of nano-TiO_2 in serum (5 mg/kg b.w.), and the tissue content of TiO_2

was determined 1, 14, and 28 days after exposure. There were no detectable levels of TiO_2 in blood cells, plasma, brain, or lymph nodes. The TiO_2 levels were highest in the liver, followed in decreasing order by the levels in the spleen, lung, and kidney, and highest on day 1 in all organs. TiO_2 levels were retained in the liver for 28 days, there was a slight decrease in TiO_2 levels from day 1 to days 14 and 28 in the spleen, and there was a return to control levels by day 14 in the lung and kidney. There was no detectable inflammatory response or organ toxicity [129].

6.3.4 Nanosilver

Due to their inherent antimicrobial activity, silver nanoparticles have become the most common nanomaterial used in manufactured consumer products such as disinfectant sprays and powders, deodorants, laundry detergents, shampoos, fabrics, filters, and washers. Various methods can synthesize nanosilver of different size and shape, including spheres, rods, cubes, wires, and films. Some of them have surface coatings to prevent agglomeration or to add functional characteristics for intended use. In addition, some are hybridized with other materials to form nanocomposites or colloids. Therefore, different silver nanoparticles can have different physicochemical properties, which can influence their surface reactivity, solubility, absorption, deposition, cellular uptake, interaction with biomolecules, translocation and clearance in the body, and, subsequently, their biological and toxicological activities.

Silver is known for its high affinity for the thiol group, so nanosilver particles may interact with proteins and enzymes with thiol groups. Silver nanoparticles, therefore, may deplete the antioxidant defense mechanism, leading to ROS accumulation by binding to proteins and enzymes like GSH and thioredoxin. Silver nanoparticles may also damage cell membrane by interaction with thiol-group-containing proteins in the cell membrane. Like other nanomaterials, the toxicity of nanosilver has largely been investigated by in vitro systems. Cytotoxicity and DNA damage through oxygen radical generation in mammalian cells have been reported when exposed to high levels of silver nanoparticles. In vivo toxicity data of nanosilver is limited. In view of nanosilver's diverse forms and applications, inhalation, oral, and dermal routes of human exposures are possible.

6.3.4.1 Pulmonary exposure

The pulmonary and systemic distribution of inhaled ultrafine silver nanoparticles was investigated by inhalation and intratracheal inistillation in rats [130, 131]. In the inhalation study, Fischer 344 rats were exposed by inhalation (whole body) for six hours at a concentration of 133 μg nano-Ag/m^3 (15 nm modal diameter) and were sacrificed on days 0, 1, 4, and 7. The amounts of nano-Ag in the lungs decreased rapidly with time, and by day 7 only 4% of the initial burden remained. In the blood, significant amounts of nano-Ag were detected on day 0 and thereafter decreased rapidly. In the liver, kidney, spleen, brain, and heart, low concentrations of Ag were observed. For comparison, rats were given a 150 μL aqueous suspension of 50 μg agglomerated nano-Ag particles (>100 nm) by intratracheal instillation and were sacrificed on days 1, 4, and 7 after instillation. Nano-Ag contents in the lung were constant between day 1 and day 7, indicating that no rapid particle translocation from the lung to other organs had taken place. Focal accumulation of nano-Ag-particle-laden alveolar macrophages was found. The studies suggested that particle size and the tendency of particles to form agglomerates affect the distribution pathway in the lungs. After inhalation, rapid clearance of nano-Ag from the lungs occurred and significant amounts were detected in the liver, kidney, spleen, brain, and heart, indicating a systemic distribution of nano-Ag by circulating blood. It is likely that because of their nanoscale size and higher rate of dissolution, the silver nanoparticles were not efficiently phagocytized by macrophages and instead were cleared rapidly through the circulatory system. It has been suggested a substantial portion of nano-Ag toxicity is caused by the dissolved ions released from the particles [132].

To investigate the inflammatory responses and pulmonary function changes in rats, groups of Sprague–Dawley rats were exposed to silver nanoparticles by inhalation (whole body) at concentrations of 0.6×10^6 particles/cm^3 (49 μg/m^3), 1.4×10^6 particles/cm^3 (113 μg/m^3), and 3.0×10^6 particles /cm^3 (515 μg/m^3) 6 hours/day, 5 days/week, for 90 days [133, 134]. The test materials in the aerosols were spherical in shape with both agglomerated and nonaggregated forms and diameters between 6 nm and 55 nm (median diameter 18–19 nm). As was observed in the acute inhalation and intratracheal inistillation studies in rats [130, 131], silver nanoparticles distributed in all tissues examined, including the lung, blood, liver, kidney and brain. The results indicated that lungs and the liver were the major target tissues for prolonged silver nanoparticle exposure by inhalation. Effects on the lungs were a decrease in the tidal volume and minute volume and other inflammatory responses, such as a mixed inflammatory cell infiltrate and chronic alveolar inflammation and small granulomatous lesions. Furthermore, exposure-related bile duct hyperplasia was noted in both male and female rats. On the basis of the effects in lungs and the liver, the NOAEL was estimated to be 100 μg/m^3 in rats. The authors [133, 134] noted that doubling the dosage increased the surface area fivefold, and no effects were seen from the change in mass, suggesting that the toxicity was related to the surface area of nano-Ag.

6.3.4.2 Oral exposure

The oral toxicity of silver nanoparticles is of concern as they are present in many consumer products such as toothpastes, nursing bottles, water disinfectants, and cooking utensils. To investigate the oral toxicity of silver nanoparticles, groups of male and female Sprague–Dawley rats were given silver nanoparticles (60 nm), oral doses of 30 mg/kg, 300 mg/kg, and 1000 mg/kg over a period of 28 days. After 28 days of exposure, the rats did not show any significant changes in body weight. However, the tissue distribution showed a dose-dependent accumulation of silver content in all the tissues (e.g., liver, kidney, spleen, brain) examined. In particular, a gender-related difference in the accumulation of silver was noted in the kidneys, with a twofold increase in the female kidneys when compared with the male kidneys. Significant dose-dependent changes were found in the activity of serum alkaline phsophatase and cholesterol values in the treated animals. As was observed in the 90-day inhalation study [134], histopathological examinations showed a similar dose-dependent increased incidence of bile duct hyperplasia [135].

6.3.4.3 Dermal exposure

Many health and consumer products containing silver nanoparticles, such as acne creams, disinfectants for wound care, and textiles for manufacturing odor-reducing or bacteria-inhibiting clothing, come into direct contact with skin. Therefore, there are concerns for their potential of skin irritation, toxicity, and penetration. Eight different forms of silver nanoparticles with different sizes (20 nm and 50 nm in diameter) and surface conditions (washed and unwashed, carbon coated and uncoated) have been studied for their dermal toxicity and penetrating potential in porcine skin in vivo [136]. Pigs were dosed on the back skin with nano-Ag solutions ranging from 0.34 μg/mL to 34.0 μg/mL for 14 days. None of the nano-Ag caused erythema or edema in any of the treated sites during the 14-day study period based on the Draize system. Microscopic examination showed a concentration-dependent increase of focal inflammation and edema in the skin treated with all nano-Ag, regardless of particle size or if washed/unwashed or coated/uncoated. Epidermal hyperplasia with lesions typical of a chronic irritation reaction was also found in the high-dosed skin. However, no penetration of any nano-Ag into the epidermis was observed.

6.3.4.4 Genotoxicity

In a subchronic toxicity study, groups of male and female Sprague–Dawley rats were given silver nanoparticles (60 nm), oral doses of 30 mg/kg, 300 mg/kg, and 1000 mg/kg over a period of 28 days. There were no statistically significant differences in the micronucleated polychromatic erythrocytes (MN PCEs) or the ratio of PCEs among the total erythrocytes after silver nanoparticle exposure when compared with the controls. Therefore, the results suggested that silver nanoparticles do not induce genetic toxicity in male and female rat bone marrow in vivo [135].

6.4 Conclusions

The increased production and uses of engineered nanomaterials in consumer products and chemical and pharmaceutical industries in recent years have compelled regulatory agencies around the world to evaluate the potential hazards/risks of these materials to humans and the environment. Although some nanomaterials have bulk-sized counterparts, experts are in agreement that the adverse effects of nanoparticles cannot be predicted from the known toxicity of their bulk materials. Currently, in vivo toxicity databases on nanomaterials are far more limited than in vitro test data, and the toxic effects of most nanoparticles have not been characterized.

It is well known that respirable, poorly soluble particulates may cause lung effects when inhaled, and by analogy to these particulates, there is concern for the pulmonary effects of nanomaterials. Hence, most in vivo toxicological studies of nanomaterials have been conducted by respiratory exposure. Since dermal and oral routes of human exposures of nanoparticles are also likely due to their applications in consumer and cosmetic products, there are a few dermal and oral studies in rodents. Methods currently used for testing and assessing chemicals are generally used for safety tests of nanomaterials. Because of the high cost and technical difficulties, conducting chronic inhalation toxicity/carcinogenicity studies in rodents for every nanomaterial with varying physicochemical properties is impractical. With appropriate dosing schemes and dispersion methods, it appears that the instillation and aspiration methods for pulmonary exposure, although not considered physiological, are reasonable alternatives to inhalation studies for hazard identification of nanomaterials. As methods currently used for testing and assessing chemicals are generally accepted for safety tests of nanomaterials, many in vivo toxicological studies on nanomaterials have used the toxicity test guidelines developed by the US EPA or the OECD for testing chemicals. Some of the main types of nanomaterials and their in vivo toxicological effects in experimental animals are summarized below.

- *CNTs/CNFs.* CNTs probably are the most extensively studied nanomaterials in recent years. Significant positive toxicity results from in vivo studies have been obtained for a number of different CNTs. These studies have shown a consistent toxicological response (e.g., pulmonary inflammation, fibrosis) independent of the study design (intratracheal, aspiration, and inhalation). Exposure to SWCNTs and MWCNTs is of special concern because of their fiber-like shape and dimensions. Their nanometer diameters and micrometer lengths closely resemble the dimensions of some pathogenic mineral fibers such as asbestos fibers. Results from animal studies with SWCNTs, MWCNTs, and CNFs show similar pulmonary responses as those reported for some respirable particles and durable fibers. In some studies, CNT-induced lung fibrosis developed more rapidly and at a lower mass burden than crystalline silica. Exposing the mesothelial lining of the body cavity of mice to long (>5 μm) MWCNTs by injection has resulted in asbestos-like, length-dependent, pathogenic behavior [46]. Evidence that MWCNTs can cause persistent pulmonary inflammation, can be translocated from the lung to the regional lymph node, and can penetrate the pleura to cause pleural inflammation has indeed been shown in mice exposed by pharyngeal aspiration [45] or by inhalation [53]. Respiratory exposure to some CNTs may cause not only pulmonary toxicity but also cardiovascular effects and immunotoxicity. Available studies indicate few or no adverse effects in rodents following acute ingestion of CNTs.

- *Fullerenes.* Fullerenes may cause biological effects due exclusively to their adsorption to biomolecules, and the ability of fullerenes to adsorb biomolecules is dependent on the size and surface area of aggregates, water solubility, lipophilicity, and surface chemistry and reactivity. The available studies demonstrated a pronounced discrepancy between in vitro and in vivo *in* the toxicity of the fullerenes [4]. This may partly be ascribed to influence of bioavailability, toxicokinetics, and the respiratory antioxidant properties of fullerenes. In general, available in vivo toxicity data shows that pristine C_{60} and most fullerene derivatives have low toxicity when tested by the inhalation, dermal, or oral route of exposure. However, some of its derivatives can be quite toxic by parenteral administration, which is highly dependent on how they are derivatized.
- *Titanium dioxide.* Pulmonary toxicology studies in rats have demonstrated that some nanoparticles administered to the lung are more toxic than larger, fine-sized particles of similar chemistry at identical mass concentrations. However, several studies of nano-TiO_2 appear to show that particle size may not be the most important physicochemical parameter for their toxicity. Rather, nano-TiO_2 toxicity is strongly influenced by crystalline forms, surface properties and reactivity, and photocatalytic activity of the particles. After inhalation exposure, some of the effects induced by nano-TiO_2 were

similar to those observed after exposure to conventional TiO_2. A number of short-term in vivo studies indicate that the pulmonary effects of TiO_2 nanoparticles in rats and mice are only transient responses and are reversible. The lung burden (in terms of surface area but not mass) and clearance rate of nano-TiO_2 appear to be important factors for their chronic toxic and carcinogenic effects on the pulmonary tract. In a four-week inhalation study in rats, Warheit et al. [137] studied the time course of pulmonary clearance impairment and persistence of inflammation following particle overload exposures to titanium dioxide or carbonyl iron particles at concentrations of 5 mg/m^3, 50 mg/m^3, and 250 mg/m^3. The findings of the study demonstrated that exposure to high, overload concentrations of innocuous dusts, such as titanium dioxide or carbonyl iron particulates, can produce sustained lung inflammation, increased pulmonary cellular proliferation, impairment of particle clearance, and reductions in macrophage function. Thus, high-dose inhalation of low-toxicity particles in rats can produce particle overload and sustained pulmonary effects, which eventually lead to lung tumors.

Oral and topical toxicity data suggests that nano-TiO_2 has low systemic toxicity and is well tolerated on the skin. Most available experimental evidence suggests that nano-TiO_2 or sunscreen formulations containing TiO_2 nanoparticles do not penetrate beyond the stratum corneum of healthy human skin. There is a discontinuity between in vitro and in vivo testing results on the toxicity of many nanomaterials, especially for the skin of nano-TiO_2. Some in vitro cultures or preparations (other than those using intact skin samples) lack the stratum corneum layer, which according to currently available data can block penetration such that in vitro tests might overstate toxicity of chemicals like TiO_2. The weight of evidence suggests that TiO_2 nanoparticles currently used in cosmetic preparations or sunscreens pose no risk to human skin or human health. However, depending on the physical/chemical properties of the ingredient and the formulation, nanosized formulations may enhance or reduce skin absorption or penetration, especially on compromised skin.

- *Nano-Ag.* Little data exists on the health effects of nanosilver particles. The small amount of in vivo toxicity data suggests the effects are similar to those seen with silver ions, which are highly toxic to bacteria, and therefore they are widely used as antimicrobial agents. After inhalation, silver nanoparticles were not efficiently phagocytized by alveolar macrophages and instead were cleared rapidly through the circulatory system because of their nanoscale size and higher rate of dissolution. It has been suggested that a substantial portion of nano-Ag toxicity in various organs/tissues is caused by the dissolved ions released from the particles. Silver is known for its high affinity for the thiol group, so nanosilver particles may interact with proteins and enzymes with thiol groups. Silver nanoparticles, therefore, may deplete the antioxidant defense mechanism, leading to ROS accumulation by binding to proteins and enzymes like GSH and thioredoxin. Silver nanoparticles may also damage cell membrane by interaction with thiol-group-containing proteins in the cell memebrane. Different silver nanoparticles can have different physicochemical properties, which can influence their surface reactivity, solubility, absorption, deposition, cellular uptake, interaction with biomolecules, translocation and clearance in the body, and, subsequently, their biological and toxicological activities.
- *Safety assessment issues.* To summarize the discussion of in vivo toxicity testing of nanoparticles, some important issues have been identified that impact the accurate evaluation of nanoparticulate hazards following exposures. There are four fundamental components of in vivo toxicity testing of nanomaterials: (1) dose–response evaluations, (2) time course experimental design, (3) appropriate physicochemical characterization of the particle-types being tested, and (4) utilization of appropriate benchmark control particulate samples. Prudent implementation and thoughtful design of the bioassays are important for obtaining hazard information on the particulate materials of interest. However, while there has been some evidence of toxicity from available in vivo studies, many of them are nonguideline research, and the materials studied have not been well characterized or reported

in regard to the four fundamental components of in vivo toxicity testing of nanomaterials identified above. Therefore, it is exceedingly difficult to identify, with certainty, the physicochemical parameter(s) attributing to the toxic effects of nanomaterials, to compare the relative hazards of different nanomaterials, and to generalize the hazard potential of all nanomaterials or even nanomaterials of the same type on the basis of the available in vivo database. For instance, a number of studies have demonstrated that nanoparticle toxicity is extremely complex and potentially being regulated by the size and shape, as well as surface charge and reactivity and other physicochemical properties, of the particles [e.g., 138, 139, 103, 19]. Not all nanomaterials under the same name/class are identical. Commercially available brands of nano-TiO_2, for instance, can vary in purity, particle size, crystalline form, photocatalytic activity, surface chemistry, and reactivity. Due to various degrees of porosity, nanoparticles with the same diameter can differ in surface area. Therefore, accurate characterization of the physicochemical properties of nanomaterials being tested is of utmost importance because they can affect whether and how particles are deposited, absorbed, transported, and cleared and, ultimately, their toxic potentials. However, attempts to characterize nanoscale particle sizes and size distributions in relation to toxicity and exposure evaluation have been prone to errors involving agglomeration/deagglomeration during sample preparation, contamination and degradation during product storage, and measurement methods. As a result, accurate data of the physicochemical properties of the nanomaterials are difficult to obtain and is often not reported correctly in toxicological studies.

Many data requirements for human risk assessment such as data on workplace air concentrations and concentrations of nanoparticles available from various consumer products and their release mechanisms are lacking. It remains uncertain whether humans are likely to be exposed to many of the nanoscale particles currently in commerce under realistic exposure scenarios. This is because of the tendency of nanoparticles to agglomerate or form large aggregates when introduced into environmental media, as well as their propensity to transform or recrystallize as they move through the environment. Some of the test results were obtained on nanomaterials that were suspended in different solvent media, sonicated or ground to smaller agglomerate sizes, primary particles, or shorter lengths with surface defects before being aerosolized for inhalation. Therefore, the hazard information generated they may bear little resemblance to the materials as human exposed. Other challenges in traditional risk assessment of nanomaterials have been discussed [1]. An innovative paradigm setting the stage for "toxicity testing and risk assessment of nanomaterials in the 21st century" has recently been proposed (2; see also Chapter 4). In this paradigm, the health hazards and molecular mechanisms of various classes/subclasses of nanomaterials are evaluated using reference nanomaterials, short-term in vivo animal studies in conjunction with high-throughput screenings, and mechanism-based short-term in vitro assays. Quantitative risk assessment will be conducted when dose–response modeling of perturbation of pathway function, toxicokinetics data, and PBPK modeling for in vitro and in vivo extrapolation are available [2].

Disclaimer

This manuscript has been reviewed by the US EPA and approved for publication. Approval does not signify that the contents necessarily reflect the views and policies of the agency, nor does the mention of trade names or commercial products constitute endorsement or recommendation for use.

References

1. Lai, D. Y., and Sayre, P. G. (2009). Toxicity testing and evaluation of nanoparticles: challenges in risk assessment, in *Nanotoxicity: From* in vivo *to* in vitro *Modelsto Human Risks*, eds. Sahu, S. C., and Casiano, D. A. (Wiley & Sons, UK), pp. 427–457.
2. Lai, D. Y. (2012). Toward toxicity testing of nanomaterials in the 21st century: a paradigm for moving forward. *WIREs Nanomed. Nanobiotechnol.*, **4**, pp. 1–15.

3. Oberdo¨rster, G, G., Maynard, A., Donaldson, K., Castranova, V., Fitzpatrick, J. W., Ausman, K., Carter, J., Karn, B., Kreyling, W., Lai, D., Olin, S. S., Monteiro-Riviere, N. A., Warheit, D. B., and Yang, H. (2005a). Principles for characterizing the potential human health effects from exposure to nanomaterials: elements of a screening strategy. A report from the ILSI Research Foundation/Risk Science Institute, Nanomaterial Toxicity Screening Working Group, Part. *Fibre Toxicol.*, **2**, p. 8.
4. Sayes, C. M., Reed, K. L., and Warheit, D. B. (2007a). Assessing toxicity of fine and nanoparticles: comparing in vitro measurements to in vivo pulmonary toxicity profiles. *Toxicol Sci.*, **97**, pp. 163–180.
5. Davis, D. A. (1994). Sunscreen oddities. *Drug Cosmet. Ind.*, **155**, p. 20.
6. FSAI. (2008). The relevance for food safety of applications of nanotechnology in the food and feed industries. Food Safety Authority of Ireland, Dublin, http://www.fsai.ie/resources and publications/scientific reports.html.
7. Limbach, L. K., Li, Y., Grass, R. N., Brunner, T. I., Hintermann, M. A., Muller, M., Gunter, D., and Starr, W. I. (2004). Oxide nanoparticle uptake 9n human lung fibroblasts: effects of particle size, agglomeration, and diffusion at low concentration. *Environ. Sci. Technol.*, **38**, pp. 5208–5216.
8. Oberdo¨rster, G., Oberdo¨rster, E., and Oberdo¨rster, J. (2005b). Nanotoxicology: an emerging discipline evolving from studies of ultrafine particle. *Environ. Health Persp.*, **113**, pp. 823–839.
9. Environmental Protection Agency (2012). *OCSPP Harmonized Test Guidelines Series 870—Health Effects Test Guidelines*, http://www.epa.gov/ocspp/pubs/frs/publications/Test_Guidelines/series870.htm.
10. Organisation for Economic Co-operation and Development (2012). Guidelines for the Testing of Chemicals, Section 4 Health Effects. http://www.oecd-ilibrary.org/environment/oecd-guidelines-for-the-testing-of-chemicals-section-4-health-effects_20745788.
11. Henderson, R. F., Driscoll, K. E., Harkema, J. R., Lindenschmidt, R. C., Chang, I. Y., Maples, K. R., and Barr, E.B. (1995). A comparison of the inflammatory response of the lung to inhaled versus instilled particles in F344 rats. *Fund. Appl. Toxicol.*, **24**, pp. 183–197.
12. Bernstein, D., Castranova, V., Donaldson, K., Fubini, B., Hesterberg, T., Kane, A., Lai, D., McConnell, E. E., Muhle, H., Oberdo¨rster, G., Olin, S., and Warheit, D. (2005). Testing of fibrous particles: short-term assays and strategies. Report of an ILSI Risk Science Institute Working Group. *Inhal. Toxicol.*, **17**, pp. 1–41.

13. Warheit, D. B., Borm, P. J., Hennes, C., and Lademann, J. (2007a). Testing strategies to establish the safety of nanomaterials: conclusions of an ECETOC workshop. *Inhal. Toxicol.*, **19**, pp. 631–643.
14. Klein, C. L. Wiench, K., Wiemann, M., Ma-Hock, L., Ravenzwaay, B., and Landsiedel, R. (2012). Hazard identification of inhaled nanomaterials: making use of short-term inhalation studies. *Arch. Toxicol.*, **86**, pp. 1137–1151.
15. Ma-Hock, L., Burkhardt, S., Strauss, V., Gamer, A. O., Wiench, K., van Ravenzwaay, B., and Landsiedel, R. (2009). Development of a short-term inhalation test in the rat using nano-titanium dioxide as a model substance. *Inhal. Toxicol.*, **21**, pp. 102–118.
16. Driscoll, K. E., Costa, D. L., Hatch, G., Henderson, R., Oberdorster, G., Salem, H., and Schesinger, R. B. (2000). Intratracheal instillation as an exposure technique for the evaluation of respiratory tract toxicity: uses and limitations. *Toxicol. Sci.*, **55**, pp. 24–35.
17. Warheit, D. B., Laurence, B. R., Reed, K. L., Roach, D. H., Reynolds, G. A., and Webb, T. R. (2004) Comparative pulmonary toxicity assessment of single-wall carbon nanotubes in rats. *Toxico. Sci.*, **77**, pp. 117–125.
18. Muller, J., Huaux, F., Moreau, N., Misson, P., Heilier, J.-F., Delos, M., Arras, M., Fonseca, A., Nagy, J. B., and Lison D. (2005). Respiratory toxicity of multi-wall carbon nanotubes. *Toxicol. Appl. Pharmacol.*, **207**, pp. 221–231.
19. Warheit, D. B., Webb, T. R., Reed, K. L., Frerichs, S., and Sayes, C. M. (2007b). Pulmonary toxicity study in rats with three forms of ultrafine-TiO_2 particles: Differential responses related to surface properties. *Toxicology*, **230**, pp. 90–104.
20. Mangum, J. B., Turpin, E. A., Antao-Menezes, A., Cesta, M. F., Bermudez, E., and Bonner, B. C. (2006). Single-Walled Carbon Nanotube (SWCNT)-induced interstitial fibrosis in the lungs of rats is associated with increased levels of PDGF mRNA and the formation of unique intercellular carbon structures that bridge alveolar macrophages *in situ*. Part. Fibre Toxicol., **3**, p. 15.
21. Porter, D., Sriram, K., Wolfarth, M., Jefferson, A., Schwegler-Berry, D., Andrew, M. E., and Castranova, V. (2008). A biocompatible medium for nanoparticle dispersion. *Nanotoxicology*, **2**, pp. 144–154.
22. Driscoll, K. E., Lindenschmidt, R. C., Murer, J. K., Perkins, L., Perkins, M., and Higgins, I. (1991). Pulmonary response to inhaled silica or titanium dioxide. *Toxicol. Appl. Pharmacol.*, **111**, pp. 201–210.
23. Henderson, R., Driscoll, K., Harkema, J., Lindenschmidt, R. C., Chang, I., Maples, K., and Barr, E. (1995). A comparison of the inflammation

response of the lungs to inhaled versus instilled particles in F344 rats. *Fund. Appl. Toxicol.*, **24**, pp. 183–197.

24. Warheit, D. B., Brook, W. J., Lee, K. P., Webb, T. R., and Reed, K. L. (2005). Comparative pulmonary toxicity inhalation and instillation studies with different TiO_2 particle formulations: impact of surface treatments on particle toxicity. *Toxicol. Sci.*, **88**, pp. 514–524.
25. Morimoto, Y., Hirohashi, M., Ogami, A., Oyabu, T., Myojo, T., Todoroki, M., Yamamoto, M., Hashiba, M., Mizuguchi, Y., Lee, B. W., Kuroda, E., Shimada, M., Wang, W. N., Yamamoto, K., Fujita, K., Endoh, S., Uchida, K., Kobayashi, N., Mizuno, K., Inada, M., Tao, H., Nakazato, T., Nakanishi, J., and Tanaka, I. (2012). Pulmonary toxicity of well-dispersed multi-wall carbon nanotubes following inhalation and intratracheal instillation. *Nanotoxicology*, **6**, pp. 587–599.
26. Li, J. G., Li, W. X., Xu, J. Y., Cai, X. Q., Liu, R. L., Li, Y. J., Zhao, Q. F., and Li, Q. N. (2007a). Comparative study of pathological lesions induced by multiwalled carbon nanotubes in lungs of mice by intratracheal instillation and inhalation. *Environ. Toxicol.*, **22**, pp. 415–421.
27. Rao, G. V. S., Tinkle, S., Weissman, D. N., Antonini, J. M., Kashon, M. L., Salmen, R., Battelli, L. A., Willard, P. A., Hubbs, A. F., and Hoover, M. D. (2003) Efficacy of a technique for exposing the mouse lung to particles aspirated from the pharynx. *J. Toxicol. Environ. Health*, **A66**, pp. 1441–1452.
28. Porter, D. W., Hubbs, A. F., Chen, B. T., McKinney, W., Mercer, R. R., Wolfarth, M. G., Battelli, L., Wu, N., Sriram, K., Leonard, S., Andrew, M., Willard, P., Tsuruoka, S., Endo, M., Tsukada, T., Munekane, F., Frazer, D. G., and Castranova, V. (2013). Acute pulmonary dose-responses to inhaled multi-walled carbon nanotubes. *Nanotoxicology*, **7**, pp. 1179–1194.
29. Porter, D. W., Hubbs, A. F., Mercer, R. R., Wu, N., Wolfarth, M. G., Sriram, K., Leonard, S. S., Batelli, L., Schwegler-Berry, D., Friend, S., Andrew, M., Chen, B. T., Tsuruoka, S., Endo, M., and Castranova, V. (2010). Mouse pulmonary dose-and time course-responses induced by exposure to multi-walled carbon nanotubes. *Toxicology*, **269**, pp. 136–147.
30. Shvedova, A. A., Kisin, E., Murray, A. R., Johnson, V. J., Gorelik, O., Arepalli, S., Hubbs, A. F., Mercer, R. R., Keohavong, P., Sussman, N., Jin, J., Stone, S., Chen, B., Deye, G., Maynard, A., Castranova, V., Baron, P. A., and Kagan, V. (2008). Inhalation versus aspiration of single walled carbon nanotubes in C57BL/6 mice: inflammation, fibrosis, oxidative stress and mutagenesis. *Am. J. Physiol. Lung Cell Mol. Physiol.*, **295**, pp. L552–L565.

31. Mercer, R. R., Scabilloni, J., Wang, L., Kisin, E., Murray, A. R., Schwegler-Berry, D., Shvedova, A. A., and Castranova, V. (2008). Alteration of deposition pattern and pulmonary response as a result of improved dispersion of aspirated single-walled carbon nanotubes in a mouse model. *Am. J. Physiol. Lung Cell Mol. Physiol.*, **294**, pp. L87–L97.

32. Oberdo¨rster, G., Cox, C., and Gelein, R., (1997). Intratracheal instillation versus intratracheal inhalation of tracer particles for measuring lung clearance function. *Exper. Lung Res.*, **23**, pp. 17–34.

33. Elder, A. C., Gelein, R., Oberdorster, G., Finketstein, J., Motter, R., and Wang, Z. (2004). Efficient depletion of alveolar macrophages using intratracheally inhaled aerosols of liposome-encapsulated clodronate. *Exp. Lung Res.*, **30**, pp. 105–112.

34. Osier, M., Baggs, R. B., and Oberdörster, G. (1997). Intratracheal instillation versus intratracheal inhalation: influence of cytokines on inflammatory response. *Environ. Health Perspect.*, **105**(Suppl 5), pp. 1265–1271.

35. Donaldson, K., Murphy, F. A., Dufin, R., and Poland, C. A. (2010). Asbestos, carbon nanotubes and the pleural mesothelium: a review of the hypothesis regarding the role of long fibre retention in the parietal pleura, inflammation and mesothelioma. Part. *Fibre Toxicol.*, **7**, p. 5.

36. Thostenson, E. T., Ren, Z. F., and Chou, T. W. (2001). Advances in the science and technolohy of carbon nanotubes and their composites: a review. *Compos. Sci. Technol.*, **61**, pp. 1899–1912.

37. Ku, B. K., Emery, M. S., Maynard, A. D., Stolzenburg, M. R., and McMurry, P. H. (2006). In situ structure characterization of airborne carbon nanofibres by a tandem mobility-mass analysis. *Nanotechnology*, **17**, pp. 3613–3621.

38. Pulskamp, K., Diabate, S., and Krug, H. F. (2007). Carbon nanotubes show no sign of acute toxicity but induce intracellular reactive oxygen species in dependence on contaminants. *Toxicol. Lett.*, **168**, pp. 58–74.

39. Lison, D., and Muller, J. (2008). Lung and systemic responses to carbon nanotubes (CNT) in mice. *Toxicol. Sci.*, **101**, pp. 179–180.

40. Lam, C. W., James, J. T., McCluskey, R., and Hunter, R. L. (2004). Pulmonary toxicity of single-walled carbon nanotubes in mice 7 and 90 days after intratracheal instillation. *Toxicol. Sci.*, **77**, pp. 126–134.

41. Shvedova, A. A., Kisin, E. R., Mercer, R., Murray, A. R., Johnson, V. J., Potapovich, A. I., Tyurina, Y. Y., Gorelik, O., Arepalli, S., Schwegler-Berry, D., Hubbs, A. F., Antonini, J., Evans, D. E., Ku, B. K., Ramsey, D.,

Maynard, A., Kagan, V. E., Castranova, V., and Baron, P. (2005). Unusual inflammatory and fibrogenic pulmonary responses to single-walled carbon nanotubes in mice. *Am. J. Physiol. Lung Cell Mol. Physiol.*, **289**, pp. L698–L708.

42. Chou, C. C., Hsiao, H. Y., Hong, Q. S., Chen, C. H., Peng, Y. W., Chen, H. W., and Yang, P. C. (2008). Single-wall carbon nanotubes can induce pulmonary injury in mouse model. *Nano. Lett.*, **8**, pp. 437–445.

43. Baron, P. A., Deye, G. J., Chen, B. T., Schwegler-Berry, D. E., Shvedova, A. A., and Castranova, V. (2008). Aerosolization of single-walled carbon nanotubes for an inhalation study. *Inhal. Toxicol.*, **20**, pp. 751–760.

44. Kobayashi, N., Naya, M., Ema, M., Endoh, S., Maru, J., Mizuno, K., and Nakanishi, J. (2010). Biological response and morphological assessment of individually dispersed multi-wall carbon nanotubes in the lung after intratracheal instillation in rats. *Toxicology*, **276**, pp. 143–153.

45. Mercer, R. R., Hubbs, A. F., Scabilloni, J. F., Wang, L., Battelli, L. A., Schwegler-Berry, D., Castranova, V., and Porter, D. W. (2010). Distribution and persistence of pleural penetrations by multi-walled carbon nanotubes. *Part. Fibre Toxicol.*, **7**, pp. 28.

46. Poland, C. A., Duffin, R., Kinloch, I., Maynard, A., Wallace, W. A. H., Seaton, A., Stone, V., Brown, S., MacNee, W., and Donaldson, K. (2008). Carbon nanotubes introduced into the abdominal cavity of mice show asbestos-like pathogenicity in a pilot study. *Nat. Nanotechnol.*, **3**, pp. 423–428.

47. Schinwald, A., Murphy, F. A., Prina-Mello, A., Poland, C. A., Byrne, F., Movia, D., Glass, J. R., Dickerson, J. C., Schultz, D. A., Jeffree, C. E., MacNee, W., and Donaldson, K. (2012). The threshold length for fiber-induced acute pleural inflammation: shedding light on the early events in asbestos-induced mesothelioma. *Toxicol. Sci.*, **128**, pp. 461–470.

48. Donaldson, K., and Poland, C. A. (2009). Nanotoxicology: new insights into nanotubes. *Nat. Nanotechnol.*, **4**, pp. 708–710, www.nature.com/naturenanotechnology.

49. Muller, J., Delos, M., Panin, N., Robolli, V., Huaux, F., and Lison, D. (2009). Absence of carcinogenici response to multiwall carbon nanotubes in a 2-year bioassay in the peritoneal cavity of the rat. *Toxicol. Sci.*, **207**, pp. 221–231.

50. Takagi, A., Hirose, A., Nishimura, T., Fukumori, N., Ogata, A., and Ohashi, N. (2008). Induction of mesothelioma in p53+/- mouse by intraperitoneal application of multi-wall carbon nanotube. *J. Toxicol. Sci.*, **33**, pp. 105–116.

51. Sakamoto, Y., Nakae, D., Fukumori, N., Tayama, K., Maekawa, A., Imai, K., Hirose, A., Nishimura, T., Ohashi, N., and Ogata, A. (2009). Induction of mesothelioma by a single intrascrotal administration of multi-wall carbon nanotube in intact male Fischer 344 rats. *J. Toxicol. Sci.*, **34**, pp. 65–76.

52. Porter, D. W., Hubbs, A. F., Chen, B. T., McKinney, W., Mercer, R. R., Wolfarth, M. G., Battelli, L., Wu, N., Sriram, K., Leonard, S., Andrew, M., Willard, P., Tsuruoka, S., Endo, M., Tsukada, T., Munekane, F., Frazer, D. G., and Castranova, V. (2013). Acute pulmonary dose-responses to inhaled multi-walled carbon nanotubes. *Nanotoxicology*, **7**, pp. 1179–1194.

53. Ryman-Rasmussen, J. P., Cesta, M. F., Brody, A. R., Shipley-Phillips, J. K., Jeffrey I., Everitt, J. I., Tewksbury, E. W., Moss, O. R., Wong, B. A., Darol, E., Dodd, D. E., Andersen, M. E., and Bonner, J. C. (2009). Inhaled carbon nanotubes reach the subpleural tissue in mice. *Nat. Nanotechnol.*, **4**, pp. 747–751. *FEBS Lett.*, **393**(1996), pp. 139–145.

54. Ellinger-Ziegelbauer, H., and Pauluhn, J. (2009). Pulmonary toxicity of multi-walled carbon nanotubes (Baytubes®) relative to α-quartz following a single 6 h inhalation exposure of rats and a 3 months post-exposure period. *Toxicology*, **266**, pp. 16–29.

55. Pauluhn, J. (2010). Subchronic 13-week inhalation exposure of rats to multi-walled carbon nanotubes:toxic effects are determined by density of agglomerate structure, not fibrillar structures. *Toxicol. Sci.*, **113**, 226–242.

56. Pauluhn, J. (2011). Poorly soluble particulates: searching for a unifying denominator of nanoparticles and fine particles for DNEL estimation. *Toxciology*, **279**, pp. 176–188.

57. Ma-Hock, L., Treumann, S., Strauss, V., Brill, S., Luizi, F., Mertler, M., Wiench, K., Gamer, A. O., van Ravenzwaay, B., and Landsiedel, R. (2009a). Inhalation Toxicity of multi-wall carbon nanotubes in rats exposed for 3 months. *Toxicol. Sci.*, **112**, pp. 468–481.

58. Patlolla, A. K., Hussain, S. M., Schlager, J. J., Patlolla, S., and Tchounwou, P. B. (2010). Comparative study of the clastogenicity of functionalized and nonfunctionalized multiwalled carbon nanotubes in bone marrow cells of Swiss-Webster mice. *Environ. Toxicol.*, **25**, pp. 608–621.

59. Patlolla, A. K., Berry, A., and Tchounwou, P. B. (2011). Study of hepatotoxicity and oxidative stress in male Swiss-webster mice exposed to functionalized multi-walled carcbon nanotubes. *Mol. Cell Biochem.*, **358**, pp. 189–199.

60. Bottini, M., Bruckner, S., Nika, K., Bottini, N., Bellucci, S., Magrini, A., Bergamaschi, A., and Mustelin, T. (2006). Multi-walled carbon nanotubes induce T-lymphocyte apoptosis. *Toxicol. Lett.*, **160**, pp. 121–126.

61. DeLorme, M. P., Muro, Y., Arai, T., Banas, D. A., Frame, S. R., Reed, K. L., and Warheit, D. B. (2012). Ninety-day inhalation toxicity study with a vapor grown carbon nanofiber in rats. *Toxicol. Sci.*, **128**, pp. 449–460.

62. Bihari, P., Holzer, M., Praetner, M., Fent, J., Lerchenberger, M., Reichel, C. A., Rehberg, M., Lakatos, S., and Krombach, F. (2010). Single-walled carbon nanotubes activate platelets and accelerate thrombus formation in the microcirculation. *Toxicology*, **269**, pp. 148–154.

63. Schladweiler, M. C., Shannahan, J. H., Thomas, R. F., Saxena, R., Gilmour, M., and Kodavanti, U. P. (2009). Single-walled carbon nanotubes induced pulmonary and vascular response following intratracheal instillation. SOT Abst. No. 2203.

64. Radomski, A., Jurasz, P., Alonso-Escolano, D., Drews, M., Morandi, M., Malinski, T., and Radomski, M. W. (2005). Nanoparticle-induced platelet aggregation and vascular thrombosis. *Br. J. Pharmacol.*, **146**, pp. 882–893.

65. Ballinger, S. W., Patterson, C., Yan, C.-N., Doan, R., Burow, D. L., Young, C. G., Yakes, F. M., Van Houten, B., Ballinger, C. A., Freeman, B. A., and Runge, M. S. (2000). Hydrogen peroxide- and peroxynitrite-induced mitochondrial DNA damage and dysfunction in vascular endothelial and smooth muscle cells. *Circ. Res.*, **86**, pp. 960–966.

66. Li, Z., Hulderman, T., Salmen, R., Chapman, R., Leonard, S. S., Young, S.-H., Shvedova, A., Luster, M. I., and Simeonova, P. P. (2007b). Cardiovascular effects of pulmonary exposure to single-wall carbon nanotubes. *Environ. Health Perspect.*, **115**, pp. 377–382.

67. Suliman, H. B., Welty-Wolf, K. E., Carraway, M. S., Tatro, L., and Piantadosi, C. A. (2004). Lipopolysaccharide induces oxidative cardiac mitochondrial damage and biogenesis. *Cardiovasc. Res.*, **46**, pp. 279–288.

68. Delfino, R. J., Sioutas, C., and Malik, M. (2005). Potential role of ultrafine particles in associations between airborne particle mass and cardiovascular health. *Environ. Health Perspect.*, **113**, pp. 934–946.

69. de Haar, C., Hassing, I., Bol, M., Bleumink, R., and Pieters, R. (2006). Ultrafine but not fine particulate matter causes airway inflammation and allergic airway sensitization to co-administered antigen in mice. *Clin. Exp. Allergy*, **36**, pp. 1469–1479.

70. Inoue, K. I., Takano, H., Yanagisawa, R., Sakurai, M., Ichinose, T., Sadakane, K., and Yoshikawa, T. (2005). Effects of nano particles on antigen-related airway inflammation in mice. *Respir. Res.*, **6**, pp. 106.

71. Løvik, M., Alberg, T., Nygaard, U. C., Samuelsen, M., Groeng, E. C., and Gaarder, P. I. (2007). Popliteal lymph node (PLN) assay to study adjuvant effects on respiratory allergy. *Methods*, **41**, pp. 72–79.

72. Nygaard, U. C., Jitka, S., Hansen, J. S., Samuelsen, M., Alberg, T., Marioara, C. D., and Martinus Løvik, M. (2009). Single-walled and multi-walled carbon nanotubes promote allergic immune responses in mice. *Toxicol. Sci.*, **109**, pp. 113–123.

73. Mitchell, L. A., Gao, J., Wal, R. V., Gigliotti, A., Burchiel, S. W., and McDonald, J. D. (2007). Pulmonary and systemic immune response to inhaled multiwalled carbon nanotubes. *Toxicol. Sci.*, **100**, pp. 203–214.

74. Mitchell, L. A., Lauer, F. T., Burchiel, S. W., and McDonald, J. D. (2009). Mechanisms for how inhaled multiwalled carbon nanotubes suppress systemic immune function in mice. *Nat. Nanotechnol.*, **4**, pp. 451–456.

75. Kolosujaj-Tabi, J., Hartman, K. B., Boudjemaa, S., Ananta, J. S., Morgant, G., Szwarc, H., Wilson, L. J., and Moussa, F. (2010). In vivo behavior of large doses of ultrashort and full-length single-walled carbon nanotubes after oral and intraperitoneal administration to Swiss mice. *ACS Nano.*, **4**, pp. 1481–1492.

76. Folkmann, J. K., Risom, L., Jacobsen, N. R., Wallin, H., Loft, S., and Mølle, P. (2009). Oxidatively damaged DNA in rats exposed by oral gavage to C_{60} fullerenes and single-walled carbon nanotubes. *Environ. Health Persp.*, **117**, pp. 703–708.

77. Murthy, B., Kishore, S., and Surekha, P. (2011). Acute toxicological effects of multi-walled carbon nanotubes (MWCNT), in *Carbon Nanotubes- Growth and Applications*, ed. Naraghi, M. (Rijeka, Croatia).

78. Lim, J. H., Kim, S. H., Shin, I. S., Park, N. H., Moon, C., Kang, S. S., Kim, S. H., Park, S. C., and Kim, J. C. (2011). Maternal exposure to multi-wall carbon nanotubes does not induce embryo fetal developmental toxicity in rats. *Birth Defects. Res. B Dev. Reprod. Toxicol.*, **92**, pp. 69–76.

79. Kroto, H. W., Heath, J. R., O'Brien, S. C., Curl, R. F., and Smalley, R. E. (1985). C_{60}: buckminsterfullerene. *Nature*, **318**, pp. 162–163.

80. Nielsen, G. D., Roursgarrd, M., Jensen, K. A., Poulsen, S. S., and Larsen, S. T. (2008). In vivo biology and toxicology of fullerenes and their derivatives. *Basic Clin. Pharmacol. Toxicol.*, **103**, pp. 197–208.

81. Johnson, H. J., Hutchison, G. R., Christensen, F. M., Aschberger, K., and Stone, V. (2010). The biological mechanisms and physicochemical characteristics responsible for driving fullerene toxicity. *Toxicol. Sci.*, **114**, pp. 162–182.

82. Sayes, C. M., Marchione, A. A., Reed, K. L., and Warheit, D. B. (2007b). Comparative pulmonary toxicity assessments of C_{60} water suspensions in rats: few differences in fullerene toxicity in vivo in contrast to in vitro profiles. *Nano Lett.*, **7**, pp. 2399–2406.

83. Roursgaard, M., Poulsen, S. S., Kepley, C. L., Hammer, M., Nielsen, G. D., and Larsen, S. T. (2008). Polyhydroxylated C60 Fullerene (Fullerenol) attenuates neutrophiliclung inflammation in mice. *Basic Clin. Pharmacol. Toxicol.*, **103**, pp. 386–388.

84. Chiang, L. Y., Lu, F. J., and Lin, J. T. (1995). Free-radical scavenging activity of water-soluble fullerenols. *J. Chem. Soc. Chem. Commun.*, **12**, pp. 1283–1284.

85. Baker, G. L., Gupta, A., Clark, M. L., Valenzuela, B. R., Staska, L. M., Harbo, S. J., Pierce, J. T., and Dill, J. A. (2008). Inhalation toxicity and lung toxicokinetics of C60 fullerene nanoparticles and microparticles. *Toxicol. Sci.*, **101**, pp. 122–131.

86. Fujita, K., Morimoto, Y., Ogami, A., Myojyo, T., Tanaka, I., Shimada, M., Wang, W.-N., Endoh, S., Uchida, K., Nakazato, T., Yamamoto, K., Fukui, H., Horie, M., Yoshida, Y., Iwahashi, H., and Nakanishi, J. (2009). Gene expression profiles in rat lung after inhalation exposure to C_{60} fullerene particles. *Toxicology*, **258**, pp. 47–55.

87. Walker, N. J., Baker, G. L., Dill, J. A., Dermolec, D. R., White, K. L., and Roycreoft, J. H. (2009). Evalaution of the effect of particle size on the toxicity and toxicokinetics of fullerene C60 in rats and mice following nose-only inhalation exposure. Abst. No. 233, Society of Toxicology.

88. Xu, J. Y., Han, K., Li, S. X., Cheng, J. S., Xu, G. T., Li, W. X., and Li, Q. N. (2009). Pulmonary responses to polyhydroxylated fullerenols, C60(OH)x. *J. Appl. Toxicol.*, **29**, pp. 578–584.

89. Rouse, G., Yang, J., Ryman-Rasmussen, J. P., Barron, A. R., and Monteiro-Riviere, N. A. (2007). Effects of mechanical flexion on the penetration of fullerene amino acid-derivatized peptide nanoparticles through skin. *Nano. Lett.*, **7**, pp 155–160.

90. Nelson, M. A., Domann, F. E., Bowden, G. T., Hooser, S. B., Fernando, Q., and Carter, D. E. (1993). Effects of acute and subchronic exposure of topically applied fullerene extracts on the mouse skin. *Toxicol. Indust. Health*, **9**, pp. 623–630.

91. Huczko, A., Lange, H., and Clako, E. (1999). Fullerene experimental evidence for a null risk of skin irritation and allergy. *Fullerene Sci. Technol.*, 7, pp. 935–939.

92. Yamago, S., Tokuyama, H., Nakamura, E., Kikuchi, K., Kananishi, S., Sueki, K., Nakahara, H., Enomoto, S., and Ambe, F. (1995). In vivo biological behavior of a water-miscible fullerene: ^{14}C labeling, absorption, distribution, excretion and acute toxicity. *Chem. Biol.*, **2**, pp. 385–389.

93. Mori, T., Takada, H., Ito, S., Matsubayashi, K., Miwa, N., and Sawaguchi, T. (2006). Preclinical studies on safety of fullerene upon acute oral administration and evaluation for no mutagenesis. *Toxicology*, **225**, pp. 48–54.

94. Chen, H. H., Yu, C., Ueng, T. H., Chen, S., Chen, B., Huang, K. J., and Chiang, L. Y. (1998). Acute and subacute toxicity study of water-soluble polyalkylsulfonated C60 in rats. *Toxicol. Pathol.*, **26**, pp. 143–151.

95. Gharbi, N., Pressac, M., Hadchouel, M., Szwarc, H., Wilson, S. R., and Moussa, F. (2005). Fullerene is a powerful antioxidant in vivo with no acute or suacute toxicity. *Nano Lett.*, **5**, pp. 2578–2585.

96. Rajagopalan, P., Wudl, F., Schinazi, R. F., and Boudinot, F. D. (1996). Pharmacokinetics of a water-soluble fullerene in rats. *Anitmicrob. Agent Chemother.*, **40**, pp. 2262–2265.

97. Ueng, T.-H., Kang, J.-J., Wang, H.-W., Cheng, Y.-W., and Chiang, L. Y. (1997). Suppression of microsomal cytochrome P450-dependent monooxygenases and mitochondrial oxidative phosphorylation by fullerenol, a polyhydroxylated fullerene C_{60}. *Toxicol. Lett.*, **93**, pp. 29–37.

98. Sumner, S., Fenner, T. R., Synder, R. W., Taylor, G. F., and Lewin, A. H. (2010). Distribution of Carbon-14 labeled([^{14}C]C_{60}) in the pregnant and in the lactation dam and the effect of C60 exposure on the biochemical profile of urine. *J. Appl. Toxicol.*, **30**, pp. 354–360.

99. Tsuchiya, T., Oguri, I., Yamakoshi, Y. N., and Miyata, N. (1996). Novel harmful effects of [60]fullerene on mouse embryos in vitro and in vivo. *FEBS Lett.*, **393**, pp. 139–145.

100. Zhu, X., Zhu, L., Li, Y., Duan, Z., Chen, W., and Alvarez, P. J. J. (2007). Developmental toxicity in zebrafish (Danio rerio) embryos after exposure to manufactured nanomaterials: Buckminsterfullerene aggregates (nC_{60}) and fullerol. *Environ. Toxicol. Chem.*, **26**, pp. 976–979.

101. Sera, N., Tokiwa, H., and Miyata, N. (1996). Mutagenicity of th efullerene C60-generated singlet oxygen dependent formation of lipid peroxides. *Carcinogenesis*, 17, pp. 2163–2169.

102. Oberdo¨rster, G., Ferin, J., and Lehnert, B. E. (1994). Correlation between particle size, in vivo particle persistence, and lung injury. *Environ. Health Persp.*, **102**(suppl. 5), pp. 173–179.

103. Warheit, D. B., Webb, T. R., Sayes, C. M., Colvin, V. L., and Reed, K. L. (2006). Pulmonary instillation studies with Nanoscale TiO_2 rods and dots in rats: toxicity is not dependent upon particle size and surface area. *Toxico. Sci.*, **91**, pp. 227–236.

104. Li, J., Li, Q., Xu, J., Li, J., Cai, X., Liu, R., Li, Y., Ma, J., and Li, W. (2007c). Comparative study on the acute pulmonary toxicity induced by 3 and 20nm TiO_2 primary particles in mice. *Environ. Toxicol. Pharmacol.*, **24**, pp. 239–244.

105. Grassian, V. H., O'Shaughnessy, P. T., Adamcakova-Dodd, A., Pettibone, J. M., and Thorne, P. S. (2007a). Inhalation exposure study of titanium dioxide nanoparticles with a primary particle size of 2 to 5 nm. *Environ. Health Perspect.*, **115**, pp. 397–402.

106. Rossi, E. M., Pylkkänen, L., Koivisto,A. J., Vippola, M., Jensen, K. A., Miettinen, M., Sirola, K., Nykäsenoja, H., Karisola, P., Stjernvall, T., Vanhala, E., Kiilunen, M., Pasanen, P., Mäkinen, M., Hämeri, K., Joutsensaari, J., Tuomi, T., Jokiniemi, J., Wolff, H., Savolainen, K., Matikainen, S., and Alenius, H. (2010). Airway exposure to silica-coated TiO_2 nanoparticles induces pulmonary neutrophilia in mice. *Toxicol. Sci.*, **113**, pp. 422–433.

107. Pott, F., and Roller, M. (2005). Carcinogenicity study with nineteen granular dusts in rats. *Eur. J. Oncol.*, **10**, pp. 249–281.

108. Grassian, V. H., Adamcakova-Dodd, A., Pettibone, J. M., O'shaughnessy, P. T., and Thorne, P. S. (2007b). Inflammatory response of mice to manufactured titanium dioxide nanoparticles: comparison of size effects through different exposure routes. *Nanotoxicology*, **1**, pp. 211–226.

109. Kobayashi, N., Naya, M., Endoh, S., Maru, J., Yamamoto, K., and Nakanishi, J. (2009). Comparative pulmonary toxicity study of nano-TiO_2 particles of different sizes and agglomerations in rats: different short- and long-term post-instillation results. *Toxicology*, **264**(1–2), pp. 110–118.

110. van Ravenzwaay, B., Landsiedel, R., Fabian, E., Burkhardt, S., Strauss, V., and Ma-Hock, L. (2009). Comparing fate and effects of three particles of different surface properties: Nano-TiO_2, pigmentary TiO_2 and quartz. *Toxicol. Lett.*, **186**, pp. 152–159.

111. Bermudez, E., Mangum, J. B., Wong, B. A., Bahman, A., Hext, P. M., Warheit, D. B., and Everitt, J. I. (2004). Pulmonary responses of mice, rats, and hamsters to subchronic inhalation of ultrafine titanium dioxide particles. *Toxicol. Sci.*, **77**, pp. 347–357.

112. Ma-Hock, L., Burkhardt, S., Strauss, V., Gamer, A. O., Wiench, K., van Ravenzwaay, B., and Landsiedel, R. (2009b). Development of a short-term inhalation test in the Rat Using Nano-Titanium Dioxide as a model substance. *Inhal. Toxicol.*, **21**, pp. 102–118.

113. Borm, P. J. A., Höhr, D., Steinfartz, Y., Zeitträger, I., and Albrecht, C. (2000). Chronic inflammation and tumor formation in rats after

intratracheal instillation of high doses of coal dusts, titanium dioxides, and quartz. *Inhal. Toxicol.*, **12**, pp. 225–231.

114. Heinrich, U., Fuhst, S., Rittinghausen, O., Creutzenberg, B., Bellmann, M., Koch, K., and Levsen, K. (1995). Chronic inhalation exposure of wistar rats and two different strains of mice to diesel engine exhaust, carbon black, and titanium dioxide. *Inhal. Toxicol.*, **7**, pp. 533-556.

115. Dunford, R., Salinaro, A., Cai, L., Serpone, N., Horikoshi, S., Hidaka, H., and Knowland, J. (1997). Chemical oxidation and DNA damage catalysed by inorganic sunscreen ingredients. *FEBS Lett.*, **418**, pp. 87–90.

116. Warheit, D., Hoke, R. A., Finlay, C., Donner, E. M., Reed, K. L., and Sayes, C. M. (2007c). Development of a base set of toxicity tests using ultrafine TiO_2 particles as a component of nanoparticle risk management. *Toxicol. Lett.*, **171**, pp. 99–110.

117. Pflücker, F., Wendel, V., Hohenberg, H., Gärtner, E., Will, T., Pfeiffer, S., Wepf, R., and Gers-Barlag, H. (2001). The human stratum corneum layer: an effective barrier against dermal uptake of different forms of topically applied micronised titanium dioxide. *Skin Pharmacol. Appl. Skin Physiol.*, **14**(Suppl. 1), pp. 92–97.

118. Lademann, J., Weigmann, H.-J., Rickmeyer, C., Barthelmes, H., Schaefer, H., Mueller, G., and Sterry, W. (1999). Penetration of titanium dioxide microparticles in a Sunscreen formulation into the horny layer and the follicular orifice. *Skin Pharmacol. Appl. Skin Physiol.*, **12**, pp. 247–256.

119. Lekki, J., Stachura, Z., Dąbroś, W., Stachura, J., Menzel, F., Reinert, T., Butz, T., Pallon, J., Gontier, E., Ynsa, M. D., Moretto, P., Kertesz, Z., Szikszai, Z., and Kiss, A. Z. (2007). On the follicular pathway of percutaneous uptake of nanoparticles: ion microscopy and autoradiography studies. *Nucl. Instrum. Meth. B*, **260**, pp. 174–177.

120. Pinheiro, T., Pallon, J., Alves, L. C., Veríssimo, A., Filipe, P., Silva, J. N., Silva, R. (2007). The influence of corneocyte structure on the interpretation of permeation profiles of nanoparticles across skin. *Nucl. Instrum. Meth. B*, pp. 119–123.

121. Nohynek, G. J., Lademann, J., Ribaud, C., and Roberts, M. S. (2007). Grey goo on the skin? Nanotechnology, cosmetic and sunscreen safety. *Crit. Rev. Toxicol.*, **37**, pp. 251–277.

122. Wang, J., Zhou, G., Chen, C., Yu, H., Wang, T., Ma, Y., Jia, G., Gao, Y., Li, B., Sun, J., Li, Y., Jiao, F., Zhao, Y., and Chai, Z. (2007). Acute toxicity and biodistribution of different sized titanium dioxide particles in mice after oral administration. *Toxicol. Lett.*, **168**, pp. 176–185.

123. Nemmar, A., Melghit, K., and Ali, B. H. (2008). The acute proinflammatory and prothrombotic effects of pulmonary exposure to rutile TiO_2 nanorods in rats. *Exp. Biol. Med.*, **233**, pp. 610–619.

124. Nurkiewicz, T. R., Porter, D. W., Hubbs, A. F., Cumpston, J. L., Chen, B. T., Frazer, D. G., and Castranova, V. (2008). Nanoparticle inhalation augments particle-dependent systemic microvascular dysfunction. Part. *Fibre Toxicol.*, **5**, p. 1.

125. Nurkiewicz, T. R., Porter, D. W., Hubbs, A. F., Stone, S., Moseley, A. M., Cumpston, J. L., Goodwill, A. G., Frisbee, S. J., Perrotta, P. L., Brock, R. W., Frisbee, J. C., Boegehold, M. A., Frazer, D. G., Chen, B. T., and Castranova, V. (2011). *Pulmonary Particulate Matter and Systemic Microvascular Dysfunction*. Research Report 164. Health Effects Institute (Blackwell, Boston, MA).

126. Wang, J., Liu, Y., Jiao, F., Lao, F., Li, W., Gu, Y., Li, Y., Ge, C., Zhou, G., Li, B., Zhao, Y., Chai, Z., and Chen, C. (2008b). Time-dependent translocation and potential impairment on central nervous system by intranasally instilled TiO_2 nanoparticles. *Toxicology*, **254**, pp. 82–90.

127. Wang, J., Chen, C., Liu, Y., Jiao, F., Li, W., Lao, F., Li, Y., Li, B., Ge, C., Zhou, G., Gao, Y., Zhao, Y., and Chai, Z. (2008a). Potential neurological lesion after nasal instillation of TiO_2 nanoparticles in the anatase and rutile crystal phases. *Toxicol. Lett.*, **183**, pp. 72–80.

128. Chen, J., Dong, X., Zhao, J., and Tang, G. (2009). In vivo acute toxicity of titanium dioxide nanoparticles to mice after intraperitoneal injection. *J. Appl. Toxicol.*, **29**, pp. 330–337.

129. Fabian, E., Landsiedel, R., Ma-Hock, L., Wiench, K., Wohlleben, W., and Ravenzwaay, B. (2008). Tissue distribution and toxicity of intravenously administered titanium dioxide nanoparticles in rats. *Arch. Toxicol.*, **82**, pp. 151–157.

130. Takenaka, S., Karg, E., Möller, W., Roth, C., Ziesenis, A., Heinzmann, U., Schramel, P., and Heyder, J. (2000). A morphological study on the fate of ultrafine silver particles: distribution pattern of phagocytized metallic silver in vitro and in vivo. *Inhal. Toxicol.*, **12**, pp. 291–299.

131. Takenaka, S., Karg, E., Roth, C., Schulz, H., Ziesenis, A., Heinzmann, U., Schramel, P., and Heyder, J. (2001). Pulmonary and systemic distribution of inhaled ultrafine silver particles in rats. *Environ. Health Perspect.*, **109**(suppl 4), pp. 547–551.

132. Kim, J., Kim, S., and Lee, S. (2011). Differentialtion of the toxicities of silver nanoparticles and silver ions to the Japanese medaka and the cladoceran Daphnia magna. *Nanotoxicology*, **5**, pp. 208–214.

133. Sung, J. H., Ji, J. H., Yoon, J. U., Kim, D. S., Song, M. Y., Jeong, J., Han, B. S., Jeong Hee Han, J. H., Chung, Y. H., Kim, J., Kim,T. S., Chang, H. K., Lee, E. J., Lee, J. H., and Yu, I. J. (2008). Lung function changes in sprague-dawley rats after prolonged inhalation exposure to silver nanoparticles. *Inhal. Toxicol.*, **20**, pp. 567–574.

134. Sung, J. H., Ji, J. H., Park, J. D., Yoon, J. U., Kim, D. S., Jeon, K. S., Song, M. Y., Jeong, J., Han, B. S., Han, J. H., Chung, Y. H., Chang, H. K., Lee, J. H., Cho, M. H., Kelman, B. J., and Yu, I. J. (2009). Subchronic inhalation toxicity of silver nanoparticles. *Toxicol. Sci.,* **108**, pp. 452–461.

135. Kim, Y. S., Kim, J. S., Cho, H. S., Rha, D. S., Kim, J. M., Park, J. D., Choi, B. S., Lim, R., Chang, H. K., Chung, Y. H., Kwon, I. H., Jeong, J., Han, B. S., and Je Yu, I. J. (2008). Twenty-eight-day oral toxicity, genotoxicity, and gender-related tissue distribution of silver nanoparticles in Sprague-Dawley rats. *Inhal. Toxicol.*, **20**, pp. 575–583.

136. Samberg, M. E., Oldenburg, S. J., and Monteiro-Riviere, N. A. (2010). Evaluation of silver nanoparticle toxicity in skin in vivo and keratinocytes in vitro. *Environ. Health Perspect.*, **118**, pp. 407–413.

137. Warheit, D. B., Hansen, J. F., Yuen, I. S., Kelly, D. P., Snajdr, S., and Hartsky, M. A. (1997). Inhalation of high concentrations of low toxicity dusts in rats results in pulmonary and macrophage clearance impairments. *Toxicol. Appl. Pharmacol.*, **145**, pp. 10–22.

138. Nemmar, A., Hoylaerts, M. F., Hoet, P. H. M., Vermylen, J., and Nemery, B. (2003). Size effect of intratracheally instilled particles on pulmonary inflammation and vascular thrombosis. *Toxicol. Appl. Pharmacol.*, **186**, pp. 38–45.

139. Sayes, C. M., Fortner, J. D., Guo, W., Lyon, D., Boyd, A. M., Ausman, K. D., Tao, Y. J., Sitharaman, B., Wilson, L. J., Hughes, J. B., West, J. L., and Colvin, V. L. (2004). The differential cytotoxicity of water-soluble fullerenes, *Nano Lett.*, **4**, pp. 1881–1887.

140. Mercer, R. R., Hubbs, A. F., Scabilloni, J. F., Wang, L., Battelli, L. A., Friend, S., Castranova, V., and Porter, D. W. (2011). Pulmonary fibrotic response to aspiration of multi-walled carbon nanotubes. *Part. Fibre Toxicol.*, **8**, **pp.** 21.

141. Sargent, L. M., Shvedova, A. A., Hubbs, A. F., Salisbury, J. L., Benkovic, S. A., Kashon, M. L., Lowry, D. T., Murray, A. R., Kisin, E. R., Friend, S., McKinstry, K. T., Battelli, L., and Reynolds, S. H. (2009). Induction of aneuploidy by single-walled carbon nanotubes. *Environ. Mol. Mutagen.*, **50**, pp. 708–717.

142. Shimadac, M., Wangc, W.-N., Endohd, S., Uchidad, K., Nakazatod, T., Yamamotoe, K., Fukuia, H., Horiea, M., and Yoshidaa, Y. (2009). Gene

expression profiles in rat lung after inhalation exposure to C60 fullerene. *Part. Toxicol.*, **258**, pp. 47–55.

143. Donaldson, K., Aitken, R., Tran, L., Stone, V., Duffin, R., Forrest, G., and Alexander, A. (2006). Carbon nanotubes: a review of their properties in relation to pulmonary toxicology and workplace safety. *Toxicol. Sci.*, **92**, pp. 5–22.

Chapter 7

Predictive Nanotoxicology: In silico Approaches

Enrico Burello
European Commission: Joint Research Centre, Systems Toxicology Unit, Institute for Health and Consumer Protection, Ispra, Italy
enricoburello@yahoo.com

7.1 Introduction

A thorough understanding of the relationships between the physicochemical properties and the behavior of nanomaterials in biological systems is mandatory for designing safe and efficacious nanomedicines. However, synthesizing and testing all materials is time consuming and resource intensive, which is why researchers are developing computational models to predict the behavior of nanomaterials in biological systems. Such predictions would allow researchers to design experiments, prioritize tests, and streamline research on real nanomaterials. The focus of this chapter is to highlight the most significant achievements and problems of modeling nanomaterials with quantitative structure–activity relationship (QSAR) and other in silico methods. More specifically,

Handbook of Safety Assessment of Nanomaterials: From Toxicological Testing to Personalized Medicine
Edited by Bengt Fadeel

ISBN 978-981-4463-36-2 (Hardcover), 978-981-4463-37-9 (eBook)
www.panstanford.com

we present and discuss a number of recent inspiring applications of modeling tools with a focus on approaches that attempt to describe the interactions that take place at the nano-/biointerface. These interactions are responsible for the biological activity of a nanomaterial in living organisms.

Historically, studies devoted to the understanding and prediction of biological activity and toxicity rely on QSAR methods. A (Q)SAR is a statistical model that relates a set of structural or property descriptors (x_n) of a chemical compound to its biological activity, for example, EC_{50} values:

$$EC_{50} = f(x_1, x_2, \ldots x_n) \tag{7.1}$$

The descriptors, which include parameters to account for hydrophobicity, topology, electronic properties, and steric effects, are determined empirically or by computational methods, whereas activities include chemical measurements and biological assays. The QSAR methodology is well known and extensively applied in the areas of drug discovery and chemical toxicity modeling for guiding the design of chemical compounds, and its growing importance for providing key information is reflected in a number of regulatory frameworks (e.g., Registration, Evaluation, Authorisation and Restriction of Chemicals [REACH]) where these approaches are considered acceptable methods under certain conditions for filling in knowledge gaps for untested chemicals. Although the usefulness of the QSAR technology to model drug-like compounds is well established, its application to predict nanomaterials biological activity is still limited. From a practical standpoint, there is a lack of sufficiently large and reliable datasets of compounds tested under the same experimental conditions (an important requisite for building a statistically significant QSAR model). On the other side, it is difficult to define the exact structure of a nanomaterial (unless in few cases such as for fullerenes), and this hampers the identification and development of descriptors for characterizing relevant physicochemical properties of nanomaterials. Moreover, nanomaterials are large and complex assemblies of inorganic and/or organic elements where the exact stoichiometry may vary from one material to another; this makes classical molecular descriptors inappropriate for this type of study or at least it requires time-consuming computational approaches. A structure characterization of a metal oxide nanoparticle would, for example, include the type and extent of exposed surfaces, the type and density of reactive sites, and the density of hydroxyl groups, as well as the changes they undergo when the size of nanomaterials changes or the environmental conditions vary.

Another approach for modeling the properties of nanomaterials, especially those made of carbon atoms, such as fullerenes, carbon nanotubes, and the like, would be that of using molecular mechanics–based modeling. Molecular mechanics is the description of molecular structures based on classical physics concepts. It uses force field equations to describe the energy that costs a molecule for deviating from its ideal geometry:

$$E = E_{\text{covalent}} + E_{\text{nonbonded}} \tag{7.2}$$

$$E_{\text{covalent}} = E_{\text{covalent}} + E_{\text{bond}} + E_{\text{dihedral}} \tag{7.3}$$

$$E_{\text{nonbonded}} = E_{\text{electrostatic}} + E_{\text{van der Waals}} \tag{7.4}$$

$$E = \sum_{\text{bonds}} k_i (b - b_0)^2 + \sum_{\text{angles}} k_\theta (\theta - \theta_0)^2 + \sum_{\text{dihedrals}} k_\varnothing [1 + \cos(n\varnothing - \delta)] + \sum_{\text{nonbonded}} \varepsilon \left[\left(\frac{R_{\text{min}_{ij}}}{r_{ij}} \right)^{12} - \left(\frac{R_{\text{min}_{ij}}}{r_{ij}} \right)^{6} \right] + \frac{q_i q_j}{\varepsilon r_{ij}} \tag{7.5}$$

Equations like 7.5 are implemented in an optimization algorithm that finds the optimum geometry of a structure corresponding to a (local) energy minimum. Molecular mechanics methods make no reference to electrons and so cannot deal with electronic properties like charge distributions or nucleophilic and electrophilic behavior of compounds. They are usually employed in QSAR studies for determining the 3D structure of organic molecules (i.e., the lowest energy conformer) as well as their nonbonding interactions with the environment, particularly with a target (macro)molecule. Three-dimensional structures are then used as input for the calculation of (3D) descriptors or for docking analyses.

Docking techniques are the computational translation of the lock-and-key concept in enzyme–inhibitor interaction studies and aim at calculating the best fit between a chemical compound and a target macromolecule.

Molecular mechanics is used in dynamic simulations as well to compute the time-dependent behavior of molecular systems; from knowledge of the force on each atom, it is possible to determine the acceleration of atoms in a system. The integration of the equations of motion then yields a trajectory that describes the positions, velocities, and accelerations of the particles as they vary with time. From this trajectory, the average values of physicochemical properties can be determined.

Another approach that is currently used to model the properties of nanomaterials, in particular those made of metals and heavy atoms, relies on density functional theory (DFT) methods, a quantum mechanical–based methodology used to investigate the electronic structure of many-body systems. A great strength of the density functional language is its appropriateness for defining and elucidating important universal concepts of molecular reactivity. In the formalism of DFT, any chemical system is characterized by an electronic chemical potential, μ, and an absolute chemical hardness, η. These quantities are reactivity descriptors of isolated chemical species that can be used to represent the electron transfer process of combined systems. The chemical potential characterizes the escaping tendency of the electron cloud from the equilibrium state and is defined as the derivative of the ground-state electronic energy, E, with respect to the number of electrons, N. E versus N plots are not straight lines but are generally convex upward; their curvature, that is, the second derivative of E, defines the absolute chemical hardness, a quantity that characterizes the resistance toward charge transfer. Using the finite difference approximation, the reactivity descriptors are computed from the vertical ionization potential (IP) and electron affinity (EA):

$$\mu = (\text{IP} + \text{EA})/2 \quad (7.6)$$

$$\eta = (\text{IP} - \text{EA})/2 \quad (7.7)$$

The chemical potential is equivalent to the negative of the electronegativity concept of classical structural chemistry, and one half of (IP + EA) is the original Mulliken formula for electronegativity. The finite difference expression for hardness is the band gap, that is, the difference between the highest occupied molecular orbital (HOMO) and the lowest unoccupied molecular orbital (LUMO) energies in a simple orbital theory. The reactivity descriptors have found widespread applications in QSAR studies, and the calculation of HOMO and LUMO energies of molecules is often used to predict their reactivity, photocatalytic, and conducting properties.

7.2 QSAR and QSPR Models for Nanomaterials

Puzyn et al. [1] have recently reported the development of a QSAR model to predict the cytotoxicity of various metal oxide nanoparticles. The authors systematically investigated the cytotoxicity of 17 metal oxide nanoparticles toward *Escherichia coli* bacteria, and on that basis of the toxicity data and computed structural descriptors, they built a model to predict the cytotoxicity of other nanomaterials. Cytotoxicity is expressed as the effective concentration at which a compound brings about a 50% reduction in bacterial viability. In the experiments, one batch of nanoparticles, containing 10 training compounds, was used to develop the QSAR model, and a second batch, consisting of 7 different nanoparticles, was used to test the validity of the model. Parallel to the toxicological testing of the first batch, the authors computed 12 descriptors to account for the reactivity and electronic properties of the nanoparticles. Because the oxide nanoparticles being tested were too large to be fully included in the calculations, the descriptors were computed from smaller representative clusters using semiempirical PM6 methods, which are quantum chemistry techniques that yield fast estimates of molecular properties. A multiple regression method to find statistically significant correlations between the measured toxicological profiles of the nanoparticles and their descriptors was then employed to find the best-fit model. The model obtained relies on only one descriptor to predict the cytotoxicity of nanoparticles, namely, the enthalpy of formation of a gaseous cation having the same oxidation state as that in the metal oxide structure tested. This key descriptor accounts for lattice energy, which describes the dissolution of nanoparticles and the sum of ionization potentials of a given metal. This parameter can be interpreted as an index of the chemical stability of the nanoparticles; the chemical stability is related to the dissolution of nanoparticles and redox reactions on the particle's surface, and these processes may release toxic ions or free radicals that can cause oxidative stress that damages the bacteria.

Predictions from the model are very close to measured cytotoxicity values, thereby confirming the reliability and relevance of the model. Overall, the model is consistent with toxicological studies reported in the literature, and this suggests that the electronic properties of these materials are responsible for their reactivity and adverse effects toward living organisms. Qualitatively, the experiments indicate that the toxicity of metal oxide nanoparticles decreases when the formal charge of the cation increases. Given that other properties of bulk oxides (e.g., isoelectric point, acidity, and lattice self-potential) also follow this trend, and that size-dependent changes in the properties of nanoparticles are expected to appear at diameters below the 15 nm minimum value reported by the authors, it seems that the nanomaterials tested in this study display bulk-like properties.

Although the QSAR methodologies are well-known and have been extensively used for organic compounds their application to nanomaterials should take into account features that do not pertain to organic and drug-like compounds: one of these features is that while organic molecules have well-defined structures, and therefore yield descriptors with one single value, nanomaterials have a structure variability that can result in different descriptor values. For example, nanomaterials are usually synthetized in ranges of sizes and shapes, and this implies that also the properties are different (especially if we consider size dependent properties). This issue has important consequences on the validity of the QSAR models: for example, predicting the toxicity of materials that lie outside the domain defined by the descriptor(s) may not be accurate because the domain, which is spanned by a particular set of training compounds, determines the type of structures that can be reliably predicted.

If a coating is covering entirely the surface of a nanoparticle then the simplest approximation one can do is to take into account only the properties of the coating and disregard the core of the material. This approximation is reasonable, as similar arguments are applied to explain the role of the protein corona in biological systems, which, by masking the nanoparticle surface, affects the bioavailability of the whole particle. In addition, if the coating is organic then molecular descriptors are relatively straightforward to calculate (or at least data is more likely to be available). It is here important, however, to underline that the coating has to mask entirely the surface and should not detach from the surface of the nanoparticle: Studies on the reactivity of nanoparticles covered with human serum albumin, for example, have shown that if the coating is partially masking the surface and/or it established dynamic interactions with the surface of the nanoparticles, such nanomaterials are still capable of producing radical oxygen species, most probably because the transient interaction with the macromolecules does not prevent the surface of the particles to catalyze the formation of radicals. Tropsha et al. [2] developed a QSAR model for predicting the cellular uptake against different cell lines of 109 nanoparticles with the same metal core but different organic coatings. Each individual nanoparticle was represented only by the structure of the organic coating, which in turn was characterized by 150 molecular descriptors. The authors used neural network techniques to correlate biological profiles with the structure of the nanomaterials and found that relevant descriptors contained information on the coating's lipophilicity, pointing out the importance of van der Waals forces for engaging interactions with biomembranes. These findings imply that the biological activity of a nanoparticle library based on a common core can be predicted using classical QSAR tools. Such models could be used to find a coating material that minimizes the interactions with proteins and/or biomembranes for controlling the cellular uptake and biodistribution of nanomaterials. In the same study, the authors performed another QSAR analysis for a set of 51 nanoparticles with different metal cores and surface modifications. The heterogeneous nature of the dataset does not allow the calculation of descriptors for all nanoparticles in this case; instead, four experimentally measured physical parameters were used: size, R1 and R2 relaxivities, and the zeta potential. The descriptors were then used to classify nanomaterials into two clusters of similar activity. Both QSAR studies led to statistically validated and externally predictive models, thereby illustrating that nanoparticle bioactivity can be predicted by either fragment or whole nanoparticle descriptors.

Fullerenes have emerged as agents that could revolutionize the treatment of many diseases. They can be active as neuroprotective agents [3, 4], antioxidants [5], antibacterials [6, 7], and inhibitors of the (HIV)-protease [8] and potentially useful as drug carrier systems and photoactive compounds inducing DNA cleavage [9, 10]. However, the low water solubility and potential toxicity limit the use of C_{60} and its derivatives for biological applications. In particular, their size and surfactant properties might be responsible for the observed cytotoxicity, which often occurs via a membrane disruption mechanism. With this hypothesis in mind, Prato et al. tested a series of water-soluble fullerene derivatives and investigated for their hemolytic properties. They observed that cationic chains introduced on the C_{60} scaffold induced significant toxicity (i.e., membrane disruption), whereas the presence of neutral or anionic moieties did not produce any response. The experimental observations were then correlated with an amphiphilicity index, defined as the ratio between the total hydrophobic and the total hydrophilic surface area of tested compounds. The descriptor was calculated on the lowest energy conformers, taking into account solvent effects on the geometry and ionization status of the C_{60} molecules. Even if the regression coefficients do not indicate a strong correlation, as the authors report, the use of a single descriptor for predicting a cytotoxic effect is a useful starting point, leaving space for further model improvement. These findings also highlight the importance of electrostatic interactions: as biomembranes are negatively charged, we can expect that positively charged nanomaterials can engage favorable electrostatic interactions with the biological membranes, resulting in membrane disruption or engulfment or even favoring uptake.

When nanomaterials come into contact with biological systems, their surface gets covered by (macro)molecules, which display high affinity for the material's surface. This interaction forms an interface, which is responsible for the fate and behavior of nanomaterials in biological systems. In a biological fluid, for example, proteins associate with nanoparticles, and the amount and type of proteins present on the surface of the particles leads to a series of in vivo responses, for example, the stimulation or suppression of the immune system, the cellular uptake, and the change in enzymatic activity. To identify and control the properties of such interfaces, Riviere et al. reported an approach called the biological surface adsorption index (BSAI) [12]. This method consists of measuring the adsorption coefficients (k_i) of different molecular probes on the surface of a nanomaterial, each probe representing a unique profile of molecular forces involved in the interaction with the nanomaterial. The team used a set of 28 diverse compounds (ranging from alcohols to different hydrocarbons) as probes and measured their adsorption on different nanomaterials in aqueous solutions using solid-phase microextraction and gas chromatography–mass spectrometry methods. The adsorption coefficients are then expressed as a function of five descriptors calculated on the molecular probes: such descriptors reflect the lone-pair electrons (R), the dipolarity/polarizability (π), the acidity/basicity (α/β), and the London dispersion (V) forces. Coulomb forces are also expected to play a fundamental role in the formation of the nano–bio interface; the model, however, does not take them into account explicitly and zeta potential measurements are instead suggested for complementing the information on the adsorption processes. Finally, the regression coefficients (c, r, p, a, b, and v) obtained are used as descriptors for nanomaterials:

$$\log k_i = c + rR_i = p\pi_i + a\alpha_i + b\beta_i + vV_i \qquad (7.8)$$

The method proposed is one of the first examples in the literature aiming at predicting the physicochemical properties and behavior of nanomaterials in complex biological systems and has the advantage of being applicable to any kind of nanomaterial, opening up the possibility to compare the biological properties of different classes of compounds. One direct application of the BSAI approach is to predict the adsorption of small molecules on the nanoparticle surface, a process that is relevant, for example, for controlling drug loading and release in biological and environmental systems. The method could also be used to correlate with membrane interaction and biodistribution parameters (such as adsorption rates, distribution coefficients, and extent of cellular uptake), which for nanomaterials are supposed to be highly dependent on the formation of the protein corona. More precisely, the adsorption of proteins and subsequent formation of the corona involve, besides the above-mentioned forces, entropic and steric contributions as well; in some cases the surface pattern is also relevant, for example, for matching surface charges or specific binding patches. It is also important to mention that soft proteins can undergo structural changes upon adsorption: this process led to an increased number of contacts between the nanomaterial's surface and the proteins, as well as the exposure of newly formed epitopes that are usually buried inside the protein's structure. Therefore, extrapolating information on the adsorption of macromolecules from small molecule adsorption coefficients might not be that straightforward, and additional efforts would be required to describe accurately the formation of the protein corona.

7.3 Density Functional Theory Approaches

There is a growing literature linking health effects of nanomaterials in cells and organisms with the ability of nanomaterials to induce oxidative stress and pro-inflammatory effects as a result of their chemical reactivity. The pro-oxidative effects observed for metal oxide nanoparticles, for example, depend on properties such as electronically reactive states, the presence of transition metals or redox cycling elements (e.g., in doped semiconductor metal oxides), photoactivation leading to the generation of electron/hole pairs (e.g., TiO_2), dissolution of the particle surface, and shedding of toxic metal ions (e.g., ZnO). Under physiological conditions, cells maintain a reduced intracellular state. This process is the result of a balance between the levels of oxidized and reduced species present in the cell. Oxidizing substances can create an imbalance in this state, for example, by depleting electrons from aqueous redox species or by acting as catalysts. The overall result is a decrease in the levels of antioxidants and/or an increase in reactive oxygen species (ROS), a cellular condition that eventually evolves into inflammatory and cytotoxic responses. ROS include hydrogen peroxide and superoxide ions; antioxidants are molecules such as vitamin C and glutathione that scavenge the unwanted oxidants. The dynamic equilibrium between antioxidant defense mechanisms that act to restore redox equilibrium and cellular responses that can lead to adverse outcomes is encapsulated in the "hierarchical oxidative stress paradigm" [13], which posits that ROS production leads to incremental cellular responses that can be classified as antioxidant defense, pro-inflammatory effects, and cytotoxicity. This paradigm is supported by studies on the adverse health effects of ambient ultrafine particles. In such studies, excellent correlation coefficients have been established between the capacity of ultrafine particles to generate ROS abiotically and their ability to induce oxidative stress responses (e.g., heme oxygenase 1 expression) in epithelial cells and macrophages. At a lower oxidative stress level the nanoparticles induce cytoprotective responses, for example, through the activation of the homeostatic antioxidant defense pathway, inducing the expression of several antioxidant and phase II metabolizing enzymes (e.g., heme oxygenase 1 and glutathione-S-transferase). If this level of protection fails, the oxidative stress will lead to cytokine and chemokine expression through transcriptional activation of the gene promoters by redox-sensitive mitogen-activated protein (MAP) kinase and the transcription factor nuclear factor kappa-B (NF-κB) signaling cascades. Further escalation will trigger disturbance of the mitochondrial function, resulting in cytotoxicity (apoptosis and necrosis).

The concept of reactivity has been recently used by Burello and Worth to rationalize with a theoretical framework the observed oxidative stress that occurs when some oxide nanoparticles come into contact with cells [14]. The authors reason that oxide nanoparticle can act like catalysts or electron donors/acceptors and therefore are potentially capable of inducing oxidative stress. The theoretical framework uses reactivity descriptors to calculate the energy structure of oxide nanoparticles and predicts their oxidative stress potential by comparing their conduction (E_C) and valence band (E_V) energy levels with relevant redox potentials (E^0) of biological reactions occurring inside cells. The relative position of E_V or E_C with respect to E^0 dictates whether an electron transfer between the oxide and an adsorbate is feasible: redox couples with E^0 above the conduction band can transfer electrons to the conduction band, whereas redox couples in the band gap can accept conduction band electrons or (photoinduced) valence band holes. The redox couples below the valence band can only be reduced by valence band electrons. The energy of the valence band edge is a measure of the ionization potential of a material. The lowest unoccupied electronic level in most semiconductors coincides with the bottom of the conduction band, and the band edge energy E_C is a measure of the electron affinity of the oxide. The standard redox potentials of couples active in biological media range from −4.12 eV to −4.84 eV relative to the absolute vacuum scale (AVS), whereas environmental aqueous redox potentials range from −3.7 eV to −5.7 eV (ref.). Assuming that oxide nanoparticles behave like bulk materials when having a diameter larger than 20–30 nm and no surface states inside the band gap, that is, between the conduction and valence bands, the authors used a set of empirical equations for calculating the energy structure of the particles instead of performing rigorous DFT calculations. The valence and conduction band energy levels are calculated from the metal oxides band gap (E_g) and electronegativity (χ). Such empirical equations are rooted in the Pauling principle of electronegativity and can be applied to a large number of oxide structures. In this way, the proposed framework, although simplified if compared with ab initio calculations, allows a larger dataset of compounds to be selected for screening and, more importantly, becomes more useful when it includes additional descriptors reflecting a variety of material's characteristics. The model predicts fairly well the conduction and valence band positions of titania nanoparticles, which have the correct energy structure to produce hydroxyl radicals and superoxide ions. Other metal oxides whose conduction band superimposes the range of biological redox potentials are WO_2, CrO_2, MoO_2, Tl_2O_3, Ag_2O, Cu_2O, Ni_2O_3, Nb_2O_5, Mn_2O_3, CoO, FeO, Ta_2O_5, TiO_2(rutile), Cr_2O_3, In_2O_3, TiO_2(anatase), and Ce_2O_3. The practical consequences of such work would be to help rationalize the implementation of toxicity-related descriptors in the nanomaterial's design process, providing a rapid tool to prioritize and streamline research.

This model has been recently verified and implemented on a high-throughput screening platform by independent experimental studies for testing a metal oxide's capability of inducing oxidative stress and acute pulmonary inflammation [15]. The authors choose 24 oxide nanomaterials among those predicted by Burello and tested their ability to induce oxidative stress. Among the materials tested, the overlap of conduction band energy levels with the cellular redox potential was strongly correlated to the ability of Co_3O_4, Cr_2O_3, Ni_2O_3, Mn_2O_3, and CoO nanoparticles to generate oxygen radicals, oxidative stress, and inflammation. In addition, Co_3O_4, Ni_2O_3, Mn_2O_3, and CoO nanoparticles were also found to oxidize cytochrome c as a representative redox couple involved in redox homeostasis. While CuO and ZnO generated oxidative stress and acute pulmonary inflammation that was not predicted by the conduction band energy levels (their conduction band energy levels do not overlap with the range of redox potentials), the adverse biological effects of these materials could be explained by their solubility. The successful implementation of the oxidative stress model on a high-throughput screening assay demonstrates that it is possible to predict the toxicity of a large series of metal oxides nanoparticles in the lung premised on reactivity properties and an integrated in vitro/in vivo hazard-ranking model. Most importantly, this work has established a predictive toxicological paradigm, which is also of considerable significance for regulatory decision making about this important class of nanomaterials.

DFT calculations can be used to study the influence of a particle's size on the structure and properties of nanomaterials. This information is crucial to determine to what size extent a material can be considered having bulk-like or nanospecific properties. Structure- and size-dependent electronic properties, surface chemistry, and atomic charges of carbonaceous nanomaterials, for example, have been investigated by several authors with ab initio calculations. Shukla and Leszczynski investigated the influence of the shape and size of small carbon (Cn; n = 40–70) species, including fullerenes, nanodisks, nanocapsules, and nanobowls, on their electronic properties [16]. They calculated the first vertical ionization potentials and electron affinities, band gaps, isolated atom energies, cohesive energies, and cohesive energies per atom using DFT. The theory predicts that the ionization potential decreases, while the electron affinity significantly increases when moving from C_{60} to larger species. Another study that demonstrates the significance of the quantum size effect for capped and uncapped carbon nanotubes was performed by Yumura et al. The authors concluded that tube lengths, edge structures, and end caps play an important role in determining properties such as band gaps and vibrational frequencies.

The same methodology was applied in two studies on carbon nanoneedles. Wang and Mezey characterized the electronic structures of open-ended and capped nanoneedles [17]. They observed that the conducting properties (expressed by the HOMO–LUMO gap) were strongly correlated to the size of the considered species. Only very long structures were characterized as semiconductors, while very short and thin structures have band gaps too large to conduct electrons. Similarly, Poater et al. have analyzed the conduction behavior of nanotubes as a function of their length and the presence of different ions (when the ring of the nanotubes is made of at least eight carbon atoms) [18]. The authors calculated energy band gaps, chemical potential, hardness, and electrophilicity of a series of nanotubes with 4-, 6-, and 8-ring atoms and possessing different lengths (from 1 to 10 layers). The band gap decreases as the length of the nanotubes increases, pointing toward a semiconductor character when size increases. The chemical hardness also decreases and the structures of carbon nanotubes become more reactive as their length increases.

7.4 Molecular Mechanics Approaches

Carbon nanotubes are another class of carbonaceous nanomaterials with interesting properties for applications in nanomedicine [19, 20]. Their shape and surface properties make them a potential candidate for new drug delivery systems with significant advantages over other methods currently in use, such as liposomes. In principle, by piercing plasma membranes and heading straight to the cytoplasm, a carbon nanotube encounters fewer biological barriers and delivers the drug more directly. The nanotubes also have a structural advantage in that they are extremely thin but very long, offering a large surface area where to graft the required drug. This again offers an improvement over liposome drug delivery and also allows the amount of drug loaded onto the nanotube to be regulated. Like fullerenes, however, carbon nanotubes have low dipersibility and display toxicity via an oxidative stress mechanism. These problems suggest that functionalization of their surface with different chemical moieties can lead to improved uptake, drug loading, and water solubility. Another problem associated with such class of compounds is related to their persistence in the body, and again their functionalization might be the right answer for designing biodegradable or more readily excretable compounds. A recent study by Kagan et al. shows, for example, that short carboxylated nanotubes are degraded when incubated in a solution where both the neutrophil myeloperoxidase enzyme and H_2O_2 are present [21]. This enzyme is known to bind negatively charged molecules and catalyze their biodegradation with the contribution of peroxidase reactive intermediates and hypochlorite, which act as oxidants. To better understand the role of surface charges on the enzyme–nanotube interaction, the authors used docking techniques to determine the binding mode of carboxylated and pristine nanotubes to the protein. The modeling reveals two potential interaction sites with the enzyme: One binding pocket can host the carboxylated nanotubes, thus allowing their biodegradation via a catalytically competent tyrosine residue. The complex formed is stabilized by electrostatic interactions between the positively charged residues on the enzyme and the negative carboxylic groups on the nanotube surface. The second binding site is far away from this residue and prefers interactions with pristine nanotubes, which are less susceptible to biodegradation. Docking methods can be used to find the optimal building blocks for designing nanomaterials with the desired characteristics and functionality, provided that the crystal structure of the target protein(s) has been determined. Currently, the availability of this information for hundreds of potential macromolecule targets opens up the possibility to quantify which nanomaterials can selectively avoid contact or interact with proteins, enzymes, immunoglobulins, and DNA.

Recent studies have demonstrated that macromolecules can be inserted into biological membranes, leading to the formation of transmembrane nanopores that can function as ion channels and antimicrobial agents. Molecular dynamics simulations on such systems suggest that the addition of a nanotube into a phospholipid membrane may change both the structure and the dynamic behavior of the bilayer. Liu and Hopfinger used membrane interaction–QSAR analysis, which relies on molecular dynamics simulation, to further explore the interaction of carbon nanotubes with a phospholipid bilayer [22]. The authors constructed a model of a fully hydrated dimyristoylphosphatidylcholine (DMPC) membrane bilayer and inserted a carbon nanotube through the bilayer. The results showed that the nanotube caused a structural rearrangement of the packing of the lipid molecules that extended 18 Å from the nanotube. Accompanying this rearrangement was a loss of flexibility of the DMPC molecules surrounding the nanotube. A particularly interesting structural feature was the development of a cylindrical ring of 2–4 Å thickness around the nanotubes that was completely unoccupied. Next, the authors monitored the diffusion of a hydrated sodium or calcium ion through the tube, as well as the diffusion of three small polar molecules—ethanol, urea, and caffeine—through the bilayer in the presence of the nanotube. The results showed that the calcium ion would be transported through the nanotube in preference to the sodium ion, even though it would cause greater disturbance in the bilayer structure during its passage. In addition, the nanotubes affected the total diffusion coefficients of the three polar molecules, increasing that of ethanol by 35% and those of urea and caffeine to a lesser extent. However, the presence of the nanotubes caused all three molecules to move more directly across the bilayer. These observations have relevant implications on the design of carbon nanotube–based nanomedicines and the assessment of their potential toxicity. However, dynamic simulations are very time consuming, and their applications are limited to a few specific cases and rather small nanomaterials.

7.5 Mathematical Modeling of Nanomaterial Bioactivity

Another approach that attempts to model the behavior of nanomaterials in biological systems and with great potential for applications in the nanomedicine sector goes under the definition of mathematical modeling. In this approach, researchers formulate a specific problem with a number of differential equations. Examples of mathematical modeling applied to the design of nanomaterials for biomedical applications are given by Decuzzi and Ferrari, who analyzed the role of specific and nonspecific interactions in receptor-mediated endocytosis of nanoparticles [23]. "Endocytosis" is the general term used to define the process used by cells to import selected extracellular molecules, viruses, microorganisms, and nanometer particles, and it is of fundamental importance in several fields as drug and gene delivery and in nanotoxicology. It includes a variety of pathways: clathrin-dependent and independent receptor-mediated endocytosis (RME), pinocytosis, and phagocytosis. Among these, the most effective mechanism for the uptake of nanometer particles is RME, where molecules (ligands) decorating the particle surface bind to countermolecules (receptors) expressed over the cell membrane. The receptors together with the bound particles enter the cell by membrane invagination. In RME, the receptors expressed over the cell membrane are collected at the site of invagination by surface diffusion, without which endocytosis would not occur or would occur over a much longer time scale. The authors presented a general formulation to predict the endocytic performances of nanometer particles where the threshold particle radius R_{th}, the optimal particle radius R_{opt}, and the characteristic wrapping time t_w are estimated as a function of specific and nonspecific particle–cell interactions. In particular, they introduced an explicit expression for $(R_{th})_{min}$, the minimum threshold radius below which no endocytosis can occur. The work showed that the contribution of the nonspecific interactions, represented by the dimensionless parameter F, is as important as the contribution of the specific interactions, represented by the binding energy factor C. As the repulsive interactions between the particle and the cell increase (F increases), the wrapping time and the threshold and optimal particle radii increase too, making particle endocytosis more difficult or even impracticable. On the other hand, attractive nonspecific interactions (negative F) would favor particle endocytosis. The dimensionless parameters having the largest influence on the endocytic performance of the nanoparticles have been shown to be C and G, the binding energy and bond elasticity factors, respectively, whose values can be controlled by changing or engineering the ligand molecules, and F, the nonspecific attractive/repulsive factor whose value can be tailored by controlling the electrostatic surface charge and the density, length, and type of surface polymeric linkers decorating the particle and its bulk dielectric properties. The formulation of this problem with mathematical modeling methods yields a set of ready-to-use formulas that could be of help in designing nanoparticles with controlled endocytic performances.

7.6 Multiscale Modelling and Other Coarse-Graining Methods

In cases where an understanding of the dual nature of the structure of matter (continuous when viewed at large length scales and discrete when viewed at an atomic scale) and its interdependences are crucial, multiscale materials modeling approaches are required to complement continuum and atomistic analyses methods. Multiscale modeling approaches relies on a systematic reduction in the degrees of freedom on the natural length scales that can be identified in the material. Usually, the following levels are distinguished: level of quantum mechanical models (information about electrons is included), level of molecular dynamics models (information about individual atoms is included), mesoscale or nanolevel (information about groups of atoms and molecules is included), level of continuum models, and level of device models. Each level addresses a phenomenon over a specific window of length and time. Connections between such scales are currently achieved either by a parameterization or by a coarse-graining procedure. Currently, the coupling between length scales has to face two main challenges: firstly, the computational complexity of seamlessly coupled simulations via the coarse-graining approach and, secondly, the inherent difficulty in dealing with system evolution stemming from time scaling, which does not permit coarse graining over temporal events. Here we introduce the work on thermodynamic cartography, which has been developed and applied by Barnard et al. to study the properties of nanomaterials as a function of size, temperature, pressure, and other environmental conditions (e.g., pH).

Controlling nanoscale features is still not a trivial matter, and dramatic changes in fundamental properties may result from small changes in the size, the shape, or the degree of dimensional anisotropy. These types of structural changes can be driven or invoked spontaneously by changes in the environment immediately surrounding any given structure. For example, minor fluctuations in the temperature and/or pressure can be sufficient to induce solid–solid transitions in the nanostructure (which do not appear in the bulk form) and change material properties entirely. For this reason, a complete understanding of the structural stability of any nanomaterial with respect to its physical and chemical environment is essential. Accurate explicit computational studies of nanoparticles are very computationally demanding and difficult to extend to sizes above 1–2 nm. This is not just because of the large computational resources required to calculate the optimized shape of nanoparticles with first principles methods but also because of the large set of structures that must be explicitly examined to effectively sample structure space. It is therefore highly desirable to use a theoretical approach that is predominantly analytical and is based on a limited set of experimental or theoretical data. A number of new shape-dependent theories (usually requiring parameterization) have been devised. A theory, based on Gibb's summation, was introduced by Barnard and Zapol for use with isolated (unsupported) nanostructures, which included both surface and edge energies, as well as corner energies, surface tensions, and bulk elastic energy:

$$G_x = G_{\text{bulk},x} + G_{\text{sirface},x} + G_{\text{edge},x} + G_{\text{corner},x} + G_{\text{defects},x} \tag{7.9}$$

This theory has so far proven very useful in examining the shape of isolated (colloidal) nanoparticles of a variety of materials. One example is TiO_2, where a broad study has been reported investigating the relative stability of quasi-spherical and faceted anatase nanoparticles and the effect of particle shape on the size-dependent phase transition between anatase and rutile nanoparticles [24]. The study also included an investigation of the effect of water on the shape of TiO_2 nanocrystals, followed by a systematic study on the effects of surface acidity on the shape and phase stability of anatase and rutile nanoparticles less than 100 nm in size, with results in good agreement with the experiment. In addition, the authors analyzed the photocatalytic activity of titania nanocrystals at different sizes and demonstrated that there is a direct relationship between the number of active sites on the crystal's surface and the experimentally measured production of radical species [25].

7.7 Conclusions

The modeling studies presented in this chapter show that currently there are different methodologies adopted by computational chemists and physics to model the behavior of nanomaterials in biological systems. Such approaches have a range of applications, and although they are still under development, they show great potential for complementing the experimental studies and drive the design of nanomaterials for medical or other applications. Future challenges are the definition and implementation of a full multiscale modeling approach, the development nanospecific descriptors for QSAR models, and the development of new testing methods, especially abiotic tests, which could be coupled with computational studies for assessing the interactions of nanomaterials with biological targets.

References

1. Puzyn, T., Rasulev, B., Gajewicz, A., Hu, X., Dasari, T. P., Michalkova, A., Hwang, H. M., Toropov, A., Leszczynska, D., and Leszczynski, J. (2011). Using nano-QSAR to predict the cytotoxicity of metal oxide nanoparticles. *Nat. Nanotechnol.*, **6**, pp. 175–178.
2. Fourches, D., Pu, D., Tassa, C., Weissleder, R., Shaw, S. Y., Mumper, R. J., and Tropsha, A. (2010). Quantitative nanostructureactivity relationship modeling. *ACS Nano*, **4**, pp. 5703–5712.
3. Huang, S. S., Tsai, S. K., Chin, C. L., Chiang, L. Y., Hsieh, H. M., Teng, C. M., and Tsai, M. C. (2001). Neuroprotective effect of hexasulfobutylated C60 on rats subjected to focal cerebral ischemia. *Free Radic. Biol. Med.*, **30**, pp. 643–649.
4. Dugan, L., Gabrielsen, J., Yu, S., Lin, T., and Choi, D. (1996). Buckminsterfullerenol free radical scavengers reduce excitotoxic and apoptotic death of cultured cortical neurons. *Neurobiol. Dis.*, **3**, pp. 129–135.
5. Dugan, L. L., Lovett, E., Cuddihy, S., Ma, B.-W., Lin, T. S., and Choi, D. W. (2000). Carboxyfullerenes as neuroprotective antioxidants, in *Fullerenes: Chemistry, Physics, and Technology*, eds. Kadish, K. M., and Ruoff, R. S. (John Wiley & Sons, New York), pp. 467–479.
6. Mashino, T., Nishikawa, D., Takahashi, K., Usui, N., Yamori, T., Seki, M., Endo, T., and Mochizuki, M. (2003). Antibacterial and antiproliferative activity of cationic fullerene derivatives. *Bioorg. Med. Chem. Lett.*, **13**, pp. 4395–4397.
7. Bosi, S., Da Ros, T., Castellano, S., Banfi, E., and Prato M. (2000). Antimycobacterial activity of ionic fullerene derivatives. *Bioorg. Med. Chem. Lett.*, **10**, pp. 1043–1045.
8. Schuster, D. I., Wilson, L. J., Kirschner, A. N., Schinazi, R. F., Schlueter-Wirtz, S., Tharnish, P., Barnett, T., Ermolieff, J., Tang, J., Brettreich, J., et al. (2000). Evaluation of the anti-HIV potency of a water-soluble dendrimeric fullerene, in *Fullerene 2000: Functionalized Fullerenes*, eds. Martin, N., Maggini, M., and Guldi, D. M. (Electrochemical Society, Pennington, NJ), vol. 9, pp. 267–270.
9. Bernstein, R., Prat, F., and Foote, C. S. (1999). On the mechanism of DNA cleavage by fullerene investigated in model systems: electron transfer from guanosine and 8-oxoguanosine derivatives to C60. *J. Am. Chem. Soc.*, **121**, pp. 464–465.
10. Boutorine, A. S., Takasugi, M., Héléne, C., Tokuyama, H., Isobe, H., Nakamura, E. (1995). Fullerene-oligonucleotide conjugates:

photoinduced sequence-specific DNA cleavage. *Angew Chem., Int. Ed. Engl.*, **33**, pp. 2462–2465.

11. Bosi, S., Feruglio, L., Da Ros, T., Spalluto, G., Gregoretti, B., Terdoslavich, M., Decorti, G., Passamonti, S., Moro, S., and Prato, M. (2004). Hemolytic effects of water-soluble fullerene derivatives. *J. Med. Chem.*, **47**, pp. 6711–6715.

12. Xia, X.-R., Monteiro-Riviere, N. A., and Riviere, J. E. (2010). An index for characterization of nanomaterials in biological systems. *Nat. Nanotechnol.*, **5**, pp. 671–675.

13. Meng, H., Xia, T., George, S., and Nel, A. E. (2009). A predictive toxicological paradigm for the safety assessment of NMs. *ACS Nano*, **3**, pp. 1620–1627.

14. Burello, E., and Worth, A. P. (2011). A theoretical framework for predicting the oxidative stress potential of oxide NPs. *Nanotoxicology*, **5**, pp. 228–235.

15. Zhang, H., Ji, Z., Xia, T., Meng, H., Low-Kam, C., Liu, R., Pokhrel, S., Lin, S., Wang, X., Liao, Y. P., Wang, M., Li, L., Rallo, R., Damoiseaux, R., Telesca, D., Mädler, L., Cohen, Y., Zink, J. I., and Nel, A. E. (2012). Use of metal oxide nanoparticles band gap to develop a predictive paradigm for oxidative stress and acute pulmonary inflammaion. *ACS Nano*, **6**(5), pp. 4349–4368.

16. Shukla, M. K., and Leszczynski, J. (2006). A density functional theory study on the effect of shape and size on the ionization potential and electron affinity of different carbon nanostructures. *Chem. Phys. Lett.*, **428**, pp. 317–320.

17. Wang, J. L., and Mezey, P. G. (2006). The electronic structures and properties of open-ended and capped carbon nanoneedles. *J. Chem. Inf. Model.*, **46**, pp. 801–807.

18. Poater, A., Saliner Gallegos, A., Carbó-Dorca, R., Poater, J., Solá, M., Cavallo, L., and Worth, A. (2009). Modelling the structure-property relationships of nanoneedles: a journey towards nanomedicine. *J. Comput. Chem.*, **30**, pp. 275–284.

19. Lacerda, L., Raffa, S., Prato, M., Bianco, A., and Kostarelos, K. (2007). Cell-penetrating CNTs for delivery of therapeutics. *Nano Today*, **2**, pp. 38–43.

20. Bianco, A., Kostarelos, K., and Prato, M. (2005). Applications of carbon nanotubes in drug delivery. *Curr. Opin. Chem. Biol.*, **9**, pp. 674–679.

21. Kagan, V. E., Konduru, N. V., Feng, W., Allen, B. L., Conroy, J., Volkov, Y., Vlasova, I. I., Belikova, N. A., Yanamala, N., Kapralov, A., Tyurina, Y.

Y., Shi, J., Kisin, E. R., Murray, A. R., Franks, J., Stolz, D., Gou, P., Klein-Seetharaman, J., Fadeel, B., Star, A., and Shvedova, A. A. (2010). Carbon nanotubes degraded by neutrophil myeloperoxidase induce less pulmonary inflammation. *Nat. Nanotechnol.*, **5**, pp. 354–359.

22. Liu, J., and Hopfinger, A. J. (2008). Identification of possible sources of nanotoxicity from carbon nanotubes inserted into membrane bilayers using membrane interaction quantitative structure-activity relationship analysis. *Chem. Res. Toxicol.*, **21**, pp. 459–466.

23. Decuzzi, P., and Ferrari, M. (2007). The role of specific and nonspecific interactions in receptor mediated endocytosis of nanoparticles. *Biomaterials*, **28**, pp. 2915–2922.

24. Barnard, A. S., and Xu, H. (2008). An environmentally sensitive phase map of titania nanocrystals. *ACS Nano*, **2**, pp. 2237–2242.

25. Barnard A. S. (2011). Mapping the photocatalytic activity or potential free radical toxicity of nanoscaletitania. *Energy Environ. Sci.*, **4**, pp. 439–443.

Chapter 8

Physiologically Based Nanomaterial Pharmacokinetic Models

Jim E. Riviere
Institute of Computational Comparative Medicine, Kansas State University, Manhattan, KS, USA
jriviere@ksu.edu

8.1 Introduction

If there is one property of nanomaterials that makes them "unique" relative to predicting their activity or toxicity relative to small molecules, it is that to a large extent their pattern of biodistribution and pharmacokinetic properties define the nature of their therapeutic and adverse effects. Recent reviews [1–5] have discussed the biodistribution and pharmacodisposition properties of nanomaterials that make this statement true. This includes, for some specific nanoparticles, detailed analyses of their biodistribution and unique protein interactions. However, surprisingly little has been published on actual pharmacokinetic models that quantitate these processes or that take specific interaction mechanisms into consideration. A recent review of nanoparticle drug delivery

Handbook of Safety Assessment of Nanomaterials: From Toxicological Testing to Personalized Medicine
Edited by Bengt Fadeel

ISBN 978-981-4463-36-2 (Hardcover), 978-981-4463-37-9 (eBook)
www.panstanford.com

strategies confirms this notion that classic pharmacokinetic parameters have not been determined for most nanomaterials [6] except for more conventional liposomes and polymeric nanoparticles [3]. The focus of this chapter will be to take into consideration what is qualitatively known about nanoparticle biodistribution and elimination and use this information to start defining the model structure and parameters needed to develop true physiologically based nanoparticle pharmacokinetic (PBNPK) models.

8.2 What Is Unique about Nanoparticle ADME?

8.2.1 Absorption

The first step in a proper adsorption, distribution, metabolism, and excretion (ADME) pharmacokinetic analysis is to define the manner in which a nanomaterial enters the body, that is, when exposure occurs by nonparenteral routes where an absorption phase is evident, usually after oral, transdermal, or inhalational administration. These topics are addressed in Chapter 12 (pulmonary effects) and Chapter 14 (dermal effects), respectively. Relative to oral exposure, early studies in rats have shown that even large micron-scale particles can be absorbed in the intestines across Peyer's patches or by intestinal enterocytes primarily into the lymphatic system [7]. This work has concentrated on vaccine delivery and targeting of local lymphatic organs and will not be extensively discussed in this chapter since it is a given that a particulate pathway exists for oral absorption.

A great deal of work in nanomedicine has to do with using nanotechnology in the pharmaceutical formulation itself. This may involve taking advantage of a nanoparticle's large surface-area-to-mass ratio to control drug release or of unique surface properties to promote drug solubility and subsequent controlled release. Alternatively, nanosized structures (e.g., needles, liposomal topical formulations, complex emulsions) may be used to facilitate delivery of small molecules [8]. Although such applications are all major drivers of nanomedicine development, from the perspective of the present chapter, they do not directly impact the fundamental nature of ADME properties of the nanoscaled materials themselves that impact their pharmacokinetics.

8.2.2 Distribution

Once absorbed into the systemic circulation, the biodistribution properties now become important. These topics have been well reviewed in recent publications related to biocorona formation [9] and mechanisms by which nanoparticles cross biological barriers [10]. Finally, elimination and metabolism of nanomaterials, or more specifically for nanoparticles probably best termed "biodegradation," are discussed in Chapter 10. These topics will briefly be discussed here, where necessary, to define highlights that are pertinent to construction of PBNPK models.

The unique and possibly even defining characteristic of nanomaterials is their strong propensity to adsorb and interact with a wide variety of biomolecules, the most important being proteins. This interaction has recently been termed biocorona formation [9] and often determines both the half-life of the nanoparticle's survival in the systemic circulation and the tissues to which it would distribute. This process by which any foreign particulate material or pathogen bound to an antibody or complement system protein in the systemic circulation was termed "opsonization." It was a prerequisite for phagocytosis and ultimate removal from the circulation through cells of the reticuloendothelial system (RES) or other mononuclear phagocytes (see Chapter 9). Many nanoparticle–protein interactions involve this classic opsonization pathway and often dictate the pattern of biodistribution.

However, the concept of biocorona formation is broader, is meant to represent a dynamic event, and involves protein interactions (e.g., apoliproteins like HDL or LDL, albumin, acute phase proteins, ferritin, coagulation factors) that target nanoparticle–protein complexes to specific cells and tissues, thereby ultimately determining their pharmacokinetic profile. There is also species specificity as to which RES cells (e.g., spleen versus liver) will clear the same nanoparticles, a topic nicely reviewed by Moghimi [3]. This is not at all surprising, yet it has not been consciously noted when extrapolating ADME data across different species. There is a great deal of biological diversity in plasma protein content as a function of species, as well as genetics, age, diet, disease, and the environment. If nanoparticle–protein interactions are a crucial determinant of biodistribution, individual differences in the plasma proteome will modulate pharmacokinetic markers of biodistribution. There is a need for a robust characterization of protein interaction patterns using biologically relevant metrics. This knowledge is a prerequisite for defining robust PBNPK models that would apply to a wide variety of nanomaterials.

It is well known that coating of nanoparticles with polyethylene glycol (PEG), block copolymers, and hyaluronic acid will suppress RES uptake and prolong circulation half-life [1–4]. In addition, the shape of surface adducts may change the nature of protein interaction and subsequent deposition [3]. In addition to just surface chemistry and adducts, particle shape, including surface area and curvature, is also an important parameter [11, 12]. Size has long been known to play a major role in determining the ability of a nanoscale particulate body (usually liposome) to circulate in the blood without passage through so-called a "leaky" fenestrated or discontinuous capillary endothelium and uptake by RES or phagocytic cells [2, 13, 14]. Liver Kupffer cells tend to take up smaller (<100 nm) particles in contrast to splenic filtration cells that capture particles >100 nm. However, larger, flexible micron-long filamentous particles may "stream through" RES-lined capillaries in the spleen, which normally would exclude rigid spheres greater than 150 nm in diameter [15]. In fact movement of any nanoparticle in the viscous medium of blood introduces rheological concerns that further dictate a nanoparticle's ability to exit a capillary bed [16]. Any model must thus encompass a multidimensional space, which could account for different properties that are known to determine this crucial phase of nanoparticle biodistribution.

The geometry, porosity, and surface chemistry of a series of silica nanoparticles were critical attributes for determining the pattern of biodistribution after intravenous injection to mice [17]. Similar to other particles discussed above, the liver and spleen were important targets. However, for particles of similar size, mesoporous silica and a high geometric aspect ratio (rods versus spheres) tended to preferentially cause deposition into pulmonary capillaries, suggesting these factors are also important determinants of biodistribution. In this four-day study, 60% to 85% of material remained in the animal and was not excreted from the body.

One aspect of biodistribution that has received attention is the movement of a therapeutic nanomaterial out of the systemic circulation to reach a targeted tumor. A paradigm of cancer chemotherapy in nanomedicine is based on the premise that tumor vasculature is "leaky" and thus becomes a mechanism for preferential targeting of systemically administered nanoparticles [18]. This paradigm implies that normal vasculature is not "leaky" yet does allow passage of some materials into select body tissues. As discussed above, most attention has been focused on the RES, which possesses "leaky" capillaries of various-sized pores and includes those residing in the liver (Kupffer cells but not hepatocytes), spleen, bone marrow, lymphatic system, and kidney (mesangium). Thus optimal targeting of a cancer chemotherapeutic nanoparticle to a tumor would require a size that would evade liver uptake but not be too large to be instead captured by the spleen, this range being somewhere between 100 nm and 200 nm.

Once a nanoparticle is circulating in the systemic circulation, some of the same physicochemical properties that modulate biocorona formation may also modulate cellular uptake. For the RES, neutral liposomes are taken up less than are negatively charged particles [19], as are many neutral iron oxide nanoparticles [20]. This is similar to that seen with quantum dot (QD) studies in human epidermal keratinocyte cultures, which showed neutral QD uptake < QD-NH_2 < QD-COOH [21]. Of significance to pharmacokinetic model development, the time course of cellular uptake also differed between QDs with different surface coatings. Of course, cellular uptake ultimately determines tissue deposition in a pharmacokinetic model. What is sorely lacking in this area is studies done using nanoparticles that have a biocorona similar to what would be found under in vivo conditions or of particles that are capable in the first place to actually cross the systemic capillary barrier and gain exposure to cell surfaces. In vitro experiments are conducted using standard cell culture techniques, which include the use of media that have protein and other biomolecule compositions optimized for cell growth. What impact do these media constituents have on biocorona formation around the tested particles? How are the concentrations used in in vitro cell culture experiments related to what could be achievable in vivo? What is the form of the nanoparticle biocorona that would migrate out of a capillary to be available for cellular uptake?

Although not a topic of the present chapter, it is instructive to realize that some nanoparticle–protein interactions are not strictly pharmacokinetic but also impact pharmacodynamics. This is when the mechanism of action of the nanoparticle involves direct interaction with proteins responsible for disease (e.g., amyloid) in Alzheimer's or Parkinson's disease. As an example, one potential approach for modulating Alzheimer's disease is to use specific nanomaterials that bind to blood-amyloid β protein in an effort to shift the blood–brain barrier equilibrium of this protein in favor of blood, which would reduce brain deposition and allow its removal from the body via the RES [22]. This is a pharmacokinetic property that determines a nanoparticle's pharmacodynamic profile and supports the hypothesis stated in the first sentence of the present chapter on the dominance of pharmacokinetic properties to predict nanoparticle activity.

The lymphatic system is where lipophilic particulate matter often transits the body yet is not normally accounted for in classic pharmacokinetic models. Intramuscular and subcutaneous injections of nanoparticles [23], as well as inhalational administration [5] and intradermal injections [24], may result in nanoparticle accumulation within the lymphatic system. In fact, QDs could be used to image lymphatic drainage to lymph nodes in surgery [25], a study that showed QDs between 15–20 nm tend to be retained in the first lymph node encountered.

8.2.3 Elimination

There are some aspects of a nanoparticle's elimination and clearance from the body that are also different from what is encountered for small molecules. Studies using QDs of different sizes have suggested that only particles smaller than 5.5 nm are capable of being cleared by the kidney [26], the primary route from where most small molecules or their metabolites are excreted from the body. This is consistent with the kidney's normal physiological role of not filtering plasma proteins or other formed elements in the blood. A lack of robust urinary or biliary secretion pathways for nanomaterials is a major component that must be taken into consideration when constructing PBNPK models.

In most cases, "elimination" from the body is not via excretion but by degradation in lysosomes of RES cells. This is highly dependent upon the particle's composition and its susceptibility to metabolism. Surface coatings and coronas may get the nanoparticle targeted to a cell, but its ultimate stability is a function of the particle's composition. Hence so-called biologically labile polymers—for example, albumin, chitosan, cellulose, milk protein, poly(lactic-co-glycolic) acid—some iron oxide particles, and nanostructured colloid liquids are easily degraded and thus are effectively eliminated from the body. For example, with iron oxide nanoparticles, once surface coatings degrade, the iron oxide core could be incorporated into the body's iron pool [20]. In contrast substances such as styrene and metallic structures such as QDs may persist due to resistance to degradation. It has recently been demonstrated that certain carbon nanotubes can be degraded in vitro by neutrophil myeloperoxidases [27], suggesting that if this occurs in vivo, even relatively "hard" nanostructures may ultimately be cleared from the body. This mechanism of degradation suggests that unlike most small molecules where the organs of elimination are the liver or the kidney, for nanomaterials the elimination organs may be dispersed throughout the body.

8.3 Pharmacokinetic Models

With this background in mind, how has pharmacokinetics been applied in general to the ADME parameters of nanomaterial disposition? It is well beyond the scope of this chapter to review the science and art of pharmacokinetics, which is adequately covered in many textbooks [28–30]. Pharmacokinetics allows quantitation of nanomaterial biodistribution parameters across different studies that facilitates cross-study comparisons. The primary parameter calculated is usually the half-life of disappearance of material in the systemic circulation. This is the time it takes for concentration in blood or plasma to decrease by 50% and is calculated as [ln $2/K$], where K is the elimination rate constant determined from the slope of the ln (concentration) versus time profile of a substance decay in blood. For nanomaterials, a longer half-life correlates to persistence of the particle in the circulation. However, unlike the interpretation for small molecules, a short half-life does not imply increased clearance from the body but rather may reflect opsonization and removal of particles by the RES. These limitations have recently been discussed in detail by this author [4].

In another approach, so called noncompartmental or stochastic models may be used where sojourn of a particle in the body is assessed by calculating mean residence times (MRTs), whose interpretation is similar to half-life. In addition, the half-life or MRT for absorption and distribution can also be calculated if enough data over proper time frames is collected to allow statistical definition of multiple slopes in the concentration versus time profile.

In classic pharmacokinetic analysis, the body is considered composed of a number of so-called compartments that consist of those regions in the body where particle movement into and out of the vasculature (reference point for all models) are similar. These compartments have no anatomical or physiological reality. When compartmental models are constructed, a simple exponential equation relating concentration (C) at any time (t) can be written as $C_t = C_0e^{-Kt}$, where K is the elimination rate constant. Since this is an equation of a first-order rate process, K is the slope of the ln C versus t plot. From this equation, pharmacokinetic parameters of volume of distribution (V_d = Dose/C_0) and clearance (Cl = KV_d) can be calculated. Some of these descriptive parameters, primarily half-life, have been occasionally calculated for a number of nanomaterials [20, 4]. If more complex distribution patterns occur (e.g., the body is not homogenous, but slow and rapid distribution phases occur), a biexponential model is needed, and these parameters can be determined using different equations.

8.3.1 PBPK Models

A particularly useful approach to pharmacokinetic modeling that takes into account the body's actual anatomical structure and physiology is termed "physiologically based pharmacokinetics" (PBPK), an area well reviewed in a recent text [31]. This modeling approach examines a molecule's time course of biodistribution and elimination in actual tissues and organs linked together by blood flow through the vascular system. An example is illustrated in Fig. 8.1. As will be seen in the section below examining how these models have been applied to nanomaterials, they also allow introduction of mechanisms of biodistribution and cell incorporation, as well as being more amenable to incorporation of in vitro data and ultimately interspecies extrapolations.

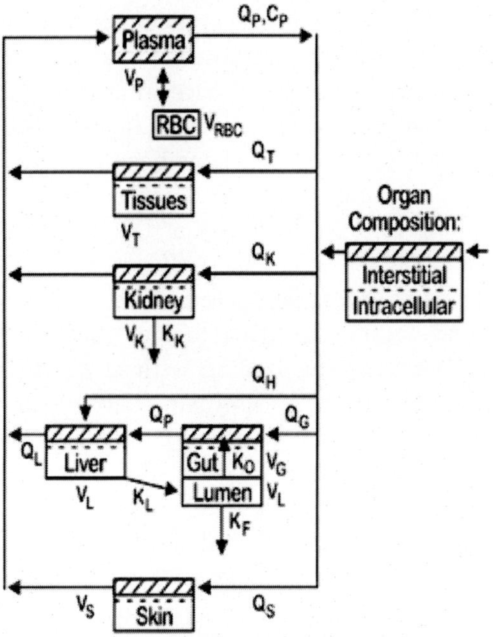

Figure 8.1 General PBPK model. Tissue blocks are connected by the vascular system with organ blood flows (Q), volumes (V), elimination rate constants (K) when an organ is a route of excretion, and absorption rates (K_0) from the gut.

8.3.2 In vitro Perfused Tissue Biodistribution Studies

Before discussing how whole-animal in vivo PBPK models have been used, it is worth briefly mentioning the application of simple pharmacokinetic models to isolated perfused tissue preparations to model nanomaterial biodistribution from the infused arterial circulation to the perfused tissue bed. Our laboratory has used the previously developed isolated perfused porcine skin flap (IPPSF) model developed for studying transdermal chemical absorption [32, 33] to study biodistribution by infusing nanomaterial into the arterial circulation perfusing this isolated skin flap in vitro, and measuring nanomaterial concentrations in arterial and venous drainage samples. Experiments are designed using an infusion phase, where the nanomaterial is included in the arterial media, followed by a washout phase, where clean media is then infused. This allows accumulation and washout phases to be modeled. Arterial/venous extraction is then determined, and simple pharmacokinetic models (Fig. 8.2) can be developed to describe nanoparticle movement from the arterial to the venous vasculature, reflecting biodistribution. Alternatively, a model similar to an organ block in the PBPK model above can be used. These models have been employed to describe QD 621 [34], the nC_{60} fullerene [35], and silver nanoparticles [36]. The studies allow direct calculation of tissue extraction, volume of distributions, and the time course of particle movement into a defined tissue bed. These parameters differed with nanoparticle composition and coating for QD infusions (anionic QD uptake > neutral QD). An interesting finding with QD infusions for both coatings was a periodicity of approximately 100 minutes in the arterial to venous extraction. If such a phenomenon occurs in vivo, tissue redistribution between vascular effects could be observed. The structure of these models allows easy incorporation into whole-animal PBPK models. Although not studied with nanoparticles, isolated perfused livers and kidneys have often been used to study mechanisms of drug and xenobiotic excretion.

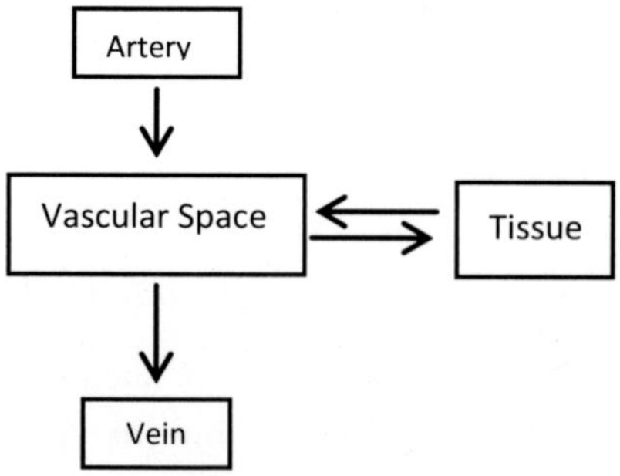

Figure 8.2 Perfused tissue pharmacokinetic model. Samples are collected from the perfusing artery and venous drainage.

An intriguing finding in some of these infusion studies was the induction of vascular toxicity with infusion of certain nanoparticles [37]. Compared to a large control series of small-molecule infusions and nC_{60} in the IPPSF, infusion with iron oxide, silver, silica, and QDs caused vascular congestion evidenced by perfused flap weight gain and a gradual increase in arterial perfusion pressure QD flaps. This is in contrast to arterial infusion of a vasoconstrictive drug, which results in an immediate increase in arterial resistance [33]. This finding of vascular congestion is consistent with recent findings in vivo in mice where silica nanoparticles caused acute toxicity due to mechanical obstruction of the vasculature in multiple organs [38]. This vascular toxicity, most dramatic in the kidney, was modulated by surface geometry and chemical characteristics, with mesoporous and high aspect ratios being more toxic. This has implications to the construction of PBNPK models since direct vascular effects will impact both clearance and distribution parameters.

8.4 Whole-Animal in vivo PBNPK Models

With this background in mind, how have PBPK models been applied to nanomaterials? To clarify this discussion, as defined above, PBPK models of nanomaterials will be termed "PBNPK" models. An excellent review of specific PBNPK models was published by Li et al. [39]. Because there have only been a few attempts to actually construct PBNPK models due to the scarcity of animal data to model, more questions arise than conclusions on definitive steps needed to formulate such models. For example, only a few types of nanomaterials have been modeled and none have included consideration of lymphatic transport. No attempts were made to monitor biocorona formation or in situ nanoparticle aggregation states in any biophase included in the models. In a manner very similar to so-called "pH partitioning" where a weak acid or base may accumulate in an organ where pH favors the ionized form of the drug, a similar phenomenon could occur with nanoparticles where the driving force would now be aggregation or agglomeration in the tissue favoring this process. This has not been addressed.

A QD 705 model in mice was described by Lin et al. [40] on the basis of a complete biodistribution study reported earlier [41]. This dataset had shown only partial and gradual QD excretion over six months. These workers proposed a one-way (blood to tissue) distribution coefficient (DC) as they argued a true partition coefficient does not exist for nanoparticles. Our group [42] reported on a single PBNPK model using the Yang data and published studies from other QDs (705, 525, 621, 800, conjugated) ranging in size from 7 nm to 80 nm in mice and rats. A single model could not describe all data, and the work by Li et al. reached a similar conclusion relative to partition coefficients and flow-limited models. In contrast, in the isolated tissue perfusion studies described above using carbon and silver nanoparticles, equilibrium was achieved in the perfused skin tissue bed, allowing for the concept of a partition coefficient to be employed [36, 35]. However, the periodicity in arterial–venous extraction observed with QD infusion [34] is more consistent with the QD mice data and strongly suggests that even in the same tissue bed, the approach used to model tissue distribution may be particle dependent. Finally, direct interaction with nanoparticles on modulating vascular function, as discussed above [37], also makes constructing tissue compartments in a PBNPK model problematic.

Lankveld et al. [43] developed a PBNPK model in rats over 16 days to describe the disposition of silver nanoparticles in sizes of 20 nm, 80 nm, and 110 nm. This study showed rapid clearance from the blood into all tissues studied independent of particle size, although smaller particles tended to accumulate in the liver, while larger particles tended toward the spleen, a finding consistent with RES distribution data discussed above for other nanomaterials. After multiple dose administration, accumulation was observed in all organs. At the end of the experiment, the largest-sized particles persisted in all tissues. These studies clearly suggest that disposition of 20 nm particles were different than the larger ones studied. A "quasi-irreversible" tissue incorporation parameter was used to account for 1.3%, 9.9%, and 6.5% of particles across all sampled tissues for 20 nm, 80 nm, and 110 nm particles, respectively. The tissue blood partition value obtained in this study for the reversible tissue concentrations (which would include skin) was the same as that reported in the isolated skin perfusion studies discussed above for 20 nm and 40 nm silver nanoparticles [36], a finding supporting the use of a perfused tissue preparation to study nanoparticle biodistribution.

This model also illustrates a number of other characteristics of a PBNPK model relative to where "elimination" occurs. As discussed earlier [4], irreversible tissue sequestration, as seen with the silver particles in this model and QDs in the models above, becomes a primary mechanism of clearance. Thus unlike many small-molecule PBPK models, the organ of elimination may be a misnomer as elimination via sequestration and degradation occurs in multiple tissues. In the Lankveld model, blood was defined as the elimination organ. It should be noted that this is somewhat similar to pharmacokinetic models describing protein and peptide disposition (e.g., kinins) [29] where enzymatic degradation by pulmonary enzymes results in effective elimination.

For particles such as iron oxide, models may get more complex as a PBNPK model may be needed to describe initial biodistribution and then a traditional PBPK model used to quantitate incorporation of iron into the body pool. The same could be postulated for any metallic particle (e.g., silver) that would leach out elemental metal. For many particles such as QDs and silica discussed earlier, particles often persist in the body at the end of an experiment. A limitation of these existing models and studies are that long-term trials have not been conducted to accurately estimate the final fate of distributed materials.

On the basis of the few PBNPK models published and the unique aspects of biodistribution and elimination discussed in the earlier parts of this chapter, there are several characteristics of nanomaterials that make the basic structure of small-molecule PBPK models, and thus data needed, potentially different from PBNPK models. These are tabulated and compared in Box 8.1, where some points where such factors could be important in a PBPK model are illustrated. These factors should be taken into consideration when nanomaterial PBNPK models are developed as data becomes available.

8.5 Need for Biological Characterization Indices

As can be seen from this analysis of data needs for developing PBNPK models, there is also a need for including parameters that quantitate a nanoparticle's interactions with biological molecules and structures, a phenomenon that seems central to a complete understanding of nanomaterial disposition. How does one take into

Box 8.1 Potential differences between small molecules and nanoparticle disposition that could modulate pharmacokinetic properties and should be accounted for in any model development

- For small molecules, the entity being modeled is normally a single molecular structure that could be modified by classical chemical or enzymatic metabolic processes. For nanoparticles, one may be dealing with an aggregate of subparticles that could change in size and surface characteristics, depending on the biological environment it is located in. Associated with this is that molecules are normally of a single molecular weight, while most nanoparticles are best described by a size distribution (mean size, size range).
- Solubility and partition coefficients are often embedded in small-molecule models to predict tissue deposition by estimating a partition coefficient. Nanoparticles have different parameters that dictate biodistribution (size, shape, complete surface chemistry, aggregation status), and thus a simple surrogate metric is not yet available.
- The concept of flow-limited and permeability-limited compartments used in small chemical models has to be modified to deal with rate-limiting blood rheology considerations operative in nanoparticle movement through capillary beds.
- Small molecules may undergo nonspecific protein binding primarily with albumin by a process well described using classic pr,otein ligand–binding models (e.g. Hill model). Nanoparticles may bind to proteins and other macromolecules in a dynamic fashion to form a biocorona that could be a primary determinant of nanoparticle distribution and elimination. These models have not yet been generalized in a mathematical presentation useful for incorporating into a pharmacokinetic model.
- For the majority of small molecules studied, systemic distribution occurs primarily through the circulatory system. For some nanoparticles, trafficking through the lymphatic system may predominate and thus dictate the structure of the pharmacokinetic model used.

- Unlike small molecules, nanomaterials have colloidal properties that may result in aggregation or agglomeration, depending upon their local microenvironment (pH, ionic strength, presence of other molecules). Particles may thus change size and surface properties when they enter other tissue sites or cellular compartments (e.g., lysosomes) in a manner that would influence their pharmacokinetic behavior (e.g., trapping on the side of the membrane promoting aggregation).
- In most cases, small molecules are taken up into cells by either diffusion or classic molecular transporter systems (organic acid transporter system [OATS], p-glycoprotein, etc.) that have been well described using saturable kinetic models (e.g., Michaelis–Menten). These are generally modeled as a reversible process. In contrast, nanoparticles seem to enter cells via slower and capacity-limited vesicular transport systems with charge and size specificity (e.g., micro- and macropinocytosis, membrane rafts) whose kinetic properties have not been well described. Cell egress via exocytosis has not often been modeled, suggesting the uptake process may be irreversible for certain particles.
- Small-molecule elimination and clearance is normally via the liver and the kidney through well-described pathways (biliary excretion, hepatic metabolism, tubular transporters, glomerular filtration, etc.). Drugs embedded or encapsulated in nanoparticles, even though ultimately released into tissues, may be protected from these normal clearance mechanisms. Clearance of a nanoparticle may primarily be accomplished by degradation in RES cells or if evading this must undergo some other particle-dependent degradation pathway or not undergo significant elimination from the body if they are resistant to metabolism.
- The pathway of absorption of a chemical into the body may have a major effect of its subsequent deposition as classically illustrated by first-pass hepatic metabolism. For nanoparticles, if an absorbed particle forms a tight association with a biomolecule (e.g., surfactant) in the process of absorption, its subsequent deposition could be affected.

account the nature of the protein corona, and what specific cellular uptake processes will be involved in a particle's deposition?

Computational approaches to describe nanomaterials using biologically relevant descriptors were recently reviewed [1] as were approaches to predict biocorona formation and composition [9]. Physicochemical properties important in understanding unique effects of nanomaterials include particle size and its distribution, agglomeration state, particle shape, crystal structure, chemical composition, surface area, surface chemistry, surface charge, electronic properties, and porosity. What differentiates nanoparticles from other molecules is the quantum effects evident in true nanosized structures. Thus as a bulk material approaches the nanoscale, properties such as conductivity, reactivity, color, strength, and solubility change. The surface-to-volume ratio dramatically increases. These changes make extrapolation of properties and biological activity difficult from bulk chemicals of the same composition.

When quantitative structure–activity relationships (QSARs) (see Chapter 6) are determined for small molecules, descriptors can be easily calculated on the basis of atomic and molecular properties. Size and surface charge can be easily measured using electron microscopy or methods based on light scattering. However, for nanomaterials, there are significant challenges in using theory to compute accurate properties. Variation in actual materials manufactured, except for all but very defined structures (e.g., C_{60} fullerenes, defined carbon nanotubes, grapheme sheets), makes calculation of molecular properties near impossible. Similarly, nanomaterials that are essentially stable colloids (e.g., QDs) also make computational approaches extremely difficult. Thus indirect approaches to probe complex nanostructure properties relevant to biological interactions have been studied. Experimental methods that indirectly probe such properties show promise.

8.5.1 Biological Surface Adsorption Index

One such approach developed in our laboratory is the multidimensional biological surface adsorption index (BSAI) developed to provide a surrogate metric describing nanoparticle interactions to the partition coefficient used for small molecules [44, 45]. The BSAI is a novel experimental-based approach to characterize the adsorption properties of nanomaterials behind such behaviors as aggregation, biocorona formation, and other interactions that dictate biological deposition. The BSAI is composed of five nanodescriptors that represent the surface adsorption forces (hydrophobicity, hydrogen bonding, polarity/polarizability, and lone-pair electrons) that dictate nanoparticle interactions under conditions encountered in biological systems. When this five-dimensional nanodescriptor fingerprint was reduced to two dimensions using principal component analysis, a series of 16 diverse nanomaterials could be classified into distinct clusters on the basis of these surface adsorption properties. The index was successful in predicting log octanol water partition coefficients as well as adsorption of a variety of small molecules onto multiwalled carbon nanotubes.

We postulate that knowledge of the surface adsorption properties embedded in an index such as the BSAI, coupled with classic characterization of the physical properties (particle size and its distribution, shape, surface area, etc.) and biochemical properties (antibody–antigen interactions, chemical reactivity: oxidation, reduction, complexion, etc.), is required if truly robust PBNPK and toxicodynamic models are to be developed. These parameters would then be embedded into the model to govern uptake or elimination for particles with these characteristics. For example, if biocorona formation targets a nanoparticle primarily to the spleen, then a spleen compartment would be added to the scheme in Fig. 8.1. If in vitro studies define uptake of particles to vesicular transport mechanisms, then instead of assuming a reversible transport process between the interstitial and intercellular compartments within a tissue, a specific one-way rate process could be included on the basis of parameters obtained from in vitro studies. If a particle traffics in the lymphatic system, lymphatic flow would be added to the basic PBNPK model structure. Another approach is to use in vitro characterization studies to classify the type of nanoparticle interactions with biomolecules expected, depending on physicochemical properties of the nanoparticle surface, to then define what organs should be included in a final model.

8.6 Conclusions

This chapter has briefly reviewed the biological properties of nanomaterials that effect their biodistribution and elimination from the context of how these properties would affect development of PBNPK models. Box 8.1 should be studied for pertinent considerations relevant to developing such models. Many of the issues involve characterizations that could be conducted in vitro and then applied to the proper PBNPK tissue module.

Because of the paucity of appropriate in vivo data available, the initial PBNPK models developed have been limited to nanosilver and QDs in laboratory rodents. Unfortunately, neither human nor large laboratory animal data is available for comparison. As discussed above, the RES shows pronounced species differences, which impact the pattern of nanomaterial biodistribution. However, one of the advantages of developing PBNPK models is that they are directly amenable to making interspecies comparisons should appropriate experiments be conducted. The fields of nanomedicine and nanotoxicology are rapidly developing and entering a more mature phase where quantitative in vivo studies must be conducted with an eye toward incorporating the latest characterization metrics and mechanisms of nanoparticle distribution and elimination and with an eye toward predicting their behavior in humans. The use of PBNPK models is a tool that would greatly facilitate this goal.

References

1. Gajewicz, A., Rasulev, B., Dinadayalane, T. C., Urbaszek, P., Puzyn, T., Leszczynska, D., and Leszczynska, J. (2012). Advancing risk assessment of engineered nanomaterials: application of computational approaches. *Adv. Drug Deliv. Rev.*, **64**, pp. 1663–1693.
2. Li, S. D., and Huang, L. (2008). Pharmacokinetics and biodistribution of nanoparticles. *Mol. Pharm.*, **5**, pp. 496–504.
3. Moghimi, S. M., Hunter, A. C., and Andresen, T. L. (2012). Factors controlling nanoparticle pharmacokinetics: an integrated analysis and perspective. *Annu. Rev. Pharmacol. Toxicol.*, **52**, pp. 481–503.
4. Riviere, J. E. (2009). Pharmacokinetics of Nanomaterials: An overview of carbon nanotubes, fullerenes and quantum dots. *WIRES Nanomed. Nanobiotechnol.*, **1**, pp. 26–34.

5. Yang, W., Peters, J. I., and Williams, R. O. (2008). Inhaled nanoparticles: a current review. *Int. J. Pharm.*, **356**, pp. 239–247.

6. Adair, J. H., Parette, M. P., Altinoglu, E. I., and Kester, M. (2010). Nanoparticulate alternatives for drug delivery. *ACS Nano*, **4**, pp. 4967–4970.

7. Des Rieux, A., Fieves, V., Garinot, M., Schneider, Y.-J., and Preat, V. (2006). Nanoparticles as potential oral delivery systems of proteins and vaccines: a mechanistic approach. *J. Controlled Rel.*, **116**, pp. 1–27.

8. Cevc, G., and Vierl, U. (2010). Nanotechnology and the transdermal route. *J. Controlled Rel.*, **141**, pp. 277–299.

9. Walkey, C. D., and Chan, W. C. W. (2012). Understanding and controlling the interaction of nanomaterials with proteins in a physiological environment. *Chem. Soc. Rev.*, **41**, pp. 2780–2799.

10. Pietroiusti, A., Campagnolo, L., and Fadeel, B. (2013). Interactions of engineered nanoparticles with organs protected by internal biological barriers. *Small*, **9**, pp. 1557–1572.

11. Cedervall, T., Lynch, I., Foy, M., Berggård, T., Donnelly, S. C., Cagney, G., Linse, S., and Dawson, K. A. (2007). Detailed identification of plasma proteins adsorbed on copolymer nanoparticles. *Angew Chem., Int. Ed. Engl.*, **46**, pp. 5754–5756.

12. Pedersen, M. B., Zhou, X., Larsen, E. K. U., Sorensen, U. S., and Kjems, J. (2010). Curvature of synthetic and natural surfaces is an important target in classical pathway complement activation. *J. Immunol.*, **184**, pp. 1931–1945.

13. Liu, D., Mori, A., and Huang, L. (1992). Role of liposome size and RES blockade in controlling biodistribution and tissue uptake of GMI-containing liposomes. *Biochim. Biophys. Acta*, **1104**, pp. 95–101.

14. Moreira, J. N., Gaspar, R., and Allen, T. M. (2001). Targeting stealth liposomes in a murine model of human small cell lung cancer. *Biochim. Biophys. Acta*, **1515**, pp. 167–176.

15. Geng, Y., Dalhaimer, P., Cai, S., Tsai, R., Tewari, M., Minko, T., and Discher, D. E. (2007). Shape effects of filaments versus spherical particles in flow and drug delivery. *Nat. Nanotechnol.*, **2**, pp. 249–255.

16. Gentile, F., Ferrari, M., and Decuzzi, P. (2008). The transport of nanoparticles in blood vessels: the effect of vessel permeability and blood rheology. *Ann. Biomed. Eng.*, **36**, pp. 254–261.

17. Yu, T., Hubbard, D., Ray, A., and Ghandehari, H. (2012). In vivo biodistribution and pharmacokinetics of silica nanoparticles as a

function of geometry, porosity and surface characteristics. *J. Controlled Rel.*, **163**, pp. 46–54.

18. Hobbs, S. K., Monsky, W. L., Yuan, F., Roberts, W. G., Griffith, L., Torchilin, V. P., and Jain, R. K. (1998). Regulation of transport pathways in tumor vessels: role of tumor type and microenvironment. *Proc. Natl. Acad. Sci. U S A*, **95**, pp. 4607–4612.
19. Levchenko, T. S., Rammohan, R., Lukyanov, A. N., Whiteman, K. R., and Torchilin, V. P. (2002). Liposome clearance in mice: the effect of a separate and combined presence of surface charge and polymer coating. *Int. J. Pharm.*, **240**, pp. 95–102.
20. Almeida, J. P. M., Chen, A. L., Foster, A., and Drezek, R. (2011). In vivo biodistribution of nanoparticles. *Nanomedicine*, **6**, pp. 815–835.
21. Ryman-Rasmussen, J. P., Riviere, J. E., and Monteiro-Riviere, N. A. (2007). Variables influencing interactions of untargeted quantum dot nanoparticles with skin cells and identification of biochemical modulators. *Nano Lett.*, **7**, pp. 1344–1348.
22. Gobbi, M., Re, F., Canovi, M., Beeg, M., Gregori, M., Sesana, S., Sonnino, S., Brogioli, D., Musicanti, C., Gasco, P., Salmona, M., and Masserini, M. E. (2010). Lipid based nanoparticles with high binding affinity for amyloid beta 1-42 peptide. *Biomaterials*, **31**, pp. 6519–6529.
23. Yim, Y. S., Choi, J. S., Jang, S. B., Kim, G. T., Park, K., Kim, C. H., Cheon, J., and Kim, D. G. (2009). Pharmacokinetic properties and tissue storage of FITC conjugated SA-MnMEIO nanoparticles in mice. *Curr. Appl. Physiol.*, **9**, pp. E304–E307.
24. Gopee, N. V., Roberts, D. W., Webb, P., Cozart, C. R., Siitonen, P. H., Warbritton, A. R., Yu, W. W., Colvin, V. L., Walker, N. J., and Howard, P. C. (2007). Migration of intradermally injected quantum dots to sentinel organs in mice. *Toxicol. Sci.*, **98**, pp. 249–257.
25. Soltesz, E. G., Kim, S., Laurence, R. G., DeGrand, A. M., Parungo, C. P., Dor, D. M., Cohn, L. H., Bawendi, M. G., Frangioni, J. V., and Mihaljevic, T. (2005). Interoperative sentinel lymph node mapping of the lung using near-infrared fluorescent quantum dots. *Ann. Thorac. Surg.*, **79**, pp. 269–277.
26. Choi, H. S., Liu, W., Misra, P., Tanaka, E., Zimmer, J. P., Ipe, B. I., Bawendi, M. G., and Frangioni, J. V. (2007). Renal clearance of quantum dots. *Nat. Biotechnol.*, **25**, pp. 1165–1170.
27. Kagan, V. E., Konduru, N. V., Feng, W., Allen, B. L., Conroy, J., Volkov, Y., Vlasova, I. I., Belikova, N. A., Yanamala, N., Kapralov, A., Tyurina, Y. Y., Sji, J., Kisin, E. R., Murray, A. R., Franks, J., Stolz, D., Gou, P., Klein-Seetharaman, J., Fedeel, B., Star, A., and Shevoda, A. A. (2010). Carbon

nanotubes degraded by neutrophil myeloperoxidase induces less pulmonary inflammation. *Nat. Nanotechnol.*, **5**, pp. 354–359.

28. Ette, E. I., and Williams, P. J. (2007). *Pharmacometrics: The Science of Quantitative Pharmacology* (Wiley, Hoboken, NJ).

29. Riviere, J. E. (2011). *Comparative Pharmacokinetics: Principles, Techniques and Applications*, 2nd ed. (Wiley-Blackwell, Ames, IA).

30. Shargel, L., Wu-Pong, S., and Yu, A. B. C. (2005). *Applied Biopharmaceutics and Pharmacokinetics*, 5th ed. (McGraw Hill, New York).

31. Reddy, M. A., Yang, R. S. H., Clewell, H. J., and Andersen, M. E. (2005). *Physiologically Based Pharmacokinetic Modeling* (Wiley, Hoboken, NJ).

32. Riviere, J. E. Bowman, K. F., Monteiro-Riviere, N. A., Dix, L. P., and Carver, M. P. (1986). The isolated perfused porcine skin flap (IPPSF). I. A novel in vitro model for percutaneous absorption and cutaneous toxicology studies. *Fundam. Appl. Toxicol.*, **7**, pp. 444–453.

33. Riviere, J. E., and Monteiro-Riviere, N. A. (1991). The isolated perfused porcine skin flap as an in vitro model for percutaneous absorption and cutaneous toxicology. *Crit. Rev. Toxicol.*, **21**, pp. 329–344.

34. Lee, H. A., Imran, M. Monteiro-Riviere, N. A., Colvin, V. L., Wu, W., and Riviere, J. E. (2007). Biodistribution of quantum dot nanoparticles in perfused skin: evidence of coating dependency and periodicity in arterial extraction. *Nano Lett.*, **7**, pp. 2865–2870.

35. Leavens, T. L., Xia, X. R., Lee, H. Y., Monteiro-Riviere, N. A., Brooks, J. D., and Riviere, J. E. (2010). Evaluation of perfused porcine skin as a model system to quantitate tissue distribution of fullerene nanoparticles. *Toxicol. Lett.*, **197**, pp. 1–6.

36. Leavens, T. L., Monteiro-Riviere, N. A., Inman, A. O., Brooks, J. D., Oldenburg, S. J., and Riviere, J. E. (2012). In vitro biodistribution of silver nanomaterial in isolated perfused porcine skin flaps. *J. Appl. Toxicol.*, **32**, pp. 913–919.

37. Riviere, J. E., Leavens, T. L., Brooks, J. D., and Monteiro-Riviere, N. A. (2012). Acute vascular effects of nanoparticle infusion in isolated perfused skin. *Nanomed. Nanotech. Biol. Med.*, **8**, pp. 428–431.

38. Yu, T., Greish, K., McGill, L. D., Ray, A., and Ghandehari, H. (2012). Influence of geometry, porosity, and surface characteristics of silica nanoparticles on acute toxicity: their vasculature effects and tolerance threshold. *ACS Nano*, **6**, pp. 2289–2301.

39. Li, M., Al-Jamal, K., Kostarelos, K., and Reineke. J. (2010), Physiological based pharmacokinetic modeling of nanoparticles. *ACS Nano*, **11**, pp. 6303–6317.

40. Lin, P., Chen, J. W., Chang, L. W., Wu, J. P., Redding, L., Chang, H., Yeh, T. K., Yang, C. S., Tsai, M. H., Wang, H. J., Kuo, Y. C., and Yang, R. S. (2008). Computational and ultrastructural toxicology of a nanoparticle, Quantum Dot 705, in mice. *Environ. Sci. Technol.*, **42**, pp. 6264–6270.

41. Yang, R. S., Chang, L. W., Wu, J. P., Tsai, M. H., Wang, H. J., Kuo, Y. C., Yeh, T. K., Yang, C. S., and Lin, P. (2007). Persistent tissue kinetics and redistribution of nanoparticles, quantum dot 705, in mice: ICP-MS quantitative assessment. *Environ. Health Perspect.*, **115**, pp. 1339–1343.

42. Lee, H. A., Leavens, T., Mason, S. E., Monteiro-Riviere, N. A., and Riviere, J. E. (2009). Comparison of quantum dot biodistribution with blood-flow limited physiologically-based pharmacokinetic model. *Nano Lett.*, **9**, pp. 794–799.

43. Lankveld, D. P. K., Ooman, A. G., Krystek, P., Neigh, A., Troost de Jong, A., Noorlander, C. W., Van Eijkeren, J. C. H., Geertsma, R. E., and de Jong, W. H. (2010). The kinetics of the tissue distribution of silver nanoparticles of different sizes. *Biomaterials*, **31**, pp. 8350–8361.

44. Xia, X. R., Monteiro-Riviere, N. A., Mathur, S., Oldenberg, S., Fadeel, B., and Riviere, J. E. (2011). Mapping the surface adsorption forces of nanomaterials in biological systems using the biological surface adsorption index (BSAI) approach. *ACS Nano*, **5**, pp. 9074–9081.

45. Xia, X. R., Monteiro-Riviere, N. A., and Riviere, J. E. (2010). An index for characterization of nanomaterials in biological systems. *Nat. Nanotechnol.*, 5, pp. 671–675.

Chapter 9

Immunotoxicity of Nanomaterials

Barbara Lettiero,[a] **Z. Shadi Farhangrazi,**[b] **and S. Moein Moghimi**[a]
[a]*Nanomedicine Research Group, and NanoScience Centre,*
Centre for Pharmaceutical Nanotechnology and Nanotoxicology,
University of Copenhagen, Copenhagen, Denmark
[b]*Biotrends International, Denver Technology Center,*
Greenwood Village, CO, USA
moien.moghimi@sund.ku.dk

9.1 Introduction

The immune system is a complex network of blood proteins, cells, and organs capable of identifying and neutralizing dangers in the body. The defensive mechanisms are orchestrated by both the innate and the adaptive arms of the immune system [1–3]. The innate immune system consist of cells and proteins that immediately identify, clear, and eliminate danger, primarily through pattern recognition, where the main components include circulating, interstitial, and tissue macrophages, dendritic cells, natural killer cells, a plethora of circulating plasma proteins, notably the components of the complement system, and membrane-bound complement regulators [1, 2]. Adaptive immunity comprises complex immune cell processes

Handbook of Safety Assessment of Nanomaterials: From Toxicological Testing to Personalized Medicine
Edited by Bengt Fadeel

ISBN 978-981-4463-36-2 (Hardcover), 978-981-4463-37-9 (eBook)
www.panstanford.com

and communications that diversify immune responses to danger [3]. These involve activation of natural killer cells, $CD8^+$ cytotoxic T- and $CD4^+$ T helper (T_H) cells. Uncommitted $CD4^+$ cells can be induced to differentiate toward T_H-1 (accompanied by secretion of interleukin 2 and interferon g for cell-mediated responses and activation of $CD8^+$ cytotoxic T-cells), T_H-2 (characterized by secretion of interleukins 4, 5, and 6 to help B-cells and induction of IgE and IgG4 in man but IgG1, IgE, and IgA in mice), T_H-17 (interleukin 17–expressing cells that are linked to early inflammation and autoimmunity), and T_{reg} (cells for immune silencing) phenotypes according to the local cytokine profile. The adaptive immune responses are initiated in secondary lymphoid organs: the lymph nodes, mucosal-associated lymphoid tissues, and the spleen. Compelling evidence strongly suggests that stimulation of the innate immunity plays an important role in the evolution of the adaptive immune responses [1, 3].

Nanoparticulate systems in the form of liposomes, polymeric nanospheres, gold nanoshells, etc., offer many diagnostic and therapeutic applications [4–8]. These entities, however, by virtue of their size, shape, and surface characteristics (e.g., display of structures with repeated domains) may resemble microorganisms. These properties make nanoparticles prone for interception by different components of the body's defences following entry into the body [4–8]. The type and extent of immune responses, however, will depend on nanoparticle physicochemical characteristics, dose, frequency of administration, and the portal of entry [4–6, 8–14]. Immune responses may comprise safe elimination (e.g., phagocytic clearance and destruction), immunostimulation, hypersensitivity, and immunosuppression [6, 8–12, 14–17] (Fig. 9.1). Immunostimulation is considered an antigen-nonspecific or uncontrolled activation of any component of the immune system. For example, nanoparticles may display adjuvant activity resulting in chronic inflammation, cytokine release syndrome, and tissue leukocyte infiltration. Hypersensitivity refers to a set of undesirable reactions produced by the immune system, which could be damaging and even fatal. There are four types of hypersensitivity reactions, but those of interest to nanoparticles include type I or allergy, which is IgE mediated, and T_H-1-associated delayed-type hypersensitivity [9–11]. In addition to these, nanomaterials may induce acute non-IgE-mediated allergic reactions with symptoms similar to type I hypersensitivity

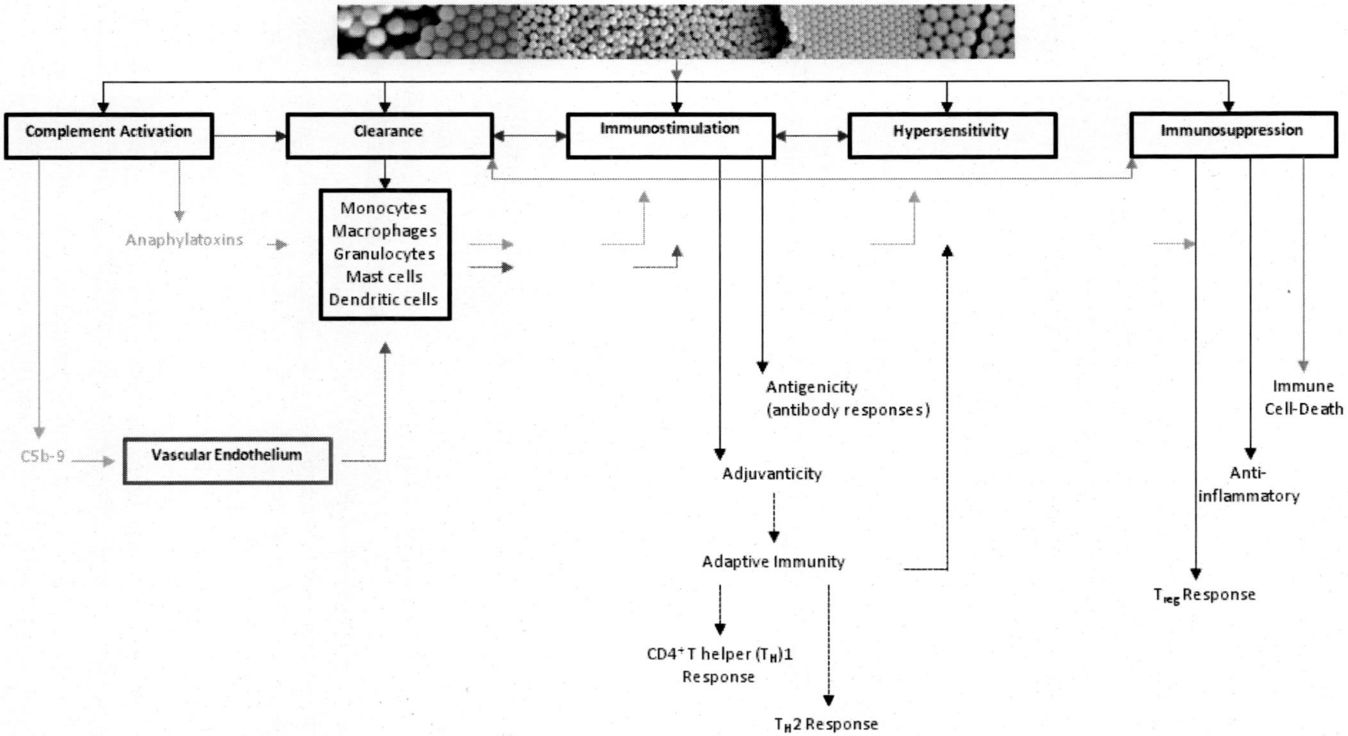

Figure 9.1 A simplified scheme of possible innate and adaptive immune responses to nanoparticles. The scheme also shows cross talk between innate and adaptive responses. Not all elements of the immune system are shown.

[12, 18, 19]. These reactions are not within the definition of standard hypersensitivity classification. Immunosuppression refers to impairment of any component of the immune systems, resulting in decreased immune functions and therefore impaired ability to cope with infections and cancer.

With increasing development of therapeutic and functional nanomaterials, and because of the high structural complexity of these entities and/or the use of peptides, proteins, or nucleic acid components as payload or targeting ligands [6], administered nanomedicines may induce a plethora of intended beneficial as well as uncontrolled adverse immune reactions. These concepts are briefly discussed where a better understanding of nanomaterial–immune system interactions could pave the way for the future of "safe-by-design" nanopharmaceuticals.

9.2 Nanoparticle Clearance by Immune Cells

Phagocytic cells are widely distributed and strategically placed in many tissues and in different body compartments and readily intercept nanoparticles [8]. In the vascular compartment, hepatic macrophages (Kupffer cells) are the predominant scavengers [8]. The task is also shared by other phagocytes in direct contact with blood, including circulating monocytes, macrophages in the marginal zone, and the red-pulp regions of the sinusoidal spleens and the persinal macrophages in the bone marrow–blood barrier [8, 20, 21]. In contrast to Kupffer cells, pulmonary intravascular macrophages are the main scavengers in pigs, goats, sheep, cats, and calves [22]. The intravenous route of entry is also a gateway for nanoparticle access to stromal and hematopoietic macrophages in the bone marrow (through transcellular and intercellular passage across the bone marrow sinus endothelium) [21] and interstitial macrophages at locations where the endothelial barrier is perturbed by inflammatory processes (e.g., atherosclerotic plagues) or by dysregulated angiogenesis (e.g., solid tumors) [4, 8, 23]. The interaperitoneal route of entry will present nanoparticles first to phagocytes in the peritoneal cavity, but some particles may reach macrophages in the lymph nodes through stomata in the diaphragm [8]. Subcutaneously exposed particles will be challenged by the local interstitial macrophages as well as dendritic cells (interstitial and dermal dendritic cells, Langerhans cells) [24], but a significant fraction of the particles will eventually drain into the initial lymphatic system and will be conveyed to the regional draining lymph nodes for extraction by macrophages located at the subcapsular floor and medullary sinuses [25, 26]. Lymph node sinus-resident immature dendritic cells may also participate in nanoparticle capture [24, 26]. On the other hand, alveolar macrophages are well placed to intercept intratracheally administered particles. Only a small fraction of nanoparticles may find themselves in close vicinity to phagocytic cells located in other organs such as the brain, gut, testis, synovial cavity, and eye, unless administered locally into these sites [8]. For example, in the gut some particles may gain access to macrophages and dendritic cells of the lamina propria through the M-cells of Peyer's patches [7, 27].

Macrophages and other phagocytes express an extensive receptor repertoire of phagocytosis receptors, including scavenger receptors, Toll-like receptors, mannose receptor, complement receptors, Fc receptors, tuftsin receptor, and Dectins [1, 2, 5, 24, 25, 27–29]. These receptors may recognize nanoparticle features either directly or through opsonisation processes in biological fluids (e.g., plasma, lymph, tears). Examples of opsonic molecules include complement proteins such as C1q, C3b, and iC3b, antibodies, and fibronectin [5, 6]. Particles of different physicochemical properties may attract a different array of opsonins [6]. Different opsonic proteins may also accumulate on the surface of a typical nanoparticle. These processes may determine which subpopulation of macrophages will host the particles and could indicate a recognition hierarchy phenomenon in phagocytic clearance [6]. Nevertheless, these natural defence strategies have offered rational approaches for selective delivery of therapeutic agents to macrophages with nanoparticulate drug carrier systems (e.g., polymeric micelles, polymeric nano- and microparticles, liposomes) (reviewed in Ref. [8]). Examples include delivery of antimicrobials (since macrophages serve as sites of proliferation of many pathogens), antigens, metabolic enzymes for enzyme replacement therapy (e.g., glucocerebrosidase), therapeutic nucleic acids, and toxins (for selective macrophage elimination).

Particulate targeting of macrophages and intracellular drug release processes can be optimized through modifications of the drug carrier physicochemical properties, including size, shape, composition, and surface characteristics [5, 6]. For example, nanoparticle targeting of macrophages may be improved through surface modification with macrophage receptor ligands such as mannose and tuftsin [8]. The macrophage phagocytic/endocytic pathway will direct nanoparticulate drug carriers to lysosomes, where local degradation processes will release the entrapped cargo from the carrier into phagosome-lysosome vesicles [8]. Efficient cargo delivery to cytosol, however, may be achieved by triggered release mechanisms in late endosomes. Examples include pH-sensitive and fusogenic drug carriers [4, 5]. On the other hand, nanoparticle surface modification with methoxypoly(ethylene glycol) (mPEG) can suppress opsonization processes (as in complement activation and fibronectin binding) and sterically prevent nanoparticle binding to the macrophage receptors [30]. Through appropriate nanoparticle engineering and together with an understanding of macrophage microheterogeneity, targeting may be aimed at a particular subset of macrophages. Examples include liposomes and polymeric nanospheres capable of targeting newly recruited monocytes in the liver [31, 32] and the splenic red-pulp macrophages [33], respectively. Through the control of nanoparticle surface features, such as the architectural arrangements of their projected polymers, the extent of nanoparticle presentation to both interstitial phagocytic cells and lymph node macrophages can also be controlled, as discussed elsewhere [26, 34].

Similarly, particle size plays an important role in the efficiency of cross presentation of exogenous antigens to dendritic cells [35]. For instance, antigens bound to 50 nm polystyrene nanoparticles home to late acidic (pH 4.5–5.0) endosomes and lysosomes in dendritic cells [35]. This results in rapid and unregulated antigen degradation and inefficient cross presentation. Dendritic cells, however, shuttle antigens bound to larger particles (500 nm) to early endosomal compartments with pH close to neutrality [35]. This preserves the majority of antigens for subsequent generation of peptides for presentation to the major histocompatibility complex I pathway.

In contrast to phagocytic processes, the neutrophil extracellular trap (NET) is a nonphagocytic mechanism that further plays a role in elimination of particulate intruders [36–38]. NETs are sticky filaments of DNA and proteins decorated with vesicles containing many proteases. NETs are released by dying polymorphonuclear phagocytes and are capable of physically entangling intruders such as microorganisms as well as man-made nanoparticles, notably the cationic species [38].

9.3 Nanoparticle Modulation of Immune Responses

Nanoparticle-mediated immune responses is a poorly understood set of complex interfacial and molecular processes that involve many elements of both innate and adaptive immunity, depending on nanoparticle physicochemical characteristics, dose, and the route of exposure [9–11]. For example, nanoparticle-mediated inadvertent activation of the complement system through any of the three complement initiation pathways may trigger consequential secondary responses with hemodynamic, respiratory, cutaneous, and subjective manifestations [12, 18, 19]. This is due to liberation of potent complement bioactive products (e.g., C3a, C5a, and C5b-9) with the ability to modulate the function of a variety of immune cells and vascular endothelial cells. Indeed, excessive production of C5a may down-regulate immune responses in some leukocytes, while overactivating other cell types [39, 40]. Triggering of mast cells and basophils by anaphylatoxins may lead to secretion of a cocktail of vasoactive meditors (e.g., histamine, thromboxanes, leukotrienes) and induce anaphylaxis and other undesirable effects [19]. On the other hand, C5b-9 complexes may elicit nonlytic stimulatory responses from vascular endothelial cells and modulate endothelial regulation of hemostasis and inflammatory cell recruitment [41]. Other complement activation products such as iC3b could induce up-regulation of certain adhesion molecules on neutrophils and endothelial cells [42]. Complement activation by nanoparticles—and its consequence—is discussed further in Chapter 9.

Direct interaction of nanoparticles with a variety of immune cells such as M1, M2, and regulatory macrophages, mast cells, basophils, eosinophils, neutrophils, natural killer cells, and different subsets of lymphocytes and dendritic cells may initiate many immunostimulatory or immunosuppressive reactions [5, 9–11]. Nanoparticle (e.g., various metallic and polymeric particles, carbon nanotubes) interaction and uptake by some immune cells may generate reactive oxygen species and trigger production of inflammatory cytokines, such as interleukin 1β, causing severe tissue damage [9–11]. Cytokine release by nanoparticles and potential immunotoxic effects can sometimes be ascribed to the type of surfactant used, presence of impurities, and instability (aggregation) of the formulation. Nanoparticles can also induce immunosuppression, either inadvertently through cytokine production (as in transforming growth factor β), T-cell impairment, and complement activation or intentionally through delivery of immunosuppressive agents [9–11].

Here we limit our discussion to the application of nanoparticles in therapeutics and medicine.

9.3.1 Immunostimulation

Two interesting immunostimulatory properties of nanoparticles in medicine are antigenicity and adjuvanticity.

9.3.1.1 Antigenicity

Antigenicity is usually referred to a specific antibody response. The immune response to a typical nanoparticle may result in generation of antibodies (IgG and IgM classes) against some components of a nanoparticle. The presence of these antibodies may affect nanoparticle stability as well as pharmacokinetics [5, 6, 43–45]. For example, there are many naturally occurring antibodies to phospholipid headgroups and cholesterol with different titers, specificity, and interindividual variations [43, 44]. The binding of such antibodies to liposomes may trigger complement activation, which in turn will generate the lytic complex C5b-9 for insertion into the liposomal bilayer [44, 45]. This could result in substantial leakage of vesicular encapsulated aqueous cargo before it reaches its cellular targets. Others have shown that repeated injection of long-circulating polyethylene glycol (PEG)ylated liposomes can generate anti-PEG antibodies, since PEG acts as a type 2 T-cell-independent antigen and elicits a potent IgM response by direct stimulation of B-cells [46]. Consequently, in the presence of circulating anti-PEG antibodies PEGylated liposomes will no longer exhibit prolonged circulation profiles and become prone to sequestration by hepatic and splenic macrophages as a result of antibody (IgM)-mediated complement fixation. Induction of anti-PEG antibodies can be prevented if PEGylated liposomes carry a cytotoxic drug such as doxorubicin [47]. This is due to doxorubicin-induced macrophage death and the inhibition of B-cell proliferation and/or killing of proliferated B-cells. In some cases, liposomal cargo, such as therapeutic nucleic acids, may act as a potent enhancer of immunogenicity. Antibody induction by therapeutic liposomes may also generate new complications such as infusion-related reactions or anaphylaxis [19].

Apart from liposomes, there are limited studies that have assessed antibody generation against nanoparticles. For example, there are inconsistent reports on antibody generation against fullerene nanoparticles [48–50]. While some reports have shown antibody generation against C_{60} and C_{70} fullerenes [48, 49], others did not detect fullerene antigenicity, even in the presence of a complete Freund's adjuvant [50]. These discrepancies may be accounted for by differences in fullerene surface properties and animal choice.

9.3.1.2 Adjuvanticity

Nanoparticles may act as immune potentiators or adjuvants triggering early innate immune responses that subsequently assist the generation of potent and persistent adaptive immune responses. Accordingly, a wide variety of nanoparticles have been used as adjuvants for vaccine formulations, particularly to enhance the immunogenicity of subunit vaccines, not only through antigen protection and targeting to antigen-presenting cells, but also through direct immunostimualtion [16, 24]. For instance, cationic liposomes are potent activators of the innate immunity when combined with nucleic acid agonists of endosomally located Toll-like receptors 3, 7, 8, and 9 [51]. Nanoparticle surface decoration with appropriate ligands against a plethora of plasma membrane and internal dendritic cell receptors can augment antigen delivery, signalling events and intracellular processing (reviewed in Ref. [24]). These include ligands that target DEC-205, mannose receptor MRC1/CD206, DC-SIGN (CD209), LOX-1, langerin (CD207), CLEC4A, DECTIN-2, DORA, and Fcg receptor type I. However, these receptors are differently expressed by distinct subsets of dendritic cells, and their expression may vary with the state of dendritic cell maturation. Adjuvanticity may also depend on the ability of particles to activate the complement system as in the case of liposomes, carbon nanotubes, and a variety of polymeric nanospheres and microspheres [12, 52–55]. Indeed, the complement split-product C3d can induce B-cell activation, whereas anaphylatoxins C3a and C5a may serve as "danger signals" [56]. C5a is also a chemoattractant to many immune cells [12, 42]. Furthermore, other complement components such as C1q may bind to anionic nanoparticles and induce dendritic cell maturation, activation (interleukin 12 and tumor necrosis factor alpha secretion), and elevation of their T-cell-stimulating capacity [12, 57].

Other potential adjuvants, such as chitosan and silica particles, as well as ISCOMS, may directly activate the NALP3 inflammasome complex (apoptosis-associated speck-like protein and caspase 1 protease), which in turn cleave and activate the immunostimulatory cytokine interleukin 1β [17, 24].

9.3.1.3 Allergenicity and hypersensitivity

Nanoparticle exposure may elicit allergic reactions, an area that is of immediate importance to occupational exposure of inorganic, organic, and composite nanomaterials and of relevance to proposed medical applications of certain multifunactional nanoparticles. For example, lung exposure to certain metal oxide nanoparticle (e.g., NiO, Co_3O_4) comprising soluble metal haptens induced chronic interstitial inflammation and pro-inflammatory T_H-1- and T_H-17-associated delayed-type hypersensitivity responses and pulmonary alveolar proteinosis in Wistar rats [58]. On the other hand, no inflammatory responses were observed following exposure to CuO and Cr_2O_3 [58]. Others have demonstrated that carbon nanotubes can promote T_H-2 allergic responses in mice [59]. For instance, both single- and multiwalled carbon nanotubes enhanced the allergenicity of ovalbumin when administered via intranasal or subcutaneous routes in BALB/c mice [59]. However, these observations must be interpreted cautiously since neither the animal model nor the allergen (ovalbumin) reflects the development of an allergic reaction in humans. Firstly, there is no direct evidence for the existence of discrete T_H-17 cells in humans [24]. Secondly, BALB/c mice are genetically prone to mount T_H-2 allergic responses [60, 61].

Acute allergic-like reactions, which are not initiated or mediated by pre-existing IgE antibodies, typically occur in 45% of human individuals within a few minutes of infusion of regulatory approved nanopharmaceuticals [18, 19]. Symptoms vary from light to severe and include flushing, skin rash, fever, chills, facial swelling, dyspnea, chest pain, and back pain. Typically, responses arise at the first exposure without prior sensitization and may lessen or disappear on later treatments. Although the molecular and genetic factors responsible for these events are still not clear, available evidence suggests that inadvertent activation of the complement system may be an important contributing factor in eliciting these reactions [18, 19]. Cardiopulmonary distress is a major physiologic consequence of complement-related acute allergic-like reactions to nanopharmaceuticals [19]. This would be of high concern in patients with cardiovascular abnormalities or where the target disease involves primary or secondary cardiac impairment.

9.3.2 Immunosuppression

Nanoparticle interception by cells of the immune system offers opportunities for intentional down-regulation of the immune response. One of the best examples is macrophage silencing and destruction in relation to many pathological conditions [8]. For instance, liposomal formulations of bisphosphonate drugs have been used for selective compartmental macrophage elimination in relation to rheumatoid arthritis, autoimmune blood disorders, spinal cord injury, restenosis, and T-cell-mediated autoimmune diabetes (reviewed in Ref. [8]). Other examples include betamethasone functionalized polymeric nanoparticles [62] and collagen type II entrapped in polymeric nanoparticles [63] that suppressed inflammation in rat and mouse models of arthritis, respectively. There are many examples of actively targeted nanoparticles to pathological endothelium for reducing leukocyte rolling and extravasation [64, 65].

Some studies have suggested that the immunogenicity of recombinant proteins may be reduced following their conjugation to nanoparticles [66]. For example, administration of the recombinant factor VIII (rFVIII) in more than 20% of hemophilia A patients triggered high titers of anti-rFVIII neutralizing antibodies [66]. Interestingly, subcutaneous administration of rFVIII associated with phosphatidylserine-containing liposomes strongly reduced the antibody response [67]. These observations may imply either the loss of recognition of rFVIII immunogenic sites by the immune cells following or an immunosuppressive role for phosphatidylserine [68]. Phosphatidylserine-incorporating liposomes may induce transforming growth factor β–mediated anti-inflammatory responses with reduction in the total amount of leukocytes [69].

Allergen-loaded nanoparticles have successfully been applied for induction of immune tolerance against environmental and food allergens as well as allergic encephalomyelitis (reviewed in Ref. [10]). Certain fullerene derivatives were further shown to inhibit type I hypersensitivity reactions to an allergen in a mouse model of anaphylaxis [70].

Conversely, therapeutic nanoparticles may induce unintentional immunosuppression, resulting in an impaired ability to cope with invading pathogens and cancer. For instance, nanoparticles carrying cytotoxic agents for cancer chemotherapy may be intercepted by Kupffer cells [71]. Following intracellular release, the cytotoxic cargo is capable of inducing macrophage destruction. Following their destruction, full restoration of Kupffer cells often takes two weeks [71]. Deficiency in the Kupffer cell number may increase the risk of bacteremia during this period.

Local complement activation may induce immunosuppressive effects. For instance, intratumoral complement activation was shown to help tumor growth and progression [72]. The process is believed to be due to the chemoattractant property of liberated C5a products, which recruits myeloid-derived suppressor cells into the tumor, deregulating or suppressing cytotoxic $CD8^+$ T-cells. These observations are of concern with the increasing use of long-circulating cancer nanomedicines and nanotechnologies that generate thrombogenesis in tumor vasculature [73]. Because of the leaky nature of tumor microvasculature, long-circulating cancer nanomedicines extravasate from the blood and accumulate passively at tumor interstitial spaces, where they may potentially trigger complement activation [4–6]. Indeed, the regulatory approved Doxil (a PEGylated liposome formulation containing doxorubicin) can activate the complement system through both classical and alternative pathways [12, 18, 19, 74]. Furthermore, cytotoxic drugs, delivered by nanoparticles, induce local apoptotic and necrotic cell death. Such cells also incite complement activation following properdin (a complement protein and stabilizer of the alternative pathway C3 convertase) binding to their plasma membrane [75]. Further complications may arise due to presence of intratumoral antitumor antibodies and tumor-resident blood clots. Cell-bound antibodies can trigger a complement through both classical and alternative pathways and generate lytic C5b-9 complexes, whereas extrinsic proteases such as kallikerin and thrombin can directly cleave C3 and C5 [42,73].

There are many other examples of unintentional immunosuppressive effects of nanoparticles, and these have been reviewed elsewhere [9, 10, 14, 69]. Of particular interest are carbon nanotubes, which have generated inconclusive results [14]. Again, these discrepancies may be ascribed to the large diversity of carbon nanotubes in terms of their geometry, aspect ratios, surface properties, associated impurities, the administered dose, and species differences.

9.4 Conclusions

Intentional targeting of nanoparticles to the immune system provides numerous opportunities for modulation of immune responses against many pathological conditions, as well as combating certain diseases and disorders of the immune system. Indeed, nanopharmaceuticals have already played significant roles in combating macrophage infections (through delivery of antimicrobials and immunomodulators) and in immunization protocols. A better understanding of biological and physicochemical factors that control nanoparticle interaction with different arms of the immune system is still required, so molecular and cellular targeting strategies involving nanoparticles may be optimized and adverse reactions eliminated [13]. Such knowledge would also be of importance for successful design and development of nanopharmaceuticals for targeting to sites other than the immune cells, thus overcoming immune system interception and initiation of adverse events such as immunogenicity and infusion-related allergic reactions [13]. Nevertheless, immunosafety profiling of nanoparticles and nanopharmaceuticals remains a key challenge. We are still in need of appropriate 3D cultures and animal models for better safety profiling and "structure–activity" mapping. As a result of the broad differences in nanoparticle properties, immunosafety assessments must be done on a case-by-case basis; extrapolation from one system to another must be done cautiously or avoided. Here, development of predictive medium- to high-throughput assays for infusion-related allergic reactions may further help with risk assessment and aid patient selection [11, 76].

Life-threatening and strongly disabling diseases such as chronic infections and cancer are responsible for more than 25% of all deaths in the Western world, and the incidences are increasing [77]. Due to the complexity of these diseases a broad spectrum of therapeutic interventions is necessary to save lives and to improve the quality of life. Disease progression is often associated with deficiencies in patient immune responses. Immune cell (dendritic and T-cell) therapy is showing great promise for patients where conventional therapies have achieved limited success [78–84]. However, the outcome of adaptive immune responses following in vivo administration of manipulated dendritic cells seems to depend on multiple factors, including the specific dendritic cell lineage, maturational stage, microenvironmental activation signals, and their favorable communication with T-cells. These factors cannot be controlled amicably by current clinical approaches. An additional challenge in the clinical use of ex vivo–generated dendritic cells is to ensure their proper migration from the point of injection to the scene of action in the lymph nodes, which is currently a strongly limiting factor for efficient dendritic cell therapy. Some of these areas are further poised for growth through nanoparticulate-based system interventions, which may aid precision immune cell manipulation and selection. For instance, dendritic cells lose the ability of cross presentation in the late stage of maturation, even though they retain their ability for antigen internalization. As pointed out earlier [35], there are strong indications that this may be associated with a change in the acidification of endosomes, which results in differences in antigen transfer to the cytosol in early and late stages of maturation. The fate of antigens presented on nanoparticles also seems to differ depending on particle size and acidification of the endosomes [35]; however, precise knowledge of how this is correlated to dendritic cell maturation is not known. Thus, there is a need to investigate differences in dendritic cell maturation and endosome acidification in relation to antigen presentation by multifunctional nanoparticle-based delivery systems.

Dendritic cells and T-cells both have tolerance-inducing potential, depending on the specific phenotype being induced. Tolerance is the very aim in treatment of autoimmune diseases, while it is highly detrimental to therapies directed toward clearance of cells. Induction of tolerance-supporting immune cells is closely linked to intercellular communication of dendritic cells and T-cells in compartments of the lymph nodes. Tolerance may have many causes, including low levels of antigen presentation by dendritic cells to T-cells, low levels of costimulatory signals during antigen presentation, overexpression of specific tolerance-inducing cell surface proteins, higher secretion of anti-inflammatory versus pro-inflammatory cytokines, and phenotypical instability during the expansion of T-cells. Some of these parameters may also be regulated with appropriate nanoparticle-based technologies [85]. For example, a very important part of cancer immunotherapy is breaking the immunological tumor tolerance, which is induced through T_{reg}. Desirable modulation of T_{reg} can perhaps be achieved through stimulation of Toll-like receptors 8 and 9 in the endocytic compartments. Here, combining nanoparticle surface functionalization with Toll-like receptor agonists together with agonists for tumor necrosis factor alpha and interferon receptors on the plasma membrane may prove beneficial. Silencing is another important tool for T_{reg} modulation, where appropriate particulate delivery systems (e.g., fusogenic liposomes) may transport to and release the desired small interfering RNA (siRNA) molecules in the cytoplasm.

Acknowledgments

S. Moein Moghimi acknowledges financial support by the Danish Agency for Science, Technology and Innovation (Det Strategiske Forskningsråd).

References

1. Dempsey, P. W., Vaidya, S. A., and Cheng, G. (2003). The art of war: innate and adaptive immune responses. *Cell Mol. Life Sci.*, **60**, pp. 2604–2621.

2. Medzhitov, R., and Janeway, Jr., C. (2000). Innate immunity. *New Engl. J. Med.*, **343**, pp. 338–344.

3. Delves, P. J., and Roitt, I. M. (2000). The immune system. *New Engl. J. Med.*, **343**, pp. 108–117.

4. Moghimi, S. M., Hunter, A. C., and Murray, J. C. (2005). Nanomedicine: current status and future prospects. *FASEB J.*, **19**, pp. 311–330.

5. Moghimi, S. M., Hunter, A. C., and Murray, J. C. (2001). Long-circulating and target-specific nanoparticles: theory to practice. *Pharmacol. Rev.*, **53**, pp. 283–318.

6. Moghimi, S. M., Hunter, A. C., and Andresen, T. L. (2012). Factors controlling nanoparticle pharmokinetics: an integrated analysis and perspective. *Annu. Rev. Pharmacol. Toxicol.*, **52**, pp. 481–503.

7. Hunter, A. C., Elsom, J., Wibroe, P. P., and Moghimi, S. M. (2012). Polymeric particulate technologies for oral drug delivery and targeting: a pathophysiological perspective. *Nanomed. Nanotechnol. Biol. Med.*, **8**, pp. S5–S20.

8. Moghimi, S. M., Parhamifar, L., Ahmadvand, D., Wibroe, P. P., Farhangrazi, Z. S., and Hunter, A. C. (2012). Particulate systems for targeting of macrophages: basic and therapeutic concepts. *J. Innate Immun.*, **4**, pp. 509–528.

9. Dobrovolskaia, M. A., and McNeil, S. E. (2007). Immunological properties of engineered nanomaterials. *Nat. Nanotechnol.*, **2**, pp. 469–478.

10. Zolnik, B. A., Gonzalez-Fernandez, A., Sadrieh, N., and Dobrovolskaia, M. A. (2010). Nanoparticles and the immune system. *Endocrinology*, **151**, pp. 458–465.

11. Boraschi, D., Costantino, L., and Italiani, P. (2012). Interaction of nanoparticles with immunocompetent cells: nanosafety considerations. *Nanomedicine (UK)*, **7**, pp. 121–131.

12. Moghimi, S. M., Andersen, A. J., Ahmadvand, D., Wibroe, P. P., Andresen, T. L., and Hunter, A. C. (2011). Material properties in complement activation. *Adv. Drug Delivery Rev.*, **63**, pp. 1000–1007.

13. Moghimi, S. M., Peer, D., and Langer, R. (2011). Reshaping the future of nanopharmaceuticals: ad iudicium. *ACS Nano*, **5**, pp. 8454–8458.

14. Andersen, A. J., Wibroe, P. P., and Moghimi, S. M. (2012). Perspectives on carbon nanotube-mediated adverse immune effects. *Adv. Drug Delivery Rev.*, **64**, pp. 1700–1705.

15. Pashine, A., Vaiante, N. M., and Ulmer (2005). Targeting the innate immune responses with improved vaccine adjuvants. *Nat. Med.*, **11**, pp. S63–S68.

16. O'Hagan, D. T., Singh, M., and Ulmer, J. B. (2004). Microparticles for the delivery of DNA vaccines. *Immunol. Rev.*, **199**, pp. 191–200.

17. Li, H., Willingham, S. B., Ting, J. P., and Re, F. (2008). Cutting edge: inflammasome activation by alum and alum's adjuvant effects are mediated by NLRP3. *J. Immunol.*, **181**, pp. 17–21.

18. Moghimi, S. M., Andersen, A. J., Hashemi, S. H., Lettiero, B., Ahmdvand, D., Hunter, A. C., Andresen, T. L., Hamad, I., and Szebeni, J. (2010). Complement activation cascade triggered by PEG-PL engineered nanomedicines and carbon nanotubes: the challenges ahead. *J. Controlled Rel.*, **146**, pp. 175–181.

19. Szebeni, J., Muggia, F., Gabizon, A., and Barenholz, Y. (2011). Activation of complement by therapeutic liposomes and other lipid excipient-based therapeutic products: prediction and prevention. *Adv. Drug Delivery Rev.*, **63**, pp. 1020–1030.

20. Moghimi, S. M. (1995). Mechanisms of splenic clearance of blood cells and particles: towards development of new splenotropic agents. *Adv. Drug Delivery Rev.*, **17**, 103–115.

21. Moghimi, S. M. (1995). Exploiting bone marrow microvascular structure for drug delivery and future therapies. *Adv. Drug Delivery Rev.*, **17**, pp. 61–73.

22. Winkler, G. (1988). Pulmonary intravascular macrophages in domestic animal species: a review of structural and functional properties. *Am. J. Anat.*, **181**, pp. 217–234.

23. Williams, H. J., Fisher, E. A., and Greaves, D. R. (2012). Macrophage differentiation and function in atherosclerosis: opportunities for therapeutic intervention? *J. Innate Immun.*, **4**, pp. 498–508.

24. Moghimi, S. M. (2009). The innate immune responses, adjuvants and delivery systems, in *Delivery Technologies for Biopharmaceuticals. Peptides, Proteins, Nucleic Acids and Vaccines*, eds. Jorgensen, L., and Nielsen, H. M. (Wiley, Chichester), pp. 113–127.

25. Gray, E. E., and Cyster, J. G. (2012). Lymph node macrophages. *J. Innate Immun.*, **4**, pp. 424–436.

26. Moghimi, S. M. (2010). Nanoparticle engineering for the lymphatic system and lymph node targeting, in *Polymer-Based Nanostructures. Medical Applications*, ed. Broz, P. (RSC, Cambridge), *RSC Nanosci. Nanotechnol.*, **9**, pp. 81–97.

27. Mowat, A. M., and Bain, C. C. (2011). Mucosal macrophages in intestinal homeostasis and inflammation. *J. Innate Immun.*, **3**, pp. 550–564.

28. Taylor, P. P., Martinez-Pomares, L., Stacey, M., Lin, H-H., Brown, G. D., and Gordon, S. (2005). Macrophage receptors and immune recognition. *Annu. Rev. Immunol.*, **23**, pp. 901–944.

29. Kanazawa, N. (2007). Dendritic cell immunoreceptors: C-type lectin receptors for pattern-recognition and signaling on antigen-presenting cells. *J. Dermatol. Sci.*, **45**, pp. 77–88.

30. Moghimi, S. M., and Szebeni, J. (2003). Stealth liposomes and long circulating nanoparticles: critical issues in pharmacokinetics, opsonization and protein-binding properties. *Prog. Lipid Res.*, **42**, pp. 463–478.

31. Moghimi, S. M., and Patel, H. M. (2002). Modulation of murine liver macrophage clearance of liposomes by diethylstilbestrol. The effect of vesicle surface charge and a role for the complement receptor Mac-1 (CD11b/CD18) of newly recruited macrophages in liposome recognition. *J. Controlled Rel.*, **78**, pp. 55–65.

32. Moghimi, S. M., Hedeman, H., Christy, N. M., Illum, L., and Davis, S. S. (1993). Enhanced hepatic clearance of intravenously administered sterically stabilized microspheres in zymosan-stimulated rats. *J. Leukoc. Biol.*, **54**, pp. 513–517.

33. Moghimi, S. M., Hedeman, H., Muir, I. S., Illum, L., and Davis, S. S. (1993). An investigation of the filtration capacity and the fate of large filtered sterically-stabilized microspheres in rat spleen. *Biochim. Biophys. Acta*, **1157**, pp. 233–240.

34. Moghimi, S. M. (2006). The effect of methoxy-PEG chain length and molecular architecture on lymph node targeting of immune-PEG liposomes. *Biomaterials*, **27**, pp. 136–144.

35. Tran, K. K., and Shon, H. (2009). The role of phagosomal pH on the size-dependent efficiency of cross-presentation by dendritic cells. *Biomaterials*, **30**, pp. 1356–1362.

36. Brinkmann, V., Reichard, U., Goosmann, C., Fauler, B., Uhlemann, Y., Weiss, D. S., Weinrauch, Y., and Zychlinsky, A. (2004). Neutrophil extracellular traps kill bacteria. *Science*, **303**, pp. 1532–1535.

37. Brinkmann, V., and Zychlinsky, A. (2007). Beneficial suicide: why neutrophils die to make NETs. *Nat. Rev. Microbiol.*, **5**, pp. 577–582.

38. Bartneck, M., Keul, H. A., Zwadlo-Klarwasser, G., and Groll, J. (2010). Phagocytosis independent extracellular nanoparticle clearance by human immune cells. *Nano Lett.*, **10**, pp. 59–63.

39. Huber-Lang, M. S., Younkin, E. M., Sarma, J. V., McGuire, S. R., Lu, K. T., Guo, R. F., Padgaonkar, V. A., Curnutte, J. T., Erickson, R., and Ward, P. A. (2002). Complement-induced impairment of innate immunity during sepsis. *J. Immunol.*, **169**, pp. 3223–3231.

40. Riedemann, N. C., Guo, R. F., Bernacki, K. D., Ruben, J. S., Laudes, I. J., Neff, T. A., Gao, H., Speyer, C., Sarma, V. J., Zetoune, F.S., and Ward, P. A.

(2003). Regulation by C5a of neutrophil activation during sepsis. *Immunity*, **19**, pp. 193–202.

41. Hattori, R., Hamilton, K. K., McEver, R. P., and Sims, P. J. (1989). Complement proteins C59-b induce secretion of high molecular-weight multimers of endothelial von Willebrand-factor and translocation of granule membrane-protein GMP-140 to the cell-surface. *J. Biol. Chem.*, **264**, pp. 9053–9060.
42. Markiewski, M. M., Nilsson, B., Ekdahl, K. N., Mollnes, T. E., and Lambris, J. D. (2007). Complement and coagulation: strangers or partners in crime? *Trends Immunol.*, **28**, pp. 184–192.
43. Alving, C. R., and Swartz, Jr., G. M. (1991). Antibodies to cholesterol, cholesterol conjugates and liposomes: implications for atherosclerosis and autoimmunity. *CRC Crit. Rev. Immunol.*, **10**, pp. 441–453.
44. Haxby, J. A., Gotze, O., Muller-Eberhard, H. J., and Kinsky, S. C. (1969). Release of trapped marker from liposomes by the action of purified complement components. *Proc. Natl. Acad. Sci. U S A*, **64**, pp. 290–295.
45. Moghimi, S. M., and Hunter, A. C. (2001). Recognition by macrophages and liver cells of opsonized phospholipid vesicles and phospholipid headgroups. *Pharm. Res.*, **18**, pp. 1–18.
46. Ishida, T., Kashima, S., and Kawada, H. (2008). The contribution of phagocytic activity of liver macrophages to the accelerated blood clearance (ABC) phenomenon of PEGylated liposomes in rats. *J. Controlled Rel.*, **126**, pp. 162–165.
47. Ishida, T., Atobe, K., Wang, X., and Kiwada, H. (2006). Accelerated blood clearance of PEGylated liposomes upon repeated injections: the effect of doxorubicin-encapsulation and high-dose first injection. *J. Controlled Rel.*, **115**, pp. 251–258.
48. Chen, B. X., Wilson, S. R., Das, M., Coughlin, D. J., and Erlanger, B. F. (1998). Antigenicity of fullerenes: antibodies specific for fullerenes and their characteristics. *Proc. Natl. Acad. Sci. U S A*, **95**, pp. 10809–10813.
49. Braden, B. C., Goldbaum, F. A., Chen, B.-X., Kirschner, A. N., Wilson, S. R., and Erlanger, B. F. (2000). X-ray crystal structure of an anti-Buckminsterfullerene antibody Fab fragment: biomolecular recognition of C(60). *Proc. Natl. Acad. Sci. U S A*, **97**, pp. 12193–12197.
50. Massalova, O. V., Shepelev, A. V., Atanadze, S. N., Parnes, Z. N., Romanaova, V. S., Vol'pina, O. M., Semiletov, Iu. A., and Kushch, A. A. (1999). Immunostimulating effect of water-soluble fullerene derivatives: perspective adjuvants for a new generation vaccine. *Dokl. Akad. Nauk.*, **369**, pp. 411–413.

51. Dow, S. W. (2008). Liposome-nucleic acid immunotherapeutics. *Exp. Opin. Drug Delivery*, **5**, pp. 11–24.

52. Hamad, I., Al-Hanbali, O., Hunter, A. C., Rutt, K. J., Andresen, T. L., and Moghimi, S. M. (2010). Distinct polymer architecture mediates switching of complement activation pathways at the nanosphere-serum interface: implications for stealth nanoparticle engineering. *ACS Nano*, **4**, pp. 6629–6638.

53. Reddy, S. T., van der Vlies, A. J., Simenoi, E., Angeli, V., Randolp, G. J., O'Niel, C. P., Lee, L. K., Swartz, M. A., and Hubbell, J. A. (2007). Exploiting lymphatic transport and complement activation in nanoparticle vaccines. *Nat. Biotechnol.*, **25**, pp. 1159–1164.

54. Moghimi, S. M., and Hunter, A. C. (2010). Complement monitoring of carbon nanotubes. *Nat. Nanotechnol.*, **5**, p. 382.

55. Andersen, A., Robinson, J. T., Dai, H., Andresen, T. L., Hunter, A. C., and Moghimi, S. M. (2013). Single-walled carbon nanotube surface control of complement recognition and activation. *ACS Nano*, **7**, pp. 1108–1119.

56. Dempsey, P. W., Allison, M. E., Akkaraju, S., Goodnow, C. C., and Fearon, D. T. (1996). C3d of complement as a molecular adjuvant: bridging innate and acquired immunity. *Science*, **271**, pp. 348–350.

57. Csomor, E., Bajtaya, Z., Sandora, N., Kristofa, K., Arlaudc, G. J., Thield, S., and Erdei, A. (2007). Complement protein C1q induces maturation of human dendritic cells. *Mol. Immunol.*, **44**, pp. 3389–3397.

58. Cho, W.-S., Duffin, R., Bradley, M., Megson, I. L., MacNee, W., Howie, S. E. M., and Donaldson, K. (2012). NiO and Co_3O_4 nanoparticles induce lung DTH-like responses and alveolar lipoprotenosis. *Eur. Resp. J.*, **39**, pp. 546–557.

59. Nygaard, U. C., Hansen, J. S., Samuelsen, M., Alberg, T., Marioara, C. D., and Lovik, M. (2009). Single-walled and multi-walled carbon nanotubes promote allergic immune responses in mice. *Toxicol. Sci.*, **109**, pp. 113–123.

60. Watanabe, H., Numata, K., Ito, T., Takagi, K., and Matsukawa, A. (2004). Innate immune response in Th1- and Th2-dominant mouse strains. *Shock*, **22**, pp. 460–466.

61. Chen, X., Oppenheim, J. J., and Howard, O. M. Z. (2005). BALB/c mice have more $CD4^+CD25^+$ T regulatory cells and shoe greater susceptibility to suppression of their $CD4^+CD25$ responder T cells than C57BL/mice. *J. Leukoc. Biol.*, **78**, pp. 114–121.

62. Higaki, M., Ishihara, T., Izumo, N., Takatsu, M. And Mizushima, Y. (2005). Treatment of experimental arthritis with poly(D,L-lactic/glycolyic

acid) nanoparticles encapsulating betamethasone sodium phosphate. *Ann. Rheum. Dis.*, **64**, pp. 1132–1136.

63. Kim, W. U., Lee, W. K., Ryoo, J. W., Kim, S. H., Kim, J., Youn, J., Min, S. Y., Bae, E. Y., Hwang, S. Y., Park, S. H., Cho, C. S., Park, J. S., and Kim, H. Y. (2002). Suppression of collagen-induced arthritis by single administration of poly(lactic-co-glycolic acid) nanoparticles entrapping type II collagen: a novel treatment strategy for induction of oral tolerance. *Arthritis. Rheum.*, **46**, pp. 1109–1120.

64. John, A. E., Lukacs, N. W., Berlin, A. A., Palecanda, A., Bargatze, R. F., Stoolman, L. M., and Nagy, J. O. (2003). Discovery of a potent nanoparticle P-selectin antagonist with anti-inflammatory effects in allergic airway disease. *FASEB J.*, **17**, pp. 2296–2298.

65. Dianzani, C., Cavalli, R., Zara, G. P., Gallicchio, M., Lombardi, G., Gasco, M. R., Panzanelli, P., and Fantozzi, R. (2006). Cholesteryl butyrate solid lipid nanoparticles inhibit adhesion of human neutrophils to endothelial cells. *Br. J. Pharmacol.*, **148**, pp. 648–656.

66. Ehrenforth, S., Kreuz, W., Scharrer, I., Linde, R., Funk, M., Güngör, T., Krackhardt, B., and Kornhuber, B. (1992). Incidence of development of factor VIII and factor IX inhibitors in haemophiliacs. *Lancet*, **339**, pp. 594–598.

67. Romani, K., Miclea, R. D., Purohit, V. S., Mager, D. E., Straubinger, R. M., and Balu-Iyer, S. V. (2008). Phosphatidylserine containing liposomes reduce immunogenicity of recombinant human factor VIII (rFVIII) in a murine model of hemophilia. *J. Pharm. Sci.*, **97**, pp. 1386–1389.

68. Hoffmann, P. R., Kench, J. A., Vondracek, A., Kruk, E., Daleke, D. L., Jordan, M., Marrack, P., Henson, P. M., and Fadok, V. A. (2005). Interaction between phospatidylserine and the phosphatidylserine receptor inhibits immune responses *in vivo*. *J. Immunol.*, **174**, pp. 1393–1404.

69. Peer, D. (2012). Immunotoxicity derived from manipulating leukocytes with lipid-based nanoparticles. *Adv. Drug Delivery Rev.*, **64**, pp. 1738–1748.

70. Ryan, J. J., Bateman, H. R., Stover, A., Gomez, G., Norton, S. K., Zhao, W., Schwartz, L. B., Lenk, R., Kepley, C. L. (2007). Fullerene nanomaterials inhibit the allergic response. *J. Immunol.*, **179**, pp. 665–672.

71. Daemen, T., Hofstede, G., Kate, M. T. T., Bakker-Woudenberg, J. A. J. M., and Scherphof, G. (1995). Liposomal doxorubicin-induced toxicity: depletion and impairment of phagocytic activity of liver macrophages. *Int. J. Cancer*, **61**, pp. 716–721.

72. Mrkiewski, M. M., DeAngelis, R. A., Benencia, F., Ricklin-Lichtsteiner, S. K., Koutoulaki, A., Gerard, C., Coukos, G., and Lambris, J. D. (2008).

Modulation of the antitumor immune response by complement. *Nat. Immunol.*, **9**, pp. 1225–1235.

73. Moghimi, S. M., and Andresen, T. L. (2009). Complement-mediated tumour growth: implications for cancer nanotechnology and nanomedicines. *Mol. Immunol.*, **46**, pp. 1571–1572.
74. Moghimi, S. M., Hamad, I., Andresen, T. L., Jorgensen, K., and Szebeni, J. (2006). Methylation of the phosphate oxygen moiety of phospholipid-methoxy (polyethylene glycol) conjugates prevents PEGylated liposome-mediated complement activation and anaphylatoxin production. *FASEB J.*, **20**, pp. 2591–2593.
75. Xu, W., Berger, S. P., Trouw, L. A., de Boer, H. C., Schlagwein, N., Mutsaers, C., Daha, M. R., and van Kooten, C. (2008). Properdin binds to late apoptotic and necrotic cells independently of C3b and regulates alternative pathway complement activation. *Proc. Natl. Acad. Sci. U S A*, **180**, pp. 7613–7621.
76. Moghimi, S. M., Wibroe, P. P., Helvig, S. Y., Farhangrazi, Z. S., and Hunter, A. C. (2012). Genomic perspectives in inter-individual adverse responses following nanomedicine administration: the way forward. *Adv. Drug Delivery Rev.*, **64**, pp. 1385–1393.
77. Jemal, A., Siegel, R., Ward, E., Murray, T., Xu, J., and Thun, M. J. (2007). Cancer statistics. *CA Cancer J. Clin.*, **57**, pp. 43–66.
78. Mills, K. H. (2004). Regulatory T cells: friend or foe in immunity to infection? *Nat. Rev. Immunol.*, **4**, pp. 841–855.
79. Curiel, T. J., Coukos, G., Zou, L., Alvarez, X., Cheng, P., Mottram, P., Evdemon-Hogan, M., Conejo-Garcia, J. R., Zhang, L., Burow, M., Zhu, Y., Wei, S., Kryczek, I., Daniel, B., Gordon, A., Myers, L., Lackner, A., Disis, M. L., Knutson, K. L., Chen, L., and Zou, W. (2004). Specific recruitment of regulatory T cells in ovarian carcinoma fosters immune privilege and predicts reduced survival. *Nat. Med.*, **10**, pp. 942–949.
80. Steinman, R. M., and Banchereau, J. (2007). Taking dendritic cells into medicine. *Nature*, **449**, pp. 419–426.
81. Dudley, M. E., and Rosenberg, S. A. (2007). Adoptive cell transfer therapy. *Semin. Oncol.*, **34**, pp. 524–531.
82. Caspi, R. R. (2008). Immunotherapy of autoimmunity and cancer: the penalty for success. *Nat. Rev. Immunol.*, **8**, pp. 970–976.
83. Chaudhuri, D., Suriano, R., Mittelman, A., and Tiwari, R. K. (2009). Targeting the immune system in cancer. *Curr. Pharm. Biotechnol.*, **10**, pp. 166–184.

84. Engell-Noerregaard, L., Hansen, T. H., Andersen, M. H., Thor Straten, P., and Svane, I. M. (2009). Review of clinical studies on dendritic cell-based vaccination of patients with malignant melanoma: assessment of correlation between clinical response and vaccine parameters. *Cancer Immunol. Immunother.*, **58**, pp. 1–14.

85. Jia, L., Kovacs, J. R., Zheng, Y., Shen, H., Gawalt, E. S., and Meng, W. S. (2008). Expansion of Foxp3-expressing regulatory T cells in vitro by dendritic cells modified with polymeric particles carrying a plasmid encoding interleukin-10. *Biomaterials*, **29**, pp. 1250–1261.

Chapter 10

Complement Activation by Nanomaterials

Janos Szebeni
Nanomedicine Research and Education Center, Semmelweis University, Budapest, Hungary
Jszebeni2@gmail.com

10.1 Introduction

It is often emphasized that a main problem in the field of nanomedicine is the unusual and unpredictable toxicities of nanoparticles (NPs), which stem from the same unique features that land these particles useful, medically exploitable functionalities. These include their large surface area, intensive interactions, unusual shapes, and increased transcapillary and transmembrane mobilities. In fact, progress in nanotoxicology, an empirical subdiscipline in the cross section of nanotechnology, toxicology, and pharmacology, represents today one of the main rate-limiting factors in the translation of nanotechnology to medical practice [1–12] (and refer to Chapter 2 for an introduction to nanotoxicology).

Handbook of Safety Assessment of Nanomaterials: From Toxicological Testing to Personalized Medicine
Edited by Bengt Fadeel

ISBN 978-981-4463-36-2 (Hardcover), 978-981-4463-37-9 (eBook)
www.panstanford.com

This chapter focuses on one particular nanotoxicity phenomenon: complement (C) activation by NPs that are used in therapy and diagnosis, that is, in nanomedicine. The C system is part of the immune system, and immune toxicity is one of the organ toxicities that are relatively unexplored for most NPs (see Chapter 8). The particular clinical relevance of inadvertent C activation by nanomedicines lies in the risks for triggering a hypersensitivity reaction, called C activation-related pseudoallergy (CARPA), and for enhancing immunogenicity, which may interfere with the clinical utility of nanomedicines, turning them into potentially dangerous vaccines. In addition, the direct C attack on nanomedicines may interfere with their efficacy by causing membrane damage and early clearance.

10.2 Complement Activation: An Overview

Considering the huge number of NP types produced by nature and man, all with unique chemical and surface properties, the diverse exposure routes, and the different organs and cells afflicted, the spectrum of abnormal NP effects is very broad, if not incomprehensible. There are numerous ways by which adverse NP effects can be categorized or classified, depending on which type of NP is examined, how they are exposed to the body, and which target organs, cells, or molecules are afflicted. The number of possible permutations of these different possibilities, in the order of hundreds, gives an idea of the potential alternatives for health risks by NPs. As for immune toxicity, this adverse effect can be branched according to the direction, timing, and duration of changes of afflicted immune cells and molecules. CARPA caused by nanomedicines represents a pseudoallergic (non-IgE-mediated) acute and transient stimulation of both the soluble and the cellular arms of innate immunity (Fig. 10.1).

10.3 Complement Activation by Nanoparticles

The C system [13, 14] represents the innate, humoral arm of the immune system, a network of some 36 glycoproteins in blood (~5%

of serum globulin) and cell surfaces (Table 10.1). Its sequential, highly regulated activation plays a key role in protecting the body from infectious agents and other environmental and inner harms. Although C was discovered 118 years ago (by Jules Bordet), even today scientists find startling new physiological roles for the cascade, most recently in conception, tissue repair, and growth [15]. Pathologic alterations of the C system have been associated with a long list of illnesses [15, 16].

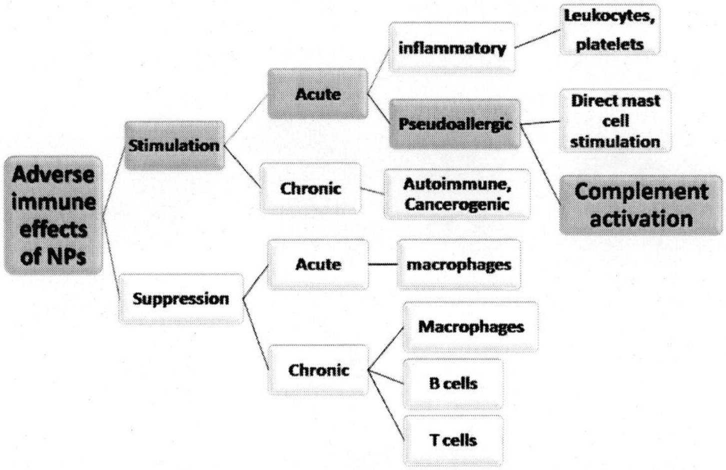

Figure 10.1 Types and routes of immune toxicity.

Pathogen- or NP-induced C activation usually starts on the surface of the agent, at the "nano–bio interface," via three independent pathways called classical pathway (CP), alternative pathway (AP), and lectin pathway (LP). The subsequent molecular and cellular changes (Fig. 10.2) can be detrimental to both the C activator and the host, via the adverse processes mentioned in the Introduction section and specified in Fig. 10.3.

As discussed later in more detail, the mechanisms by which different nanomedicines or drug carrier nanosystems trigger C3 conversion include all three activation pathways, that is, CP, AP, and LP. The process is the same as with natural pathogens, raising the question, why are nanomedicines not distinguished from harmful agents by the immune system? One likely explanation is that most nanomedicines are very similar in size, shape, and surface properties to viruses, against which the immune system developed effective protective strategies over millions of years of evolution. A remarkable example of this similarity is the morphologic resemblance of liposomal doxorubicin (Doxil) to HIV [17, 18]. It should be noted in this regard that the similarity of liposomes to viruses in the size range of 50–200 nm is the cornerstone of the concept that liposomes represent the "archetype" cell, the simplest basic supramolecular organizations from which biomembranes and, hence, life has evolved [19–21].

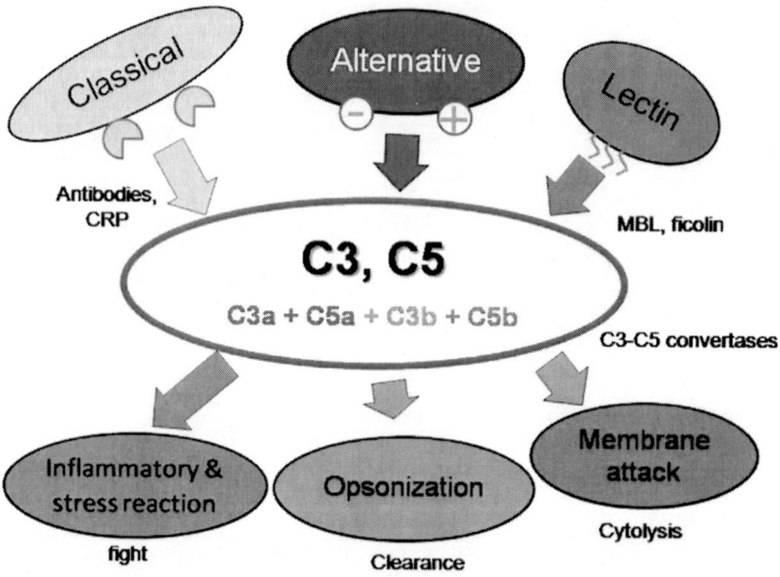

Figure 10.2 The C cascade. The central step, conversion of C3 to C3a and C3b, is enabled by the buildup of C3 convertase via three different pathways, and the by-products of this conversion perform four different effector functions for the elimination of activator pathogens or other foreign particles. The recognition molecules for CP activations are IgG, IgM, C1q, rC1r, C1s, and CRP; for AP activation C3, IgG, and properdin; and for LP activation MBL, MASP1, MASP2, and ficolin. Further details of these processes are the subject of a vast literature on C [16, 33].

Table 10.1 Complement proteins

Fluid-phase		Membrane bound	Active split product
Cascade	**Control**		
C1q	fB, B	CR1(CD35)	C2a
C1r	fD, D	CR2 (CD21)	C3a
C1s	fI, I	CR3,(CD11b/CD18 (Mac-1)	C5a
C2	fH, H, ß1H	CR4 (CD11c/CD18)	C3b
C3	fP, P, Properdin	C3aR	iC3b
C4	C1inh, C1-esterase	C5aR (CD88)	C3d
C5	C4bp, C4 binding protein	fHR	C3dg
C6	Protein S (vitronectin)	MCP (CD46)	
C7		DAF (CD55)	
C8		Protectin (CD59)	
C9		C8bp (homologous restriction factor	
MBL		Clusterin	
MASP-1		C1qR	
MASP-2			
ficolin			

Abbreviations: MBL, mannose-binding lectin, C4bp, C8bp, C4-C8 binding proteins.

The second explanation is the lack of C-binding molecules and/or structural features on the surface on NPs that normally protect the eukaryotic self and cancer cells and infectious pathogens (bacteria, viruses, and parasites) from a C attack. Table 10.2 lists some of these C-evading molecules and proteins, whose abundance and variety in nature [22–25] shed light on the biological relevance of C activation on surfaces. Once nanotechnology learns how to use C evasion molecules in drug carrier nanosystems, we will get closer to reducing the risks (Fig. 10.3) of nanomedicine-induced C activation.

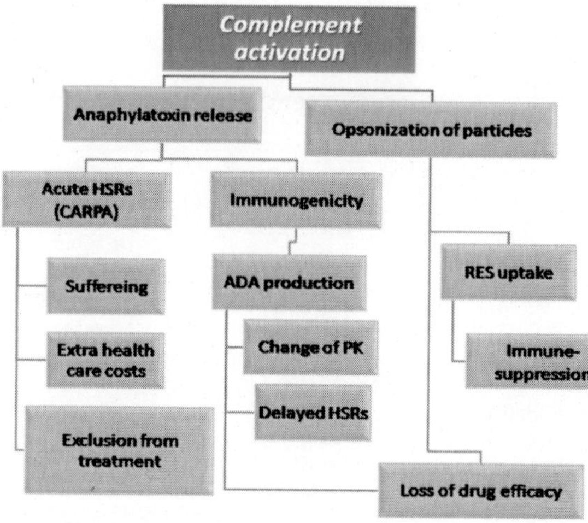

Figure 10.3 Hierarchy of adverse effects of C activation by NPs. The liberation of anaphylatoxins can lead to a severe, potentially deadly inflammatory reaction (CARPA), the associated enhancement of specific immunity can lead to immunogenicity, and the C system's eliminatory attack against the activating particle can interfere with the efficacy of nanomedicines.

Table 10.2 Complement inhibitor membrane proteins of eukaryotic cells and pathogens used for immune evasion

C inhibitor protein	Expressing cell/ pathogen	Mechanism	References
	Eukaryotic Cells		
C receptor type I (CR1, CD35	erythrocytes, monocytes, neutrophils , B and T cells mast cells and glomerular podocytes	binding and cofactor activity for inactivation of C3b and C4b by factor I	[43]
membrane cofactor protein (MCP), CD46	blood cells, sperm and cells at the maternal-fetal interface;	binding and cofactor activity for inactivation of C3b and C4b by factor I	[44]
decay accelerating factor (DAF)	hematopoietic and non-hematopoietic cells	prevents the assembly of the C3bBb complex	[45]

C inhibitor protein	Expressing cell/ pathogen	Mechanism	References
protectin (CD59)	hematopoietic and non-hematopoietic cells, endothel cells, HIV, cytomegalovirus, vaccinia	binds C5b678 and prevents further C9 binding and polymerization	[46, 47]
		Bacteria	
metalloproteinase aureolysin	Staphylococcus aureus	cleaves C3	[48]
extracellular fibrinogen binding protein (EfbP)		blocking deposition of C3	[49]
chemotaxis inhibitory protein (CHIPS)		blocks C5aR	[48, 50–54]
Staphylococcal complement inhibitor (SCIN)		prevents C3b deposition and phagocytosis	[55]
surface-associated protein (SdrE)		binds fH	[56]
alkaline protease (AprA)		blocks C3b deposition, cleaves C2	[57]
collagen binding adhesins (MSCRAMMS)	Gram-positive bacteria	inhibits CP activation	[58]
surface protein A (NspA)	serogroup B Neisseria meningitidis, Neisseria gonorrhoeae, Haemophilus influenzae	interferes with C3b deposition, inhibits AP activation	[59–61]
factor H-binding protein (fHbp), sialylated lipooligosaccharide (LOS) YadA	Yersinia enterocolitica	binds factor H,	[62]
urease	*Helicobacter pylori*	degrade bound C3b	[63]

(*Continued*)

Table 10.2 (*Continued*)

C inhibitor protein	Expressing cell/ pathogen	Mechanism	References
O-antigen	Brucella abortus	Inhibits C1q binding	[64]
protein Rck	Salmonella typhimurium	inhibits C9 polymerization	[65]
M protein	group A and B streptococci	binds factor H	[66]
streptococcal C5a inactivator		cleaves 6 amino acids from C5a	[67]
Viruses			
	VCP Vaccinia binds C3		
IMP	Cowpox		
SPICE	Variola		
gC-1, and HVSCD59	HSV	binds to C3b	[68]
gC-2, HVSCCPH	HSV		
glycoprotein C (gC)	herpesviruses-1 and HSV-2		
glycoprotein III (gIII)	pseudorabies virus		[69]
Parasites			
SCIP-1, paramyosin	Schistosoma mansoni parasite	binds to C8-9, inhibits MAC formation	[70, 71]
m28 membrane serine protease		cleaves iC3b	[72]
P70 protein		binds C3b	

It is important to note about membrane-bound natural C inhibitors that a part of them, those on the surface of eukaryotic cells (CR1, MCP, DAF, CD59), contain tandem repeating motifs of some 60 amino acids termed "short consensus repeats" (SCRs), or "Sushi" domains [26]. It has been established that a minimum of three such domains are able to inhibit C activation via binding C3b and C4b. Attempts to formulate a membrane-binding derivative of SCR1's first three Sushi domains yielded a drug candidate against a range of C-mediated diseases (Mirococept, APT070) [27–32].

Taken together, the above-discussed C-evading molecules, Mirococept, Soliris (a C5 antagonist monoclonal antibody already in the market [34]), and several other C inhibitors under research an development today [23, 24, 35, 36] represent a new class of anti-inflammatory drugs that could also be utilized against the adverse consequences of nanomedicine-induced C-activation. However, (re)producing C evasion molecules as soluble or membrane-bound C inhibitors is not the only goal in the pharmaceutical application of these molecules, as inhibiting C evasion by pathogens is also a potentially useful application in strengthening the body's immune response against these agents. Accordingly, C evasion molecules are also important molecular targets in anticancer and anti-inflammatory chemotherapy [22–24, 37, 38].

10.4 Mechanisms of Complement Activation by Nanoparticles

In theory, all NPs can activate the C system whose surface area is large enough to allow the deposition of C3 convertase or that binds C3 directly or indirectly. Nevertheless, only a few NP types are known to cause C activation, including liposomes, micelles, dendrimers, polymeric micelles, and carbon nanotubes (CNTs).

10.4.1 Complement Activation by Liposomes

The information available on C activation by NPs is by far most extensive in the case of liposomes. It was discovered as early as in the late sixties [39–42] that liposomes, as a membrane model, can be used to study the mechanism of C activation on cell membranes. A great number of subsequent studies, utilizing different experimental setups, different endpoints, liposome membrane constituents, fluid phase triggers, and serum/plasma/blood ratios (Table 10.3) led to some general conclusions as outlined below.

Table 10.3 Experimental variables in studies on C activation by liposomes

End point/ Readout of C activation	C consumption; CH50; liberation of C split products in the medium (C3a, C5a, C4d, SC5b-9, Bb); Release from the vesicles of glucose, galactose, enzymes, fluorophores (CF), isotopes (Rb^{86}), spin labels (TCC)
C binding membrane constituents	Cardiolipin; CBS; Cetylmannoside; DCP; DTP-DPPE; GalCer; gangliosides; lysolecithin; N-Ac-glucosamine; paragloboside; PE-conj. haptens; PGPI; PI, PIP; PIP2; PS-PS; SPDP-PE-IgG; TNP-Cap-DPPE; TNP-PE; Torpedo AChR; CRP

Abbreviations: Ab, antibody; ABA-Tyr-, azobenzenarsonate-tyrosyl-; AChR, acetylcholine receptor; CBS, cerebroside sulfate (acidic glycolipid); CF, carboxyfluorescein; CRP, C reactive protein; DCP, dicetyl phosphate; DNP-, dinitrophenyl-; DPPE, dipalmitoylphosphatidylethanolamine; DTP-, dithiopyridyl-; Gal, galactose; GalNac, n-acetylgalactosamine; MB-, maleimidobenzoyl-; nat., naturally occurring; NHS, normal human serum; PG, phosphatidylglycerol; PI, phosphatidylinositol; PIP, phosphatidylinositolphosphate; PIP_2, PI, phosphatidylinositol-diphosphate; spec., specific; SPDP, N-hydroxysuccinimidyl 3-(2-pyridyldithio)propionate; TCC, tempocholine chloride; TNP-, trinitrophenyl-; TNP-Cap-, TNP-aminocaproyl-. Modified from Ref. [73].

Studies on liposomal C activation revealed numerous experimental conditions and physicochemical factors influencing the process. Thus, C activation by different liposomes was shown to depend on the (phospho)lipid composition and resultant fluidity and phase transition of the bilayer membrane; its surface features (curvature, presence of charge, coating layers, ligands); the size, shape, and homogeneity (polydispersity) of vesicles; and, finally, the contamination of the preparation with lipopolysaccharide (LPS) and aggregated components (proteins, lipids, etc.). Complement activation by liposomes is concentration dependent; however, in addition to the (phospho)lipid level in the medium, the dilution of serum or plasma is also critical. Among these variables the presence of membrane surface electric charges, whether positive or negative, enhances C activation, as do large size, homogeneicity, the high cholesterol content (above 50%) of vesicles, their LPS contamination, and the presence of aggregates [73, 75, 76].

The pathways by which liposomes activate C also vary, depending on experimental system and conditions. Liposomal C activation can proceed via two basic pathways (classical and alternative), as well as through two additional "atypical" mechanisms (Fig. 10.4). CP activation is typically induced by negatively charged liposomes, and it involves the binding of naturally occurring IgG and IgM antibodies to phospholipid head groups and the hydroxyl moiety of cholesterol [77–79]. Additional epitopes for antiphospholipid antibody binding are adducts of oxidized phospholipid and beta2 glycoprotein 1 and liposome-adsorbed plasma proteins, for example, apolipoprotein H. [80]. Liposomes can further trigger CP activation through direct binding of C1q, which is promoted by a negative surface charge [81], or through complement regulatory protein (CRP) binding [82, 83]. The subsequent activation of the C1 complex, C2, C4, and C3, leads to the formation of C3 and C5 convertases.

AP activation of C, when it is the primary process and not the amplification loop, can be induced by positively charged liposomes, for example, stearylamine or 1,2-bis(oleoyloxy)-3-(trimethylammonio)propane containing vesicles [74]. In this case, formation of the C3 convertase is triggered by nonspecific binding to the membrane of C3b-C3b-IgG or intact C3.

The unconventional activation sequences include 1) CP activation, which proceeds without the involvement of antibodies, through direct or through CRP-mediated binding of C1q to the vesicles (Fig. 10.4C), and 2) AP activation, which is initiated and/or augmented by antibodies. Activation of the terminal sequence (C5b, C6, C7, C8, C9), leading to the formation of a membrane attack complex (MAC), is identical with all liposomes. Certain liposomes can activate C via the LP, for example, phosphatidylinositol-containing liposomes [84] and mannose- or other carbohydrate-decorated liposomes, depending on the surface density of the sugar [76]. Nonspecific adsorption of abnormally glycosylated proteins, such as apolipoprotein variants and IgG-G0, may further aid LP activation [76].

In addition to the large number of protein interactions involved in C activation by liposomes, what lends these studies particularly high individual variation is the variable surface area on the membranes available for protein deposition and the competition of C proteins against each other for membrane-binding sites (e.g., C3b, C1q, CRP, C4b). Finally, it should be emphasized that C proteins are heat labile, and the conditions of their storage, freezing, and thawing also have major impacts on experimental results in this field.

10.4.2 Complement Activation by Micelles

Micelles, that is, assemblies of amphiphilic molecules into 8–30 nm solid core-shell structures, are widely used to dissolve and deliver water-insoluble drugs, nutrients, or other agents into the body. Micelles can be ionic or nonionic, and they can contain natural, semisynthetic, or synthetic lipids or surfactant polymers. Among the semisynthetic micellar systems the best known perhaps is Cremophor EL (CrEL), which is polyethoxylated castor (ricinus) oil containing a large variety of polyethylene glycols (PEGs) and amphiphilic polyethoxylated glycerols, polyethoxylated fatty acids (mostly ricinoleic acid) and polyethoxylated glycerol esters differing in acyl chain and/or polyethyleneoxide length. Another similar product is Emulphor, that is, polyethoxylated vegetable oil. Synthetic surfactant polymers include nonionic block copolymers, such as poloxamer 188 (Fluosol-DA), polysorbate 80 (PS80), polysorbate 20 (PS20), and sorbitan monooleate 80.

In previous studies we have investigated C activation by CrEL as a likely cause of hypersensitivity reactions (HSRs) paclitaxel (Taxol®), one of the best-known anticancer drugs. Complement activation was shown to involve both the CP and the AP, as was shown for PS80 and PS20 [85]. Nevertheless, considering the study of Perkins and Sim (1986) showing the radius (of gyration, R_G) of C3 to be 5.2 nm, with molecular dimensions of 18 × 2 × 10 nm [86], it is very difficult to explain how the 8–22 nm CrEL (Fig. 10.4A) and other similar micelles bind and provide surface for the deposition of C3 convertases (C4b2a, C3bBb), not to talk about the much bigger C5 convertases (C2a4bC3b, C3bBbC3b). Although theoretically it cannot be excluded that micelles cause C scissions without convertase formation on surfaces, the identification of 50–300 nm lipid-like microdroplets in human plasma incubated with CrEL (Fig. 10.4B), taken together with information on the massive transformation of lipoproteins in patients treated with Taxol, with the formation of positively charged sudanofilic (lipid containing) microdroplets [87], as well as with lipid exchange between micelles and lipoproteins following their interaction in blood, it was hypothesized that CrEL micelles interact with plasma lipoproteins to form polyoxyethylen-coated mixed micelle–lipoprotein complexes, which bind C3 similarly to the block copolymer surfactants (L101, L121) adhered to oil droplets [88] (Fig. 10.4C). To the author's knowledge, however, no study has yet proved or disproved this hypothesis, and the mechanism by which PS20 and PS80 activate C [85] is entirely obscure.

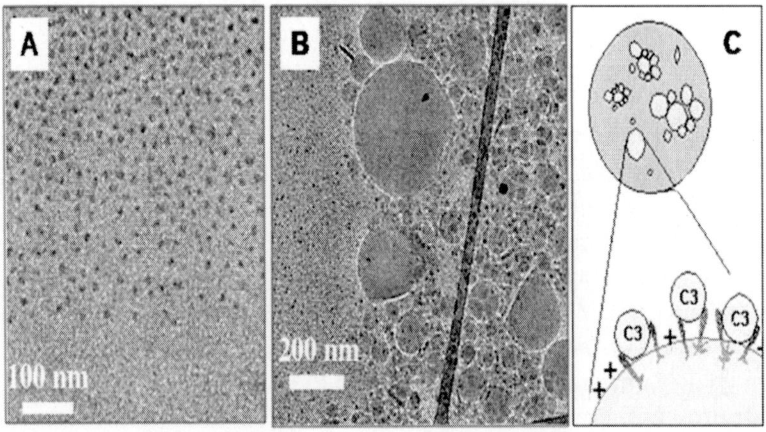

Figure 10.4 Cryo-TEM images of (A) micelles formed from Cremophor EL in water, (B) plasma samples following incubation with CrEL, and (C) scheme of C activating NPs in blood, illustrating the direct binding of C3 to the NPs (modified from [88, 89]). *Abbreviation*: TEM, transmission electron microscopy.

Among other types of nonionic amphiphilic surfactants with micellar buildup, poloxamers (traded as Pluronics or Kolliphor) are also known to activate C. These are triblock copolymers composed of a central hydrophobic chain of polyoxypropylene (POP) blocks flanked by two hydrophilic polyoxyethylene (POE), and one widely used member of the family, poloxamer 188 (Fluosol DA), was shown to cause C3 conversion, decreased hemolytic C levels, and generation of C5a-related leukocyte-aggregating activity in plasma, which was prevented by ethylenediaminetetraacetic acid (EDTA) but not ethyleneglycoltetraacetic acid (EGTA), suggesting alternative C activation.

As for ionic polymers, Plank et al. [90] examined the C-activating properties of synthetic cationic molecule complexes (polylysine, polyethyleneimine [PEI]), which activated C to varying extents, which activation was most likely due to the positive charges, as complexing the polymer with DNA reduced C activation. Complement activation by polycationic PEI was also shown by Merkel et al. [91], for 25 kDa PEI. This activation was reduced or completely blocked when the polymer was functionalized with 2K- or 20K-PEG.

10.4.3 Complement Activation by PEG

PEG, also referred to as polyethylene oxide (PEO) or POE, depending on its molecular weight, has the formula H-(O-CH_2-CH_2)$_n$-OH, with the element in parentheses being repeated in different products in the 7–2300-fold range, producing 0.3–100 kDa PEG. PEG has been used as an excipient in many pharmaceutical products and solvents in oral liquids and eyedrops; as an ointment base, tablet binder, film coating, and lubricant; and also as a therapeutic agent against spinal cord and brain injuries in dogs [92, 93]. PEG is also commonly used to fuse B-cells with myeloma cells in monoclonal antibody production. In clinical pharmacotherapy, however, PEG is best known as a ligand for protein therapeutics, extending their circulation time. Examples include PEG-interferon alpha, which is used to treat hepatitis C, and PEGfilgrastim (Neulasta), which is used to treat neutropenia.

In nanomedicne PEG is the gold standard as coating polymer for NPs, with the purpose of extending their life span for drug delivery. The different derivative types are monofunctional linear branched, multiarm, bifunctional, and heterofunctional. PEG is conjugated to different acceptor molecules or particles via maleimide, aldehyde, oxyamine, or other end groups. The effects of PEG on C activation by NPs vary in different studies. Mosqueira et al. [94], for example, measured C3 cleavage by poly(d,l-lactide) NPs to which different-chain-length PEGs were covalently bound at different densities. Consistent with its function as a steric barrier to protein binding, PEG inhibited C activation in proportion with its chain length and surface density. In another study, Peracchia et al. [95] demonstrated the importance of PEG conformation on suppression of C activation by polyisobutylcyanoacrylate NPs, which finding was extended by Hamad et al. [96], who showed that not only the intensity but also the pathway of C activation may depend on the spatial "architecture" of surface-projected PEG chains. Namely, the authors showed that C activation by poloxamine 908-coated (a PEO) polystyrene NPs proceeded via C1q-dependent CP when the PEO chains were in the PEG-dilute, narrow "mushroom" configuration, while activation switched to the LP and decreased in the case of PEG-dense, wider "brush" coating type [97]. In addition, AP activation also showed differences with the two types of coats inasmuch as properdin-mediated activation of the AP was restricted to particles displaying transitional mushroom-brush configuration of PEO [96].

PEG is also widely applied to extend the circulation time of liposomes in blood, for example, in liposomal doxorubicin, that is, Doxil, the first Food and Drug Administration (FDA)-approved anticancer nanodrug [98]. Doxil contains 5 mole% of 2K-PEG, which does not prevent C activation by these vesicles [99–101]. In a study from our laboratories analyzing PEGylated liposome-induced C activation in human serum we varied the concentration of 2K-PEG-PE in liposomes in the 2%–5% range and also looked at the impact of the PEG chain length in the 0.3–12 K range. Complement activation by these liposomes was weak and not significantly affected by any PEG-PE, suggesting that PEG neither promotes nor protects against C activation under the conditions examined [101]. Thus, the stealth character of Doxil [102] is unlikely to be due to interference by PEG with C activation.

10.4.4 Complement Activation on Polymer-Coated Nanoparticles

NPs are often coated with different polymers in order to reduce their interaction with plasma proteins and, hence, increase their circulation time. Most studies on this matter, wherein the biocompatibility-enhancing effect of polymer coats is assessed by measuring C activation and/or deposition, give account of modulations of (particle-induced) C activation rather than actual activation by the polymer coat. In one of the early studies on this subject Norman et al. [103] described the critical dependence of C3 absorption to poloxamer and poloxamine-coated polystyrene microspheres on the chain lengths of POE and POP blocks. The total amount and type of proteins adsorbed to the microspheres (IgG, C3, transferring, fibronectin) showed differential dependence on the polymer chain length and also depended on the plasma or serum concentration, highlighting the complexity and individuality of NP–plasma protein interactions.

10.4.5 Complement Activation by Dendrimers, Other Polymers

Dendrimers are hyperbranched polymers with highly symmetric, spherical structures in the 2–5 nm range. Their properties are determined by various functional groups on their surface or in their interior. Charged or hydrophilic groups on their outer shell make them water soluble and, thus, usable as drug carriers. Dendrimers are classified according to the number of dendron layers they display, formed by sequential branching cycles during their synthesis. Each successive generation results in a dendrimer roughly twice the molecular weight of the previous generation and about a 1 nm increase of size. Higher (4–6) generation dendrimers also have more exposed functional groups on the surface, which can later be used to customize the dendrimer for a given application. Given the size of the even highest (sixth) dendrimer generation being smaller than a C3 molecule (see above), just as with micelles, it is difficult to imagine how dendrimers would activate C via C3 and C5 convertase deposition. Nevertheless, charged dendrimers do activate C, as attested to by an exhaustive study by Plank et al. [90], who examined the C-activating properties of synthetic cationic molecule complexes with DNA (polylysine, PEI) and the fifth-generation poly(amidoamine) (PAMAM) dendrimers. All these agents activated C to varying extents, and this activation was clearly due to the positive charges, as the presence of polyanionic DNA reduced C activation. Complement activation by polycationic PEI was confirmed for a 25 kDa polymer by Merkel et al. (2011), which was reduced or absent when the polymer was functionalized with 2K- or 20K-PEG [91].

10.4.6 Complement Activation by Carbon Nanotubes

Multiwalled carbon nanotubes (MWCNTs) were shown to activate human C via both the CP and the AP [104], with direct binding of C1q triggering CP activation. Covalent functionalization of MWCNTs with epsilon-caprolactam or L-alanine significantly reduced this activation, suggesting that functionalization represents a viable approach for making MWCNTs more biocompatible [12]. Later it was shown that PEGylated single-walled carbon nanotubes (SWCNTs) also activated C via the CP and the AP, but the AP split-product Bb was not increased, suggesting activation exclusively through C4 cleavage. Furthermore, C4d elevation in C1q-depleted serum was inhibited by *N*-acetylglucosamine and by anti-MASP-2 antibodies, providing evidence for LP activation [10, 105]. Most recently the same authors showed that SWCNTs can bind MBL and ficolin, thus better explaining C activation via the LP [106]. In particular, the study demonstrated that coating CNTs with human serum albumin activated C through C1q-mediated CP and AP, while surface coating with PEG led to a different activation mechanism, depending on the amphiphile structure. CNTs with linear PEG triggered the LP through both l-ficolin and mannan-binding lectin recognition. The LP activation, however, did not trigger the amplification loop of the AP. A branched PEG also activated the LP but only through l-ficolin recognition. Importantly, this mode of activation neither generated anaphylatoxins nor induced triggering of the effector arm of the complement system [106].

10.5 Consequences of Complement Activation

10.5.1 The CARPA Concept

The concept that C activation may underlie non-IgE-mediated infusion reactions to intravenous (i.v.) liposomes was proposed in 1999, when evidence was obtained that the hemodynamic changes in pigs that mimicked the cardiovascular changes of patients following the infusion of liposomal doxorubicin (Doxil) were due to C activation [107]. Later studies provided a long list of evidence for the causal role of C activation, among which the specific inhibition by C blockers, the mimicking of symptoms by the C activator zymosan, and the correlation between C activation and reactivity in pigs are the most important. These factors, taken together with the fact that the reaction develops within minutes after starting the infusion of liposomes at first administration, thus ruling out a role of IgE, led to the proposal of the term "C activation-related pseudoallergy," or CARPA. Considering that major hemodynamic, bronchopulmonary, and skin changes in pigs, as well as cardiac arrhythmia and heart arrest, correspond to the symptoms of liposome-induced hypersensitivity syndrome in humans, this suggests furthermore that the porcine CARPA model represents a faithful animal model of the human hypersensitivity reaction to liposomes and other drugs, also known as an infusion or anaphylactoid or idiosyncratic reaction. On the basis of the prevalence of adverse events (AEs) during pharmacotherapy in US hospitals (which happens in about 30% of all treated patients) and the estimated ratio of non-IgE-mediated adverse immune effects, 19%, or ~400,000, of severe pseudoallergic reactions may occur each year in US hospitals. Considering 0.1% as the top ratio of fatal reactions, CARPA represents a major health risk for thousands of patients each year who need i.v. medication with reactogenic drugs. Among these, those suffering from cardiovascular disease are at increased risk for a fatal outcome. The frequency of HSRs reported for different individual reactogenic drugs varies between 3% and 45% [108]. Out of these, the reactions to Doxil have been studied in most detail in humans, whose frequency is reported to be up to 10%. The best-known symptoms of CARPA are flushing and a rash with dyspnea, followed by, in decreasing percentages, back pain, chest pain, abnormal blood pressure, chills, and tachycardia. In rare cases the consequences are more severe and might include myocardial infarction, shock, cardiac fibrillation, arrest, and death. There are major differences between true allergy and pseudoallergy in several respects; namely, CARPA arises upon the first exposure, the latency period increases, and the strength of the reaction decreases upon repeated exposures.

10.5.2 The Effector Arm of CARPA

The anaphylatoxins released upon C activation activate the immune cells, which release a long array of vasoactive mediators, which, in the next step, act upon smooth muscle and endothelial cells in the lung, heart, gut, and many other organs in the body. The responsive cells include mast cells, basophils, and macrophages that express a group of G-protein-coupled receptors, which bind anaphylatoxins (i.e., C3a/C5a/C5L2 receptors). Binding of C activation by-products, C3a and C5a, to these receptors can trigger essentially the same intracellular signal transduction system that is activated upon the engagement of an allergen to membrane-bound IgE, triggering the release of a battery of secondary vasoactive mediators, including histamine, tryptase, platelet-activating factor (PAF), and leukotrienes (LTB2, LTB4, LTC4, LTD4, LTE4, TXA2, PGD2, and TXD4). Some of these "allergomedins" (e.g., PAF, histamine, tryptase, and TXA2) are preformed and liberated from the cells immediately upon activation, while others are de novo synthesized and, hence, liberate slower. In the next step of CARPA allergomedins bind to their respective receptors on endothelial and smooth muscle cells, modifying their function in ways that lead to the symptoms of CARPA (Fig. 10.5).

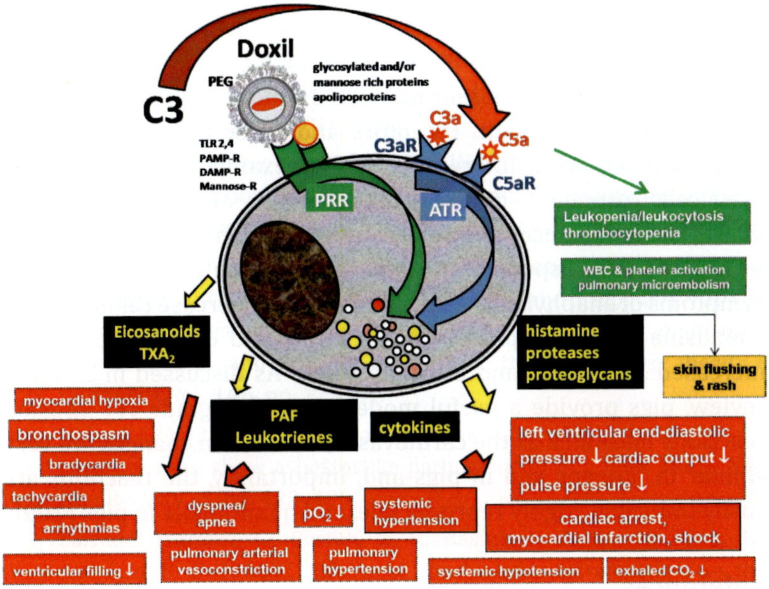

Figure 10.5 Schematic illustration of the molecular and cellular events leading to CARPA. Complement (C) is activated by NPs (illustrated here by Doxil), leading to the liberation of anaphylatoxins. The binding of anaphylatoxins to their respective receptors (ATRs) on mast cells (or PIM cells in pigs [109]) lead to a release reaction, that is, secretion of vasoactive secondary mediators (yellow text, black background). The particles also bind to the cells via PRRs, possibly TLRs 2 and 4, PAMP-Rs, DAMP-Rs, or mannose-Rs. In addition to binding to PRR, apolipoproteins and glycosylated and/or mannose-rich proteins can mediate the binding of particles to mast cells. According to the "double hit" hypothesis [109], simultaneous stimulation of mast cells by anaphylatoxin binding to ATRs and particle binding to PRRs amplifies the release reaction and production of secondary mediators. The latter lead to a HSRs with the listed hemodynamic and cardiopulmonary (white text, red background), skin (black text, yellow background), and blood cell (white text, green background) changes. The hemodynamic and cardiopulmonary changes are intricately related, as detailed elsewhere [109]. *Abbreviations*: ATR, anaphylatoxin receptor; PIM, pulmonary intravascular macrophage; PRR, pattern recognition receptor; TLR, Toll-like receptor; PAMP-R, pathogen-associated molecular pattern receptor; DAMP-R, danger-associated molecular pattern receptor; mannose-R, mannose receptor.

20. Deamer, D., and Weber, A. L. (2010). Bioenergetics and life's origins. *Cold Spring Harb. Perspect. Biol.*, **2**.

21. Deamer, D. W. (2008). Origins of life: how leaky were primitive cells? *Nature*, **454**(7200), pp. 37–38.

22. Harris, C. L., and Morgan, B. P. (2004). The many faces of the membrane regulators of complement, in *The Complement System: Novel Roles in Health and Disease*, ed. Szebeni, J. (Kluwer, Boston, MA).

23. Mollnes, T. E. (2004). Therapeutic manipulation of thecomplement system, in *The Complement System: Novel Roles in Health and Disease*, ed. Szebeni, J. (Kluwer, Boston, MA), pp. 483–516.

24. Mollnes, T. E., and Kirschfink, M. (2006). Strategies of therapeutic complement inhibition. *Mol. Immunol.*, **43**, pp. 107–21.

25. Zipfel, P. F., and Skerka, C. (2009). Complement regulators and inhibitory proteins. *Nat. Rev. Immunol.*, **9**, pp. 729–40.

26. Campbell, I. D., Baron, M., Day, A. J., Sim, R. B., Norman, D. G., and Barlow, P. N. (1991). Three-dimensional structure of a complement control protein module in solution. *J. Mol. Biol.*, **219**, pp. 717–725.

27. Linton, S. M., Williams, A. S., Dodd, I., Smith, R., Williams, B. D., and Morgan, B. P. (2000). Therapeutic efficacy of a novel membrane-targeted complement regulator in antigen-induced arthritis in the rat. *Arthritis Rheum.*, **43**, pp. 2590–2597.

28. Smith, G. P., and Smith, R. A. G. (2001). Membrane-targeted complement inhibitors. *Mol. Immunol.*, **38**, pp. 249–255.

29. Smith, R. A. (2002). Targeting anticomplement agents. *Biochem. Soc. Trans.*, **30**, pp. 1037–1041.

30. Smith, R. A. G., Esser, D., Ridley, S. H., and Bradford, R. (2004). Engineering control of complement inhibition at the cell surface, in *The Complement System: Novel Roles in Health and Disease*, ed. Szebeni, J. (Kluwer, Boston, MA), pp. 531–546.

31. Souza, D. G., Esser, D., Bradford, R., Vieira, A. T., and Teixeira, M. M. (2005). APT070 (Mirococept), a membrane-localised complement inhibitor, inhibits inflammatory responses that follow intestinal ischaemia and reperfusion injury. *Br. J. Pharmacol.*, **145**, pp. 1027–1034.

32. De Silva, R. J., Vuylsteke, A., Fritchley, S. J., Trull, A. K., Dunning, J. J., and Wallwork, J. (2006). APT070 inhibits complement activation during in vitro cardiopulmonary bypass. *Eur. J. Cardiothorac. Surg.*, **30**, pp. 72–76.

33. Janeway, C. A. (2001). The complement system and innate immunity—Immunobiology, in *The Immune System in Health and Disease, Immunobiology*, 5th ed., eds. Janeway, C. A., Travers, J. P., Walport, M., and Shlomchik, M. J. (Garland Science, New York).

34. Hillmen, P., Young, N. S., Schubert, J., Brodsky, R. A., Socié, G., Muus, P., Röth, A., Szer, J., Elebute, M. O., Nakamura, R., Browne, P., Risitano, A. M., Hill, A., Schrezenmeier, H., Fu, C. L., Maciejewski, J., Rollins, S. A., Mojcik, C. F., Rother, R. P., and Luzzatto, L. (2006). The complement inhibitor eculizumab in paroxysmal nocturnal hemoglobinuria. *N. Engl. J. Med.*, **355**, pp. 1233–1243.

35. Kirschfink, M. (1997). Controlling the complement system in inflammation. *Immunopharmacology*, **38**, pp. 51–62.

36. Kirschfink, M., and Mollnes, T. E. (2004). The role of complement in transplantation, in *The Complement System: Novel Roles in Health and Disease*, ed. Szebeni, J. (Kluwer, Boston, MA), pp. 397–418.

37. Fishelson, Z., Donin, N., Zell, S., Schultz, S., and Kirschfink, M. (2003). Obstacles to cancer immunotherapy: expression of membrane complement regulatory proteins (mCRPs) in tumors. *Mol. Immunol.*, **40**, pp. 109–23.

38. Donin, N., Ziporen, L., Masarwa, M., Ben-Moshe, O., Kirschfink, M., and Fishelson, Z. (2012). Silencing of mortalin, a regulator of the complement membrane attack complex, as anticancer adjuvant therapy. *Immunobiology*, **217**, p. 1203.

39. Alving, C. R., Kinsky, S. C., Haxby, J. A., and Kinsky, C. B. (1969). Antibody binding and complement fixation by a liposomal model membrane. *Biochemistry*, **8**, pp. 1582–1587.

40. Haxby, J. A., Kinsky, C. B., and Kinsky, S. C. (1968). Immune response of a liposomal model membrane. *Proc. Natl. Acad. Sci. U S A*, **61**, pp. 300–307.

41. Haxby, J. A., Gotze, O., Muller-Eberhard, H. J., and Kinsky, S. C. (1969). Release of trapped marker from liposomes by the action of purified complement components. *Proc. Natl. Acad. Sci. U S A*, **64**, pp. 290–295.

42. Kinsky, S. C., Haxby, J. A., Zopf, D. A., Alving, C. R., and Kinsky, C. B. (1969). Complement-dependent damage to liposomes prepared from pure lipids and Forssman hapten. *Biochemistry*, **8**, pp. 4149–4158.

43. Ahearn, J. M., and Fearon, D. T. (1989). Structure and function of the complement receptors, CR1 (CD35) and CR2 (CD21). *Adv. Immunol.*, **46**, pp. 183–219.

44. Liszewski, M. K., Post, T. W., and Atkinson, J. P. (1991). Membrane Cofactor Protein (MCP or CD46): newest member of the regulators of complement activation gene cluster. *Annu. Rev. Immunol.*, **9**, pp. 431–455.

45. Medof, M. E., Lublin, D. M., Holers, V. M., Ayers, D. J., Getty, R. R., Leykam, J. F., Atkinson, J. P., and Tykocinski, M. L. (1987). Cloning and characterization of cDNAs encoding the complete sequence of decay-accelerating factor of human complement. *Proc. Natl. Acad. Sci. U S A*, **84**, pp. 2007–2011.

46. Huang, Y., Qiao, F., Abagyan, R., Hazard, S., and Tomlinson, S. (2006). Defining the CD59-C9 binding interaction. *J. Biol. Chem.*, **281**, pp. 27398–27404.

47. Holmes, C. H., Simpson, K. L., Okada, H., Okada, N., Wainwright, S. D., Purcell, D. F., and Houlihan, J. M. (1992). Complement regulatory proteins at the feto-maternal interface during human placental development: distribution of CD59 by comparison with membrane cofactor protein (CD46) and decay accelerating factor (CD55). *Eur. J. Immunol.*, **22**, pp. 1579–1585.

48. Laarman, A. J., Ruyken, M., Malone, C. L., van Strijp, J. A., Horswill, A. R., and Rooijakkers, S. H. (2011). Staphylococcus aureus metalloprotease aureolysin cleaves complement C3 to mediate immune evasion. *J. Immunol.*, **186**, pp. 6445–6453.

49. Lee, L. Y., Hook, M., and Haviland, D. (2004). Inhibition of complement activation by a secreted Staphylococcus aureus protein. *J. Infect. Dis.*, **190**, pp. 571–579.

50. Sharp, J. A., Echague, C. G., Hair, P. S., Ward, M. D., Nyalwidhe, J. O., Geoghegan, J. A., Foster, T. J., and Cunnion, K. M. (2012). Staphylococcus aureus surface protein SdrE binds the complement regulator factor H to evade the immune response. *Immunobiology*, **217**, p. 1204.

51. Stapels, D. A. C., Ramyar, K. X., Ricklin, D. F., Milder, J. J., Bischoff, M., Herrmann, M., Geisbrecht, B. V., Lambris, J. D., von Köckritz-Blickwede, M., van Kessel1, K. P. M., and Rooijakkers, S. H. M. (2012). Extracellular adherence protein (Eap) of Staphylococcus aureus evades innate immunity by inhibiting complement activation and neutrophil elastase. *Immunobiology*, **217**, p. 1170.

52. Jusko, M., Potempa, J., Kantyka, T., Bielecka, E., Miller, H. K., Kalinska, M., Dubin, G., Garred, P., Shaw, L. N., and Blom, A. M. (2012). Staphylococcal proteases aid evasion of human complement system. *Immunobiology*, **217**, p. 1204.

53. Postma, B., Kleibeuker, W., Poppelier, M. J., Boonstra, M., Van Kessel, K. P., Van Strijp, J. A., and de Haas, C. J. (2005). Residues 10–18 within the C5a receptor N terminus compose a binding domain for chemotaxis inhibitory protein of Staphylococcus aureus. *J. Biol. Chem*, **280**, pp. 2020–2027.

54. Boehm, S., Reuter, M., and Zipfel, P. F. (2012). The surface binding region of complement Factor H guides the staphylococcal Efb to cell surfaces to allow local complement inhibition. *Immunobiology*, **217**, p. 1213.

55. Rooijakkers, S. H., Ruyken, M., and Roos, A. (2005). Immune evasion by a staphylococcal complement inhibitor that acts on C3 convertases. *Nat. Immunol.*, **6**, pp. 920–927.

56. Sharp, J. A., Echague, C. G., Hair, P. S., Ward, M. D., Nyalwidhe, J. O., Geoghegan, J. A., Foster, T. J., and Cunnion, K. M. (2012). Staphylococcus aureus surface protein SdrE binds complement regulator factor H as an immune evasion tactic. *PLOS ONE*, **7**, p. e38407.

57. Laarman, A. J., Bardoel, B. W., Ruyken, M., Fernie, J., Milder, F. J., van Strijp, J. A., and Rooijakkers, S. H. (2012). Pseudomonas aeruginosa alkaline protease blocks complement activation via the classical and lectin pathways. *J. Immunol.*, **188**, pp. 386–393.

58. Ya-Ping, K., Mingsong, K., Liang, X., Ross, C. L., Barbara, L. Q., Murray, B. E., and Höök, M. (2012). Collagen-binding MSCRAMMS of Gram-positive bacteria inhibit complement activation via the classical pathway. *Immunobiology*, **217**, p. 1151.

59. Vogel, U., Weinberger, A., Frank, R., Müller, A., Köhl, J., Atkinson, J. P., and Frosch, M. (1997). Complement factor C3 deposition and serum resistance in isogenic capsule and lipooligosaccharide sialic acid mutants of serogroup B Neisseria meningitidis. *Infect. Immun.*, **65**, pp. 4022–4029.

60. Lewis, L. A., Vasudhev, S., Shaughnessy, J., and Ram, S. (2012). Factor H-dependent regulation of the alternative pathway of complement by Meningococci that lack FHBP, NSPA, capsule and lipooligosaccharide sialic acid. *Immunobiology*, **217**, p. 1170.

61. Lewis, L. A., Carter, M., and Ram, S. (2012). The relative roles of factor H binding protein, neisserial surface protein A, and lipooligosaccharide sialylation in regulation of the alternative pathway of complement on meningococci. *J. Immunol.*, **188**, pp. 5063–5072.

62. China, B., Sory, M. P., N'Guyen, B. T., De Bruyere, M., and Cornelis, G. R. (1993). Role of the YadA protein in prevention of opsonization of

Yersinia enterocolitica by C3b molecules. *Infect. Immun.*, **61**, pp. 3129–3136.

63. Rokita, E., Makristathis, A., Presterl, E., Rotter, M. L., and Hirschl, A. M. (1998). Helicobacter pylori urease significantly reduces opsonization by human complement. *J. Infect. Dis.*, **178**, pp. 1521–1525.
64. Eisenschenk, F. C., Houle, J. J., and Hoffmann, E. M. (1999). Mechanism of serum resistance among Brucella abortus isolates. *Vet. Microbiol.*, **68**, pp. 235–244.
65. Heffernan, E. J., Reed, S., Hackett, J., Fierer, J., Roudier, C., and Guiney, D. (1992). Mechanism of resistance to complement-mediated killing of bacteria encoded by the Salmonella typhimurium virulence plasmid gene rck. *J. Clin. Invest.*, **90**, pp. 953–964.
66. Horstmann, R. D., Sievertsen, H. J., Knobloch, J., and Fischetti, V. A. (1988). Antiphagocytic activity of streptococcal M protein: selective binding of complement control protein factor H. *Proc. Natl. Acad. Sci. U S A*, **85**, pp. 1657–1661.
67. Wexler, D. E., Chenoweth, D. E., and Cleary, P. P. (1985). Mechanism of action of the group A streptococcal C5a inactivator. *Proc. Natl. Acad. Sci. U S A*, **82**, pp. 8144–8148.
68. Lubinski, J., Wang, L., Mastellos, D., Sahu, A., Lambris, J. D., and Friedman, H. M. (1999). In vivo role of complement-interacting domains of herpes simplex virus type 1 glycoprotein gC. *J. Exp. Med.*, **190**, pp. 1637–1646.
69. Huemer, H. P., Larcher, C., and Coe, N. E. (1992). Pseudorabies virus glycoprotein III derived from virions and infected cells binds to the third component of complement. *Virus Res.*, **23**, pp. 271–280.
70. Parizade, M., Arnon, R., Lachmann, P. J., and Fishelson, Z. (1994). Functional and antigenic similarities between a 94-kD protein of Schistosoma mansoni (SCIP-1) and human CD59. *J. Exp. Med.*, **179**, pp. 1625–1636.
71. Marikovsky, M., Parizade, M., Arnon, R., and Fishelson, Z. (1990). Complement regulation on the surface of cultured schistosomula and adult worms of Schistosoma mansoni. *Eur. J. Immunol.*, **20**, pp. 221–227.
72. Fishelson, Z. (1995). Novel mechanisms of immune evasion by Schistosoma mansoni. *Mem. Inst. Oswaldo Cruz.*, **90**, pp. 289–292.
73. Szebeni, J. (1998). The interaction of liposomes with the complement system. *Crit. Rev. Ther. Drug Carrier Syst.*, **15**, pp. 57–88.
74. Chonn, A., Cullis, P. R., Devine, D. V. (1991). The role of surface charge in the activation of the classical and alternative pathways of complement by liposomes. *J. Immunol.*, **146**, pp. 4234–4241.

75. Szebeni, J. (2001). Complement activation-related pseudoallergy caused by liposomes, micellar carriers of intravenous drugs and radiocontrast agents. *Crit. Rev. Ther. Drug Carr. Syst.*, **18**, pp. 567–606.

76. Andersen, A. J., Hashemi1, S. H., Andresen, T. L., Hunter, A. C., and Moghimi, S. M. (2009). Complement: alive and kicking nanomedicines. *J. Biomed. Nanotechnol.*, **5**, pp. 364–372.

77. Alving, C. R., and Swartz, G. M. (1991). Antibodies to cholesterol, cholesterol conjugates, and liposomes: implications for atherosclerosis and autoimmunity. *Crit. Rev. Immunol.*, **10**, pp. 441–453.

78. Alving, C. R., and Wassef, N. M. (1999). Naturally occurring antibodies to cholesterol: a new theory of LDL cholesterol metabolism. *Immunol. Today*, **20**, pp. 362–366.

79. Sherer, Y., and Shoenfeld, Y. (2001). Antiphospholipid syndrome, antiphospholipid antibodies, and atherosclerosis. *Curr. Atheroscler Rep.*, **3**, pp. 328–333.

80. Horkko, S., Miller, E., Branch, D. W., Palinski, W., and Witztum, J. L. (1997). The epitopes for some antiphospholipid antibodies are adducts of oxidized phospholipid and beta2 glycoprotein 1 (and other proteins). *Proc. Natl. Acad. Sci. U S A*, **94**, p. 10356.

81. Bradley, A. J., Brooks, D. E., Norris-Jones, R., and Devine, D. V. (1991). C1q binding to liposomes is surface charge dependent and is inhibited by peptides consisting of residues 14–26 of the human C1qA chain in a sequence independent manner. *Biochim. Biophys. Acta*, **1418**, p. 19.

82. Richards, R. L., Gewurz, H., Osmand, A. P., and Alving, C. R. (1977). Interactions of C-reactive protein and complement with liposomes. *Proc. Natl. Acad. Sci. U S A*, **74**, p. 5672.

83. Volanakis, J. E., and Wirtz, K. W. (1979). Interaction of C-reactive protein with artificial phosphatidylcholine bilayers. *Nature*, **281**, p. 155.

84. Kuroki, Y., Honma, T., Chiba, H., Sano, H., Saitoh, M., Ogasawara, Y., Sohma, H., and Akino, T. (1997). A novel type of binding specificity to phospholipids for rat mannose-binding proteins isolated from serum and liver. *FEBS Lett.*, **414**, p. 387.

85. Weiszhár, Z., Czúcz, J., Révész, C., Rosivall, L., Szebeni, J., and Rozsnyay, Z. (2012). Complement activation by polyethoxylated pharmaceutical surfactants: Cremophor-EL, Tween-80 and Tween-20. *Eur. J. Pharm. Pharm.*, **45**, pp. 492–498.

86. Perkins, S. J., and Sim, R. B. (1986). Molecular modelling of human complement component C3 and its fragments by solution scattering. *Eur. J. Biochem.*, **157**, pp. 155–168.

87. Kessel, D., Woodburn, K., Kecker, D., and Sykes, E. (1995). Fractionation of Cremophor EL delineates components responsible for plasma lipoprotein alterations and multidrug resistance reversal. *Oncol. Res.*, **7**, pp. 207–212.

88. Szebeni, J., Alving, C. R., Savay, S., Barenholz, Y., Priev, A., Danino, D., and Talmon, Y. (2001). Formation of complement-activating particles in aqueous solutions of Taxol: possible role in hypersensitivity reactions. *Intern. Immunopharm.*, **1**, pp. 721–735.

89. Szebeni, J. (2004). Complement activation-related pseudoallergy: mechanism of anaphylactoid reactions to drug carriers and radiocontrast agents, in *The Complement System: Novel Roles in Health and Disease*, ed. Szebeni, J. (Kluwer, Boston, MA), pp. 399–440.

90. Plank, C., Mechtler, K., Szoka, C. F., and Wagner, E. (1996). Activation of the complement system by synthetic DNA complexes: a potential barrier for intravenous gene delivery. *Hum. Gene Ther.*, **7**, pp. 1437–1446.

91. Merkel, O., Urbanics, R., Bedõcs, P., Rozsnyay, Z., Rosivall, L., Toth, M., Kissel, T., and Szebeni, J. (2011). In vitro and in vivo complement activation and related anaphylactic effects associated with polyethylenimine and polyethylenimine-graft-poly(ethylene glycol) block copolymers. *Biomaterials*, **32**, pp. 4936–4942.

92. Laverty, P., Leskovar, A., Breur, G. J., Coates, J. C., Bergman, R. L., Widmer, W. R., Toombs, J. P., Shapiro, S., and Borgens, R. B. (2004). A preliminary study of intravenous surfactants in paraplegic dogs: polymer therapy in canine clinical SCI. *Neurotrauma J.*, **21**, pp. 1767–1777.

93. Krause, T. L., and Bittner, G. D. (1990). Rapid morphological fusion of severed myelinated axons by polyethylene glycol. *PNAS*, **87**, pp. 1471–1475.

94. Mosqueira, V. C., Legrand, P., Gulik, A., Bourdon, O., Gref, R., Labarre, D., and Barratt, G. (2001). Relationship between complement activation, cellular uptake and surface physicochemical aspects of novel PEG-modified nanocapsules. *Biomaterials*, **22**, pp. 2967–2979.

95. Peracchia, M. T., Vauthier, C., Passirani, C., Couvreur, P., and Labarre, D. (1997). Complement consumption by poly(ethylene glycol) in different conformations chemically coupled to poly(isobutyl 2-cyanoacrylate) nanoparticles. *Life Sci.*, **61**, pp. 749–761.

96. Hamad, I., Al-Hanbali, O., Hunter, A. C., Rutt, K. J., Andresen, T. L., and Moghimi, S. M. (2010). Distinct polymer architecture mediates switching of complement activation pathways at the nanosphere-

serum interface: implications for stealth nanoparticle engineering. *ACS Nano*, **4**, pp. 6629–6638.

97. Georgiev, G. A., Al-Hanbali, S. D. K. O., Georgiev, G. D., and Lalchev, Z. (2007). Effects of poly (ethylene glycol) chains conformational transition on the properties of mixed DMPC/DMPE-PEG thin liquid films and monolayers. *Colloids Surf., B*, **59**, pp. 184–193.

98. Barenholz, Y. (2012). Doxil—The first FDA-approved nano-drug: from and idea to product, in *Handbook of Harnessing Biomaterials in Nanomedicine*, ed. Peer, D. (Pan Stanford, Singapore), pp. 335–398.

99. Szebeni, J., Baranyi, B., Savay, S., Lutz, L. U., Jelezarova, E., Bunger, R., and Alving, C. R. (2000). The role of complement activation in hypersensitivity to pegylated liposomal doxorubicin (Doxil®). *J. Liposome Res.*, **10**, pp. 347–361.

100. Szebeni, J., Baranyi, L., Savay, S., Milosevits, J., Bunger, R., Laverman, P., Metselaar, J. M., Storm, G., Chanan-Khan, A., Liebes, L., Muggia, F. M., Cohen, R., Barenholz, Y., and Alving, C. R. (2002). Role of complement activation in hypersensitivity reactions to Doxil and HYNIC-PEG liposomes: experimental and clinical studies. *J. Liposome Res.*, **12**, pp. 165–172.

101. Szebeni, J., Bedõcs, P., Rozsnyay, Z., Weiszhár, Z., Urbanics, R., Rosivall, L., Cohen, R., Garbuzenko, O., Báthori, G., Tóth, M., Bünger, R., and Barenholz, Y. (2012). Liposome-induced complement activation and related cardiopulmonary distress in pigs: factors promoting reactogenicity of Doxil and AmBisome. *Nanomed. NBM*, **8**, pp. 176–184.

102. Gabizon, A., Catane, R., Uziely, B., Kaufman, B., Safra, T., Cohen, R., Martin, F., Huang, A., and Barenholz, Y. (1994). Prolonged circulation time and enhanced accumulation in malignant exudates of doxorubicin encapsulated in polyethylene-glycol coated liposomes. *Cancer Res.*, **54**, pp. 987–992.

103. Norman, M. E., Williams, P., and Illum, L. (1993). Influence of block copolymers on the adsorption of plasma proteins to microspheres. *Biomaterials*, **14**(3), pp. 193–202.

104. Salvador-Morales, C., Flahaut, E., Sim, E., Sloan, J., Green, M. L., and Sim, R. B. (1996). Complement activation and protein adsorption by carbon nanotubes. *Mol. Immunol.*, **43**, pp. 193–201.

105. Hamad, I., Hunter, A. C., Rutt, K. J., Liu, Z., Dai, H., and Moghimi, S. M. (2008). Complement activation by PEGylated single-walled carbon nanotubes is independent of C1q and alternative pathway turnover. *Mol. Immunol.*, **45**, pp. 3797–3803.

106. Andersen, A. J., Robinson, J. T., Dai, H., Hunter, A. C., Andresen, T. L., and Moghimi, S. M. (2013). Single-walled carbon nanotube surface control of complement recognition and activation. *ACS Nano*, **7**, pp. 1108–1119.

107. Szebeni, J., Fontana, J. L., Wassef, N. M., Mongan, P. D., Morse, D. S., Dobbins, D. E., Stahl, G. L., Bünger, R., and Alving, C. R. (1999). Hemodynamic changes induced by liposomes and liposome-encapsulated hemoglobin in pigs: a model for pseudo-allergic cardiopulmonary reactions to liposomes. Role of complement and inhibition by soluble CR1 and anti-C5a antibody. *Circulation*, **99**, pp. 2302–2309.

108. Chanan-Khan, A., Szebeni, J., Savay, S., Liebes, L., Rafique, N. M., Alving, C. R., and Muggia, F. M. (2003). Complement activation following first exposure to pegylated liposomal doxorubicin (Doxil): possible role in hypersensitivity reactions. *Ann. Oncol.*, **14**, pp. 1430–1437.

109. Szebeni, J., Bedõcs, P., Csukás, D., Rosivall, L., Bunger, R., and Urbanics, R. (2012). A porcine model of complement-mediated infusion reactions to drug carrier nanosystems and other medicines. *Adv. Drug Delivery Rev.*, **64**, pp. 1706–1716.

110. SCENIHR. (2009). *Risk Assessment of Products of Nanotechnologies*, http://ec.europa.eu/health/ph_risk/committees/04_scenihr/docs/scenihr_o_023.pdf.

111. Hugli, T. E., Chenoweth, D. E. (1980). Biologically active peptides of complement. Techniques and significance of C3a and C5a measurements, in *Laboratory and Research Methods in Biology and Medicine*, eds. Nakamura, R. M., Dito, W. R., and Tucher, E. S. (Alan R. Liss, New York), p. 443.

112. Hugli, T. E., and Morgan, E. L. (1984). Mechanisms of leukocyte regulation by complement-derived factors. *Contemp. Top. Immunobiol.*, **14**, pp. 109–153.

113. Hugli, T. E. (1984). Structure and function of anaphylatoxins. *Spring Semin. Immunopathol.*, **7**, pp. 193–219.

114. Marceau, F., Lundberg, C., and Hugli, T. E. (1987). Effects of anaphylatoxins on circulation. *Immunopharmacology*, **14**, pp. 67–84.

115. Chenoweth, D. E., Cooper, S. W., Hugli, T. E., Stewart, R. W., Blackstone, E. H., and Kirklin, J. W. (1981). Complement activation during cardiopulmonary bypass: evidence for generation of C3a and C5a anaphylatoxins. *N. Engl. J. Med.*, **304**, pp. 497–503.

116. Heideman, M., and Hugli, T. E. (1984). Anaphylatoxin generation in multisystem organ failure. *J. Trauma*, **24**, pp. 1038–1043.

Chapter 11

Biodegradation of Carbon-Based Nanomaterials

Cyrill Bussy,[a] Alberto Bianco,[b] Maurizio Prato,[c] and Kostas Kostarelos[a]
[a]*Nanomedicine Laboratory, Institute of Inflammation and Repair, Faculty of Medical and Human Sciences, and National Graphene Institute, The University of Manchester, Manchester, UK*
[b]*CNRS, Institut de Biologie Moléculaire et Cellulaire, Laboratoire d'Immunologie et Chimie Thérapeutiques, Strasbourg, France*
[c]*Department of Pharmaceutical Sciences, University of Trieste, Trieste, Italy*
k.kostarelos@ucl.ac.uk

11.1 Introduction

Among the various carbon-based nanomaterials (CNMs), carbon nanotubes (CNTs) and graphene have in the last few years emerged as two material types with the potential to be further developed in the field of nanomedicine. Indeed, modifying and engineering their basic graphitic structure in order to improve their biocompatibility has led to the demonstration of their possible use as delivery systems, biosensors, or composites for tissue engineering. However, while chemically functionalized CNMs present reduced toxicity and great

Handbook of Safety Assessment of Nanomaterials: From Toxicological Testing to Personalized Medicine
Edited by Bengt Fadeel

ISBN 978-981-4463-36-2 (Hardcover), 978-981-4463-37-9 (eBook)
www.panstanford.com

biomedical promise, they are still viewed with scepticism, owing to the paradigm that their physicochemical characteristics make them nonbiodegradable. Recently, different studies have uncovered that peroxidase enzyme–based processes could lead to their oxidation and biodegradation. This chapter provides a review of current knowledge on this topic, including the proposed mechanism for enzymatically catalyzed biodegradation. In the context of biomedical use, these new findings offer novel perspectives for CNMs and also stress the need for future investigations that could reveal how to promote or inhibit their biodegradation, depending on the biomedical application desired.

11.2 Carbon-Based Nanomaterials

In the last decade, a novel class of nanomaterials based exclusively on carbon (Fig. 11.1) has been explored in the fields of nanobiotechnology and nanomedicine. Several proof-of-concept studies have demonstrated that fullerenes, CNTs, carbon nanohorns (CNHs), nanodiamonds, and, more recently, graphene may offer new tools for diagnosis and treatment of various diseases [1–4]. Amongst those CNMs, CNTs and graphene have both attracted a lot of interest from biomedical researchers, owing to their unique combination of chemical and physical properties.

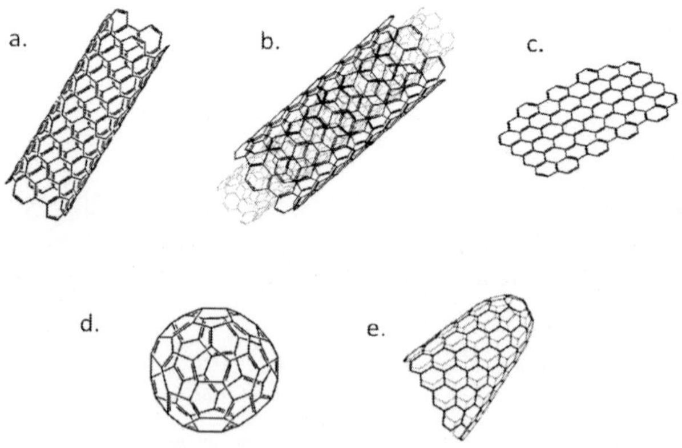

Figure 11.1 Different CNMs. (a) SWCNT, (b) MWCNT, (c) graphene, (d) fullerene, and (e) CNH.

CNTs are made of sp^2-hybridized carbon atoms, which are organized in one (single-walled carbon nanotubes [SWCNTs]) or more (multiwalled carbon nanotubes [MWCNTs]) concentric graphene sheets rolled up into thin, hollow cylinders. While the diameter of SWCNTs is in the 0.47–2.0 nm range, the MWCNTs' diameter can reach up to 100 nm. Both CNT types have lengths of several micrometers, making them a 1D nanomaterial with a characteristic high aspect ratio and high surface area. Likewise, graphene is also an sp^2-hybridized CNM but is a 2D material due to its one-atom-thick planar graphitic structure. Lateral dimensions of graphene flakes can be tailored according to specific applications. Compared to CNTs, both sides of the planar axis of graphene sheets are available, providing a platform for potentially higher degree of functionalization.

Both CNTs and graphene have outstanding features (electronic, mechanical, electrical, optical, thermal) that have been explored for a wide variety of biological and medical applications such as biosensors, tissue engineering, and targeted delivery systems [2, 4, 5]. They are both considered as unique materials for chemical or photothermal therapies but also molecular imaging [6–10]. Some of these prospective biomedical applications have been made possible by extensive chemical studies on methods to modify and functionalize the inherently hydrophobic surface of CNMs so as to make them compatible with aqueous-based environments [8,11–14]. Interestingly, functionalization (by covalent or noncovalent approaches) has also been shown to improve dramatically the biocompatibility of CNMs constructs compared to nonfunctionalized materials. Numerous studies have demonstrated how functionalized CNMs are able to deliver drugs, genes, or antigens to various types of cells without inducing toxicological effects or immune responses [15–17]. Pharmacokinetic studies have moreover revealed that the nature and degree of functionalization are essential parameters not only for the tissue absorption and distribution of CNMs but also for their excretion [7, 18, 19]. Similarly, for tissue engineering purposes functionalized CNM-based substrates have been shown to support and improve the growth and differentiation of neuron or osteoblast progenitors and also stem cells without deleterious biological responses [20–23]. Nevertheless, an essential issue that needs elucidation is the long-term fate of CNMs that will stay inside the body. In relation to that, the question of the possible metabolism of CNMs by the body over time (i.e., biodegradation) was for a long time disclaimed due to their strong and chemically inert *sp*^{2} graphitic structure that was believed to make them nondegradable under physiological conditions.

Recent work has reported that the perceived nondegradable CNMs may undergo enzyme catalyzed oxidation and biodegradation under specific conditions [24, 25]. In this chapter, we will review the current knowledge on this topic and discuss the implications for the biomedical field and some potential directions for future work. Indeed, if CNMs are naturally biodegradable, it is rational to believe that tailoring the CNM structure in a certain manner could accelerate or inhibit their inherent biodegradability, resulting in significant implications for their biomedical use. For instance, while durability is more likely to be suited for long-term tissue engineering products, highly biodegradable nanovectors for short-time intervention purposes seem more appropriate. Although clinical applications are still immature, prospective research that would focus on engineering the CNM structure and surface to control degradability will surely increase their chances for translation into the clinic.

11.3 Oxidation of Carbon-Based Nanomaterials

CNMs such as CNTs and graphene are by definition materials made exclusively of *sp*^{2} carbon atoms that are difficult to oxidize. Only strong oxidative (i.e., nitric acid, sulfuric acid in combination with hydrogen peroxide [H_2O_2]) or severe mechanical (e.g., probe sonication or milling) treatments have been shown to introduce defects in the form of oxygen-containing groups onto the carbon lattice of pristine materials [26]. In addition to these chemical modifications, oxidation of CNMs also leads to structural alterations such as length shortening for nanotubes and hole opening in the planar sheets for graphene, producing polycyclic aromatic hydrocarbons and CO_2 as by-products [27–29].

Given such knowledge, CNMs were considered as nonbiodegradable material at least in physiological environment until Star et al. hypothesized and proved that enzyme-catalyzed oxidation processes are able to degrade them [30]. Two peroxidase enzymes were primarily selected to study that hypothesis, horseradish peroxidase (HRP) [30, 31] and myeloperoxidase (MPO) [28]. While HRP, a plant enzyme, could be seen as a proof of concept not relevant to biomedical applications, MPO is the enzyme responsible for the production of strong oxidants (i.e., hypochlorite) able to destroy pathogens in immune cells such as neutrophils and macrophages, which are likely to encounter CNMs after administration in the body. Both enzymes contain in their active sites a heme group that catalyzes enzymatic reactions in the presence of hydrogen peroxide. Following the same principle, it was also demonstrated that CNMs could undergo degradation when incubated with either lactoperoxidase (LPO) [32] or eosinophil peroxidase (EPO) [33], both enzymes being able to produce another strong oxidative agent, hypobromite, in the presence of H_2O_2. Cyclo-oxygenases, catalases, or cytochrome C oxidases are other heme-containing peroxidases that may be studied in the future in order to extend the knowledge on CNM degradability in various parts of the body where those nanomaterials might accumulate or stay after environmental or biomedical exposure.

Alternatively, other research groups have used oxidizing and acidic electrolyte solutions to mimic the macrophage phagolysosome milieu in which CNMs have been described to be accumulating after cellular uptake [34–36]. In all cases, the concept in these studies did not involve active processes of degradation based on enzymatic activity (no hydrolytic enzymes added) but rather questioned the durability of CNMs over time in chemically aggressive physiological environments. As for enzymatic processes, other simulated biological fluids will need to be tested to consolidate the understanding of CNM fate in the body [37]. For instance, environments simulating the intraluminal (e.g., gastric or lung fluid) or interstitial (e.g., body or synovial fluid) fluids seem also relevant.

11.4 Ex vivo Biodegradation of CNMs

Both CNTs and graphene have been studied following the action of peroxidase enzymes or phagolysosomal-mimicking environments (Table 11.1). Even though various possible models of structural degradation of the material can be considered, only a few have been experimentally shown (Fig. 11.2).

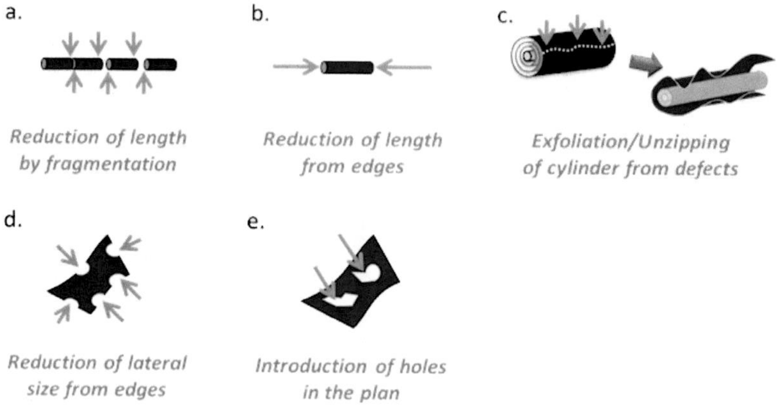

Figure 11.2 Theoretical (a–e) and experimentally demonstrated (b, c, and e) models of biological degradation of CNTs (a, b, and c) and graphene (d and e) (without taking into account dissolution of the *sp*^{2}-hybridized carbon backbone structure).

11.4.1 Ex vivo Biodegradation of SWCNTs

The first study that reported strong oxidative enzymes able to degrade CNTs appeared in 2008 [30]. In that study, Star et al. described how carboxylated SWCNTs can be degraded at 4°C by the oxidative activity of HRP in the presence of low concentrations of hydrogen peroxide. Repeating the same experiment but at room temperature, they showed that carboxylated SWCNTs can degrade more quickly than at 4°C, while pristine SWCNTs could not [31]. Oxidized polycyclic aromatic hydrocarbons were detected as by-products of partial or incomplete degradation, and carbon dioxide was the ultimate product of complete degradation after 10 days. From this work, it was concluded that structural defects (presence of oxygen-containing groups) or functional groups in the graphitic lattice may act as initiator sites for enzymatic action.

In a second set of studies, the human MPO enzyme was used [28]. The primary hypothesis was that reactive radical intermediates and hypochlorite generated by MPO from H_2O_2 would catalyze the oxidation of carboxylated SWCNTs. It was observed that degradation occurred when SWCNTs were incubated either in a mixture of MPO-H_2O_2 or in hypochlorite alone, but there was no degradation in MPO or H_2O_2 solutions alone. Moreover, nanotube degradation by MPO-H_2O_2

treatment was enhanced in the presence of NaCl, suggesting a role for both hypochlorous acid and peroxidase reactive intermediates in the MPO-catalyzed degradation. Notably, the biodegradation of pristine noncarboxylated SWCNTs was reduced under the same conditions (i.e., MPO-H_2O_2 with NaCl). It was concluded that degradation by oxidation of pristine SWCNTs is mediated by hypochlorite, whereas degradation of carboxylated SWCNTs involved both hypochlorite and the MPO enzyme. Going further, the authors hypothesized a mechanism in which hypochlorite would be introducing oxidative defects in the form of oxygen-containing groups that could in a second step favor interaction with the enzyme, leading subsequently to degradation by peroxidase reactive radical intermediates.

Similarly, Vlasova et al. investigated the potential of degradation by different oxidants produced by MPO (i.e., active peroxidase intermediates, hypochlorite, and reactive free radicals) on carboxylated SWCNTs and compared the effectiveness of MPO degradation with the LPO action [32]. Using hypochlorite scavengers, they confirmed the major role of hypochlorite in SWCNT degradation by MPO. Both MPO and LPO were found to be more effective in the presence of chloride and bromide ions, respectively, demonstrating that hypochlorite and hypobromite accelerate the enzyme-mediated degradation. The same group more recently demonstrated that covalently polyethylene glycol (PEG)ylated SWCNTs can also undergo hypochlorite-/hypobromite-mediated oxidation [38].

The durability of SWCNTs was also assessed in a phagolysosomal simulant fluid medium, at pH 4.5 [34]. As for HRP, oxidized SWCNTs were prompted to degradation following addition of hydrogen peroxide, compared to pristine SWCNTs, which appeared to be resistant to these mild physiological oxidative conditions. In another study, also mimicking an acidic phagolysosomal environment (pH 4.5), an electrolyte solution (Gamble's solution) was used to assess whether SWCNTs can undergo structural modifications (diameter and length), but no such changes were observed [35].

11.4.2 Ex vivo Biodegradation of MWCNTs

The capacity of carboxylated MWCNTs to biodegrade when mixed with HRP or phagolysosomal simulating fluid (PSF) was also investigated and compared to carboxylated SWCNTs under the same conditions (Fig. 11.3) [36]. In less than 30 days, the carboxylated SWCNTs were fully degraded in both environments, whereas MWCNTs were still not fully degraded after 60 days, albeit their structure was clearly modified. On the basis of TEM analysis, it was concluded that degradation of SWCNTs occurred by an unzipping mechanism, and the lengths of MWCNTs were significantly shortened compared to the starting material when both HRP and PSF were used. The comparison between HRP (enzyme-based) and PSF (nonenzyme-based) degradation conditions demonstrated that there was no clear difference in terms of degradation potential using the two types of oxidative environments.

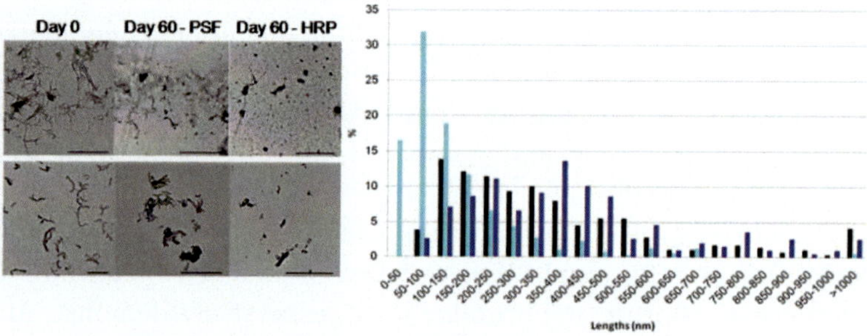

Figure 11.3 Ex vivo biodegradation of SWCNTs and MWCNTs. TEM images of ox-SWCNTs and ox-MWCNTs during the degradation process. The samples were collected at time 0 and 60 days in PSF and HRP oxidizing conditions. Scale bars correspond to 500 nm. Statistical distribution of the length of the ox-MWCNTs at the end of the degradation process (60 days). Blue bars correspond to starting material. Black and cyan bars correspond to PSF and HRP treatment, respectively. *Abbreviation*: TEM, transmission electron microscopy. Adapted from Russier et al., 2010, copyright Royal Society of Chemistry.

Zhao et al. also investigated the degradation of different MWCNTs (purified, oxidized, or nitrogen doped) under HRP/H_2O_2 conditions and offered a tentative explanation on the mechanism that was taking place [39]. In agreement with the previous study, the authors showed a decrease in length and diameter and proposed a layer-by-layer exfoliation process due to accumulation of structural defects on the outer wall of the tubes. The degree of carboxylation was correlated with the rate of degradation. Interestingly, oxidized

MWCNTs were less degraded than nitrogen-doped nanotubes, which completely disappeared after 90 days, suggesting that the amount of pre-existing defects present on the graphene backbone (which is higher for N-doped CNTs) is also an important parameter in CNM biodegradation. Overall, oxidative defects present in the outer layer seemed to act as initiator of further oxidation by reactive intermediates (such as $HOCl^-$) produced by the enzyme.

In another study, three kinds of MWCNTs were incubated in Gamble's solution (pH 4.5) to mimic the long-term degradation that may take place in phagolysosomes of macrophages [35]. Only the material identified as "long" reported a significant decrease in length after 10 weeks of incubation, with no change in diameter observed. Moreover a 30% decrease in weight for the same sample was recorded after 24 weeks of incubation. No mechanistic interpretations regarding degradation were offered in that study.

11.4.3 Ex vivo Biodegradation of Graphene

Only a single study has so far reported biodegradation of CNMs that do not belong to the family of CNTs but graphene [29]. In this work, graphene oxide (GO) and reduced graphene oxide (RGO) were exposed to a mixture of HRP and H_2O_2. HRP-catalyzed oxidation was efficient on GO but not on RGO, demonstrating, as mentioned before, the role of functional oxygen-containing groups as initiator sites for enzymatic activity. Degradation by enzymatic oxidation of the graphene lattice led to formation of holes in the planar sheet, with the diameter of these holes increasing with time. As for CNTs, the ultimate by-product of biodegradation was CO_2. Gel electrophoresis and modeling studies were conducted to understand the interactions of the GO or RGO with the HRP enzyme at the molecular level, suggesting that the HRP active site can achieve closer proximity to GO compared to RGO. Moreover, molecular modeling results indicated that HRP could oxidize GO sheets across their planar axis but not from their edges.

Putting together the different results generated by the ex vivo degradation of graphene, SWCNTs, and MWCNTs, the kinetics of degradation of one MWCNT can now be more precisely envisioned (Fig. 11.4).

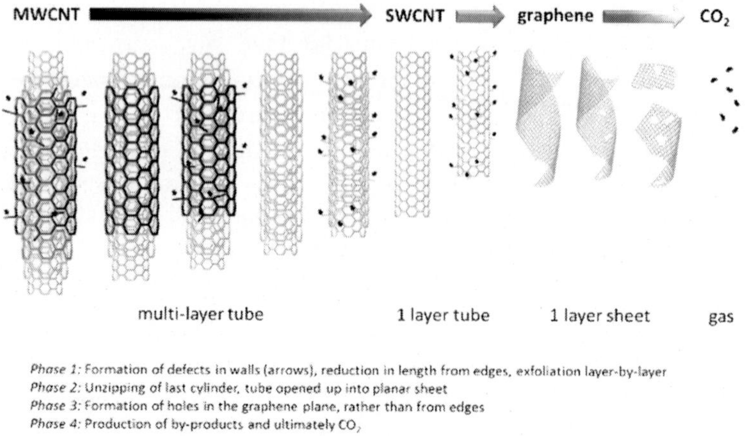

Figure 11.4 The continuum of graphene materials' degradation. The schematic is representing a proposed kinetic of MWCNT degradation into CO_2 through SWCNT and graphene successive states according to the current knowledge. Phase 1: Layer-by-layer exfoliation and shortening by tips of a multilayered tube to form a monolayered tube. Phase 2: Unzipping of the damaged one-layer tube to form a planar sheet. Phase 3: Formation of holes in the one-layer sheet. Phase 4: "Dissolution" into by-products, polycyclic aromatic hydrocarbons, and ultimately CO_2.

11.5 Biodegradation of CNMs in Living Systems

Beside the demonstration of ex vivo degradability, evidence has been reported that CNMs, and especially CNTs, are also degradable in vitro and in vivo (Table 11.2). At the cellular level, Kagan et al. were the first to report that carboxylated and IgG-functionalized SWCNTs targeted to human isolated neutrophils, could undergo MPO-related degradation after stimulated uptake [28]. In the same work, human monocyte-derived macrophages were also proven to be able to engulf carboxylated SWCNTs and to degrade them, though to a lesser extent compared to high-concentration MPO-containing neutrophils (13% and 100% of the original dose, respectively). Raman spectroscopy used as a mean to assess the evolution of CNT structure within neutrophils revealed that the amount of defects increased over time. In another study, PEGylated SWCNTs were shown to activate isolated neutrophils, which in response were

Table 11.2 In vitro and in vivo biodegradation studies

Type of CNT	Type of functionalization	Biological model	Time points	Techniques	Main results	References
MWCNT	pristine	Intratracheal instillation in rat	15 days	Transmission electron microscopy	Decrease in length of CNT	[41]
DWCNT	COOH	Human prostate adenocarcinoma cells PC3	0.5 to 24 hrs	Raman spectroscopy	Loss of Raman signal from the outer layer of the DWCNT over time; loss of Raman signal from inner layer of DWCNT appeared at 24 hrs only. Proposal: Exocytosis of CNT rather than degradation.	[40]
SWCNT	IgG-COOH-CNT complex vs COOH-CNT	Stimulated human neutrophils, human monocyte derivated macrophages	6h and 12h (48h for macrophage)	Raman spectroscopy, UV-Vis-NIR spectra	100% IgG-CNT degraded by neutrophil and to a lesser extent by macrophage (13%-up to 50% at 48hrs); non-IgG-functionalized CNT underwent 30% degradation by neutrophil; hypothesis: myeloperoxidase and NADPH oxidase are essential in CNT degradtion by neutrophils.	[28]
SWCNT	COOH	Pharyngal aspiration in wild-type and MPO knock-out mice	28 days	Hyperspectral imaging methodology and Photoacoustic imaging technology complete with Raman and NIR spectroscopies, and TEM	Inflammatory response is higher in myeloperoxidase knock-out compare to wild-type mice, similarly SWCNT oxidation and clearance from the lungs are reduced in those KO mice. Myeloperoxidase is essential in the lung biodegradation of SWCNT.	[42]
MWCNT	NH_3^+	Mouse cortical brain injection	2 and 14 days	Transmission electron microscopy and Raman spectroscopy	Uptake by microglia; partial loss of carbon tubular structure and evolution of Raman signature over time.	[43]

producing more MPO and hypochlorite [38]. Following the same pathway, biodegradation of SWCNTs was also demonstrated when incubated with primary eosinophils, which can produce HOBr via EPO [33].

Cellular degradation was also mentioned in a different study to explain the alteration of the graphene structure of oxidized RNA-coated double-walled CNTs probed by Raman mapping [40]. In that study, following internalization by human prostate adenocarcinoma cells, defects in the outer layer of CNTs increased and the outer diameter of the tubes was reduced. No experimental evidence was, however, provided to elucidate these observations.

At the tissue level, early suggestions of in vivo degradation of MWCNTs (i.e., reduction in length) in lung 15 days after intratracheal instillation were reported in 2008 [41]. However, detailed evidence of in vivo digestion of carboxylated SWCNTs via an MPO-mediated process was only provided in 2012 [42]. In that work, MPO knock-out mice were exposed to carboxylated SWCNTs via pharyngeal aspiration and compared to wild-type mice. The results showed that SWCNTs disappeared significantly from the lungs in the wild-type mice but not in the MPO knock-out mice. Raman spectroscopy and TEM were moreover used to confirm that degradation of SWCNTs was more pronounced in the wild-type mice than in the knock-out mice. This work clearly confirms the role of the enzyme in the degradation of CNTs, at least in the lung area where neutrophils are commonly the first responders toward exogenous threats.

In a relevant study for biomedical application, we have also reported that ammonium-functionalized MWCNTs (NH_3^+-CNTs) undergo degradation over time (from 2 to 14 days after injection) following an intracranial injection into the brain cortex of mice [43]. TEM analysis demonstrated that NH_3^+-CNTs were uptaken by microglia and that their cylindrical structure was altered. Moreover, Raman spectroscopy confirmed an evolution of the characteristic Raman signal for CNTs with modification of the D/G band intensity ratio, showing an increase of defects over time. Interestingly, the same NH_3^+-CNT compounds were used previously by our group to mitigate neurodegeneration after local ischemic damage via a local delivery of caspase 3 short interfering RNA (siRNA)-related sequence [44].

11.6 Biological Effects of Biodegraded CNMs

As mentioned before, biodegradation by oxidation of the CNM graphitic lattice could lead to the formation of polycyclic aromatic hydrocarbons, which have been identified as carcinogenic, mutagenic, or teratogenic chemicals for some of them. The biological effects (e.g., immunogenic profile) and potential toxicity of the by-products of CNM biodegradation have therefore to be investigated with close attention.

In the first study to explore this question, Kagan et al. demonstrated that products of CNTs, predegraded ex vivo by MPO treatment induce neither inflammation—in the form of lung neutrophil accumulation or increase of proinflammatory cytokine levels—nor formation of granulomas in pharyngeal instilled mice [28]. Those products of degradation were identified as short-chain carboxylated alkanes and alkenes. On the contrary, nondegraded CNTs (or nanotubes incubated with MPO or H_2O_2 alone that were not extensively modified) were inflammogenic and led to the formation of tissue granulomas. Moreover, partially degraded CNTs still induced inflammation, though to a lesser extent compared to nondegraded nanotubes. In a second work, they showed that lung inflammation response to CNT pharyngeal aspiration was stronger in MPO knock-out mice compared to wild-type mice, where CNTs can still undergo MPO-mediated degradation to alleviate their effects [42]. These results suggest that MPO-degraded CNTs were less toxic than the nondegraded tubes and that controlling degradation (i.e., triggering) could be a key toward safer use of CNMs for nanomedicine.

In another study, Donaldson et al. treated four different kinds of CNTs for up to 24 weeks in a modified Gamble's solution, pH 4.5 [35]. For one sample of long MWCNTs, they observed a 30% decrease of mass after 24 weeks' incubation and a shortening of the length of the tubes was reported at 10 weeks. Importantly, when intraperitoneally injected, the degraded CNTs (10 weeks' incubation) were less inflammogenic and fibrogenic and did not induce formation of granulomas compared to nonincubated (long) CNTs. Those results demonstrate that biological effects can be mitigated by degradation on a sample-specific basis.

11.7 Conclusions

For a long time, CNMs were perceived as materials with great promises for the biomedical field but with a major drawback: their nonbiodegradability due to a chemically inert graphitic structure. This fact has led to the conclusion that their translation into the clinic will never be achievable with the CNM products available. Recent reports have, however, demonstrated that CNMs undergo biodegradation under specific conditions and that these processes might also take place in living systems, offering new perspectives for the development of CNMs in nanomedicine.

At the moment, research on the biodegradation of CNMs is in its infancy: only the biodegradation of CNTs and graphene has been explored so far, and the knowledge and understanding of the mechanisms involved are still limited. Moreover, the CNMs used were in most cases not directly relevant for biomedical applications and were not studied in a realistic physiological environment (e.g., in the organ of accumulation after administration or after uptake by phagocytic cells). Therefore, there are still various investigations to be conducted to complete the understanding of the potency of CNM degradation in living systems and to decipher how to use this biodegradation property for the benefit of CNM biomedical applications.

Among those studies, the main questions to address will be (i) the inherent degradability of CNMs, (ii) the inherent degradability of biomedical constructs (i.e., CNMs modified to optimize their biocompatibility and biological activity), (iii) the different possibilities of CNM degradation in living systems (via enzymes, oxidizing agents, pH, or others), and (iv) how to control (i.e., enhance or inhibit) the biodegradation of carbon biomedical constructs. The knowledge that could generate such investigations will greatly improve the design of CNMs for biomedical applications and also participate to their potential clinical translation.

References

1. Chen, Z., Ma, L., Liu, Y., and Chen, C. (2012). Applications of functionalized fullerenes in tumor theranostics. *Theranostics*, **2**(3), pp. 238–250.

2. Bianco, A., Kostarelos, K., and Prato, M. (2008). Opportunities and challenges of carbon-based nanomaterials for cancer therapy. *Expert Opin. Drug Delivery*, **5**(3), pp. 331–342.
3. Mochalin, V. N., Shenderova, O., Ho, D., and Gogotsi, Y. (2012). The properties and applications of nanodiamonds. *Nat. Nanotechnol.*, **7**(1), pp. 11–23.
4. Feng, L., and Liu, Z. (2011). Graphene in biomedicine: opportunities and challenges. *Nanomedicine*, **6**(2), pp. 317–324.
5. Liu, Y., Dong, X., and Chen, P. (2012). Biological and chemical sensors based on graphene materials. *Chem. Soc. Rev.*, **41**(6), pp. 2283–2307.
6. Liu, Z., Tabakman, S., Welsher, K., and Dai, H. (2009). Carbon nanotubes in biology and medicine: in vitro and in vivo detection, imaging and drug delivery. *Nano Res.*, **2**(2), pp. 85–120.
7. Liu, Z., Davis, C., Cai, W., He, L., Chen, X., and Dai, H. (2008). Circulation and long-term fate of functionalized, biocompatible single-walled carbon nanotubes in mice probed by Raman spectroscopy. *Proc. Natl. Acad. Sci. U S A*, **105**(5), pp. 1410–1415.
8. Pastorin, G., Wu, W., Wieckowski, S., Briand, J. P., Kostarelos, K., Prato, M., and Bianco, A. (2006). Double functionalization of carbon nanotubes for multimodal drug delivery. *Chem. Commun.*, **11**, pp. 1182–1184.
9. Prato, M., Kostarelos, K., and Bianco, A. (2008). Functionalized carbon nanotubes in drug design and discovery. *Acc. Chem. Res.*, **41**(1), pp. 60–68.
10. Zhang, Y., Nayak, T. R., Hong, H., and Cai, W. (2012). Graphene: a versatile nanoplatform for biomedical applications. *Nanoscale*, **4**(13), pp. 3833–3842.
11. Georgakilas, V., Tagmatarchis, N., Pantarotto, D., Bianco, A., Briand, J. P., and Prato, M. (2002). Amino acid functionalisation of water soluble carbon nanotubes. *Chem. Commun.*, **24**, pp. 3050–3051.
12. Bianco, A. (2004). Carbon nanotubes for the delivery of therapeutic molecules. *Expert Opin. Drug Delivery*, **1**(1), pp. 57–65.
13. Tasis, D., Tagmatarchis, N., Bianco, A., and Prato, M. (2006). Chemistry of carbon nanotubes. *Chem. Rev.*, **106**(3), pp. 1105–1136.
14. Menard-Moyon, C., Fabbro, C., Prato, M., and Bianco, A. (2011). One-pot triple functionalization of carbon nanotubes. *Chem. Eur. J.*, **17**(11), pp. 3222–3227.
15. Gaillard, C., Duval, M., Dumortier, H., and Bianco, A. (2011). Carbon nanotube-coupled cell adhesion peptides are non-immunogenic: a

12.1.2 Exogenous DNA Damage

- *Environmental toxins.* Carcinogen-induced DNA damage is associated with increased cancer risks in humans. Chronic exposure to chemical carcinogens has been shown to be associated with various human cancers, and many of these chemicals result in the occurrence of DNA adducts that can lead to DNA mutations. One class of compounds, polycyclic aromatic hydrocarbons, is found as an airborne contaminant in soot, resulting from the incomplete combustion of hydrocarbons. Benzo[a]pyrene diolepoxide (BPDE) is derived from benzo[a]pyrene (BP), a polycyclic aromatic hydrocarbon that is ubiquitous in the environment. BPDE binds to deoxyguanosine and induces point mutations that are mainly GC $\rightarrow$ TA transversions. Other environmental contaminants includes acetaldehyde (AA) and ethylene oxide (EtO), which can generate large genomic deletions in human fibroblasts [5].
- *Radiation.* DNA DSBs can be generated from various endogenous processes (e.g., DNA replication, meiotic recombination, immunoglobin rearrangements, DNA repair), which involve the accidental or programmed generation of DSBs as intermediates, or from exogenous DNA damage as a result of radiation or exposure to chemicals/mutagens. Ionizing radiation can also result in single-stranded breaks and base changes, which interfere with DNA replication and will potentially give rise to DSBs as a result of replication "runoff." In particular, ionizing radiation in cancer therapy and topoisomerase inhibitors used in chemotherapy can give rise to DSBs. Other than direct damage, ionizing radiation can damage the cell indirectly by generating oxygen-free radicals and stimulation of oxidases and nitric oxide synthases. ROS and reactive nitrogen species (RNS) result in oxidative damages, including protein carbonylation, lipid peroxidation, and increased rate of spontaneous DNA mutations. ROS and RNS can generate extensive damages to DNA, such as DNA breaks, base changes, DNA crosslinks, and telomere dysfunction. For example, ROS can oxidize deoxyguanosine to 8-hydroxyguanine (8-oxodG). Failure of the repair of oxidized

bases carries a risk for mutation during replication, since 8-oxodG can mispair with adenine instead of cytosine, resulting in a C → A point mutation. Other than oxidation of bases, ROS can also result in increased depurination or other modification of bases. Cytosine reacts with S-adenosylmethionine, forming 5-methylcytosine, and this occurs mainly in the CpG sequences in the genome. Spontaneous deamination of 5-methylcytosine generates thymine, which results in a C→T point mutation in the next round of DNA replication. Therefore, if unrepaired, many of these DNA damages can potentially lead to neoplastic transformation of cells. Some of the most toxic lesions that can result from UV radiation are the formation of cyclobutane pyrimidine dimers (CPDs) and pyrimidine 6-4 pyrimidone photoproducts (6-4PPs), which require the nucleotide excision repair (NER) pathway for repair. UV-induced ROS can also result in severe oxidative damage and the oxidation of DNA bases, giving rise to 8-oxodG and 2,6-diamino-4-hydroxy-5-formamidoguanine (FapyGua) products.

- *Dietary factors.* Many interstrand crosslinking (ICL) agents result in a wide variety of DNA damages, including DNA monoadducts, intrastrand crosslinks, and DNA–protein adducts. ICL-inducing agents exist in natural food sources, that is, in certain edible plants, and are also produced endogenously. Metabolites such as malondialdehyde, generated from lipid peroxidation and prostagladin biosynthesis, crosslinks two deoxyguanosine residues in dCpG sequences. Malondialdehyde also crosslinks proteins and DNA, contributing to its toxicity in cells, and has been shown to be highly mutagenic [6]. Malondiadehyde has also been found in fresh and processed meats and heated edible oils [7–9]. It is evident that the genotoxicity of lipid oxidation products contributes to increase tumor frequency and atherosclerosis [9]. ICL is one of the most cytotoxic lesions, demonstrating selectivity against proliferating cells. Therefore, ICLs are commonly used in cancer therapy. Some of the common agents used include mitomycin C, cisplatin, and psoralens. ROS is necessary for lipid peroxidation, and the generation of malondialdehyde arises from endogenous metabolic pathways (mitochondria respiration) as well as during prostagladin biosynthesis and respiratory burst in immune cells. Dietary exposure to aflatoxin B1 was associated with increased liver cancer risk. Aflatoxin B1 binds to deoxyguanosine, forming AFB(1))-N(7)-guanine as the predominant DNA adduct, which is thought to be repaired by the NER pathway [10].

12.1.3 Repair of Various DNA Lesions by Specific DNA Repair Pathways

- *BER.* The BER pathway is one of the major pathways that a cell utilizes to defend against oxidation, alkylation, deamination, and depurination. BER is initiated by a group of DNA glycosylases that recognize and excise the modified bases, leaving AP sites in the DNA backbone. AP endonuclease (APE1) then nicks the DNA at the AP site, and DNA polymerase beta (POLB) then fills the gap and DNA ligase seals the DNA nicks to complete the repair process. The first DNA glycosylase identified is uracil DNA glycosylase (UNG), which removes the mutagenic uracil DNA base [11]. Uracil arises from the deamination of cytosine, and base-pairs with adenine, resulting in a C→T transition mutation. Another commonly modified base is 8-oxo-guanine (8-oxoG), which results from the oxidation of the guanosine base, causing it to base-pair with adenine as well as with cytosine, resulting in a C→T transition. OGG1 recognizes and removes 8-oxoG as well as another modified base, FapyG, 2,6-diamino-4-hydroxy-5-formamidopyridine. 3-methylpurine DNA glycosylases (MPG) recognize and process a variety of the modified DNA bases, including deaminated adenine (hypoxanthine), guanine (xanthine), and N3- and N7-alkylated purines, as well as cyclic etheno adducts of adenine and guanine, which are induced during lipid peroxidation. There are many other DNA glycosylases that can initiate BER, including SMUG1, TDG, MBD4, MUTYH, NTHL1, NEIL1, NEIL2, and NEIL3. The primary function of DNA glycosylase is to cleave the *N*-glycosidic bond, removing a free (modified) base and creating an AP site. The one-nucleotide short-patch BER involves a DNA glcosylase, AP endonuclease, DNA POLB, and DNA ligase 3. In the alternative long-patch BER pathway, AP endonuclease action recruits

DNA POLB, and DNA polymerase delta (POLD), flap-structure specific endonuclease 1 (FEN1), DNA replicative ligase 1 (Lig1), and PCNA. The coordinated action of these enzymes leads to the polymerization of long tracts of DNA, allowing the repair of more than one nucleotide.

- *NER.* One of the important DDR pathways triggered by exposure to UV light is the NER pathway. UV radiation causes the formation of CPDs and 6-4 photoproducts (6-4 PPs), which are bulky lesions that distort the DNA double helix. NER recognizes any bulky adducts that distort the DNA helix and repairs these lesions through the coordinated action of endonucleases, which incise and remove a single-stranded region containing the lesion, and polymerases, which perform the repair synthesis using the intact template strand. ICLs also require NER for repair, and many platinum-based drugs in chemotherapy generate platinum adducts, which are substrates for the NER pathway. The importance of NER is exemplified by the severity of the genetic diseases carrying mutations of the NER repair genes, xeroderma pigmentosum (XP) and Cockayne's syndrome (CS). There are at least two main subpathways of NER, the global genome repair (GGR) and the TCR. The key difference between the two pathways lies in the set of proteins that recognize the damage. In GGR, XPC-Rad32B, DDB1, and DDB2 proteins are involved, and in TCR, CSA and CSB proteins recognize and bind to types of DNA damage. Following damage recognition, both pathways utilize XPB/XPD/TFIIH to unwind the DNA at the sites of damage and XPG and XPF-ERCC1 endonucleases to incise the DNA, removing the damaged single-stranded DNA. Polymerases delta or eta then fills the gap, followed by DNA ligase to seal the nicks. Recently, hMus81-Eme1 structure-selective endonuclease involved in HR was found to be required for DSB formation after mitomycin C and cisplatin treatment (both compounds results in ICLs), thus implicating hMus81-Eme1 in the NER pathway.
- *Mismatch repair (MMR).* MMR is the main pathway for the repair of base–base mismatches and insertion or deletion loops that forms during DNA replication. Defects in the genes that control the MMR pathway have been detected in

cancers and are associated with microsatellite instability. Microsatellite are short tandem DNA repeats that are more prone to replication slippage, resulting in increase or decrease in the number of DNA residues in each microsatellite. These postreplication errors are normally recognized by the MMR pathway. The canonical pathway of MMR comprises several key players, including MutSa (MSH2/MSH6), which recognizes the distortion in the DNA containing the mismatch and recruits the heterodimer MutL, which forms a complex that recruits a DNA exonuclease, EXO1, to remove the short stretch of DNA along with the DNA mismatch. The MutL heterodimer exists in several forms: MutLa (MLH1 and PMS2), MutLb (MLH1 and PMS1), or MutLg (MLH1 and MLH3). A DNA polymerase then synthesizes the DNA to replace the excised sequence, and a DNA ligase joins any remaining gaps in the DNA. In addition, MMR also mediates the response to some DNA damage-inducing agents that alter DNA bases, such as thioguanine, methylmethanesulfonate, 5-fluorouracil, and oxidizing and methylating agents. Activation of the MMR pathway also leads to the activation of ATM/ATR-dependent cell cycle checkpoints eliciting cell cycle arrest, repair, or apoptosis. MMR also has suggested roles in mitotic and meiotic recombination and antibody class-switch recombination pathways.

- *HR.* One of the most common DNA lesions is DNA DSBs. DSBs can arise from radiation, environmental toxins, or endogenous free-radical attacks, and are closely associated with mutagenic chromosomal DNA rearrangements and chromosomal loss. Chromosomal rearrangements have been observed in radiation-induced thyroid carcinomas and in patients who developed thyroid cancer after the Chernobyl incident. Therefore, radiation-induced DSBs are potential mutagenic lesions that could give rise to DNA recombination that result in deleterious events. HR provides a high-fidelity, template-dependent repair of DNA damages, including DNA double- and single- stranded breaks and DNA interstrand crosslinks (ICLs). HR is thought to predominate in the S and G2 phases of the cell cycle and competes with the nonhomologous end-joining (NHEJ) pathway for DSB repair and TLS in the DNA damage tolerance pathway.

HR requires ATM, an important checkpoint kinase that coordinates the activity of DSB repair with cell cycle checkpoint activation. The autophosphorylation of ATM precedes the phosphorylation of a large number of important substrates involved in the checkpoint response (e.g., p53, Chk2), the amplification of the DNA damage signal (e.g., γ-H2AX), and the recruitment of repair proteins to the DSB (e.g., SMC1, KAP1) [12]. Prior to ATM recruitment, the Mre11-Nbs1-Rad50 complex is recruited to the DSB, which functions to resect the DNA ends and prepare for repair synthesis. The central reaction in HR involves the search for homology, which is catalyzed by Rad51, which binds to the processed DNA DSB, forming a nucleoprotein filament on the single-stranded 3' end. DNA strand invasion proceeds with the Rad51-ssDNA filament invading a template duplex DNA to initiate repair synthesis. A "D-loop" forms, which primes synthesis from the 3'-invading strand off the template DNA. Rad52 catalyzes the second end-capture event, resulting in a double Holiday junction (dHJ), which is processed by dissolution, catalyzed by BLM/TopIIIa, into noncrosssover products or resolution, and catalyzed by structure-specific endonucleases (Mus81/Mms4, Slx4/Slx1, Gen1), into crossover and noncrossover products.

Several repair pathways compete for the repair of DNA DSBs, including HR, NHEJ, and single-strand annealing (SSA). Besides the classical HR-mediated DSB repair pathway just described, other variations of HR utilized for the repair of DSB include synthesis-dependent strand annealing (SDSA) and break-induced replication (BIR), which also involve homology search and template recognition.

NHEJ consists of at least two biochemically distinct subpathways: the classical NHEJ (C-NHEJ), which requires the Ku-heterodimeric protein, and the alternative NHEJ (A-NHEJ), which generates microhomology at the repair junctions. C-NHEJ involves the use of Ku70 and Ku86 heterodimers to protect the DSB ends from exonucleases and to recruit DNA protein kinase (DNA-PKcs. The trimeric complex, known as the DNA-dependent protein kinase complex (DNA-PK), activates the kinase activity of DNA-PKcs, resulting in the phosphorylation of Artemis. Ligation of the DNA ends is then catalyzed by the trimeric DNA lig4-XRCC4 complex, consisting of the catalytic DNA ligase IV and its accessory factors, XLF and XRCC4. In A-NHEJ, the broken ends are thought to be resected to reveal the 3′ single-stranded overhangs containing regions of microhomology, and this process involves MRN and CtIP. Ligation of DNA ends is dependent on DNA ligase III instead of DNA ligase IV [13–15].

12.2 Evidence for Nanomaterial-Induced Genotoxicity and Carcinogenicity

Chemically, nanomaterials can be categorized into three distinct groups [16]:

- Carbon-based materials, including fullerenes, single-walled carbon nanotubes (SWCNTs), multiwalled carbon nanotubes (MWCNTs), and their derivatives
- Metal-based materials encompassing quantum dots (QDs), rare-earth materials, pristine metal nanoparticles (NPs) such as silver and gold, and metal oxides such as TiO_2, ZnO, and Fe_2O_3
- Organic dendrimers consisting of homo- and copolymers assembled to contain branched units, which can be tailored to have specific and multifunctional capabilities

These categories of nanomaterials have been heavily studied for various applications such as biodiagnostics, therapeutics, cosmetics, electronics, automobiles, construction, etc. For this reason, there is also growing interest in evaluating their toxicity. In the following sections, studies on the genotoxic and carcinogenic potential of these materials are reviewed.

12.2.1 Carbon-Based Nanomaterials

Carbon-based nanomaterials are dominated by those made up entirely of carbon (fullerenes). These come in a number of different forms and shapes such as sheets (graphene), spheres (C_{60}), ellipsoids (C_{70}), and tubes (SWCNTs and MWCNTs, respectively) [17, 18]. These nanostructures are popular for their enhanced structural, mechanical, and electrical properties, as well as their flexibility to be tuned for various applications. As a result, fullerenes are being widely explored in biomedical applications in areas such as cell biology and drug delivery in nanomedicine [19, 20].

The first important family of carbon-based nanomaterials is spherical fullerenes. There are many conflicting reports in the literature regarding the genotoxicity of these. Pristine C_{60} and its derivatives showed potential for genotoxicity with irradiation [21, 22] and without [23] in a number of cell lines. Dhawan et al. [24] reported DNA damage in lymphocytes using the comet assay. Gene mutation assays by Sera et al. [25] on *Salmonella* strains TA102,

TA104, and YG3003 also showed positive albeit weak mutagenicity of C_{60} dispersed in polyvinylpyrrolidone (PVP) under visible light irradiation. In vivo investigation was carried out by Totsuka et al. [26], who showed DNA damage using the comet assay in the lungs of C57BL/6J mice following single intratracheal instillation of 0.2 mg/kg of C_{60}, but this effect was not observed at a lower concentration of 0.05 mg/kg. Similarly Folkman et al. [27] showed oxidative DNA damage and increase levels of 8-oxoG in the lung and liver tissue of female Fischer rats after a single intragastric dose of 0.64 mg/kg of C_{60}. However, a host of studies also reported contrary results [28]. Shinohara et al. [29] concluded that C_{60} NPs had no genotoxic influence on bacteria (*Ames Salmonella* and *Escherichia coli* [*E. coli*]), Chinese hamster lung fibroblasts (CHL/IU) cells, and imprinting control region (ICR) mice, using the bacterial reverse mutation assay, imprinting control region in vitro chromosome aberration assay, and the in vivo micronuclei assay, respectively.

The second broad category of carbon-based nanomaterials is nanotubes. These are further separated into two main classes depending upon the number of carbon atom layers that make up the wall, SWCNTs and MWCNTs (Fig. 12.1). Both have been studied rigorously for their toxic effects in vivo and in vitro. However, the jury is still out as many varying observations have been reported. Most studies do report toxic responses, especially in the context of inhalation as the point of entry [30]. Specific kinds of SWCNTs and MWCNTs are cyto- and genotoxic and may induce mesothelioma in mice similar to asbestos [31–33]. Genotoxicity has enjoyed its fair share of attention, and the reports are as contradictory as those focused on cytotoxicity. Using high purity SWCNTs, Naya et al. [34] were recently able to show that *E. coli* and *Salmonella typhimorium* experienced no mutagenicity to a dosage of 5–500 μg/plate, while CHL/IU cells experienced no genotoxicity in in vitro chromosomal aberration tests with a dosage of 300–1000 μg/mL SWCNTs. The same group also carried out in vivo studies on mice administered orally with a maximum dosage of 200 mg/kg for two days. Micronucleus (MN) assays on the bone marrow showed no genotoxic potential of the SWCNTs. On the other hand, Cicchetii et al. [33] showed that commercially purchased SWCNTs functionalized with a carboxylic group were internalized by human gingival fibroblasts within two hours and enhanced the frequency of chromosomal aberrations at 50–100 μg/mL, using the cytokinesis-block micronuclei (CBMN) test.

This frequency was decreased at 125–150 μg/mL. However, comet assay results from the same study showed significant DNA damage at all concentrations of SWCNTs used, which suggested that the drop in micronuclei frequency at higher concentrations was due to a higher level of DNA damage, which CBMN could not detect accurately. On the same note, significant DNA damage induced by SWCNTs had also been reported in murine macrophages (RAW 264.7) [35], Chinese hamster lung fibroblasts (V79) [36], normal and malignant human mesothelial cells [37], and human bronchial epithelial cells (BEAS-2B) [38]. An interesting study reported recently by Wang et al. [39] looked at chronic exposure to SWCNTs with the rationale that carcinogenesis in itself is a chronic process and requires persistent stimulus. Within 12 weeks of exposure to SWCNTs at a subtoxic concentration of 0.02 μg/cm^2, BEAS-2B cells displayed morphological alterations and transformed into tumor cells, which were resistant to apoptosis and displayed increased angiogenic activity coupled with cell migration and invasive behavior that is uncharacteristic of untreated cells. The complementary in vivo study showed tumor formation within a week in SWCNT-treated mice. These tumors were growing and showed classic cancer cell traits upon histological investigation. Furthermore a several-fold decrease in the expression of the phosphorylated p53 tumor suppressor was observed.

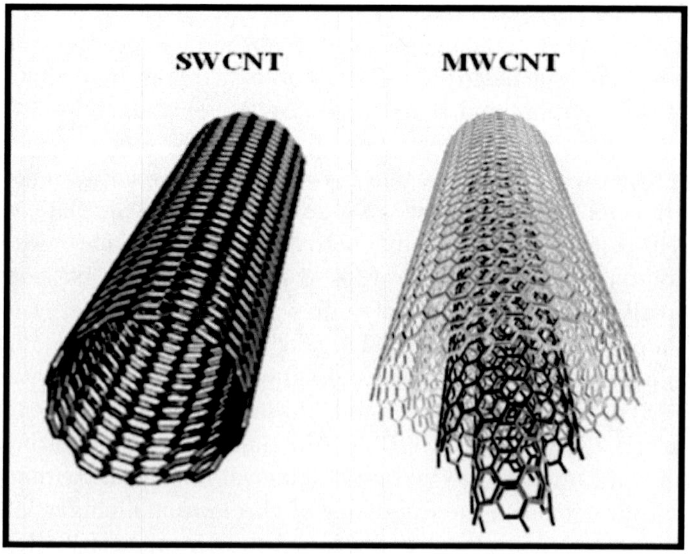

Figure 12.1 SWCNTs vs. MWCNTs [40].

MWCNTs, on the other hand, have been shown to induce relatively lower levels of genotoxicity. Cavollo et al. [32] performed an in vitro study on adenocarcinomic human lung alveolar epithelial cells (A549) using commercial MWCNTs and reported no significant DNA damage despite observing strong cytotoxicity, cell membrane damage, and cellular uptake. Separately, a lack of mutagenic effect on bacteria regardless of metabolic activation using chemical vapor deposition–prepared MWCNTs with <0.1% amorphous carbon residue and metal contaminants had also been reported [41]. Nagai et al. [42] performed a study on MWCNTs of different diameters and suggested that thinner CNTs can pierce a cell and are therefore more toxic. This could explain the difference between the reported genotoxicity of SWCNTs and MWCNTs, although there is currently no direct evidence. Nonetheless, reports suggesting that MWCNTs can be genotoxic are not uncommon. Increased DNA damage in RAW 264.7 cells using the CBMN test and the comet assay has been shown [35]. A comparative study between MWCNTs and asbestos from a toxicogenomic point of view had also been carried out, which delineated more than 1000 genes up-regulated and more than 1500 genes down-regulated by MWCNTs as well as asbestos [43]. Up-regulated genes included 11 mesothelioma-related and 22 lung-cancer related genes as well as others related to signal transduction, cell morphogenesis, protein complex assembly, and angiogenesis. Genes involved in neuropathy, schizophrenia, and Alzheimer's, as well colorectal, prostrate, and bladder cancer, were also up-regulated. Down-regulated genes included those implicated in signal transduction, cytokine interaction, colorectal lung, and gastric cancer and regulatory genes affecting brain, lung, and liver functions. A separate study on the ability of MWCNTs to produce reactive species found that these can induce cytokine-inducible nitric oxide synthase (iNOS) expression and reactive nitrogen species (RNS) generation, which could lead to DNA damage and carcinogenesis [44] through an alternate route to commonly studied ROS-induced DNA damage.

In an attempt to mitigate the contradictory reports on the roles of carbon-based nanomaterials in genotoxicity and carcinogenicity, Jacobsen et al. [45] conducted a comparative study between different carbon-based nanomaterials, including carbon black, fullerenes (C_{60}), and SWCNTs, as well as nanogold and QDs in ApoE–/– C57 mice. It was shown that SWCNTs were more genotoxic than spherical fullerenes but less genotoxic than CB. However, Yang et al. [46] studied the cyto- and genotoxicity of a pool of inorganic NPs (CB, SWCNTs, SiO_2 NPs, and ZnO NPs) on primary mouse embryonic fibroblasts and observed that "pure" SWCNTs induced significant dose-dependent cytotoxicity and significantly high levels of ROS, second only to ZnO, and the highest levels of DNA damage measured by the comet assay. The study suggested that due to their high aspect ratio, SWCNTs can penetrate cell membranes and even nuclear envelops, causing DNA damage therein.

12.2.2 Metal-Based Nanomaterials

Metal-based nanomaterials are one the most important classes of nanomaterials, one that is infiltrating the market of consumer products at a phenomenal rate. As of March 2011, the major materials used in nanomaterial-based consumer products were dominated by metals and their oxides with silver, leading the charts followed by carbon (including fullerenes and carbon nanotubes), titanium (including titanium oxide), silicon/silica, zinc (including zinc oxide), and gold.

While nanomaterial-driven toxicity is discussed in detail elsewhere in this volume (see, for instance, Chapter 2), it is worthwhile to mention a few relevant parameters that govern metal and metal oxide toxicity. For instance, ions that have leached out from the metal-based NPs surface or core, and/or reactive species that may be formed on the surface of the metal-based NP through the overcoming of the crystal's band gap, are known culprits. The former is more evident when sparsely soluble nanomaterials are considered. With the enhanced surface area, even sparsely soluble materials can experience a greater solubility and therefore ions are readily formed and can disturb the osmotic pressure as well as the normal homeostasis of cells, leading to stress. The latter is more dominant in semiconductors, for example, oxides, which produce electron–hole pairs when provided with energy (through light, for example) enough to overcome its band gap. These charged species can cause the formation of reactive species like radicals, which can cause stress (oxidative, etc.) to the cells. This stress can then cause DNA damage. By no means are these two the only mechanisms, but they are the most studied and documented ones. Neither are these two mechanisms, or others, mutually exclusive, and a whole cocktail of concurrent processes may be responsible for the overall toxicity of nanomaterials. This will be delved into in closer detail in the following section.

Studies on the genotoxicity and clastogenicity of metal-based nanomaterials are plentiful. In fact most of these materials have been studied with much attention to the influence of physicochemical parameters and underlying mechanisms. Silver (Ag) NPs have shown the ability to cause DNA damage in numerous studies. For example, Li et al. [48] showed that 5 nm uncoated Ag NPs were genotoxic to human lymphoblastoid TK6 cells. However, such genotoxicity can be mitigated with a coating of protein or surfactant. Uygur et al. [49] compared the toxicity of dialyzed pristine Ag NPs (15 nm), polyethylene glycol (PEG)-coated Ag NPs (PEGylated Ag NPs), and Ag nanorods in a number of cell lines and showed that they were genotoxic to peripheral blood cells when tested using the CBMN assay. Foldbjerg et al. [50] studied polyvinylpyrrolidone (PVP)-coated Ag NPs with a primary size of above 50 nm using human adenocarcinomic human lung alveolar epithelial cells (A549) and observed bulky DNA adducts in the nucleus through (32)P postlabeling, which strongly correlated to oxidative stress readings. Singh et al. [51], however, used a novel glycolipid, sophorolipid, to conjugate 10 nm Ag NPs and observed no genotoxicity on human liver carcinoma cells (Hep G2) up to a concentration of 100 μg/mL. Kim et al. [52], on the other hand, concluded that Ag NPs caused DNA damage in BEAS-2B and mouse lymphoma cells (L5178Y) but did not cause mutagenicity. In yet another recent study, starch-coated Ag NPs of primary sizes between 6 nm and 20 nm were found to be cytotoxic, genotoxic, and antiproliferative when introduced to human fibroblasts and glioblastomas [53]. From this study, the authors suggested that Ag NPs can cause perturbations in mitochondria respiration, leading to oxidative stress, affecting ATP production, and resulting finally in DNA damage. This is believed to be a classical pathway for metal nanomaterial–induced DNA damage, especially when they are soluble.

Gold (Au) is another important nanomaterial with strong potential in biodiagnostics and therapeutics [54]. Au NPs are generally considered as inert as many comparative studies have suggested [45]. Comet and CBMN assays done on lung, liver, and blood cells and the micronuclei assay done on circulating reticulocytes of male Wistar rats after three consecutive instillations of Au NPs showed negative genotoxic effects [55]. A similar study looking at polychromatic erythrocytes of the bone marrow of Wistar rats corroborated this result [56]. In *Drosophila*, however, Vecchio et al. [57] showed that Au NPs caused phenotypic changes, which can be hereditarily passed down.

Titanium dioxide (TiO_2) is one of the most popular nanomaterials used in consumer products, along with zinc oxide (ZnO). Both of these have found a wide range of applications as photoactive and UV-absorptive materials. It appears that TiO_2 imparts little cytotoxicity in its ground state (without photoactivation), as alluded to by a number of studies. However, genotoxicity influence of TiO_2 has been shown. Wang et al. [58] found that at 65 μg/mL of TiO_2 NPs showed a significant tail moment in the comet assay and micronuclei formation in the CBMN assay when exposed to human lymphoblastoid cells. The same study also showed positive genotoxicity using hypoxanthine-guanine phosphoribosyltransferase (HPRT) mutation assays at a TiO_2 NP concentration of 130 μg/mL [58]. Xu et al. [23] showed an increase in the number of mutations in transgenic mouse primary embryo fibroblasts (MEFs) in a dose-dependent manner due to interactions with anatase TiO_2 NPs of 5 nm and 40 nm. Correlation to $ONOO^-$ formation due to TiO_2 NPs was also shown. Kang et al. [59] showed a significant increase in micronuclei formation and DNA breakages in peripheral blood lymphocytes, as well as p53 accumulation, while Bernardeschi et al. [60] showed that both anatase- and rutile-phase TiO_2 cause DNA fragmentation in human leukocytes.

ZnO NPs, on the other hand, are typical definers of nanomaterials. The bulk form of ZnO was considered inert and safe and has been used in consumer products for decades, but the nanometric form has showed acute toxicity and significant genotoxic influence. DNA damage in human nasal mucosa cells has been detected upon exposure to ZnO NPs at concentrations as low as 10 μg/mL [61], while the same in human epidermal keratinocytes exposed to 8–14 μg/mL ZnO NPs for six hours has also been demonstrated [62]. The same group also showed DNA damage in HepG2 cells exposed to ZnO NPs at similar concentrations using the Fpg-modified comet assay, which is more sensitive to single-strand DNA breaks [63]. It was shown that *N*-acetyl cysteine (NAC), a common antioxidant, afforded some protection against the DNA damage potential of ZnO. This indicates that DNA damage was due to increased oxidative stress, which was confirmed using 2′, 7′-dichlorofluorescein (DCF) fluorescence as the indicator. Similarly, Sharma et al. [64] showed significant increase in DNA damage in human epidermal cells (A431) exposed to ZnO

NPs at concentrations as low as 0.8 μg/mL using the comet assay. In vivo oral exposure of 300 mg/kg of ZnO NPs in male Swiss albino mice also showed significant DNA damage in liver cells compared to the control, measured by the Fpg-modified comet assay [65]. A mechanistic study by Ng et al. [66] showed using BEAS-2B cells that p53, a major tumor suppressor, was a mediator in DNA damage caused by exposure to ZnO NPs, leading to apoptotic responses. Cells with the p53 gene knocked down were shown to be less susceptible to the genotoxic potential of ZnO NPs.

Silicon dioxide (SiO_2) or silica NPs, in contrast, are found to be relatively safe, with most studies showing little cyto- and genotoxicity effects. Barnes et al. [67] declared SiO_2 NPs free from the genotoxic effect on 3T3-L1 fibroblasts, while Downs et al. [55] suggested that SiO_2 NPs cause very small damage in vivo, which was probably secondary in nature through the release of inflammatory cell–derived oxidants.

Iron oxide (FeO_2, Fe_2O_3) NPs are thought to be genotoxic depending on their redox state, as suggested by Singh et al. [68], but most studies agree that the genotoxic influence is mediated via ROS generation or oxidative stress induction [68, 69]. Konczol et al. showed ROS-mediated genotoxicity in A549 cells [69]. Using bacterial assays, however, Szalay et al. [70] concluded that 10 nm iron oxide NPs were devoid of the mutagenic effect on *Salmonella typhimulium*. Coating the iron oxide can influence genotoxicity, as shown by Zhang et al. [71], who observed no genotoxic responses of L-glutamic acid–coated FeO_2 NPs on Chinese hamster lung fibroblasts using comet and CBMN assays. However, Doak et al. [72] showed induction of gross chromosomal change and significant clastogenicity by dextran-coated Fe_2O_3 NPs using CBMN and HPRT assays.

Aluminum oxide (Al_2O_3) NPs have shown slight potential for genotoxicity but not mutagenicity. Kim et al. [73] employed comet and mouse lymphoma thymidine kinase gene mutation (MLA) assays in BEAS-2B and L5178Y cells and showed that Al_2O_3 NPs could cause primary DNA damage (and cytotoxicity) but not mutagenicity. Balasubramanyam et al. evaluated, in three different studies, the effects of Al_2O_3 NPs on peripheral blood cells of female Wistar rats [74] using comet and CBMN assays, on bone marrow of Wistar rats undergoing oral exposure [75], and on *Salmonella typhimurium* [76]. Both rat studies confirmed dose- and size-dependent genotoxicity, where Al_2O_3 NPs were more genotoxic compared to their bulk forms and controls [74, 75]. The Ames test on bacteria, however, suggested that Al_2O_3 was devoid of mutagenicity [76]. Simon et al. [77] conducted a comparative study between nanomaterials, including Al_2O_3, TiO_2, and Au using A549, HepG2, and NRK-52E cells to represent the lungs, liver, and kidneys, respectively. Using the MN assay, the comet assay, and gamma-H2AX histone foci quantification, genotoxicity induced was shown to be weak and DNA damage was limited to single-strand breaks and/or alkali-labile sites.

12.2.3 Polymeric Nanoparticles

Polymers have significant potential in nanomedicine applications, primarily due to their established compatibility with various biological systems, their versatile processibility, and the possibility of producing blends and copolymers of varied properties to suit different needs. Specifically, polymeric nanomaterials could find important uses as vehicles for the delivery of bioactive agents, including drugs, DNA, short interfering RNA (siRNA), and growth factors. There are numerous polymers that have been approved by the US Food and Drug Administration (FDA), including water-soluble ones such as PEG [78], PVP [79], copolymers of *N*-(2-hydroxypropyl) methacrylamide (PHMPA) [80], and Pluronic® block copolymers [81], as well as biodegradable ones such as poly(lactic-co-glycolic acid) (PLGA) [82], chitosan [83], poly(orthoesters) (POE) [84], and polyisohexylcyanoacrylate (PIHCA) [85].

As with other nanomaterials discussed in this chapter, the toxicity of these polymers in any nanoconfiguration needs to be independently evaluated, even with proven compatibility records when used in bulk form. Generally, the bulk of reported studies suggest that genotoxicity, and in fact the overall toxicity, of polymeric nanomaterials is much less significant than the other nanomaterial types discussed in this chapter. For example, chitosan and PMAA NPs induced genotoxicity in human lymphocytes only at very high dosages [86]. PLGA NPs (95 nm) showed no genotoxicity influence in mouse fibroblasts (3T3) and human lymphocytes, even up to a concentration of 540 μg/mL [87]. Poly(epsilon-caprolactone)-poly(ethylene glycol)-poly(epsilon-caprolactone) (PCL-PEG-PCL, PCEC) nanomaterials also showed no genotoxicity potential when tested both in vitro and in vivo [88]. More complicated copolymers, such as a tri-block copolymer of monomethoxy (polyethylene glycol)-poly(D,L-lactic-co-glycolic acid)-monomethoxy(polyethylene glycol) (MeO-

PEG-poly[D,L-lactic-co-lycolic acid]-PEG-OMe, PELGE) resulted in only very weak clastogencity when tested with various PEG ratios [89].

Nonetheless there are many polymers that have not fulfilled their remarkable potential for nanomedicine applications due to concerns over their genotoxicity. One such example is polyethyleneimine (PEI), which is considered the most effective cationic polymer for pulmonary gene delivery [90]. However, cationic polymers are more toxic than water-soluble, nonionic, or even anionic polymers [91]. This is evident as a host of studies had observed toxic responses to PEIs. Chollet [92] reported lethal toxicity at a dosage as low as 85 mg of L-PEI/DNA complexes in nude mice. Using luciferase reporters in Balb/c mice, Regnström et al. [93] showed that between 24 to 72 hours after intratracheal administration, PEI up-regulated genes in the lungs involved in inflammatory responses, such as cyclooxygenases 1 and 2, suggesting the material was possibly involved in the development of adverse reactions. In comparison chitosan up-regulated genes that protect cells from oxidative stress and inflammation, such as heme oxygenase-1 and catalase. However, Beyerle et al. [94] observed that PEI-based polymers had no mutagenicity or genotoxicity effects in MutaMouse lung epithelial cells.

Collectively, current data available from the literature suggests that the major types of nanomaterials can cause varying degrees of genotoxicity in different biological systems, both in vitro and in vivo. Among them, polymeric nanomaterials are the least genotoxic and could be the most promising material type to be used in nanomedicine applications (see Chapter 1). Nonetheless, conflicting results from reported studies have risen due to the multitude and interplay of various physical and chemical parameters that affect the mechanism of interaction between nanomaterials and biological systems. These are further discussed in the subsequent sections.

12.3 Mechanisms of Nanomaterial-Induced Genotoxicity and Carcinogenicity

12.3.1 Physicochemical Properties

It is the unique physical and chemical properties of nanomaterials that have brought about the nanotechnology revolution. However, these same unique properties are also responsible for the plethora of biochemical phenomena arising from the physiological and biomolecular interactions between nanomaterials and biological systems. Whether these effects turn out to be beneficial or catastrophic will hinge on our understanding of the mechanisms of such interactions. Characterization of the physical and chemical properties of nanomaterials is therefore important before exploring possible applications in nanomedicine (see Chapter 3). Nanomaterial characterization usually refers to an array of tests to measure primary and secondary particle size, size distribution, shape, chemical composition, surface area, surface charge, etc. In this section, the influence of these parameters of potential genotoxicity of nanomaterials is discussed.

- *Size.* Nanomaterials are defined as materials with at least one physical dimension smaller than 100 nm [95]. Materials of that size can potentially penetrate the body more readily through different routes of entry (inhalation, ingestion, dermal penetration). They can contact and interact with living cells at the molecular level and initiate size-dependent biochemical reactions. In terms of cellular uptake, pristine or aggregates of nanomaterials up to 200 nm can be internalized into cells by diffusion or endocytosis [96, 97]. Evidence has clearly demonstrated that size has a prominent impact on toxicological consequences of nanomaterials [98, 99]. Generally, NPs can penetrate deeper into a variety of organs such as skin, lungs, and spleen and localize in different cell types in comparison to their microsized counterparts [100–103], thereby inducing genotoxicity and carcinogenicity.
- *Morphology.* Very few studies have focused on the shape of NPs in connection with their genotoxicity, but shape is an important influence. A recent finding reported by Muller et al. suggested that the inflammatory response and toxic effects of MWCNTs remarkably decreased after the removal of their structural defects [104]. Needle- and plate-shaped hydroxyapatite NPs were also found to induce more significant cytotoxicity and inflammatory IL-6 expression in human lung epithelial cells (BEAS-2B) compared to spherical hydroxyapatite NPs [105]. In addition, the morphology of nanomaterials determines their rate of cellular uptake. It has been demonstrated that the internalization of spherical NPs is faster than that of

nanorods, while the speed of internalization of cylindrical-shaped nanomaterials increases with their aspect ratio [96, 98].

- *Surface area.* In addition to their direct role in influencing nanomaterial toxicity, size and morphology of the particles also determine their surface area. The total surface area per unit mass of particles increases with decreasing size of the particles; the surface area of spheres is smaller than that of octagonal structures of the same size/volume. Surface area is very important because it governs, to a large extent, the reactivity and catalytic activity of nanomaterials. Surface atoms have less binding partners than atoms in the bulk of the material, and the electrons in their valence orbitals are not in the energetically most favorable configuration. Therefore, they exhibit higher tendency to react with other molecules and stabilize their electron configuration by formation of bonds. Hydroxyapatite NPs with the smallest surface area per unit mass registered the lowest level of ROS in BEAS-2B cells, compared to similar NPs with a higher surface area per unit mass [106]. Besides the increase of reactivity with decreasing particle size caused by the increase in surface area per unit mass, smaller particles tend to have more edges and corners per unit surface area. Atoms at such sites have even less binding partners than those forming planar surfaces, and therefore, reactivity per unit surface area also tends to increase with decreasing particle size [107]. Once they have penetrated the cellular environment, NPs are, due to their larger surface area, more likely to cause damage by direct chemical interactions with biomolecules or by promoting oxidative reactions compared to microsized particles of the same total mass.
- *Surface charge.* NPs having positively charged surfaces seem to be more cytotoxic than particles with anionic surfaces. It, however, remains unclear whether their toxicity is related to direct interaction of the positive charges on their surface with anionic biomolecules, such as nucleic acids, or whether their positive charge enhances their uptake by cells [108]. Since both the plasma membrane and the intracellular environment have a net electronegative charge, endocytosis of cationic nanomaterials is likely to be enhanced by electrostatic forces in comparison with anionic nanomaterials.

- *Aggregation*. Nanomaterials show a tendency to aggregate in a physiological, aqueous environment. This is a phenomenon that needs to be taken into account when studying the toxicological influence of a nanomaterial, whether in vitro or in vivo. It has been observed that TiO_2 NPs dispersed with aggregates smaller than 200 nm exhibited no genotoxicity in human lymphoblast cells (Tk6), human embryonic epithelial cells (EUE), and monkey kidney fibroblasts (Cos-1). However, larger aggregates of the same TiO_2 NPs induced double-stranded DNA breaks and/or DNA oxidation lesions in the same cell types [109]. Proper characterization of the state of NP aggregation is therefore critical in establishing accurate understanding of their genotoxic influence. However, instead of focusing efforts on achieving highly monodispersed suspensions of NPs, the key should be to develop methodologies to accurately measure the state of aggregation of the NPs in the environment relevant to the intended application or study condition.

Collectively, several physicochemical features are relevant to nanomaterial-induced genotoxicity and carcinogenicity. Inevitably, these features will also influence the initial adsorption of proteins and other biomolecules onto the nanomaterials, which will affect the mechanisms of further downstream interactions with the relevant biological system (cells or tissues). In the context of cellular interactions, Fubini et al. [110] have postulated that once internalized, nanomaterials may either cause direct damage to cells or generate ROS or be engulfed by immune cells, leading to inflammation and indirectly damaging surrounding cells (Fig. 12.2).

12.3.2 Primary and Secondary Genotoxicity

12.3.2.1 Primary genotoxicity

Primary genotoxicity of nanomaterials is described as the induction of genetic damage without the presence of inflammation [111]. There are two forms of primary genotoxicity, direct and indirect primary genotoxicity. A mechanism in which nanomaterials directly interact with genomic DNA and/or affect DNA integrity results in direct primary genotoxicity, while indirect primary genotoxicity is caused by the elicited ROS, which is produced by damaged organelles, including mitochondria and membrane-bound NADPH oxidases corresponding to their interaction with nanomaterials.

Direct primary genotoxicity requires nuclear penetration of nanomaterials that directly interfere with genomic DNA or cellular biomass processes such as chromosome segregation during mitosis. Free diffusion of nanomaterials within cells potentially enables them to enter the nucleus. It has been demonstrated that gold NPs can gain entry into the nucleus, especially when their surfaces have been modified [112, 113]. A size-dependent nuclear penetration pattern has also been observed in silicon dioxide NPs [114]. Additionally, micronuclei formation has been shown in specific cells exposed to SWCNTs, MWCNTs, silver, or TiO_2 NPs, suggesting that NPs are capable of causing chromosomal breakages or disrupting mitotic spindle function during cell division [115–117]. In fact, SWCNTs have been found to localize inside the nucleus in association with DNA, centrosomes, and microtubules of mitotic spindles (Fig. 12.3).

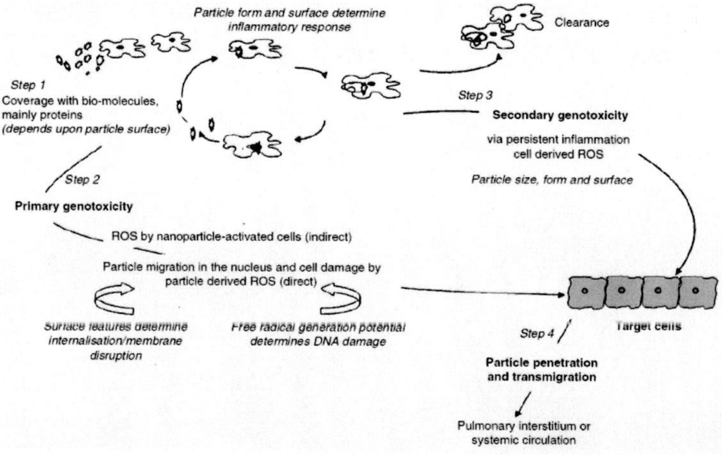

Figure 12.2 Hypothesized mechanisms of NP-induced genotoxicity. As a first step the NPs will be partially or fully covered by endogenous molecules, mostly proteins (protein corona). Primary genotoxicity results from either direct (e.g., physical interaction between NPs and DNA after migration into the nucleus or DNA interaction with NP-derived ROS) or indirect pathways (e.g., ROS generated by NP-activated target cells). Secondary genotoxicity results from DNA damage caused by persistent inflammation through activation of macrophages due to the presence of NPs (reprinted with permission from [110]).

Nanomaterials are capable of depleting antioxidants in the environment and thus increasing the potential for generating free radicals (ROS/RNS) that can damage DNA [118, 119]. Oxidative DNA damage is referred to as indirect primary genotoxicity. Impaired mitochondrial activity has been found to be responsible for oxidative DNA damage in lung epithelial cells exposed to quartz, asbestos, silver, and TiO_2 NPs, arising from mitochondria-derived ROS [120–122]. It has been postulated that indirect genotoxic effects are caused by the following mechanism:

- Superoxide anions leaking from the mitochondrial respiratory chain are subject to spontaneous or enzyme-mediated dismutation whereby

  $$2O_2 + 2H^+ \rightarrow H_2O_2 + O_2,$$

  giving rise to relatively stable oxidant H_2O_2.
- H_2O_2 is then converted by the Fenton reaction:

  $$M^{n+} + H_2O_2 \rightarrow M^{(n+1)} + OH + OH^- \qquad \text{(M: transition metal)}$$

  into hydroxyl radicals upon its diffusion into the nucleus [123].

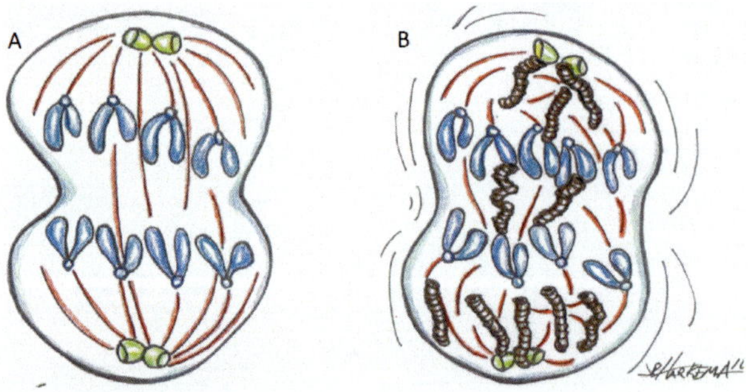

Figure 12.3 Potential interactions between NPs and DNA in a dividing cell. (A) In normal mitosis, the centrosomes (green) and microtubules (red) of the mitotic spindle ensure that each daughter cell receives the same number of chromosomes. (B) NPs disrupt the orderly segregation of chromosomes into daughter cells. SWCNTs (brown) form nanoropes that may vary in diameter but can be similar to the diameter of the microtubules of the mitotic spindle. SWCNT nanoropes can potentially interact with centrosomes, microtubules, and DNA, thereby disrupting chromosomal segregation and causing aneuploidy (reprinted with permission from [115]).

One study suggested the impairment of mitochondria in a placental cell line by chromium-cobalt NPs [124]. However, there is currently insufficient evidence to prove this mitochondria-mediated mechanism for indirect primary genotoxicity. Another possible mechanism for metallic nanomaterial–induced indirect genotoxicity is the inhibition of DNA repair in cells [125], which will also need further validation.

12.3.2.2 Secondary genotoxicity

Secondary genotoxicity is believed to be driven by inflammation, which is an important physiological response to cell and tissue injury. The keys to initiation of inflammation are the recruitment of inflammatory cells to the site of nanomaterial deposition and their subsequent secretion of various factors in response to invading foreign bodies and/or injury (refer to Chapter 8). The protective mechanism of inflammation, however, will inevitably generate oxidants such as O^{-}, H_2O_2, RNS, and ROS, which can lead to unintended adverse effects in cells, including chromosomal fragmentation, genetic mutations, formation of DNA adducts, abnormal methylation patterns, and alteration of gene expression profiles (Fig. 12.4) [107].

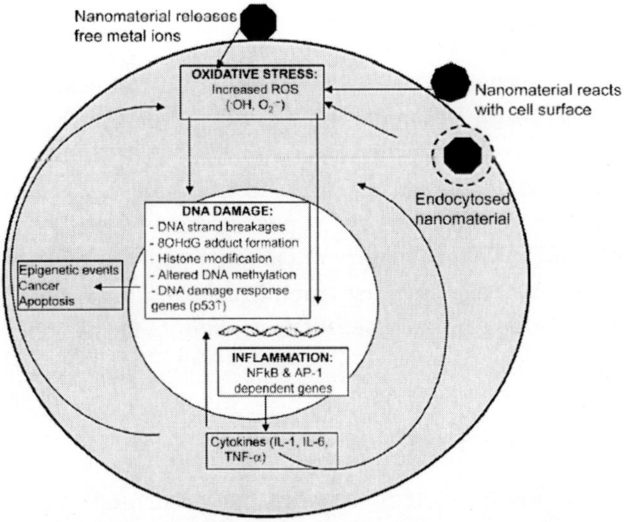

Figure 12.4 Potential inflammation-mediated mechanisms for nanomaterial-induced secondary genotoxicity (reprinted with permission from [107]).

CNTs, silica, TiO_2, and ZnO NPs have all been demonstrated to exert inflammation both in vitro and in vivo [126–131]. CNTs, for instance, induced inflammatory response in the abdominal cavities of mice in a length-dependent manner [130], whilst TiO_2 NPs were also found to be potent inflammation triggers through pulmonary exposure in rats [131] (see Chapter 5 for a more detailed discussion). Therefore, the activation of inflammation with increasing levels of oxidative stress contributes to secondary genotoxicity of some nanomaterials.

12.3.3 Oxidative Stress

Oxidative stress has been documented to be a significant contributor for nanomaterial induced genotoxicity, which typically stems from increasing levels of intracellular ROS and decreasing levels of antioxidants. Genetic damage resulting from such oxidative stress levels can occur regardless of whether NPs penetrate into the cell nucleus. Besides mitochondria-derived ROS, other initiators of genotoxicity where nuclear penetration of NPs is not necessary include the release of small molecules such as surface metal ions. In addition small-molecular-weight oxidants, generated due to interactions with NPs, can freely diffuse into the nucleus and induce oxidative DNA damage. An example is the hemolytic decomposition of lipid hyperperoxides to release α/β-unsaturated aldehyde genotoxins (4-oxo-2-nonenal, 4,5-epoxy-2(E)-decenal or 4-hydroxy-2-nonenal) from or near the surfaces of NPs [115]. In the same light, oxidative stress induced by TiO_2 and fullerene C_{60} NPs has been attributed to the production of peroxynitrite ($ONOO^-$), a potent oxidant [132]. Taken together, oxidative attack will eventually lead to DNA damage such as strand breakages, base modifications, 8-hydroxy-2-deoxyguanosine (8-OHdG) adduct formation, and DNA chain crosslinks.

12.3.4 Carcinogenicity

In vivo studies have demonstrated the occurrence of tumors in some animal models subjected to high dose exposure of TiO_2 NPs, carbon black NPs, and MWCNTs via inhalation or intratracheal administration [133]. However, the mechanism of nanomaterial-induced carcinogenesis is largely unknown [134]. Nonetheless, chronic inflammation could be an important perpetuator, bringing about elevated levels of oxidants in the system. Sustained DNA damage due to oxidative stress could overwhelm DNA repair mechanisms, potentially leading to carcinogenesis [135].

12.4 Methods to Study Nanomaterial-Induced Genotoxicity and Carcinogenicity

Cells are constantly bombarded by events that damage DNA, including ionizing radiation, UV light, and chemical agents such as methyl methanesulfonate, cisplatin, and mitomycin C. A healthy cell responds to DNA damage by first attempting to repair it and, if irreparable, undergoes either apoptosis (programmed cell death) or senescence (aging). This natural protective mechanism prevents any damaged DNA from progressing to deleterious mutations that would be passed down to progenies. Any dysfunction in these defensive mechanisms can lead to uncontrollable proliferation, which ultimately presents itself as cancer. The study of nanomaterial induced genotoxicity and carcinogencity is therefore of significant interest to the scientific community. The most commonly used models in current genotoxicity studies include the Ames bacterial mutagenesis assay and a number of in vitro and in vivo genotoxicity assays [136], which are described in this section.

12.4.1 Ames Bacterial Mutagenesis

The genotoxicity and carcinogenicity of nanomaterials can be evaluated by using a bacterial reverse mutation assay (Ames test) [137–139]. The Ames test is widely used to identify reagents that can cause gene mutations [140]. Normally, mutated *Salmonella* strains that are unable to produce some certain substances, such as histidine, are used because these bacteria cannot survive in the environment without histidine. Therefore, if certain chemicals or nanomaterials are mutagenic, new mutations can help restore the synthesis of histidine and enable the bacteria to survive in an environment without histidine. As such, this method is also known as the "reversion assay" or the "*Salmonella* test" [140]. By using the Ames test, Hasegawa et al. [138] found that nanosized Dy_2O_3, In_2O_3, and WO_3 had a higher mutagenic potential as compared to their corresponding microsized particles, indicating the potential of nanosized materials in potentially causing genotoxicity.

12.4.2 In vitro and in vivo Genotoxicity Assays

Besides the use of mutated bacteria, other assays are similarly available for genotoxicity testing. DNA break tests, chromosomal aberration, MN assays, formation of 8-OHdG, and tumor suppressor assays are the most commonly used in vitro and in vivo techniques in current genotoxicity and carcinogenicity studies.

12.4.3 DNA Breakage Assays

There are several assays that can be used to qualify and quantify genotoxicity caused by nanomaterials:

- *Comet assay.* The comet assay is also known as the single-cell gel electrophoresis assay. It is a well-established method to measure DNA damage in single eukaryotic cells, which is based on the mechanism of electrophoresis [141]. It can detect both double-stranded and single-stranded DNA breaks [142]. The rationale of the comet assay is that if negatively charged DNA contains damaged breaks, the resulting shorter strands will be able to migrate faster toward the anode during electrophoresis compared to intact DNA of a larger molecular size [143]. The result is an image showing a trail of migrating DNA in a sample containing DNA breaks, resembling that of a comet in flight and thus the name. The comet assay is a rapid, easy, and sensitive way to measure and analyze DNA damage [144], and Olive et al. [143] have published a detailed protocol on the use of the comet assay in measuring DNA damage in individual cells. Ng et al. [66] have used this technique to ascertain the potential genotoxicity of ZnO NPs before conducting further tests to verify this in vitro observation.
- *Formation of γ-H2AX.* DNA breaks can also be detected by probing the biomarker γ-H2AX. When a DNA double strand is damaged, the histone H2AX in cells will be phosphorylated into γ-H2AX [145]. The γ-H2AX assay is believed to have

much higher sensitivity toward DNA breakage compared to the comet assay [146]. Setyawati et al. [147] probed the formation of γ-H2AX and found that TiO_2 and Tb-Gd_2O_3 NPs could induce dose-dependent genotoxic effects on human skin fibroblasts (BJ). An optimized protocol of this assay can be found in Muslimovic et al.'s report [148]. Trouiller et al. [149] used both the comet assay and γ-H2AX immunostaining to demonstrate the genotoxic potential of TiO_2 NPs in mice.

- *Chromosomal aberration assay.* Chromosome aberration means an abnormal number and/or structure of chromosomes. Chromosome aberration is another crucial test in evaluating the potential genotoxicity of nanomaterials and has been reported in different studies [150–152]. For example, silver NPs were found to cause chromosomal aberration [152, 153] in different cell lines in in vitro studies. Balasubramanyam et al. [154] orally exposed female Wistar rats with different-sized aluminum oxide (Al_2O_3) particles. They found Al_2O_3 NPs (30 nm and 40 nm) were able to cause chromosomal aberrations in the bone marrow of rats. In brief, the cells arrested at the metaphase were fixed, stained, and observed under a light microscope to identify chromosomal gaps and breaks [153, 155, 156].
- *MN assay.* In the early 1970s, Boller et al. put forth the definition of a MN [157], which is one-third to one-sixteenth the diameter of the main nucleus and located in the cytoplasm of cells [158]. Normally, an MN forms because of chromosome breakage and/or whole chromosomes not being able to arrive at the right position in dividing cells [159]. The MN assay has been widely used in nanotoxicity studies to evaluate the genotoxicity of nanomaterial in different cell lines [160, 161], because it can detect clastogens, aneugens, mitotic delay, apoptosis, chromosome breakage, chromosome loss, and also nondisjunction [162–164]. The disadvantage of the MN assay is that it cannot distinguish different types of structural chromosome aberrations [163]. However, the MN assay is also widely considered as a reliable, reproducible, transferable, and predictive test to evaluate genotoxic effects of different chemicals or materials [161–163]. Studies have proven that NPs, such as silver NPs [48] and TiO_2 NPs [149], are able to

induce micronuclei in cells. QDs, CdSe, and 1% cobalt ion–doped CdSe were found to be able to induce the formation of an MN in the bone marrow and liver of rats after the rats were treated with doped or undoped CdSe NPs for seven days [165]. The qualitative study can be automotively done by flow cytometry and/or image analysis [159, 163]. The detailed protocol of this assay can be found in Kirsch-Volders et al.'s report [159].

- *Formation of 8-OHdG.* 8-OHdG is one of the main forms of oxidative lesions caused by oxidative stress in the nucleus and mitochondria [166, 167]. 8-OHdG, which indicates genomic damage, is utilized as a marker for carcinogenesis [167]. The 8-OHdG generated can be quantified by using high-performance liquid chromatography (HPLC) with electrochemical detection (EC) or gas chromatography–mass spectrometry (GC-MS) [167]. This biomarker has been used in both in vitro and in vivo tests to prove the genotoxicity and carcinogenicity of NPs such as TiO_2 NPs [168, 169], SiO_2 NPs [170], magnetic doped CdSe QDs [165], gold NPs, and gold–cobalt nanoalloy [166]. A detailed protocol can be found in Bogdanov et al.'s report [171].
- *Expression of tumor suppressor genes.* Normally, tumor suppressors are activated when cells are faced with irreparable DNA damage. Tumor suppressors can trigger cells to undergo apoptosis or senescence to prevent the formation of tumors [66]. Tumor suppressors, such as p53 and retinoblastoma protein (Rb), are crucial intracellular regulators to ensure the correct functioning of cells [66]. Techniques such as reverse transcriptase polymerase chain reaction (RT-PCR) [172], western blotting [66], and enzyme-linked immunosorbent assays (ELISAs) [173] can be utilized to indicate the activation or inactivation of tumor suppressors. The inactivation of tumor suppressor could lead to carcinogenicity in different cells. By detecting the expression of p53 and Rb in both wild-type skin fibroblast BJ cells and BJ cells with p53 knocked down, Ng et al. [66] found that ZnO NPs might be able to initiate the carcinogenesis process or increase a pre-existing carcinogenic effect by failing to trigger cell death.

12.5 Conclusions

In summary, it has been demonstrated by various research groups that many of the nanomaterials currently being explored for nanomedicine applications can exert genotoxic pressures on different biological systems under specific conditions. However, it should be pointed out that proof of genotoxicity does not equate to cancer progression. Whether cells are able to cope with the genotoxic effects, through the arsenal of endogenous mechanisms as described in this chapter, will be an important consideration for future research. The challenge will also be to develop effective experimental methodologies to clearly demonstrate the carcinogenic potential of genotoxic nanomaterials. At the molecular level, it will also be interesting to study how various nanomaterials can affect/interfere with the native DDR mechanisms, even if the nanomaterials do not exert any genotoxicity. From the cancer biology perspective, there could be mechanistic influences from nanomaterials that can give rise to novel cancer therapy approaches.

From the literature presented in this chapter, it is obvious that there is much conflicting data on the potential genotoxic effects of various nanomaterials. This is somewhat expected, given the multitude of parameters that can affect the physical and biochemical behavior of nanomaterials even of the same composition. For accurate comparisons across different genotoxic studies, scientists therefore need to report as many details as possible about the physicochemical properties of the nanomaterials under investigation. It is also important that these properties be measured under conditions that mirror the relevant biological system.

References

1. Narayan, S., and Roy, D. (2003). Role of APC and DNA mismatch repair genes in the development of colorectal cancers. *Mol. Cancer*, **2**, p. 41.

2. Jaiswal, A. S., Armas, M. L., Izumi, T., Strauss, P. R., and Narayan, S. (2012). Adenomatous polyposis coli interacts with flap endonuclease 1 to block its nuclear entry and function. *Neoplasia*, **14**(6), pp. 495–508.

3. Klapacz, J., Lingaraju, G. M., Guo, H. H., Shah, D., Moar-Shoshani, A., Loeb, L. A., and Samson, L. D. (2010). Frameshift mutagenesis

and microsatellite instability induced by human alkyladenine DNA glycosylase. *Mol. Cell*, **37**(6), pp. 843–853.

4. Lyons, D. M., and O'Brien, P. J. (2010). Human base excision repair creates a bias toward -1 frameshift mutations. *J. Biol. Chem.*, **285**(33), pp. 25203–2512.
5. Lambert, B., Andersson, B., Bastlova, T., Hou, S. M., Hellgren, D., and Kolman, A. (1994). Mutations induced in the hypoxanthine phosphoribosyl transferase gene by three urban air pollutants: acetaldehyde, benzo[a]pyrene diolepoxide, and ethylene oxide. *Environ. Health Perspect.*, **102**(Suppl 4), pp. 135–138.
6. Niedernhofer, L. J., Daniels, J. S., Rouzer, C. A., Greene, R. E., and Marnett, L. J. (2003). Malondialdehyde, a product of lipid peroxidation, is mutagenic in human cells. *J. Biol. Chem.*, **278**(33), pp. 31426–31433.
7. Hartman, P. E. (1983). Putative mutagens and carcinogens in foods. IV. Malonaldehyde (malondialdehyde). *Environ. Mutagen.*, **5**(4), pp. 603–607.
8. Marnett, L. J. (1999). Chemistry and biology of DNA damage by malondialdehyde. *IARC Sci. Publ.*, **150**, pp. 17–27.
9. Esterbauer, H. (1993). Cytotoxicity and genotoxicity of lipid-oxidation products. *Am. J. Clin. Nutr.*, **57**(5 Suppl), pp. 779S–785S; discussion 785S–786S.
10. Bedard, L. L., and Massey, T. E. (2006). Aflatoxin B1-induced DNA damage and its repair. *Cancer Lett.*, **241**(2), pp. 174–183.
11. Lindahl, T. (1974). An N-glycosidase from Escherichia coli that releases free uracil from DNA containing deaminated cytosine residues. *Proc. Natl. Acad. Sci. U S A*, **71**(9), pp. 3649–3653.
12. Goodarzi, A. A., Noon, A. T., Deckbar, D., Ziv, Y., Shiloh, Y., Lobrich, M., and Jeggo, P. A. (2008). ATM signaling facilitates repair of DNA double-strand breaks associated with heterochromatin. *Mol. Cell*, **31**(2), pp. 167–177.
13. Ma, J. L., Kim, E. M., Haber, J. E., and Lee, S. E. (2003). Yeast Mre11 and Rad1 proteins define a Ku-independent mechanism to repair double-strand breaks lacking overlapping end sequences. *Mol. Cell Biol.*, **23**(23), pp. 8820–8828.
14. Audebert, M., Salles, B., and Calsou, P. (2004). Involvement of poly(ADP-ribose) polymerase-1 and XRCC1/DNA ligase III in an alternative route for DNA double-strand breaks rejoining. *J. Biol. Chem.*, **279**(53), pp. 55117–55126.

15. Guirouilh-Barbat, J., Huck, S., Bertrand, P., Pirzio, L., Desmaze, C., Sabatier, L., and Lopez, B. S. (2004). Impact of the KU80 pathway on NHEJ-induced genome rearrangements in mammalian cells. *Mol. Cell*, **14**(5), pp. 611–623.

16. Lin, D., and Xing, B. (2007). Phytotoxicity of nanoparticles: Inhibition of seed germination and root growth. *Environ. Pollut.*, **150**(2), pp. 243–250.

17. Basiuk, E. V., Ochoa-Olmos, O. E., and De la Mora-Estrada, L. F. (2011). Ecotoxicological effects of carbon nanomaterials on algae, fungi and plants. *J. Nanosci. Nanotechnol.*, **11**(4), pp. 3016–3038.

18. Babu, S. S., Mohwald, H., and Nakanishi, T. (2010). Recent progress in morphology control of supramolecular fullerene assemblies and its applications. *Chem. Soc. Rev.*, **39**(11), pp. 4021–4035.

19. Canas, J. E., Long, M. Q., Nations, S., Vadan, R., Dai, L., Luo, M. X., Ambikapathi, R., Lee, E. H., and Olszyk, D. (2008). Effects of functionalized and nonfunctionalized single-walled carbon nanotubes on root elongation of select crop species. *Environ. Toxicol. Chem.*, **27**(9), pp. 1922–1931.

20. Ke, P. C., and Qiao, R. (2007). Carbon nanomaterials in biological systems. *J. Phys.: Condens. Matter*, **19**(37).

21. Arbogast, J. W., Darmanyan, A. P., Foote, C. S., Diederich, F. N., Whetten, R. L., Rubin, Y., Alvarez, M. M., and Anz, S. J. (1991). Photophysical properties of sixty atom carbon molecule (C60). *J. Phys. Chem.*, **95**(1), pp. 11–12.

22. Guldi, D. M., and Prato, M. (2000). Excited-state properties of C60 fullerene derivatives. *Acc. Chem. Res.*, **33**(10), pp. 695–703.

23. Xu, A., Chai, Y., Nohmi, T., and Hei, T. K. (2009). Genotoxic responses to titanium dioxide nanoparticles and fullerene in gpt delta transgenic MEF cells. *Part. Fibre Toxicol.*, **6**(1), p. 3.

24. Dhawan, A., Taurozzi, J. S., Pandey, A. K., Shan, W., Miller, S. M., Hashsham, S. A., and Tarabara, V. V. (2006). Stable colloidal dispersions of C60 fullerenes in water: evidence for genotoxicity. *Environ. Sci. Technol.*, **40**(23), pp. 7394–7401.

25. Sera, N., Tokiwa, H., and Miyata, N. (1996). Mutagenicity of the fullerene C60-generated singlet oxygen dependent formation of lipid peroxides. *Carcinogenesis*, **17**(10), pp. 2163–2169.

26. Totsuka, Y., Higuchi, T., Imai, T., Nishikawa, A., Nohmi, T., Kato, T., Masuda, S., Kinae, N., Hiyoshi, K., Ogo, S., Kawanishi, M., Yagi, T., Ichinose, T., Fukumori, N., Watanabe, M., Sugimura, T., and Wakabayashi, K. (2009).

Genotoxicity of nano/microparticles in in vitro micronuclei, in vivo comet and mutation assay systems. *Part. Fibre Toxicol.*, **6**, p. 23.

27. Folkmann, J. K., Risom, L., Jacobsen, N. R., Wallin, H., Loft, S., and Moller, P. (2009). Oxidatively damaged DNA in rats exposed by oral gavage to C-60 fullerenes and single-walled carbon nanotubes. *Environ. Health Persp.*, **117**(5), pp. 703–708.

28. Babynin, E. V., Nuretdinov, I. A., Gubskaya, V. P., and Barabanshchikov, B. I. (2002). Study of mutagenic activity of fullerene and some of its derivatives using his+ reversions of salmonella typhimurium as an example. *Russ. J. Genet.*, **38**(4), pp. 359–363.

29. Shinohara, N., Matsumoto, K., Endoh, S., Maru, J., and Nakanishi, J. (2009). In vitro and in vivo genotoxicity tests on fullerene C60 nanoparticles. *Toxicol. Lett.*, **191**(2–3), pp. 289–296.

30. Lam, C.-W., James, J. T., McCluskey, R., and Hunter, R. L. (2004). Pulmonary toxicity of single-wall carbon nanotubes in mice 7 and 90 days after intratracheal instillation. *Toxicol. Sci.*, **77**(1), pp. 126–134.

31. Yamashita, T., Yamashita, K., Nabeshi, H., Yoshikawa, T., Yoshioka, Y., Tsunoda, S.-I., and Tsutsumi, Y. (2012). Carbon nanomaterials: efficacy and safety for nanomedicine. *Materials*, **5**(2), pp. 350–363.

32. Cavallo, D., Fanizza, C., Ursini, C. L., Casciardi, S., Paba, E., Ciervo, A., Fresegna, A. M., Maiello, R., Marcelloni, A. M., Buresti, G., Tombolini, F., Bellucci, S., and Iavicoli, S. (2012). Multi-walled carbon nanotubes induce cytotoxicity and genotoxicity in human lung epithelial cells. *J. Appl. Toxicol.*, **32**(6), pp. 454–464.

33. Cicchetti, R., Divizia, M., Valentini, F., and Argentin, G. (2011). Effects of single-wall carbon nanotubes in human cells of the oral cavity: geno-cytotoxic risk. *Toxicol. In Vitro*, **25**(8), pp. 1811–1819.

34. Naya, M., Kobayashi, N., Mizuno, K., Matsumoto, K., Ema, M., and Nakanishi, J. (2011). Evaluation of the genotoxic potential of single-wall carbon nanotubes by using a battery of in vitro and in vivo genotoxicity assays. *Regul. Toxicol. Pharmacol.*, **61**(2), pp. 192–198.

35. Migliore, L., Saracino, D., Bonelli, A., Colognato, R., D'Errico, M. R., Magrini, A., Bergamaschi, A., and Bergamaschi, E. (2010). Carbon nanotubes induce oxidative DNA damage in RAW 264.7 cells. *Environ. Mol. Mutagen.*, **51**(4), pp. 294–303.

36. Kisin, E. R., Murray, A. R., Keane, M. J., Shi, X. C., Schwegler-Berry, D., Gorelik, O., Arepalli, S., Castranova, V., Wallace, W. E., Kagan, V. E., and Shvedova, A. A. (2007). Single-walled carbon nanotubes: Geno- and cytotoxic effects in lung fibroblast V79 cells. *J. Toxicol. Environ. Health A Curr. Issues*, **70**(24), pp. 2071–2079.

37. Pacurari, M., Yin, X. J., Zhao, J. S., Ding, M., Leonard, S. S., Schwegier-Berry, D., Ducatman, B. S., Sbarra, D., Hoover, M. D., Castranova, V., and Vallyathan, V. (2008). Raw single-wall carbon nanotubes induce oxidative stress and activate MAPKs, AP-1, NF-kappa B, and Akt in normal and malignant human mesothelial cells. *Environ. Health Persp.*, **116**(9), pp. 1211–1217.

38. Lindberg, H. K., Falck, G. C. M., Suhonen, S., Vippola, M., Vanhala, E., Catalan, J., Savolainen, K., and Norppa, H. (2009). Genotoxicity of nanomaterials: DNA damage and micronuclei induced by carbon nanotubes and graphite nanofibres in human bronchial epithelial cells in vitro. *Toxicol. Lett.*, **186**(3), pp. 166–173.

39. Wang, L., Luanpitpong, S., Castranova, V., Tse, W., Lu, Y., Pongrakhananon, V., and Rojanasakul, Y. (2011). Carbon nanotubes induce malignant transformation and tumorigenesis of human lung epithelial cells. *Nano Lett.*, **11**(7), pp. 2796–2803.

40. Veena, C., and Anju, G. (2011). Polymer/Carbon Nanotube Nanocomposites. In *Carbon Nanotubes - Polymer Nanocomposites,* Siva Yellampalli (Ed.), ISBN: 978-953-307-498-6, InTech, DOI: 10.5772/18423.

41. Di Sotto, A., Chiaretti, M., Carru, G. A., Bellucci, S., and Mazzanti, G. (2009). Multi-walled carbon nanotubes: lack of mutagenic activity in the bacterial reverse mutation assay. *Toxicol. Lett.*, **184**(3), pp. 192–197.

42. Nagai, H., Okazaki, Y., Chew, S. H., Misawa, N., Yamashita, Y., Akatsuka, S., Ishihara, T., Yamashita, K., Yoshikawa, Y., Yasui, H., Jiang, L., Ohara, H., Takahashi, T., Ichihara, G., Kostarelos, K., Miyata, Y., Shinohara, H., and Toyokuni, S. (2011). Diameter and rigidity of multiwalled carbon nanotubes are critical factors in mesothelial injury and carcinogenesis. *Proc. Natl. Acad. Sci.*, **108**(49), pp. E1330–E1338.

43. Kim, J. S., Song, K. S., Lee, J. K., Choi, Y. C., Bang, I. S., Kang, C. S., and Yu, I. J. (2011). Toxicogenomic comparison of multi-wall carbon nanotubes (MWCNTs) and asbestos. *Arch. Toxicol.*, **86**(4), pp. 553–562.

44. Guo, F., Ma, N., Horibe, Y., Kawanishi, S., Murata, M., and Hiraku, Y. (2012). Nitrative DNA damage induced by multi-walled carbon nanotube via endocytosis in human lung epithelial cells. *Toxicol. Appl. Pharmacol.*, **260**(2), pp. 183–192.

45. Jacobsen, N., Møller, P., Jensen, K., Vogel, U., Ladefoged, O., Loft, S., and Wallin, H. (2009). Lung inflammation and genotoxicity following pulmonary exposure to nanoparticles in ApoE-/- mice. *Part Fibre Toxicol.*, **6**(1), p. 2.

46. Wu, J., Liu, W., Xue, C., Zhou, S., Lan, F., Bi, L., Xu, H., Yang, X., and Zeng, F. D. (2009). Toxicity and penetration of TiO2 nanoparticles in hairless mice and porcine skin after subchronic dermal exposure. *Toxicol. Lett.*, **191**(1), 1–8.

47. *The Project on Emerging Nanotechnologies (PEN)* [online] 2012, http://www.nanotechproject.org/consumerproducts.

48. Li, Y., Chen, D. H., Yan, J., Chen, Y., Mittelstaedt, R. A., Zhang, Y. B., Biris, A. S., Heflich, R. H., and Chen, T. (2012). Genotoxicity of silver nanoparticles evaluated using the Ames test and in vitro micronucleus assay. *Mutat. Res.: Gen. Toxicol. Environ. Mutagen.*, **745**(1–2), pp. 4–10.

49. Uygur, B., Craig, G., Mason, M. D., and Ng, A. K. (2009). *Cytotoxicity and Genotoxicity of Silver Nanomaterials. Nanotech Conference & Expo 2009*, Vol. 2, *Technical Proceedings: Nanotechnology 2009: Life Sciences, Medicine, Diagnostics, Bio Materials and Composites*, eds. Laudon, M., and Romanowicz, B. pp. 383–386.

50. Foldbjerg, R., Dang, D. A., and Autrup, H. (2011). Cytotoxicity and genotoxicity of silver nanoparticles in the human lung cancer cell line, A549. *Arch. Toxicol.*, **85**(7), pp. 743–750.

51. Singh, S., D'Britto, V., Prabhune, A. A., Ramana, C. V., Dhawan, A., and Prasad, B. L. V. (2010). Cytotoxic and genotoxic assessment of glycolipid-reduced and -capped gold and silver nanoparticles. *N. J. Chem.*, **34**(2), pp. 294–301.

52. Kim, Y. J., Yang, S. I., and Ryu, J. C. (2010). Cytotoxicity and genotoxicity of nano-silver in mammalian cell lines. *Mol. Cell. Toxicol.*, **6**(2), pp. 119–125.

53. AshaRani, P. V., Low Kah Mun, G., Hande, M. P., and Valiyaveettil, S. (2009). Cytotoxicity and genotoxicity of silver nanoparticles in human cells. *ACS Nano*, **3**(2), pp. 279–290.

54. Giljohann, D. A., Seferos, D. S., Daniel, W. L., Massich, M. D., Patel, P. C., and Mirkin, C. A. (2010). Gold nanoparticles for biology and medicine. *Angew. Chem., Int. Ed.*, **49**(19), pp. 3280–3294.

55. Downs, T. R., Crosby, M. E., Hu, T., Kumar, S., Sullivan, A., Sarlo, K., Reeder, B., Lynch, M., Wagner, M., Mills, T., and Pfuhler, S. (2012). Silica nanoparticles administered at the maximum tolerated dose induce genotoxic effects through an inflammatory reaction while gold nanoparticles do not. *Mutat. Res.: Gen. Toxicol. Environ. Mutagen.*, **745**(1–2), pp. 38–50.

56. Schulz, M., Ma-Hock, L., Brill, S., Strauss, V., Treumann, S., Groters, S., van Ravenzwaay, B., and Landsiedel, R. (2012). Investigation on the

genotoxicity of different sizes of gold nanoparticles administered to the lungs of rats. *Mutat. Res.: Gen. Toxicol. Environ. Mutagen.*, **745**(1–2), pp. 51–57.

57. Vecchio, G., Galeone, A., Brunetti, V., Maiorano, G., Rizzello, L., Sabella, S., Cingolani, R., and Pompa, P. P. (2012). Mutagenic effects of gold nanoparticles induce aberrant phenotypes in Drosophila melanogaster. *Nanomed.: Nanotechnol. Biol. Med.*, **8**(1), pp. 1–7.

58. Wang, J. J., Sanderson, B. J. S., and Wang, H. (2007). Cyto- and genotoxicity of ultrafine TiO2 particles in cultured human lymphoblastoid cells. *Mutat. Res.: Gen. Toxicol. Environ. Mutagen.*, **628**(2), pp. 99–106.

59. Kang, S. J., Kim, B. M., Lee, Y. J., and Chung, H. W. (2008). Titanium dioxide nanoparticles trigger p53-mediated damage response in peripheral blood lymphocytes. *Environ. Mol. Mutagen.*, **49**(5), pp. 399–405.

60. Bernardeschi, M., Guidi, P., Scarcelli, V., Lucchesi, P., Nigro, M., and Frenzilli, G. (2010). Potential genotoxicity of nano-sized TiO2 particles in isolated bottle-nose dolphin (Tursiops truncatus) and human leukocytes. *Comp. Biochem. Physiol. A Mol. Integr. Physiol.*, **157**(1), pp. S14–S14.

61. Hackenberg, S., Zimmermann, F.-Z., Scherzed, A., Friehs, G., Froelich, K., Ginzkey, C., Koehler, C., Burghartz, M., Hagen, R., and Kleinsasser, N. (2011). Repetitive exposure to zinc oxide nanoparticles induces dna damage in human nasal mucosa mini organ cultures. *Environ. Mol. Mutagen.*, **52**(7), pp. 582–589.

62. Sharma, V., Singh, S. K., Anderson, D., Tobin, D. J., and Dhawan, A. (2011). Zinc oxide nanoparticle induced genotoxicity in primary human epidermal keratinocytes. *J. Nanosci. Nanotechnol.*, **11**(5), pp. 3782–3788.

63. Sharma, V., Anderson, D., and Dhawan, A. (2012). Zinc oxide nanoparticles induce oxidative DNA damage and ROS-triggered mitochondria mediated apoptosis in human liver cells (HepG2). *Apoptosis*, **17**(8), pp. 852–870.

64. Sharma, V., Shukla, R. K., Saxena, N., Parmar, D., Das, M., and Dhawan, A. (2009). DNA damaging potential of zinc oxide nanoparticles in human epidermal cells. *Toxicol. Lett.*, **185**(3), pp. 211–218.

65. Sharma, V., Singh, P., Pandey, A. K., and Dhawan, A. (2012). Induction of oxidative stress, DNA damage and apoptosis in mouse liver after sub-acute oral exposure to zinc oxide nanoparticles. *Mutat. Res.: Gen. Toxicol. Environ. Mutagen.*, **745**(1–2), pp. 84–91.

66. Ng, K. W., Khoo, S. P. K., Heng, B. C., Setyawati, M. I., Tan, E. C., Zhao, X. X., Xiong, S. J., Fang, W. R., Leong, D. T., and Loo, J. S. C. (2011). The role of the tumor suppressor p53 pathway in the cellular DNA damage response to zinc oxide nanoparticles. *Biomaterials*, **32**(32), pp. 8218–8225.

67. Barnes, C. A., Elsaesser, A., Arkusz, J., Smok, A., Palus, J., Lesniak, A., Salvati, A., Hanrahan, J. P., de Jong, W. H., Dziubaltowska, E., Stepnik, M., Rydzynski, K., McKerr, G., Lynch, I., Dawson, K. A., and Howard, C. V. (2008). Reproducible Comet assay of amorphous silica nanoparticles detects no genotoxicity. *Nano Lett.*, **8**(9), pp. 3069–3074.

68. Singh, N., Jenkins, G. J. S., Nelson, B. C., Marquis, B. J., Maffeis, T. G. G., Brown, A. P., Williams, P. M., Wright, C. J., and Doak, S. H. (2012). The role of iron redox state in the genotoxicity of ultrafine superparamagnetic iron oxide nanoparticles. *Biomaterials*, **33**(1), pp. 163–170.

69. Konczol, M., Ebeling, S., Goldenberg, E., Treude, F., Gminski, R., Giere, R., Grobety, B., Rothen-Rutishauser, B., Merfort, I., and Mersch-Sundermann, V. (2011). Cytotoxicity and genotoxicity of size-fractionated iron oxide (Magnetite) in A549 human lung epithelial cells: role of ROS, JNK, and NF-kappa B. *Chem. Res. Toxicol.*, **24**(9), pp. 1460–1475.

70. Szalay, B., Tatrai, E., Nyiro, G., Vezer, T., and Dura, G. (2012). Potential toxic effects of iron oxide nanoparticles in in vivo and in vitro experiments. *J. Appl. Toxicol.*, **32**(6), pp. 446–453.

71. Zhang, T., Qian, L., Tang, M., Xue, Y. Y., Kong, L., Zhang, S. S., and Pu, Y. P. (2012). Evaluation on cytotoxicity and genotoxicity of the L-Glutamic acid coated iron oxide nanoparticles. *J. Nanosci. Nanotechnol.*, **12**(3), pp. 2866–2873.

72. Doak, S. H. (2010). Genotoxicity of iron oxide nanoparticles and carbon nanotubes. *Mutagenesis*, **25**(6), pp. 638–638.

73. Kim, Y. J., Choi, H. S., Song, M. K., Youk, D. Y., Kim, J. H., and Ryu, J. C. (2009). Genotoxicity of Aluminum Oxide (Al1O3) nanoparticle in mammalian cell lines. *Mol. Cell. Toxicol.*, **5**(2), pp. 172–178.

74. Balasubramanyam, A., Sailaja, N., Mahboob, M., Rahman, M. F., Hussain, S. M., and Grover, P. (2009). In vivo genotoxicity assessment of aluminium oxide nanomaterials in rat peripheral blood cells using the comet assay and micronucleus test. *Mutagenesis*, **24**(3), pp. 245–251.

75. Balasubramanyam, A., Sailaja, N., Mahboob, M., Rahman, M. F., Misra, S., Hussain, S. M., and Grover, P. (2009). Evaluation of genotoxic effects of oral exposure to Aluminum oxide nanomaterials in rat bone marrow. *Mutat. Res.: Gen. Toxicol. Environ. Mutagen.*, **676**(1–2), pp. 41–47.

76. Balasubramanyam, A., Sailaja, N., Mahboob, M., Rahman, M. F., Hussain, S. M., and Grover, P. (2010). In vitro mutagenicity assessment of aluminium oxide nanomaterials using the Salmonella/microsome assay. *Toxicol. In Vitro*, **24**(6), pp. 1871–1876.

77. Simon, A., Gouget, B., Mayne, M., Herlin, N., Reynaud, C., and Degrouard, J. (2007). In vitro investigation of TiO2, Al2O3, an nanoparticles and mutli-walled carbon nanotubes cyto- and genotoxicity on lung, kidney cells and hepatocytes. *Toxicol. Lett.*, **172**, pp. S36–S36.

78. Bhadra, D., Bhadra, S., Jain, P., and Jain, N. K. (2002). Pegnology: a review of PEG-ylated systems. *Pharmazie*, **57**(1), pp. 5–29.

79. Ajani, J. A., Hortobagyi, G. N., Frye, D., Levin, B., Boman, B. M., and Faintuch, J. S. (1987). A randomized study of two schedules of copovithane in patients with advanced colorectal carcinoma. *Am. J. Clin. Oncol.*, **10**(2), pp. 139–140.

80. Duncan, R., Gac-Breton, S., Keane, R., Musila, R., Sat, Y. N., Satchi, R., and Searle, F. (2001). Polymer-drug conjugates, PDEPT and PELT: basic principles for design and transfer from the laboratory to clinic. *J. Controlled Rel.*, **74**(1–3), pp. 135–146.

81. Maynard, C., Swenson, R., Paris, J. A., Martin, J. S., Hallstrom, A. P., Cerqueira, M. D., and Weaver, W. D. (1998). Randomized, controlled trial of RheothRx (poloxamer 188) in patients with suspected acute myocardial infarction. *Am. Heart J.*, **135**(5 I), pp. 797–804.

82. Fournier, J. B., and Durand, G. (1991). Focal conic faceting in smectic-A liquid crystals. *J. Phys. II France*, **1**(7), pp. 845–870.

83. Illum, L. (1998). Chitosan and its use as a pharmaceutical excipient. *Pharma. Res.*, **15**(9), pp. 1326–1331.

84. Einmahl, S., Capancioni, S., Schwach-Abdellaoui, K., Moeller, M., Behar-Cohen, F., and Gurny, R. (2001). Therapeutic applications of viscous and injectable poly(ortho esters). *Adv. Drug Delivery Rev.*, **53**(1), pp. 45–73.

85. Kattan, J., Droz, J. P., Couvreur, P., Marino, J. P., Boutanlaroze, A., Rougier, P., Brault, P., Vranckx, H., Grognet, J. M., Morge, X., and Sanchogarnier, H. (1992). Phase-I clinical-trial and pharmacokinetic evaluation of doxorubicin carried by polyisohexylcyanoacrylate nanoparticles. *Invest. New Drugs*, **10**(3), pp. 191–199.

86. de Lima, R., Feitosa, L., Pereira, A. D. S., de Moura, M. R., Aouada, F. A., Mattoso, L. H. C., and Fraceto, L. F. (2010). Evaluation of the genotoxicity of chitosan nanoparticles for use in food packaging films. *J. Food Sci.*, **75**(6), pp. N89–N96.

87. de Lima, R., Pereira, A. D. S., Porto, R. M., and Fraceto, L. F. (2011). Evaluation of Cyto- and Genotoxicity of Poly(lactide-co-glycolide) Nanoparticles. *J. Polym. Environ.*, **19**(1), pp. 196–202.

88. Huang, Y. N., Gao, H., Gou, M. L., Ye, H. X., Liu, Y. K., Gao, Y., Peng, F., Qian, Z. Y., Cen, X. B., and Zhao, Y. L. (2010). Acute toxicity and genotoxicity studies on poly(epsilon-caprolactone)-poly(ethylene glycol)-poly(epsilon-caprolactone) nanomaterials. *Mutat. Res.: Gen. Toxicol. Environ. Mutagen.*, **696**(2), pp. 101–106.

89. He, L., Yang, L., Zhang, Z.-r., Gong, T., Deng, L., Gu, Z., and Sun, X. (2009). In vitroevaluation of the genotoxicity of a family of novel MeO-PEG-poly(D,L-lactic-co-glycolic acid)-PEG-OMe triblock copolymer and PLGA nanoparticles. *Nanotechnology*, **20**(45), pp. 455102.

90. Densmore, C. L., Orson, F. M., Xu, B., Kinsey, B. M., Waldrep, J. C., Hua, P., Bhogal, B., and Knight, V. (2000). Aerosol delivery of robust polyethyleneimine-DNA complexes for gene therapy and genetic immunization. *Mol. Ther.*, **1**(2), pp. 180–188.

91. Kabanov, A. (2006). Polymer genomics: an insight into pharmacology and toxicology of nanomedicines. *Adv. Drug Delivery Rev.*, **58**(15), pp. 1597–1621.

92. Chollet, P., Favrot, M. C., Hurbin, A., and Coll, J.-L. (2002). Side-effects of a systemic injection of linear polyethylenimine-DNA complexes. *J. Gene Med.*, **4**(1), pp. 84–91.

93. Regnström, K., Ragnarsson, E. G. E., Fryknäs, M., Köping-Höggård, M., and Artursson, P. (2006). Gene expression profiles in mouse lung tissue after administration of two cationic polymers used for nonviral gene delivery. *Pharm. Res.*, **23**(3), pp. 475–482.

94. Beyerle, A., Long, A. S., White, P. A., Kissel, T., and Stoeger, T. (2011). Poly(ethylene imine) nanocarriers do not induce mutations nor oxidative DNA damage in vitro in MutaMouse FE1 cells. *Mol. Pharm.*, **8**(3), pp. 976–981.

95. Williams, D. (2008). The relationship between biomaterials and nanotechnology. *Biomaterials*, **29**(12), pp. 1737–1738.

96. Gratton, S. E. A., Ropp, P. A., Pohlhaus, P. D., Luft, J. C., Madden, V. J., Napier, M. E., and DeSimone, J. M. (2008). The effect of particle design on cellular internalization pathways. *Proc. Natl. Acad. Sci.*, **105**(33), pp. 11613–11618.

97. Rejman, J., Oberle, V., Zuhorn, I. S., and Hoekstra, D. (2004). Size-dependent internalization of particles via the pathways of clathrin- and caveolae-mediated endocytosis. *Biochem. J.*, **377**(1), pp. 159–169.

98. Chithrani, B. D., and Chan, W. C. W. (2007). Elucidating the mechanism of cellular uptake and removal of protein-coated gold nanoparticles of different sizes and shapes. *Nano Lett.*, **7**(6), pp. 1542–1550.

99. Chithrani, B. D., Ghazani, A. A., and Chan, W. C. W. (2006). Determining the size and shape dependence of gold nanoparticle uptake into mammalian cells. *Nano Lett.*, **6**(4), pp. 662–668.

100. Geiser, M., Rothen-Rutishauser, B., Kapp, N., Schürch, S., Kreyling, W., Schulz, H., Semmler, M., Hof, V. I., Heyder, J., and Gehr, P. (2005). Ultrafine particles cross cellular membranes by nonphagocytic mechanisms in lungs and in cultured cells. *Environ. Health Perspect.*, **113**(11), pp. 1555–1560.

101. Baggs, R. B., Ferin, J., and Oberdörster, G. (1997). Regression of pulmonary lesions produced by inhaled titanium dioxide in rats. *Vet. Pathol. Online*, **34**(6), pp. 592–597.

102. Oberdörster, G., Ferin, J., and Lehnert, B. E. (1994). Correlation between particle size, in vivo particle persistence, and lung injury. *Environ. Health Perspect.*, **102**(Suppl 5), pp. 173–179.

103. Wang, B., Feng, W.-Y., Wang, T.-C., Jia, G., Wang, M., Shi, J.-W., Zhang, F., Zhao, Y.-L., and Chai, Z.-F. (2006). Acute toxicity of nano- and micro-scale zinc powder in healthy adult mice. *Toxicol. Lett.*, **161**(2), pp. 115–123.

104. Muller, J., Huaux, F., Fonseca, A., Nagy, J. B., Moreau, N., Delos, M., Raymundo-Piñero, E., Béguin, F., Kirsch-Volders, M., Fenoglio, I., Fubini, B., and Lison, D. (2008). Structural defects play a major role in the acute lung toxicity of multiwall carbon nanotubes: toxicological aspects. *Chem. Res. Toxicol.*, **21**(9), pp. 1698–1705.

105. Zhao, X., Ng, S., Heng, B. C., Guo, J., Ma, L., Tan, T. T., Ng, K. W., and Loo, S. C. (2013). Cytotoxicity of hydroxyapatite nanoparticles is shape and cell dependent. *Arch. Toxicol*, **87**(6), pp. 1037–1052.

106. Zhao, X., Heng, B. C., Xiong, S., Guo, J., Tan, T. T., Boey, F. Y., Ng, K. W., and Loo, J. S. (2011). In vitro assessment of cellular responses to rod-shaped hydroxyapatite nanoparticles of varying lengths and surface areas. *Nanotoxicology*, **5**(2), pp. 182–94.

107. Singh, N., Manshian, B., Jenkins, G. J. S., Griffiths, S. M., Williams, P. M., Maffeis, T. G. G., Wright, C. J., and Doak, S. H. (2009). NanoGenotoxicology: the DNA damaging potential of engineered nanomaterials. *Biomaterials*, **30**(23–24), pp. 3891–3914.

108. Nan, A., Bai, X., Son, S. J., Lee, S. B., and Ghandehari, H. (2008). Cellular uptake and cytotoxicity of silica nanotubes. *Nano Lett.*, **8**(8), pp. 2150–2154.

109. Magdolenova, Z., Bilanicova, D., Pojana, G., Fjellsbo, L. M., Hudecova, A., Hasplova, K., Marcomini, A., and Dusinska, M. (2012). Impact of agglomeration and different dispersions of titanium dioxide nanoparticles on the human related in vitro cytotoxicity and genotoxicity. *J. Environ. Monitor*, **14**(2), pp. 455–464.

110. Fubini, B., Ghiazza, M., and Fenoglio, I. (2010). Physico-chemical features of engineered nanoparticles relevant to their toxicity. *Nanotoxicology*, **4**(4), pp. 347–363.

111. Donaldson, K., Poland, C. A., and Schins, R. P. F. (2010). Possible genotoxic mechanisms of nanoparticles: criteria for improved test strategies. *Nanotoxicology*, **4**(4), pp. 414–420.

112. Tkachenko, A. G., Xie, H., Coleman, D., Glomm, W., Ryan, J., Anderson, M. F., Franzen, S., and Feldheim, D. L. (2003). Multifunctional gold nanoparticleâ peptide complexes for nuclear targeting. *J. Am. Chem. Soc.*, **125**(16), pp. 4700–4701.

113. Nativo, P., Prior, I. A., and Brust, M. (2008). Uptake and intracellular fate of surface-modified gold nanoparticles. *ACS Nano*, **2**(8), pp. 1639–1644.

114. Chen, M., and von Mikecz, A. (2005). Formation of nucleoplasmic protein aggregates impairs nuclear function in response to SiO2 nanoparticles. *Exp. Cell Res.*, **305**(1), pp. 51–62.

115. Hubbs, A. F., Mercer, R. R., Benkovic, S. A., Harkema, J., Sriram, K., Schwegler-Berry, D., Goravanahally, M. P., Nurkiewicz, T. R., Castranova, V., and Sargent, L. M. (2010). Nanotoxicology: a pathologist's perspective. *Toxicol. Pathol.*, **39**(2), pp. 301–324.

116. AshaRani, P. V., Low Kah Mun, G., Hande, M. P., and Valiyaveettil, S. (2008). Cytotoxicity and genotoxicity of silver nanoparticles in human cells. *ACS Nano*, **3**(2), pp. 279–290.

117. Falck, G. C. M., Lindberg, H. K., Suhonen, S., Vippola, M., Vanhala, E., Catalán, J., Savolainen, K., and Norppa, H. (2009). Genotoxic effects of nanosized and fine TiO2. *Hum. Exp. Toxicol.*, **28**(6–7), pp. 339–352.

118. Li, K. G., Chen, J. T., Bai, S. S., Wen, X., Song, S. Y., Yu, Q., Li, J., and Wang, Y. Q. (2009). Intracellular oxidative stress and cadmium ions release induce cytotoxicity of unmodified cadmium sulfide quantum dots. *Toxicol. In Vitro*, **23**(6), pp. 1007–1013.

119. Wang, F., Gao, F., Lan, M., Yuan, H., Huang, Y., and Liu, J. (2009). Oxidative stress contributes to silica nanoparticle-induced cytotoxicity in human embryonic kidney cells. *Toxicol. In Vitro*, **23**(5), pp. 808–815.

120. Driscoll, K. E., Carter, J. M., Howard, B. W., Hassenbein, D., Janssen, Y. M., and Mossman, B. T. (1998). Crocidolite activates NF-kappa B and MIP-2

gene expression in rat alveolar epithelial cells. Role of mitochondrial-derived oxidants. *Environ. Health Perspect.*, **106**(Suppl 5), pp. 1171–1174.

121. Li, H., Haberzettl, P., Albrecht, C., Höhr, D., Knaapen, A. M., Borm, P. J. A., and Schins, R. P. F. (2007). Inhibition of the mitochondrial respiratory chain function abrogates quartz induced DNA damage in lung epithelial cells. *Mutat. Res.: Fund. Mol. Mech. Mutag.*, **617**(1–2), pp. 46–57.

122. Panduri, V., Weitzman, S. A., Chandel, N. S., and Kamp, D. W. (2004). Mitochondrial-derived free radicals mediate asbestos-induced alveolar epithelial cell apoptosis. *Am. J. Physiol. Lung Cell. Mol. Physiol.*, **286**(6), pp. L1220–L1227.

123. Valko, M., Rhodes, C. J., Moncol, J., Izakovic, M., and Mazur, M. (2006). Free radicals, metals and antioxidants in oxidative stress-induced cancer. *Chem.: Biol. Interact.*, **160**(1), pp. 1–40.

124. Bhabra, G., Sood, A., Fisher, B., Cartwright, L., Saunders, M., Evans, W. H., Surprenant, A., Lopez-Castejon, G., Mann, S., Davis, S. A., Hails, L. A., Ingham, E., Verkade, P., Lane, J., Heesom, K., Newson, R., and Case, C. P. (2009). Nanoparticles can cause DNA damage across a cellular barrier. *Nat. Nano*, **4**(12), pp. 876–883.

125. Beyersmann, D., and Hartwig, A. (2008). Carcinogenic metal compounds: recent insight into molecular and cellular mechanisms. *Arch. Toxicol.*, **82**(8), pp. 493–512.

126. Donaldson, K., and Poland, C. A. (2012). Inhaled nanoparticles and lung cancer: what we can learn from conventional particle toxicology. *Swiss Med. Wkly.*, **142**, p. w13547.

127. Morishige, T., Yoshioka, Y., Inakura, H., Tanabe, A., Narimatsu, S., Yao, X., Monobe, Y., Imazawa, T., Tsunoda, S.-i., Tsutsumi, Y., Mukai, Y., Okada, N., and Nakagawa, S. (2012). Suppression of nanosilica particle-induced inflammation by surface modification of the particles. *Arch. Toxicol.*, pp. 1–11.

128. Roursgaard, M., Jensen, K. A., Poulsen, S. S., Jensen, N.-E. V., Poulsen, L. K., Hammer, M., Nielsen, G. D., and Larsen, S. T. (2011). Acute and subchronic airway inflammation after intratracheal instillation of quartz and titanium dioxide agglomerates in mice. *ScientificWorldJournal*, **11**, pp. 801–825.

129. Sun, J., Wang, S., Zhao, D., Hun, F., Weng, L., and Liu, H. (2011). Cytotoxicity, permeability, and inflammation of metal oxide nanoparticles in human cardiac microvascular endothelial cells. *Cell Biol. Toxicol.*, **27**(5), pp. 333–342.

130. Poland, C. A., Duffin, R., Kinloch, I., Maynard, A., Wallace, W. A. H., Seaton, A., Stone, V., Brown, S., MacNee, W., and Donaldson, K. (2008). Carbon nanotubes introduced into the abdominal cavity of mice show asbestos-like pathogenicity in a pilot study. *Nat. Nano*, **3**(7), pp. 423–428.

131. Nemmar, A., Melghit, K., and Ali, B. H. (2008). The acute proinflammatory and prothrombotic effects of pulmonary exposure to rutile TiO2 nanorods in rats. *Exp. Biol. Med.*, **233**(5), pp. 610–619.

132. Xu, A., Chai, Y., Nohmi, T., and Hei, T. (2009). Genotoxic responses to titanium dioxide nanoparticles and fullerene in gpt delta transgenic MEF cells. *Part. Fibre Toxicol.*, **6**(1), p. 3.

133. Tsuda, H., Xu, J., Sakai, Y., Futakuchi, M., and Fukamachi, K. (2009). Toxicology of engineered nanomaterials: a review of carcinogenic potential. *Asian Pac. J. Cancer Prevent.*, **10**, pp. 975–980.

134. Roller, M. (2009). Carcinogenicity of inhaled nanoparticles. *Inhal. Toxicol.*, **21**(s1), pp. 144–157.

135. Klaunig, J. E., Xu, Y., Isenberg, J. S., Bachowski, S., Kolaja, K. L., Jiang, J., Stevenson, D. E., and Walborg, E. F. (1998). The role of oxidative stress in chemical carcinogenesis. *Environ. Health Perspect.*, **106**(Suppl 1), pp. 289–295.

136. Mahadevan, B., Snyder, R. D., Waters, M. D., Benz, R. D., Kemper, R. A., Tice, R. R., and Richard, A. M. (2011). Genetic toxicology in the 21st century: reflections and future directions. *Environ. Mol. Mutagen.*, **52**(5), pp. 339–354.

137. Maron, D. M., and Ames, B. N. (1983). Revised methods for the Salmonella mutagenicitytest. *Mutat. Res.*, **113**(3–4), pp. 173–215.

138. Hasegawa, G., Shimonaka, M., and Ishihara, Y. (2012). Differential genotoxicity of chemical properties and particle size of rare metal and metal oxide nanoparticles. *J. Appl. Toxicol.*, **32**(1), pp. 72–80.

139. Pan, X., Redding, J. E., Wiley, P. A., Wen, L., McConnell, J. S., and Zhang, B. (2010). Mutagenicity evaluation of metal oxide nanoparticles by the bacterial reverse mutation assay. *Chemosphere*, **79**(1), pp. 113–116.

140. Mortelmans, K., and Zeiger, E. (2000). The Ames Salmonella/microsome mutagenicity assay. *Mutat. Res.: Fund. Mol. Mech. Mutag.*, **455**, pp. 29–60.

141. Collins, A. R., Oscoz, A. A., Brunborg, G., Gaivao, I., Giovannelli, L., Kruszewski, M., Smith, C. C., and Stetina, R. (2008). The comet assay: topical issues. *Mutagenesis*, **23**(3), pp. 143–151.

142. Walmsley, R. M., and Billinton, N. (2011). How accurate is in vitro prediction of carcinogenicity? *Br. J. Pharmacol.*, **162**(6), pp. 1250–1258.

143. Olive, P. L., and Banath, J. P. (2006). The comet assay: a method to measure DNA damage in individual cells. *Nat. Protoc.*, **1**(1), pp. 23–29.

144. McKelvey-Martin, V. J., Green, M. H. L., Schmezer, P., Pool-Zobel, B. L., De Meo, M. P., and Collins, A. (1993). The single cell gel electrophoresis assay (comet assay): a European review. *Mutat. Res.: Fund. Mol. Mech. Mutag.*, **288**(1), pp. 47–63.

145. Sedelnikova, O. A., Redon, C. E., Dickey, J. S., Nakamura, A. J., Georgakilas, A. G., and Bonner, W. M. (2010). Role of oxidatively induced DNA lesions in human pathogenesis. *Mutat. Res.: Rev. Mutat. Res.*, **704**, pp. 152–159.

146. Ismail, I. H., Wadhra, T. I., and Hammarsten, O. (2007). An optimized method for detecting gamma-H2AX in blood cells reveals a significant interindividual variation in the gamma-H2AX response among humans. *Nucl. Acids Res.*, **35**(5), p. e36.

147. Setyawati, M. I., Khoo, P. K., Eng, B. H., Xiong, S., Zhao, X., Das, G. K., Tan, T. T., Loo, J. S., Leong, D. T., and Ng, K. W. (2013). Cytotoxic and genotoxic characterization of titanium dioxide, gadolinium oxide and poly(lactic-co-glycolic acid) anoparticles in human fibroblasts. *J. Biomed. Mater. Res. A.*, **101**(3), pp. 633–640.

148. Muslimovic, A., Ismail, I. H., Gao, Y., and Hammarsten, O. (2008). An optimized method for measurement of gamma-H2AX in blood mononuclear and cultured cells. *Nat. Protoc.*, **3**(7), pp. 1187–1193.

149. Trouiller, B., Reliene, R., Westbrook, A., Solaimani, P., and Schiestl, R. H. (2009). Titanium dioxide nanoparticles induce DNA damage and genetic instability in vivo in mice. *Cancer Res.*, **69**(22), pp. 8784–8789.

150. Shinohara, N., Matsumoto, K., Endoh, S., Maru, J., and Nakanishi, J. (2009). In vitro and in vivo genotoxicity tests on fullerene C60 nanoparticles. *Toxicol. Lett.*, **191**, pp. 289–296.

151. Maenosono, S., Yoshida, R., and Saita, S. (2009). Evaluation of genotoxicity of amine-terminated water-dispersible FePt nanoparticles in the Ames test and in vitro chromosomal aberration test. *J. Toxicol. Sci.*, **34**(3), pp. 349–354.

152. Lee, B. M., Kwak, E. H., Bae, J. Y., Lim, S. K., Kwon, M. J., Cho, M. C., Ahn, I. Y., and Kim, S. M. (2011). Chromosomal aberration for colloidal silver nanoparticles (AgNPs). *Toxicol. Lett.*, **205**(Supplement 0), p. S291.

153. Hackenberg, S., Scherzed, A., Kessler, M., Hummel, S., Technau, A., Froelich, K., Ginzkey, C., Koehler, C., Hagen, R., and Kleinsasser, N.

(2010). Silver nanoparticles: evaluation of DNA damage, toxicity and functional impairment in human mesenchymal stem cells. *Toxicol. Lett.*, **201**(1), pp. 27–33.

154. Balasubramanyam, A., Sailaja, N., Mahboob, M., Rahman, M. F., Misra, S., Hussain, S. M., and Grover, P. (2009). Evaluation of genotoxic effects of oral exposure to Aluminum oxide nanomaterials in rat bone marrow. *Mutat. Res.: Gen. Toxicol. Environ. Mutagen.*, **676**, pp. 41–47.

155. Sonoda, E., Sasaki, M. S., Buerstedde, J.-M., Bezzubova, O., Shinohara, A., Ogawa, H., Takata, M., Yamaguchi-Iwai, Y., and Takeda, S. (1998). Rad51-deficient vertebrate cells accumulate chromosomal breaks prior to cell death. *EMBO J.*, **17**(2), pp. 598–608.

156. Levy, D., Reeder, C., Loucas, B., Hlatky, L., Chen, A., Cornforth, M., and Sachs, R. (2007). Interpreting chromosome aberration spectra. *J. Comput. Biol.*, **14**(2), pp. 144–155.

157. Kalantari, H., Gourabi, H., and Baharvand, H. (2012). Adapted cytokinesis-block micronucleus assay (CBMn) for mouse embryonic stem cells. *Protocol Exchange,* DOI 10.1038/protex.2012.011.

158. Thomas, P., Holland, N., Bolognesi, C., Kirsch-Volders, M., Bonassi, S., Zeiger, E., Knasmueller, S., and Fenech, M. (2009). Buccal micronucleus cytome assay. *Nat. Protoc.*, **4**(6), pp. 825–837.

159. Kirsch-Volders, M., Plas, G., Elhajouji, A., Lukamowicz, M., Gonzalez, L., Loock, K. V., and Decordier, I. (2011). The in vitro MN assay in 2011: origin and fate, biological significance, protocols, high throughput methodologies and toxicological relevance. *Arch. Toxicol.*, **85**(8), pp. 873–899.

160. Poma, A., Limongi, T., Pisani, C., Granato, V., and Picozzi, P. (2006). Genotoxicity induced by fine urban air particulate matter in the macrophages cell line RAW 264.7. *Toxicol. In Vitro*, **20**(6), pp. 1023–1029.

161. Gonzalez, L., Sanderson, B. J. S., and Kirsch-Volders, M. (2011). Adaptations of the in vitro MN assay for the genotoxicity assessment of nanomaterials. *Mutagenesis*, **26**(1), pp. 185–191.

162. Kirsch-Volders, M., Decordier, I., Elhajouji, A., Plas, G., Aardema, M. J., and Fenech, M. (2011). In vitro genotoxicity testing using the micronucleus assay in cell lines, human lymphocytes and 3D human skin models. *Mutagenesis*, **26**(1), pp. 177–184.

163. Corvi, R., Albertini, S., Hartung, T., Hoffmann, S., Maurici, D., Pfuhler, S., van Benthem, J., and Vanparys, P. (2008). ECVAM retrospective validation of in vitro micronucleus test (MNT). *Mutagenesis*, **23**(4), pp. 271–283.

164. Bolognesi, C., and Fenech, M. (2012). Mussel micronucleus cytome assay. *Nat. Protoc.*, **7**(6), pp. 1125–1137.

165. Khalil, W. K. B., Girgis, E., Emam, A. N., Mohamed, M. B., and Rao, K. V. (2011). Genotoxicity evaluation of nanomaterials: DNA damage, micronuclei, and 8-hydroxy-2-deoxyguanosine induced by magnetic doped CdSe quantum dots in male mice. *Chem. Res. Toxicol.*, **24**(5), pp. 640–650.

166. Girgis, E., Khalil, W. K. B., Emam, A. N., Mohamed, M. B., and Rao, K. V. (2012). Nanotoxicity of gold and gold-cobalt nanoalloy. *Chem. Res. Toxicol.*, **25**(5), pp. 1086–1098.

167. Valavanidis, A., Vlachogianni, T., and Fiotakis, C. (2009). 8-hydroxy-2′-deoxyguanosine (8-OHdG): a critical biomarker of oxidative stress and carcinogenesis. *J. Environ. Sci. Health C Environ. Carcinog. Ecotoxicol. Rev.*, **27**(2), pp. 120–139.

168. Bhattacharya, K., Davoren, M., Boertz, J., Schins, R. P. F., Hoffmann, E., and Dopp, E. (2009). Titanium dioxide nanoparticles induce oxidative stress and DNA-adduct formation but not DNA-breakage in human lung cells. *Part. Fibre Toxicol.*, **6**, p. 11.

169. Shi, Y., Zhang, J. H., Jiang, M., Zhu, L. H., Tan, H. Q., and Lu, B. (2009). Synergistic genotoxicity caused by low concentration of titanium dioxide nanoparticles and p,p′-DDT in human hepatocytes. *Environ. Mol. Mutagen.*, **51**(3), pp. 192–204.

170. Gong, C. M., Tao, G. H., Yang, L. Q., Liu, J. J., He, H. W., and Zhuang, Z. X. (2011). The role of reactive oxygen species in silicon dioxide nanoparticle-induced cytotoxicity and DNA damage in HaCaT cells. *Mol. Biol. Rep.*, **39**(4), pp. 4915–4925.

171. Bogdanov, M. B., Beal, M. F., McCabe, D. R., Griffin, R. M., and Matson, W. R. (1999). A carbon column-based liquid chromatography electrochemical approach to routine 8-hydroxy-2′-deoxyguanosine measurements in urine and other biologic matrices: a one-year evaluation of methods. *Free Rad. Biol. Med.*, **27**(5/6), pp. 647–666.

172. Wu, J., and Sun, J. A. (2010). Effect of iron oxide nanoparticles on PC12 Cells by MTS and p53, in *Inec: 2010 3rd International Nanoelectronics Conference, Vols 1 and 2* (IEEE, New York), pp. 842–843.

173. Jia, C.-P., Zhong, X.-Q., Hua, B., Liu, M.-Y., Jing, F.-X., Lou, X.-H., Yao, S.-H., Xiang, J.-Q., Jin, Q.-H., and Zhao, J.-L. (2009). Nano-ELISA for highly sensitive protein detection. *Biosensors Bioelectron.*, **24**(9), pp. 2836–2841.

Chapter 13

Pulmonary and Cardiovascular Toxicity of Nanomaterials

Flemming R. Cassee[a] and Vincent Castranova[b]
[a]*National Institute for Public Health and the Environment Netherlands, Bilthoven, The Netherlands*
[b]*National Institute for Occupational Safety and Health, Centers for Disease Control and Prevention, Morgantown, WV, USA*
flemming.cassee@rivm.nl

13.1 Introduction

Nanoparticles are being incorporated at a rapid rate into new structures, materials, and devices with a wide range of commercial applications. Since nanoparticles exhibit physicochemical properties strikingly different than fine particles of the same composition, there is every reason to suspect that nanoparticles could exhibit unique bioactivity. Therefore, the field of nanotoxicology has emerged to evaluate health effects of exposure to nanoparticles, determine the dose dependency and time course of these responses, and identify mechanisms of action. Environmental assessment studies indicate that nanoparticles can be aerosolized during weighing, blending,

Handbook of Safety Assessment of Nanomaterials: From Toxicological Testing to Personalized Medicine
Edited by Bengt Fadeel

ISBN 978-981-4463-36-2 (Hardcover), 978-981-4463-37-9 (eBook)
www.panstanford.com

mixing, sonication, and spraying procedures [1–3]. In addition, high levels of aerosolized nanoparticles can be generated during cleaning of ovens used for nanoparticle synthesis [4]. Therefore, inhalation is believed to be a primary route of nanoparticle exposure in the workplace. Inhalation is also a relevant route of administration of certain drugs. The nanotoxicology literature has grown rapidly in the last decade. The largest number of studies has evaluated the toxicological potential of titanium dioxide (TiO_2) nanospheres and carbon nanotubes (CNTs). Therefore, this chapter will summarize data concerning the response of the respiratory and cardiovascular systems after pulmonary exposure to nano-TiO_2 and CNTs.

13.2 Respiratory and Cardiovascular Effects of Pulmonary Exposure to Nanoparticles/Nanotubes

Epidemiological studies have consistently reported a positive relationship between levels of ambient particular matter (PM) and both respiratory and cardiovascular morbidity and premature mortality [5–12]. Respiratory mortality associated with ambient PM includes chronic obstructive pulmonary disease, pneumonia, and acute lower respiratory disease. Respiratory causes for increased hospital visits include asthma and pulmonary infections. Causes of increased cardiovascular mortality include chronic coronary artery disease, congestive heart failure, ischemic heart disease, and cerebrovascular disease. Morbidity was noted as increased hospital admissions for myocardial infarction and ventricular fibrillation. The associations between PM and cardiopulmonary dysfunction are stronger for $PM_{2.5}$ than PM_{10}, that is, smaller particles appear to exhibit greater respiratory and cardiovascular dysfunction on an equal mass basis [13, 14]. It is possible that on an equal mass basis nanoparticles may be more potent that fine particles. There is increasing, yet limited, epidemiological evidence on the effects of ultrafine (<0.1 μm) particles on cardiorespiratory health and the central nervous system. Clinical and toxicological studies have shown that ultrafine particles in part act through mechanisms not shared with larger particles that dominate mass-based metrics, such as $PM_{2.5}$ or PM_{10}. Ultrafine particles or nanoparticles in ambient air consist mainly of carbonaceous combustion-derived particles and exhibit a stronger association with respiratory and cardiovascular dysfunction than that observed for $PM_{2.5}$ and PM_{10} [15, 16]. Chapter 5 provides an overview of animal data on toxicity of various classes of nanoparticles upon different routes of exposure. Here, we focus on the respiratory and cardiovascular impact after pulmonary exposure to nano-TiO_2 and CNTs.

13.2.1 Respiratory Response to Pulmonary Exposure to TiO_2 Nanoparticles

Although fine TiO_2 is known as a relatively low toxic chemical, bioactivity does seem to increase with decreasing size. This was clearly demonstrated studies by Oberdorster et al. [17, 18], who exposed rats and mice by intratracheal instillation of nanoparticles of TiO_2 (20 nm particle diameter) and larger pigment-grade particles (250 nm), establishing dose–response relationships. These studies indicate that the nanoparticles lead to a greater acute pulmonary inflammatory response as well as fibrosis at a later stage than pigment-grade material. Although, on a mass basis nanoparticles are more toxic, it was also shown that when using total surface area as the metric for dose, the dose–response curves for both ultrafine and fine TiO_2 overlapped for the studies performed in mice and rats [19]. On the other hand, Warheit et al. [20] reported that neither ultrafine nor fine TiO_2 causes long-term toxicity after intratracheal instillation in rats. Furthermore, they found that acute inflammation was not related to particle surface area as measured by gas absorption. However, dynamic light-scattering data indicates that structure sizes of the ultrafine and fine samples as delivered to the rats were identical, being greater than 2 μm. Similarly, Horie et al. [21] reported that ultrafine and fine TiO_2 did not differ in toxicity one to seven days after intratracheal instillation in rats. However, the exposure dose used in this study was so low that neither particle caused significant pulmonary effects. Therefore, one cannot discern a size-dependent difference in potency. In addition, structure diameters of the ultrafine and fine samples only differed by threefold. In contrast, when an effort was made to disperse the particulate samples, ultrafine TiO_2 was found to be 40-fold more toxic and inflammatory than fine TiO_2 1–42 days after intratracheal instillation in rats [22]. Another study in which rats were exposed by inhalation [23] indicated that very little difference in effects was observed between ultrafine or fine TiO_2. Rats were exposed six hours/day, seven days/week, for three weeks up to 45 mg/m^3 with 21 μm or 300 μm particles [23]. Subtle if any effects were observed with no clear difference between the two types of TiO_2. However, given the fact that the mass median aerodynamic diameters were much more comparable (0.7 μm vs. 1.1 μm) than the primary size, more or less similar dose levels will have been obtained in terms of particle numbers and particle mass. This most likely explains the relative small differences that were observed compared to the study by Sager et al. [22], where particle diameters of the ultrafine and fine samples as delivered were substantially different.

The molecular structure of TiO_2 is fixed, yet the physical aspects can be altered. Particle shape affects the performance of TiO_2 for various applications (e.g., photocatalyst, solar energy). Therefore, the toxicological profile may also change due to differences in dimensions. A recent study demonstrated that nanospheres behave differently from high-aspect-ratio nanobelts after pharyngeal aspiration in mice [24]. On an equivalent mass dose basis, TiO_2 nanobelts induced greater pulmonary inflammation and damage than nanospheres, with longer nanobelts being more toxic than short nanobelts. In addition, nanobelts induced interstitial fibrosis at 112 days postexposure (long belts being more fibrogenic than short), while TiO_2 nanospheres did not. Furthermore, high-aspect-ratio TiO_2, but no nanospheres, resulted in impaired lymphatic clearance of these particles. In general, oxidative stress and the expression of inflammatory cytokines are recognized as the underlying mechanisms of adverse health effects reported at relatively high exposure levels of nano-TiO_2. Results indicate that the size and shape of delivered TiO_2 structures influence bioactivity.

13.2.2 Respiratory Response to Pulmonary Exposure to Carbon Nanotubes

CNTs comprise a large group of carbonaceous cylindrical structures that are characterized by the high length-to-diameter ratio of up to more than a billion. CNTs are recognized for unusual thermal conductivity and mechanical and electrical properties. The applications are far reaching, from construction materials and electronics to medical applications, such as drug delivery. In general, two types of CNTs can be distinguished, multiwalled carbon nanotubes (MWCNTs) and single-walled carbon nanotubes (SWCNTs). The literature indicates that exposure to well-dispersed CNTs will lead to oxidative stress, inflammatory responses, malignant transformation, DNA damage and mutation, granulomas, and interstitial fibrosis [25–27]. Exposure to CNTs causes concentration-dependent pulmonary inflammogenicity with evidence of regression over time [27, 28]. There are a number of critical aspects related to pulmonary toxicity, such as the length and rigidity of the fibers. For example, length is a critical parameter in the development of pleural inflammation, which has been associated with initial stages of fiber-induced mesothelioma [29]. However, length alone is not the ultimate predictor. Well-dispersed CNTs can cause thickening of the alveolar wall and fewer granulomatous lesions than agglomerated CNTs, which in turn resulted in more pronounced granulomatous inflammation [30, 31]. In addition to the greater lung collagen levels, indicating fibrosis, increased dispersion of MNCNTs resulted in greater transforming growth factor beta-1 (TGF-β1) and platelet-derived growth factor (PDGF) levels in bronchoalveolar lavage fluid in a mouse model [32]. Rigid, straight, long fibers are needed to cause severe pulmonary responses as compared to tangled fibers [29]. Also, surface modifications will affect the toxicity of CNTs. For example, in a study in which mice were exposed by pharyngeal aspiration to a single dose (2 mg/kg body weight) of MWCNTs carboxylation tended to decrease the production of the fibrogenic mediators TGF-β1 and PDGF [32]. Lam et al. [33] have evaluated the effect of metallic impurities on pulmonary response of mice to intratracheal instillation of SWCNTs and reported that metals do not play a major role in the granulomatous response. Therefore, although raw CNTs contain metal impurities, the principal reaction of the lung to CNTs appears more related to the structure (physical aspects) and not catalyst impurities. Despite the range of differences in physicochemical properties, CNTs to date have been treated as one entity in guidance documents from Japan, Europe, and the United States.

13.2.3 Cardiovascular Response to Pulmonary Exposure to TiO_2 Nanoparticles

Nurkiewicz et al. [34] reported that the ability of arterioles in the shoulder muscle (spinotrapezius muscle) of rats to response to dilators is inhibited 24 hours after intratracheal instillation of fine residual oil fly ash (ROFA) or fine TiO_2. This cardiovascular effect of pulmonary exposure appears to be a particulate effect rather than a response to soluble metals, because ROFA (containing soluble metals) and TiO_2 (no soluble metals) exhibited similar dose–response curves. Inhalation of nano-TiO_2 caused significantly greater systemic microvascular inhibition of responsiveness to dilators on an equal mass lung burden basis than fine TiO_2 [35]. Indeed, nano-TiO_2 was ~6.7-fold more potent than fine TiO_2, inhibiting dilation of arterioles in the shoulder muscle in response to calcium ionophore (A23187) by 37% at lung burdens of 10 μg versus 67 μg, respectively. Normalizing the pulmonary dose as surface area (measured as nitrogen absorption by the Brunauer–Emmett–Teller [BET] technique) delivered to the lungs overestimated the potency of fine TiO_2. It should be noted that aerosolized nano-TiO_2 exhibited a count median diameter of 150 nm in contrast to a primary particle size of 22 nm. If one calculated the physical surface area of aerosolized structures, assuming a spherical shape, to normalize the structure surface area delivered, then the potency of nano- versus fine TiO_2 was identical. Therefore, this data indicates the surface of structures delivered to the lung directly affects bioactivity when comparing particles of similar chemical composition.

Inhalation of nano-TiO_2 also inhibited the ability of arterioles from the subepicardial, left anterior descending artery to respond to acetylcholine, A23187, or increased shear stress [36]. As with inhibition of dilation of shoulder muscle arterioles, substantial microvascular dysfunction was noted 24 hours after an inhaled lung burden of 10 μg/rat. Of interest, markers of lung inflammation and injury in bronchoalveolar lavage samples were not significantly elevated above filtered air controls at this low exposure dose. This indicates that the cardiovascular system is very sensitive to pulmonary exposure to particles.

Inhalation of nano-TiO_2 (lung burden 10 μg/rat) also increased phosphorylation of p38 mitogen-activated protein kinase and troponin I in cardiac tissue 24 hours postexposure [37]. Such phosphorylation has been reported to affect cardiac muscle contractility and myocardial function [38–39]. Furthermore, inhalation of nano-TiO_2 (lung burden 10 μg/rat) increased the heart rate in response to isoproterenol and elevated mean and diastolic blood pressure in response to norepinephrine 24 hours postexposure [40].

13.2.4 Cardiovascular Response to Pulmonary Exposure to Carbon Nanotubes

Stapleton et al. [41] reported in a rat model that inhalation of MWCNTs at 5 mg/m^3 for five hours (lung burden of 13.5 μg/lung) inhibited the ability of coronary arterioles to respond normally to dilators, which act via an endothelial-dependent mechanism, such as acetylcholine or the calcium ionophore (A23187). At 24 hours postexposure, maximal inhibition of dilation was greater than 100%, indicating complete inhibition of dilation and an additional contractile effect of pulmonary exposure to MWCNTs. This inhibition of microvascular responsiveness to dilators decreased with time postexposure (24–168 hours) but did not return to control levels even at 168 hours postexposure. Responsiveness of coronary arteriolar smooth muscle to nitric oxide was not affected. In addition, pulmonary exposure to MWCNTs did not alter microvascular dilation in response to transmural pressure (myogenic responsiveness) or contraction in response to an α-adrenergic agonist (phenylephrine).

Li et al. [42] reported enhanced aortic plaque formation in atherosclerotis-susceptible mice (ApoE−/−) after pulmonary exposure to SWCNTs. In this study ApoE−/− mice were exposed by pharyngeal aspiration to SWCNTs (20 μg, once every two weeks for eight weeks) and the number and area of aortic and arterial plaques measured. Pulmonary exposure to SWCNTs significantly increased the aortic plaque lesion area by 69%, while increasing the plaque lesion area of the brachiocephalic artery by 114%.

Legramante et al. [43] measured arterial pressure and heart rate in rats 24 hours after intratracheal instillation of SWCNTs (300 μg). No changes in blood pressure, heart rate, or heart rate variability were noted after pulmonary exposure to SWCNTs. However, a significant reduction baroreceptor sensitivity (a 72%

change from controls) was noted after SWCNT exposure. A decrease in baroreceptor sensitivity is an indicatory of the degree of vagal control and has been associated with enhanced risk of ventricular fibrillation [44].

13.3 Mechanisms by Which Pulmonary Exposure to Nanoparticles/Nanotubes Affects Cardiovascular Function

Several mechanisms, including particle translocation from the lung to the cardiovascular tissue, elevation of inflammatory or thrombotic mediators in blood, oxidant production at systemic vessels, and alteration of automatic control of the cardiovascular system, have been proposed to explain how particle inhalation results in cardiovascular dysfunction [45]. Evidence supporting or discounting these mechanisms to explain animal study results after pulmonary TiO_2 or CNT exposure follows.

Oberdorster et al. [46] reported movement of ^{13}C ultrafine particles from the lung to the liver 24 hours after inhalation in a rat model, indicating that systemic translocation of nanoparticles from the lung is possible. However, Kreyling et al. [47] found the extent of this systemic translocation to be very low, that is, less that 1% over seven days after inhalation of ^{192}Ir nanoparticles in rats. Additionally, the translocation rate decreased significantly as particle size increased from 15 nm to 80 nm. To date, no studies have evaluated the translocation of nano-TiO_2; however, one report for MWCNTs is available [41]. In this study, rats were exposed by inhalation to MWCNTs (lung burden 13.5 μg/lung) and MWCNT counts in the lung, heart, kidney, and liver determined by morphometric analysis at 24 hours postexposure. Results indicate that some MWCNTs were observed in systemic organs. However, the extent of such translocation was extremely low (0.00009% of the lung burden had moved to the heart, 0.0002% to the kidney, and 0.0007% to the liver). Therefore, approximately 1500 MWCNTs had migrated to the heart at a time when coronary arteriole responsiveness to dilators was completely blocked [41]. This particle burden seems too low to explain the microvascular dysfunction resulting from pulmonary exposure to MWCNTs.

A second proposed mechanism is that pulmonary exposure to TiO_2 or CNTs causes lung inflammation, which enhances inflammatory or thrombotic mediators in the blood to alter cardiovascular response. Nurkiewicz et al. [48] reported complete inhibition of the ability of arterioles in the shoulder muscle to respond to dilators at a lung burden of fine TiO_2 in the rat, which did not cause a significant elevation of inflammatory markers in bronchoalveolar lavage fluid. Similarly, they failed to find any elevations of mRNA for inflammatory mediators in peripheral blood samples. Likewise, Kan et al. [37] failed to find elevated levels of tumor necrosis factor alpha (TNF-α) or interleukin-1 (IL-1) either immediately or 24 hours after inhalation of nano-TiO_2 (rat lung burden of 10 μg). However, this pulmonary lung burden of nano-TiO_2 has been reported to significantly inhibit the ability of arterioles in the shoulder muscle [35] or the heart [36] to respond to dilators. Li et al. [42] also failed to observe an increase in plasma levels of inflammatory cytokines and chemokines in mice exposed to SWCNTs at pulmonary exposure levels that induced arterial plaque formation. Likewise, Inoue et al. [49] found serum levels of inflammatory cytokines and chemokines unchanged 24 hours after intratracheal instillation of 120 μg of SWCNTs or MWCNTs in mice. However, they did report significant increases in plasma fibrinogen and von Willebrand factor (coagulatory mediators) 24 hours after pulmonary exposure to SWCNTs but not MWCNTs. Meng et al. [50] reported similar results in hypertensive rats after pulmonary exposure to SWCNTs, that is, no changes in blood inflammatory mediators but an elevation in plasma endothelin-1- and angiotensin-1-converting enzyme. Erdely et al. [51] reported that serum inflammatory markers (IL-6, IL-5, CCL11, CCL22, and CXCL1) were elevated in mice 4 hours after pulmonary exposure to CNTs (40 μg/lung) but returned to normal by 24 hours postexposure. At this time, elevations of acute-phase proteins (C-reactive protein, haptoglobin, and serum amyloid) were noted. Taken as a whole, currently available data does not strongly support the hypothesis that pulmonary exposure to nano-TiO_2 or CNTs results in blood mediator levels sufficient to explain the cardiovascular effects reported in the literature.

Nurkiewicz et al. [48] documented that pulmonary exposure to fine TiO_2 increases adherence of polymorphonuclear leukoctyes (PMNs) to microvessel walls in the shoulder muscle of rats 24 hours postexposure. A later study from this group documented that inhalation of nano-TiO_2 (10 μg lung burden in rats) potentiated the generation of reactive oxygen species (ROS) from blood PMNs 24 hours postexposure and that increased ROS generation was observed at the microvessel wall in the shoulder muscle [52]. They proposed that this ROS generation scavenged dilator-induced nitric oxide produced by endothelial cells and, thus, inhibited microvascular dilation. Indeed, antioxidants reversed the depletion of nitric oxide in arteriolar tissue and the dilator dysfunction seen after inhalation of nano-TiO_2. Similar results were reported for coronary arterioles after inhalation of nano-TiO_2 (10 μg lung burden) [53]. The local oxidant stress mechanism for the transduction of pulmonary CNT exposure to cardiovascular effects is supported by data from Li et al. [42]. These results indicate that a significant elevation in heme oxygenase-1(HO-1) in cardiac and aortic tissue and depression of glutathione in aortic mitochondria were associated with pulmonary aspiration of SWCNTs (40 μg/mouse).

Data also supports a neurogenic mechanism linking pulmonary exposure to nano-TiO_2 and cardiovascular dysfunction. Kan et al. [37] have reported that inhalation of nano-TiO_2 significantly increases substance P levels (a neurotransmitter) in the nodose ganglia (the synapase of sensory efferents from the lung) immediately after exposure. This sensory neuron activation increased further 24 hours after exposure. Inhibition of sensory receptors in the lung with a transient receptor potential blocker (ruthenium red) not only inhibited the up-regulation of substance P but also inhibited the phosphorylation of cardiac troponin I and changes in heart rate and blood pressure observed 24 hours after inhalation of nano-TiO_2 (10 μg lung burden in rats) [40]. Furthermore, inhibition of neuronal input to arterioles in the shoulder muscle with tetrodotxin partially restored microvascular dilation in rats exposed to nano-TiO_2 [54]. Additionally, inhalation of nano-TiO_2 increased sensitivity to adrenergic blockade, indicating enhanced sympathetic input to the microvasculature due to the pulmonary exposure [55]. Evidence that the cardiovascular effects of pulmonary exposure to CNTs may be mediated via the neurogenic pathway includes the following: 1) aspiration of MWCNTs induces mRNA for inflammatory mediators in the olfactory bulb, midbrain, hippocampus, and frontal cortex of the mouse brain 24 hours postexposure [56], and 2) pulmonary exposure to SWCNTs affects baroreceptor sensitivity [43].

The following is a schema that summarizes a proposed mechanistic pathway for induction of cardiovascular responses to pulmonary exposure to nano-TiO_2 or CNTs, which is supported by currently available data.

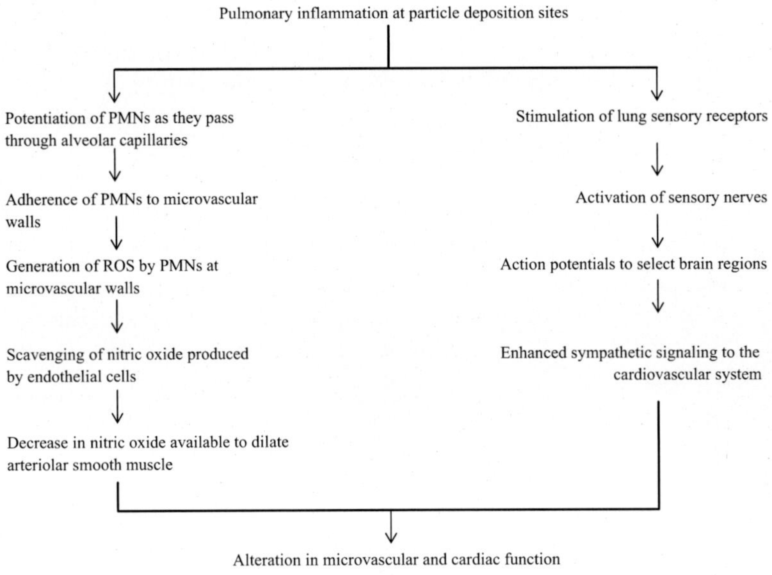

13.4 Conclusions

To date, data indicates that pulmonary exposure to nano-TiO_2 induces a greater degree of pulmonary inflammation and damage and a more robust cardiovascular effect than an equal mass of fine TiO_2 [18, 19, 35]. The dispersion status appears to affect the potency of TiO_2 nanoparticles [20, 22]. The dispersion status also affects the deposition site and pulmonary response to CNTs, that is, agglomerates deposit at the terminal bronchioles and proximal alveoli and induce granulomatous lesions, while smaller CNT structures can reach distal alveoli and induce interstitial fibrosis [30]. The fibrous shape and hydrophobic nature of CNTs appear to influence the ability of CNTs to penetrate epithelial cells and rapidly enter the interstitium of alveolar septa, where they directly stimulate proliferation and collagen formation of fibroblasts [26, 57, 58]. On an equal mass basis, CNTs appear to be more fibrogenic than asbestos

[59]. In addition, inhaled MWCNTs can induce cardiovascular effects at low lung burdens (<15 mg/rat) and appear more potent on a mass basis than nano-TiO_2 [35, 36, 41].

Disclaimer

The findings and conclusions in this chapter are those of the authors and do not necessarily represent the view of the National Institute for Occupational Safety and Health.

References

1. Maynard, A. D., Baron P. A., Foley, M., Shvedova, A. D., Kisin, E. R., and Castranova, V. (2004). Exposure to carbon nanotube material: aerosol release during the handling of unrefined single walled carbon nanotube material. *J. Toxicol. Environ. Health A*, **67**, pp. 87–107.
2. Han, J. H., Lee, E. J., Lee, J. H., So, K. P., Lee, Y. H., Bae, S. N., Lee, S. B., Ji, J. H., Cho, M. H., and Yu, I. J. (2008). Monitoring multiwalled carbon nanotube exposure in carbon nanotube research facility. *Inhal. Toxicol.*, **20**, pp. 741–749.
3. Lee, J. H., Lee, S. B., Bae, G. N., Jeon, K. S., Yoan, J. U., Ji, J. H., Sung, J. H., Lee, B. G., Lee, J. H., Yang, J. S., Kim, H. Y., Kung, C. S., and Yu, S. J. (2010). Exposure assessment of carbon nanotube manufacturing workplaces. *Inhal. Toxicol.*, **22**, pp. 369–381.
4. Methner, M. M. (2008). Effectiveness of local exhaust ventilation (LEV) in controlling engineered nanomaterial emissions during reactor cleanout operations. *J. Occup. Environ. Hyg.*, **5**, pp. D63–D69.
5. Pope, C. A. 3rd, and Dockery, D. W. (2006). Health effects of fine particulate air pollution: lines that connect. *J. Air Waste Manag. Assoc.*, **56**, pp. 709–742.
6. Brook, R. D., Rajagopalan, S., Pope, C. A., Brook, J. R., Bhatnagar, A., Diez-Roux, A. V., Holguin, F., Hong, Y., Luepker, R. V., Mittleman, M. A., Peters, A., Siscovick, D., Smith, S. C., Whitsel, L., and Kaufman, J. D. (2010). Particulate matter air pollution and cardiovascular disease: an update to the scientific statement from the American heart association. *Circulation*, **121**, pp. 2331–2378.
7. Dockery, D. W., Pope, C. A., Xu, X., Spengler, J. D., Ware, J. H., Fay, M. E., Ferris, B. G., and Speizer, F. E. (1993). An association between air pollution and mortality in six US cities. *New Engl. J. Med.*, **329**, pp. 1753–1759.

8. Pope, C. A., Thun, M. J., Namboordiri, M. M., Dockery, D. W., Evans, J. S., Speizer, F. F., and Hath, C. W. (1995). Particulate air pollution as a predictor of mortality in a prospective study of US adults. *Am. J. Respir. Crit. Care Med.*, **151**, pp. 669–674.
9. Abbey, D. E., Nishino, N., McDonnell, W. F., Burchette, R. J., Knutsen, S. F., Beeson, W. L., and Yang, J. X. (1999). Long-term inhalable particles and other air pollutants related to mortality in nonsmokers. *Am. J. Respir. Crit. Care Med.*, **159**, pp. 373–382.
10. Samet, J. M., Dominici, F., Curriero, F. C., Coursue, I., and Zeger, S. L. (2000). Fine particulate air pollution and mortality in 20 US cities. *New Engl. J. Med.*, **343**, pp. 1742–1749.
11. Peters, A., Liu, E., Verrier, R. L., Schwartz, J., Gold, D. R., Mittleman, M., Baliff, J., Oh, J. A., Allen, G., Monuhan, K., and Dockery, D. W. (2000). Air pollution and incidence of cardiac arrhythmia. *Epidemiology*, **11**, pp. 11–17.
12. Pope, C. A., Dockery, D. W., Spengler, J. D., and Raizenne, M. E. (1991). Respiratory health and PM10 pollution: a daily time series analysis. *Am. Rev. Respir. Dis.*, **144**, pp. 668–674.
13. Dockery, D. W., and Pope, C. A. (1994). Acute respiratory effects of particulate air pollution. *Annu. Rev. Pub. Health*, **15**, pp. 107–132.
14. Dockery, D. W. (2001). Epidemiologic evidence of cardiovascular effects of particulate air pollution. *Environ. Health Perspect.*, **109**(Suppl 4), pp. 438–486.
15. Janssen, N. A., Hoek, G., Simic-Lawson, M., Fischer, P., van Bree, L., ten Brink, H., Keuken, M., Atkinson, R. W., Anderson, H. R., Brunekreef, B., and Cassee, F. R. (2011). Black carbon as an additional indicator of the adverse health effects of airborne particles compared with PM_{10} and $PM_{2.5}$. *Environ. Health Perspect.*, **119**(12), pp. 1691–1699.
16. Peters, A., Rückerl, R., and Cyrys, J. (2011). Lessons from air pollution epidemiology for studies of engineered nanomaterials. *J. Occup. Environ. Med.*, **53**(6 Suppl), pp. S8–S13.
17. Baggs, R. B., Ferin, J., and Oberdörster, G. (1997). Regression of pulmonary lesions produced by inhaled titanium dioxide in rats. *Vet. Pathol.*, **34**(6), pp. 592–597.
18. Oberdörster, G., Finkelstein, J. N., Johnston, C., Gelein, R., Cox, C., Baggs, R., and Elder, A. C. (2000). Acute pulmonary effects of ultrafine particles in rats and mice. *Res. Rep. Health Eff. Inst.*, **96**, pp. 5–74; disc. 75–86.
19. Oberdörster, G., Oberdörster, E., and Oberdörster, J. (2005). Nanotoxicology: an emerging discipline evolving from studies of ultrafine particles. *Environ. Health Perspect.*, **113**(7), pp. 823–839.

20. Warheit, D., Webb, T., Reed, K., Frerichs, S., and Sayes, C. (2007). Pulmonary toxicity study in rats with three forms of ultrafine-TiO_2 particles: differential response related to surface properties. *Toxicology*, **230**, pp. 90–104.

21. Horie, M., Fukui, H., Endoh, S., Maru, J., Miyauchi, A., Shichiri, M., Fujita, K., Niki, E., Hagihara, Y., Yoshida, Y., Morimoto, Y., and Iwahashi, H. (2012). Comparison of acute oxidative stress on rat lung induced by nano and fine-scale, soluble and insoluble metal oxide particles: NiO and TiO_2. *Inhal. Toxicol.*, **24**(7), pp. 391–400.

22. Sager, T. M., Kommineri, C., and Castranova, V. (2008). Pulmonary response to intratracheal instillation of ultrafine versus fine titanium dioxide: role of surface area. *Part. Fibre Toxicol.*, **5**, p. 17.

23. Eydner, M., Schaudien, D., Creutzenberg, O., Ernst, H., Hansen, T., Baumgärtner, W., and Rittinghausen, S. (2012). Impacts after inhalation of nano- and fine-sized titanium dioxide particles: morphological changes, translocation within the rat lung, and evaluation of particle deposition using the relative deposition index. *Inhal. Toxicol.*, **24**(9), pp. 557–569.

24. Porter, D. W., Wu, N., Hubbs, A. F., Mercer, R. R., Funk, K., Meng, F., Li, J., Wolfarth, M. G., Battelli, L., Friend, S., Andrew, M., Hamilton, R. Jr., Sriram, K., Yang, F., Castranova, V., and Holian, A.(2013). Differential mouse pulmonary dose and time course responses to titanium dioxide nanospheres and nanobelts. *Toxicol. Sci.*, **131**(1), pp. 179–193.

25. Liu, Y., Zhao, Y., Sun, B., and Chen, C. (2013). Understanding the toxicity of carbon nanotubes. *Acc. Chem. Res.*, **46**(3), pp. 702–713.

26. Porter, D. W., Hubbs, A., Mercer, R., Wu, N., Wolfarth, M. G., Sriram, K., Leonard, S., Battelli, L., Schwegler-Berry, D., Friend, S., Andrew, M., Chen, B. T., Isuruoka, S., Endo, M., and Castranova V. (2010). Mouse pulmonary dose- and time course-responses induced by exposure to multi-walled carbon nanotubes. *Toxicology*, **269**, pp. 136–147.

27. Porter, D. W., Hubbs, A. F., Chen, B. T., McKinney, W., Mercer, R. R., Wolfarth, M. G., Battelli, L., Wu, N., Sriram, K., Leonard, S., Andrew, M., Willard, P., Tsuruoka, S., Endo, M., Tsukada, T., Munekane, F., Frazer, D. G., and Castranova, V. (2013). Acute pulmonary dose-responses to inhaled multi-walled carbon nanotubes. *Nanotoxicology*, **7**(7), pp. 1179–1194.

28. Ellinger-Ziegelbauer, H., and Pauluhn, J. (2009). Pulmonary toxicity of multi-walled carbon nanotubes (Baytubes) relative to alpha-quartz

following a single 6h inhalation exposure of rats and a 3 months post-exposure period. *Toxicology*, **266**(1–3), pp. 16–29.

29. Poland, C. A., Duffin, R., Kinloch, I., Maynard, A., Wallace, W. A., Seaton, A., Stone, V., Brown, S., Macnee, W., and Donaldson, K. (2008). Carbon nanotubes introduced into the abdominal cavity of mice show asbestos-like pathogenicity in a pilot study. *Nat. Nanotechnol.*, **3**, pp. 423–428.
30. Mercer, R. R., Scabilloni, J., Wang, L., Kisin, E., Schwegler-Berry, D., Shvedova, A. A., and Castranova, V. (2008). Alteration of deposition patter and pulmonary response as a result of improved dispersion of aspirated single-walled carbon nanotubes in a mouse model. *Am. J. Physiol.: Lung Cell Mol. Physiol.*, **294**, pp. L87–L97.
31. Morimoto, Y., Horie, M., Kobayashi, N., Shinohara, N., and Shimada, M. (2013). Inhalation toxicity assessment of carbon-based nanoparticles. *Acc. Chem. Res.*, **46**(3), pp. 770–781.
32. Wang, X., Xia, T., Ntim, S. A., Ji, Z., Lin, S., Meng, H., Chung, C.-H., George, S., Zhang, H., Wang, M., Li, N., Yang, Y., Castranova, V., Mitra, S., Bonner, J. C., and Nel, A. E. (2011). The dispersal state of multi-walled carbon nanotubes elicits pro-fibrotic cellular responses that correlate with fibrogenesis biomarkers and fibrosis in the murine lung. *ACS Nano*, **5**, pp. 9772–9787.
33. Lam, C.-W., James, J. T., McCluskey, R., and Hunter, R. L. (2004). Pulmonary toxicity of single-wall carbon nanotubes in mice 7 and 90 days after intratracheal instillation. *Toxicol. Sci.*, **77**, pp. 125–134.
34. Nurkiewicz, T. R., Porter, D. W., Barger, M., Castranova, V., and Boegehold, M. A. (2004). Particulate matter exposure impairs systemic microvascular endothelium-dependent dilation. *Environ. Health Perspect.*, **112**, pp. 1299–1306.
35. Nurkiewicz, T. R., Porter, D. W., Hubbs, A. F., Cumpston, J. L., Chen, B. T., Frazer, D. G., and Castranova, V. (2008). Nanoparticle inhalation augments particle-dependent systemic microvascular dysfunction. *Part. Fibre Toxicol.*, **5**, p. 1.
36. LeBlanc, A. J., Cumpston, J. L., Chen, B. T., Frazer, D., Castranova, V., and Nurkiewicz, T. R. (2009). Nanoparticle inhalation impairs endothelium-dependent vasodilation in subepicardial arterioles. *J. Toxicol. Environ. Health A*, **72**, pp. 1576–1584.
37. Kan, H., Wu, Z. X., Young, S. H., Chen, B. T., Cumpston, J. L., Chen, F., Kashon, M. L., and Castranova, V. (2012). Pulmonary exposure of rates

to ultrafine titanium dioxide enhances cardiac protein phosphorylation and substance P synthesis in nodose ganglia. *Nanotoxicology*, **6**, pp. 736–745.

38. Bellahcene, M., Jacquet, S., Cao, X. B., Tanno, M., Haworth, R. S., Layland, J., Kabir, A., Gaestel, M., Davis, R. J., Flavell, R. A., Shah, A. M., and Aukiran, M. (2006). Activation of p38 mitogen-activated protein kinase contributes to the early cardiodepressant action of tumor necrosis factor. *J. Am. Coll. Cardiol.*, **48**, pp. 545–555.

39. Messer, A. E., Jacques, A. M., and Marston, S. B. (2007). Troponin phosphorylation and regulatory function in human heart muscle: dephosphorylation of Ser23/24 on troponin I could account for the contractile defect in end-stage heart failure. *J. Mol. Cell Cardiol.*, **42**, pp. 245–259.

40. Kan, H., Wu, Z., Lin, Y., Chen, B. T., Cumpston, J. L., Kashon, M. L., Munson, A. E., and Castranova, V. (2012). Pulmonary inhalation of ultrafine TiO_2 and cardiovascular effects: a neuron-regulated pathway. *Toxicologist*, **126**, p. A1495.

41. Stapleton, P. A., Minarchick, V. C., Cumpston, A. M., McKinney, W., Chen, B. T., Sager, T. M., Frazer, D. G., Mercer, R. R., Scabillioni, J., Andrew, M. E., Castranova, V., and Nurkiewicz, T. R. (2012). Impairment of coronary arteriolar endothelium-dependent dilation after multi-walled carbon nanotube inhalation: a time-course study. *Int. J. Mol. Sci.*, **13**, pp. 13781–13803.

42. Li, Z., Hulderman, T., Salmen, R., Chapmen, R., Leonard, S. S., Young, S.-H., Shvedova, A., Luster, M. I., and Simeonova, P. (2007). Cardiovascular effects of pulmonary exposure to single-wall carbon nanotubes. *Environ. Health Perspect.*, **115**, pp. 377–382.

43. Legramente, J. M., Sacco, S., Crobeddu, P., Magrini, A., Valentini, F., Palleschi, G., Pallante, M., Balocchi, R., Iavicolli, I., Bergamaschi, A., Galante, A., Campagnolo, L., and Pietroiusti, A. (2012). Changes in cardiac autonomic regulation after acute lung exposure to carbon nanotubes: implications for occupational exposure. *J. Nanomater.*, doi:10.115/2012/397206.

44. Billman, G. E., Schwartz, P. J., and Stone, H. L. (1982). Baroreceptor reflex control of heart rate: a predictor of sudden cardiac death. *Circulation*, **66**, pp. 874–880.

45. Utell, M. J., Frampton, M. W., Zareba, W., Devlin, R. B., and Cascio, W. E. (2002). Cardiovascular effects associated with air pollution: potential mechanisms and methods of testing. *Inhal. Toxicol.*, **14**, pp. 1231–1247.

46. Oberdorster, G., Sharp, Z., Atudoriei, V., Eleder, A., Gelein, R., Lunts, A., Kreyling W., and Cox, C. (2002). Extrapulmonary translocation of ultrafine carbon particles following whole body inhalation exposure of rats. *J. Toxicol. Environ. Health A*, **65**, pp. 1531–1543.

47. Kreyling, W. G., Semmler, M., Erbe, F., Mayer, P., Takenaka, S., Schulz, H., Oberdorster, G., and Ziesenis, A. (2002). Translocation of ultrafine insoluble iridium particles from lung epithelium to extrapulmonary organs is size dependent but very low. *J. Toxicol. Environ. Health A*, **65**, pp. 1513–1530.

48. Nurkiewicz, T. R., Porter, D. W., Barger, M., Millecchia, L., Rao, K. M., Marvar, P. J., Hubbs, A. F., Castranova, V., and Boegehold, M. A. (2006). Systemic microvascular dysfunction and inflammation after pulmonary particulate matter exposure. *Environ. Health Perspect.*, **114**, pp. 412–419.

49. Inoue, K.-I., Takano, H., Koike, E., Yanugisawa, R., Sakurai, M., Tasaka, S., Ishizaku, A., and Shimada, A. (2008). Effects of pulmonary exposure to carbon nanotubes on lung and systemic inflammation with coagulatory disturbance induced by lipolysaccharide in mice. *Exp. Biol. Med.*, **233**, pp. 1583–1590.

50. Meng, G. C., Xu, L., Bai, R., Du, J., Zhang, L., Li, Y., Chang, Y., Zhao, Y., and Chen, C. (2012). Acute pulmonary and moderate cardiovascular responses of spontaneously hypertensive rats after exposure to single-wall carbon nanotubes. *Nanotoxicology*, **6**, pp. 526–542.

51. Erdely, A., Liston, A., Salmen-Muniz, R., Hulderman, T., Young, S.-H., Zeidler-Erdely, P. C., Castranova, V., and Simeonova, P. P. (2011). Identification of systemic markers from a pulmonary carbon nanotube exposure. *J. Occup. Environ. Med.*, **53** (Supp 16), pp. 580–586.

52. Nurkiewicz, T. R., Porter, D. W., Hubbs, A. F., Stone, S., Chen, B. T., Frazer, D. G. Boegehold, M. A., and Castranova, V. (2009). Pulmonary nanoparticle exposure disrupts systemic microvascular nitric oxide signaling. *Toxicol. Sci.*, **110**, pp. 191–203.

53. LeBlanc, A. J., Moseley, A. M., Chen, B. T., Frazer, D., Castranova, V., and Nurkiewicz, T. R. (2010). Nanoparticles inhalation impairs coronary microvascular reactivity via a local reactive oxygen-dependent mechanism. *Cardiovasc. Toxicol.*, **10**, pp. 27–36.

54. Nurkiewicz, T. R., Porter, D. W., Hubbs, A. F., Stone, S., Moseley, A. M., Cumpston, J. L., Goodwill, A. G., Frisbee, S. J., Perrotta, P. L., Brock, R. W., Frisbee, J. C., Boegehold, M. A., Frazer, D. G., Chen, B. T., and Castranova,

V. (2011). Pulmonary particle matter and systemic microvascular dysfunction. *Health Eff. Inst. Res. Rep.*, **164**, pp. 1–64.

55. Knuckles, T. L., Yi, J., Frazer, D. G., Leonard, H. D., Chen, B. T., Castranova, V., and Nurkiewicz, T. R. (2012). Nanoparticle inhalation alters systemic arteriolar vasoreactivity through sympathetic and cyclooxygenase-mediated pathways. *Nanotoxicology*, **6**, pp. 724–735.
56. Sriram, K., Porter, D. W., Jetterson, A. M., Lin, G. X., Wolfarth, M. G., Chen, B. T., McKinney, W., Frazer, D. G., and Castranova, V. (2009). Neuro inflammation and blood brain barrier changes following exposure to engineered nanomaterials. *Toxicologist*, **108**, A2197.
57. Wang, L., Mercer, R. R., Rojanasakul, Y., Qui, A., Lu, Y., Scabilloni, J. F., Wu, N., and Castranova V. (2010). Direct fibrogenic effects of dispersed single-walled carbon nanotubes on human lung fibroblasts. *J. Toxicol. Environ. Health A*, **73**, pp. 410–422.
58. Wang, X., Xia, T., Ntim, S. A., Ji, Z., George, S., Meng, H., Zhang, H., Castranova, V., Mitra, S., and Nel, A. E. (2010). Quantitative techniques for assessing and controlling the dispersion state and biological effects of multi-walled carbon nanotubes in mammalian tissue culture cells. *ACS Nano*, **4**, pp. 7241–7252.
59. Murray, A. R., Kisin, E. R., Tkach, A. V., Yanamala, N., Mercer, R., Young, S.-H., Fadell, B., Kagan, V. E., and Shvedova, A. A. (2012). Factoring in agglomeration of carbon nanotubes and nanofibers for better prediction of their toxicity versus asbestos. *Part. Fibre Toxicol.*, **9**, p. 10.

Chapter 14

Neurotoxicity of Nanomaterials

Hari Shanker Sharma and Aruna Sharma
Department of Surgical Sciences, University Hospital, Uppsala University, Uppsala, Sweden
sharma@surgsci.uu.se, aruna.sharma@surgsci.uu.se

> All substances are poisons: there is none which is not a poison. The right dose differentiates a poison and a remedy.
>
> —Paracelsus (1493–1541)

14.1 Human Exposure to Nanoparticles

Nanoparticles (NPs) are present in the air, water, or soil from various sources and could affect human health functions or exacerbate disease processes (see Table 14.1) [1, 2]. NPs may enter food materials through water and/or soil [2]. When food or water containing NPs is ingested then these NPs enter into the body fluid microenvironment [2–5]. Other routes of NPs entry into the human body are through inhalation or dermal contacts [2, 6–9]. Once NPs enter into the body they could cross different cellular barriers to enter the bloodstream [3–5]. From circulation the NPs could enter almost every organ or tissue in

Handbook of Safety Assessment of Nanomaterials: From Toxicological Testing to Personalized Medicine
Edited by Bengt Fadeel

ISBN 978-981-4463-36-2 (Hardcover), 978-981-4463-37-9 (eBook)
www.panstanford.com

the body, including the brain [1, 5, 7, 9, 10]. NPs, once having entered cells or tissues, are not metabolized easily. Thus, these NPs stay inside the cells or tissues for several months to years, causing abnormal cellular and tissue reactions, including organ dysfunction, cancer or cell death [1, 5]. Thus, further studies on NP-induced human health factors are needed to expand our knowledge in this area.

This chapter is focused on the potential neurotoxicity in vivo following exposure of NPs from various sources. Furthermore, pathophysiological effects of NPs in influencing CNS trauma or stress are also discussed. Finally, drug delivery using nanowires to achieve neuroprotection is described on the basis of our own investigations on CNS injuries in animal models.

14.1.1 NP Exposure Affects Disease Pathology

Available evidence suggests that people living in highly polluted areas and breathing contaminated air with different NPs are prone to respiratory, cardiac, renal, and/or liver diseases [7, 9]. These NP-intoxicated people with cardiovascular, endocrine, or metabolic disease also require frequent hospitalization [7]. This suggests that NP intoxication exacerbates the magnitude and severity of chronic diseases [3–5, 11–15]. Thus, it is likely that human populations suffering from common diseases, for example, hypertension or diabetes, may also be adversely affected by NP intoxication [16–18]. However, further studies are needed to confirm this hypothesis.

14.1.2 Military Personnel and NPs Exposure

Military personnel are normally more susceptible to NP exposure because of intense stress associated with them during either combat operations or peacekeeping activities across the world (Table 14.1) [1, 19–22]. These personnel are also exposed to additional stress from the environment, for example, heat, dust, cold, and/or gunpowder explosion [19, 20]. Use of military equipment and missile explosion are the sources of various NPs to which military personnel are exposed during their day-to-day activities [1, 6–9]. Dust storms in Middle East countries cause military personnel to SiO_2 NP exposure, whereas gunpowder explosion or handling of explosive materials and other strategic equipment results in C, S, Cu, Ag, or Ti NP exposure [1, 3–5]. In addition, the geographical locations where these personals are engaged in combat activities are also often heavily polluted with several magnetite fine particles (MfPs), ultrafine particles (UfPs), and NPs [9]. Breathing the polluted and/or contaminated air for a long period of time poses serious health hazards to the military [7, 9]. As a result, prolonged exposure of military personnel to an environment polluted with NPs is harmful (see Ref. [5]). Furthermore, when NP-intoxicated military personnel are injured during combat operations, their brain pathology could be exacerbated due to the NPs' involvement [13–15]. In such situations, it is still unclear whether known drugs are able to thwart central nervous system (CNS) injury in their standard doses to induce neuroprotection.

Table 14.1 Common NP exposure to humans from various sources

Type of NPs	Source of exposure
Carbon black, C	Motor vehicle exhaust, fire, wood burning
Silver, Ag	Water, food, air, soil, industrial waste
Copper, Cu	Water, air, food, industrial waste, warfare, military equipment
Titanium, Ti	Cosmetics, toothpastes, military aircraft, medicines
Aluminum, Al	Food, water, soil, industrial waste, aircraft, military equipment
Manganese, Mn	Mines, welding works
Iron, Fe	Food, water, industrial waste

Data modified after Sharma and Sharma [2–5].

14.2 Neurotoxicity of Nanoparticles

There is evidence that NPs could induce respiratory, cardiac, renal, liver, skin, and spleen toxicity [1, 6, 7, 9]. However, NP-induced neurotoxicity, in especially in vivo situations, is still not well known (see Refs. [1, 5]). Few sporadic studies showed NP-induced changes in brain function in animal models, but systematic investigation on neuronal, glial, or endothelial cell toxicity after NP exposure is still lacking.

Nowadays NPs or nanowires are used to deliver drugs across the blood–brain barrier (BBB) to treat neurological diseases in model situations (see Refs. [2, 12, 23–25]). Thus, a detailed knowledge of neurotoxicity of NPs in relation to dose, type, or their sizes is needed before nanodrug delivery becomes a reality in nanomedicine.

Our laboratory has initiated detailed investigations on NP-induced neurotoxicity in several in vivo animal models during the last decade (see Refs. [1, 5, 11, 12]). Our investigations are focused on neurotoxicity of different engineered NPs either from metals, for example, Ag, Al, or Cu, or from other sources, namely, Si, Mn, and carbon nanotubes (CNTs) [3–5, 13–15, 23]. We also evaluated the neurotoxic effects of Ti nanowires, if any are that are often used for drug delivery [3–5, 11, 12, 23, 25].

14.3 Concepts of Neurotoxicity

The term "neurotoxicity" is often used to denote neuronal damages caused by a variety of chemicals, toxins, or related materials [1, 26–28]. The term "neurotoxicity" also includes anatomical, biochemical, pathological, electrophysiological, or even psychological malfunction of neuronal, glia, and axons, as well as endothelial cells [1, 2, 13, 29]. However, the basic mechanisms of neurotoxicity in relation to brain damage and behavioral dysfunction are still not well understood.

14.3.1 Blood–Brain Barrier Disruption: A Gateway to Neurotoxicity

The CNS is a privileged organ and receives the highest amount of blood supply as compared to any other organ [2, 13, 26, 30]. Neurons are highly vulnerable to oxygen supply, and a lack of oxygen and glucose for a minute will have devastating consequences on neuronal survival [2, 11, 13, 30, 32]. Thus, a constant blood supply to the neurons is essential for their normal functioning [30]. To provide uninterrupted oxygen and glucose supply to the nerve cells and other components of the CNS, cerebral circulation is regulated strictly from changes in peripheral circulation [2, 30]. The homeostasis of the neuronal microenvironment is maintained by the endothelial cells of the cerebral or spinal cord capillaries that are specialized to regulate the transport of materials from blood to brain or vice versa within a strict limit [2, 29–32]. These specialized endothelial cells in the CNS constitute the basis of the BBB [26, 30]. The spinal cord microvessels are also equipped with a similar barrier that is commonly referred to as the blood–spinal cord barrier (BSCB) [2].

The BBB and the BSCB are collectively referred to as blood–CNS barrier (BCSNB) [11–13, 29].

Anatomically, the BCNSB is specialized endothelial cells that are connected with the tight junctions (Fig. 14.1). Thus these endothelial

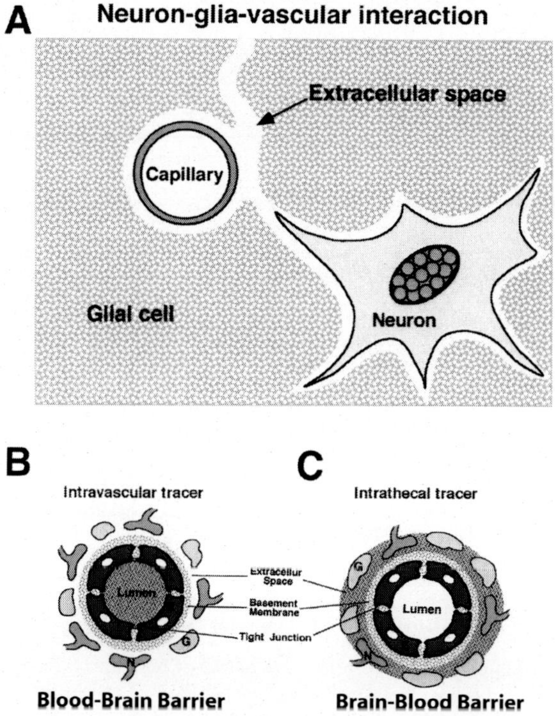

Figure 14.1 Cerebral endothelial cells and their surroundings. (A) Diagram showing spatial relationships between a cerebral capillary, a neuron, glial cells, and the extracellular space around them [26, 29]. (B) Schematic diagram showing endothelial cells (E) in the brain or spinal cord are connected with a tight junction and are surrounded by a thick basement membrane. The glial cells (G) and nerve cells (N) around the CNS endothelial cells are clearly seen. The intravascular tracer is normally stopped at the tight junction and does not penetrate the luminal endothelial cell membrane to reach the extracellular space, indicating a very tight BBB. (C) Likewise, the intrathecal tracer does not normally pass the abluminal endothelial cell membrane and/or the tight junction to reach the vascular compartment (cf B), suggesting that the *bbb* also effectively regulates the exchange of substances between the CNS microenvironment and the vascular system. Modified after Refs. [26, 29, 31].

cells behave like an extended plasma membrane in the CNS [13, 29–31]. The tight junctions do not allow even the tiniest particle lanthanum (La^{3+}), the molecular diameter of which is as about 12 Å, to pass (Fig. 14.2) [11–13]. Thus, the BCNSB could also prevent the entry of NPs under normal conditions into the CNS from the vascular compartment [14, 15]. However, NPs may enter the brain or spinal cord from blood by breaking the cellular membrane of endothelial cells.

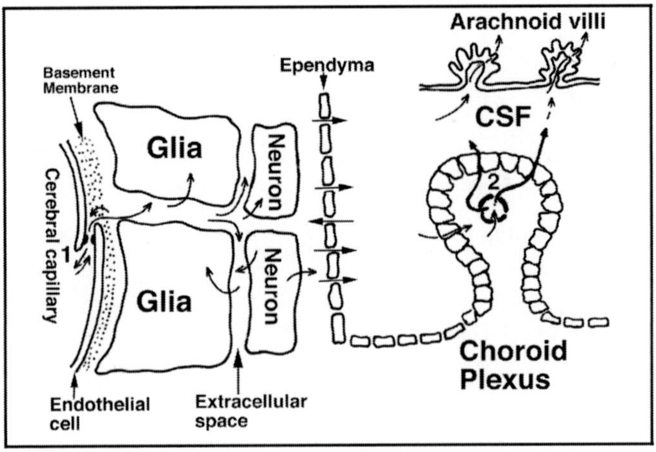

Anatomical basis of Blood-Brain (1) and Blood-CSF (2) Barrier System

Figure 14.2 Schematic representation of the BCNSB and the communication and/or exchange routes in between. The BBB (1) is located within the cerebral endothelial cells and exchanges essential nutrients or waste materials between neurons, glia, and endothelial cells for effective homeostasis. The BSCB is localized within the choroid plexus epithelial cells (2) that are connected with tight junctions. The microvessels supplying the choroid plexus are fenestrated. However, exchanges between the extracellular brain and ependymal linings (BCSFB) are quite frequent without any strict barrier between them. Modified after [26, 29, 31].

The endothelial cells of the CNS microvessels are surrounded by a thick basement membrane, or basal lamina (Fig. 14.1), and lack vesicular transport [29–31]. More than 85% of the endothelial cell surfaces are covered by astrocytic end feet, and neurons also make close contacts with the capillaries [13, 26, 29, 30, 33, 34]. This neuronal-glial-endothelial cell complex constitutes the physiological aspects of the BCNSB [13].

The luminal and abluminal surfaces of the endothelial cells constitute the identical anatomical barrier between the blood and brain interfaces [13, 29]. The abluminal sides of the endothelial cells also strictly regulate the transport of materials from brain to blood, construing a *brain–blood barrier* (*bbb*) (see Fig. 14.1) [13, 26, 27, 29, 31, 35, 36]. However, details on the functional properties of the *bbb* are not well known [13, 26, 36]. Whether the *bbb* is also broken down equally in neurological or neurodegenerative diseases or following CNS injuries like the BBB is still uncertain and requires additional investigations [11, 13, 26]. Obviously, perturbation of the neuronal microenvironment due to BBB or *bbb* breakdown results in neurotoxicity.

Another important barrier exists between the blood and the cerebrospinal fluid (CSF), known as the blood–CSF barrier (BCSFB) [26, 30, 37]. However, the anatomical site of the BCSFB is lying within the apical sides of the choroid plexus epithelia cells that are connected with tight junctions (Fig. 14.2) [26, 30]. The microvessels of the choroid plexuses that are responsible for CSF secretion are fenestrated and do not contain any tight junctions [11, 13, 26, 29, 30, 37]. The BCSFB is less strict as compared to the BBB due to the fact that the choroid plexuses are responsible for CSF production [13, 37]. Thus, transport of key substances from the blood to the CSF is necessary to maintain the CSF composition [13, 26, 30, 37]. However, the CSF is not an ultrafiltrate of the plasma and has a specific composition that is regulated by the BCSFB [30, 37]. The CSF is devoid of proteins in healthy situations [13, 26, 30, 37]. However, the CSF protein concentrations are increased several fold in almost all neurological or neurodegenerative diseases, indicating a breakdown of the BCSFB [13, 37]. However, data on NP-induced BCSFB breakdown or their presence in the CSF in disease conditions is still lacking.

14.3.2 BBB Breakdown to Proteins: Cause of Brain Edema Formation

The fluid microenvironment of the CNS is devoid of proteins due to the effective BBB and BCSFB [11, 13, 26, 27, 29–31]. However, breakdown of the BCNSB results in penetration of serum proteins within the CNS microfluid compartments [11, 26, 29, 31]. Passage of proteins within the extracellular fluid (ECF) compartment (Fig. 14.1) induces widespread immunological, biochemical, or pathological reactions [13, 26, 29]. In addition, passage of serum proteins across the vascular compartment alters the osmotic balance between the blood and the brain interface [26, 30]. This results in bulk flow of water from the vascular compartment to the brain ECF compartment, causing vasogenic edema formation [11, 13, 26, 27, 29, 30]. Swelling of the brain in a close cranial compartment causes compression of vital brain regions, leading to brain dysfunction and/or death [38–41]. Thus it is quite likely that NPs could induce breakdown of the BCNSB, resulting in brain edema formation and cellular injury.

14.3.3 Brain Pathology and Neurotoxicity

Brain pathology is the abnormal morphological changes in the neurons, glial cells, axons, and/or endothelial cells, caused by either external factors, for example, toxins, chemicals, drugs, or trauma, or internal factors, viz., neurodegenerative, vascular, or metabolic abnormalities, that is, Alzheimer's, Parkinson's, or Huntington's disease [13, 26, 29, 30]. Moreover chronic endocrine or vascular anomalies, for example, hypertension, diabetes, cirrhosis, hypoxia, anoxia, stroke, infarction, and/or bacterial and viral infections also induced brain pathology [11, 13, 26, 30]. This is interesting to note that in these situations BBB breakdown to proteins is quite common [11, 13, 26, 29, 30]. This suggests that BBB breakdown is instrumental in brain pathology or neurotoxicity. In addition, pathological or toxicological changes in the brain also include over- or underexpression of different genes within the CNS [11, 13, 26, 30]. It is likely that breakdown of the BBB results in exposure of serum components to the brain cells, causing alterations in gene expression [38, 39].

14.3.4 Pharmacology of Neuroprotection and Neurotoxicity

The hypothesis that brain pathology or neurotoxicity caused by various insults to the brain is working through BBB breakdown can be examined further using a pharmacological approach. The BBB and edema formation could be regulated through neurochemical mediators, and thus pharmacological manipulations of brain chemistry will alter the BBB function in disease conditions [11, 26, 30]. In addition, neuroregenerative agents comprising various neurotrophic factors and/or a combination of them with other neuroprotective agents could repair brain damage [35, 41–44]. Thus, to counteract damage of neuronal, vascular, and glial elements of the CNS, neurotrophic factors from nerve cells, glial cells, or endothelial cells in combination could be more powerful in exerting their effects than any individual growth factor alone. Accordingly, a combination of brain-derived neurotrophic factor (BDNF), glial-derived neurotrophic factor (GDNF), and vascular endothelial growth factor (VEGF) is needed to enhance regeneration of all brain components in the CNS after brain damage more effectively [35]. Thus, in our hands a mixture of BDNF, GDNF, and VEGF with other growth factors, for example, nerve growth factor (NGF), ciliary neurotrophic factor (CNTF), and insulin-like growth factor (IGF), is found useful in promoting neuroregeneration in the damaged CNS and BBB repair rapidly [11, 13, 26, 35]. In this context, cerebrolysin (Ever NeuroPharma, Austria) is an ideal combination of the different growth factors together with some active peptide fragments needed for neuronal plasticity, neuroregeneration, and neuroprotection [41–44]. Studies in our laboratory show that cerebrolysin attenuates BBB breakdown and reduces edema formation and thus prevents brain damage in different models of CNS injury [17, 41–44, 46]. This suggests that BBB breakdown and edema formation are instrumental in causing neurotoxicity.

14.4 Neurotoxicity of Engineered Metal Nanoparticles

To expand our knowledge on neurotoxicity of NPs, our laboratory is focused on investigations of engineered NPs from metal on brain damage in rodent models in vivo [4, 5, 12]. To check the possibility whether NPs of same sizes from different metals could induce selective and specific brain pathology in rats or mice when administered through different systemic or intracerebral (i.c.)

routes [3–5, 12, 13–15, 23], we used engineered NPs from Al, Ag, or Cu (50–60 nm) either given through the intraperitoneal (i.p.), intravenous (i.v.), or intracarotid (i.c.a.) route or administered into the intracerebroventricular (i.c.v.) or i.c. infusion in rats or mice [14, 15]. The animals are allowed to survive either 4 or 24 hours after NP intoxication (see Table 14.2). The dosages were calculated on the basis of the possible level of exposure of NPs from the environment to human cases following missile or gunpowder explosion. However, mimicking human exposure of NPs from the environment in any animal models is not realistic [1, 14, 26]. In these animals, the BBB permeability to serum albumin using immunohistochemistry and leakage of Evans blue albumin (EBA) and radioiodine ([131]-iodine) in different brain regions were examined biochemically [29, 31, 36, 47]. To understand the role of BBB breakdown in edema formation, changes in brain water content and electrolytes (Na^+ and K^+) were measured [14, 48]. The cellular changes in the brain were examined by morphological techniques using light microscopy (LM) and transmission electron microscopy (TEM) [5, 29, 36, 38, 47, 49–51].

14.4.1 Engineered NPs Induce BBB Breakdown

Metal NPs administered either systemically or into the brain ventricles significantly disrupted the BBB to EBA and radioiodine in rats but was most pronounced in mice [14, 15]. The i.c. or i.c.v. administration of NPs induced the most marked disruption of the BBB, followed by i.c.a. or i.v. injection [14]. The i.p. NPs were the least effective in BBB dysfunction in rats or mice [14]. Interestingly, the vehicles used for NP suspension, for example, saline or Tween 80, did not alter the BBB function. Equimolar concentrations of saline or Tween 80 (comparable to the NPs' concentration in solution) administered either systemic or through the i.c.v. route also did not influence the BBB to EBA or radioiodine. This suggests that NPs may have specific effects on the microvessels in disrupting the BBB function [4, 14]. The i.p. NPs (50 mg/kg) resulted in only a mild leakage of tracers in the rat or mice brain after 4 hours but not after 24 hours' administration (Table 14.2). However, i.v. (30 mg/kg) NPs resulted in marked BBB disruption that was most pronounced after 24 hours as compared to a 4-hour period (Table 14.2). Mice brains showed greater extravasation of tracers

Table 14.2 Neurotoxicity of nanomaterials examined in rats or mice in vivo studies

Experiment type	Nanomaterials	Dose/route	Species	Survival period
BBB	Ag, Cu, Al	50 mg/kg, i.p.	Rats, mice	4/24 hours
Brain edema		30 mg/kg, i.v.		
Brain pathology		2.5 mg/kg, i.c.a.		
		20 μg i.c.v. or i.c.		
BBB	SiO_2, Ag, Cu,	50 mg/kg, i.p.	Rats	8 days
Brain edema		daily for 7 days		
Brain pathology	MnO_2	10 or 20 mg/kg,	Rats	8 days
		i.p. daily for 7 days		
	TiO_2			
	SWCNTs	5 or 10 μg/kg, i.v.	Rats	4/24 hours
Neuroproetction	TiO_2 nanowires+			
	H-290/51;	50 mg/kg, i.p.	Rats	5 hours
	cerebrolysin;	2.5 or 5 mL/kg, i.v.	Rats	4 hours
	SCI-1 to SCI-5	10 μg/10 μL local	Rats	5 hours
Neuroproetction	Ag, Cu+ heat stress	50 mg/kg/day for 7 days	Rats	8 days
	SiO_2+SCI	50 mg/kg/day for 7 days	Rats	8 days

Compiled after Sharma et al. [14, 15] and Sharma and Sharma [3–5].

as compared to rats after i.v. NP administration in identical doses (Table 14.2) [4, 14, 15]. Interestingly, Ag and Cu exerted the most pronounced BBB breakdown as compared to Al NPs. No differences in the magnitude or intensity of the BBB breakdown was seen after NP administration in saline or Tween 80 in rats or mice [4, 14]. Internal carotid administration of NPs (2.5 mg/kg) disrupted the BBB to tracers largely on the ipsilateral side of the brain in rats or mice (Table 14.2). This effect was most marked at 24 hours in mice as compared to rats after Ag and Cu NP administration [14]. The i.c.v. or i.c. administration of NPs (20 μg in 10 μL) into the right lateral cerebral ventricle (i.c.v.) or right cerebral cortex (i.c.) resulted in tracer extravasation into the infused side of the cerebral hemisphere [14]. This effect was most prominent in mice brains after 24 hours following Ag and Cu NPs as compared to Al infusion [2, 4, 14]. These observations support the idea that NPs, depending on their inherent property and concentration, could induce BBB breakdown [4]. A higher magnitude of BBB disruption following the i.v., i.c.a., or i.c. route as compared to i.p. administration is in line with this idea [4, 14].

14.4.1.1 Regional distribution of Evans blue albumin in the CNS

BBB disruption following NP treatment is very selective and specific. Thus, widespread leakage of EBA was seen in the CNS of rats and mice after NP intoxication. The cerebellum, hypothalamus, piriform cortex, and brain stem showed moderate EBA leakage [14]. Ag-treated mice exhibited profound leakage of EBA at 24 hours in the brain and spinal cord after i.v., i.c.a., or i.c.v. administration. EBA staining was also present in the contralateral side of the brain after i.c.a. NP administration [4, 14]. The i.c.v. administration of Cu NPs induced EBA leakage in both dorsal and ventral surfaces of the CNS. However, the Al-treated mice did not show much leakage of EBA in the CNS (Fig. 14.1). In rats, the most marked blue staining was seen following 24 housr after Ag NP treatment intravenously. Moderate blue staining is seen in many parts of the dorsal and ventral surfaces of the brain [4, 14]. The deeper brain structures also showed profound blue staining in various areas of the brain. The i.v. Cu NPs caused EBA leakage in the rat CNS. The dorsal surface of the brain was moderately stained as compared to the ventral surface. The spinal cord showed moderate blue staining in both dorsal and ventral sides. Al NPs did not induce much EBA leakage in the CNS at any time point [14].

14.4.1.2 Immunostaining of serum albumin in the CNS

Leakage of endogenous albumin in NP-treated rats or mice further confirms the BBB breakdown to proteins [14]. Albumin immunostaining was most prominent following Cu and Ag NP administration though the i.v., i.c.a., or i.c.v. route. The leakage was most pronounced in mice as compared to rats in several brain regions. Albumin leakage caused by Cu or Ag NPs in the rat cerebral cortex is reported in Ref. [14]. Several albumin-labeled neurons are present 24 hours after i.v. administration of Cu or Ag in rats. These albumin-positive neurons are largely present in the edematous areas of the neuropil showing cellular damages [4, 14].

14.4.1.3 Ultrastructural changes in the BBB permeability

The passage of tracer transport across the endothelial cells by NPs was examined at the ultrastructural level using the electron dense tracer lanthanum (La^{3+}). Ag and Cu NPs induced profound exudation of La^{3+} across the cortical microvessels in rats or mice. La^{3+} was largely confined within the endothelial cell cytoplasm but in some cases was also seen within the basal lamina [14]. In large microvessels, the tracer was also located within the microvesicle profiles. However, the tight junctions are not disrupted to La^{3+} [14, 15]. These observations confirm that NPs are able to disrupt the BBB function by enhancing the cell membrane permeability and could be instrumental for neurotoxicity.

14.4.2 Nanoparticles Induce Brain Edema Formation

The i.v., i.c.a., or i.c.v. administration of Ag or Cu NPs induced profound increase in the brain water content and volume swelling that was most marked in mice (f = 3% to 14%) than rats (f = 2% to 8%). This increase in brain water is tightly correlated with EBA leakage following NP intoxication irrespective of route or types used [4, 14].

14.4.2.1 NPs alter brain electrolyte content

Movement of water from the vascular compartment to the brain ECF is accompanied by changes in tissue electrolytes content [30, 13]. Thus, i.v., i.c.a., or i.c.v. NP treatment altered Na^+ and K^+ contents in the brain tissues of rats and mice. However, i.p. injection of NPs did not alter tissue electrolytes significantly [14]. A significant increase in Na^+ and a marked decrease in K^+ at the time of Ag or Cu NP-induced BBB breakdown and edema formation were seen. These changes in electrolytes are most marked in mice brains (Fig. 14.2). Al NPs were not effective in altering brain electrolyte contents (see Fig. 14.2) [4, 14].

14.4.3 Nanoparticles Induce Brain Pathology

LM and TEM investigations showed profound brain pathology in rats or mice after NP administration through various routes [14]. Thus, damage to nerve cells, glial cells, endothelial cells, and axons is seen in NP-treated animals. In addition, marked alterations in neuronal, glials and myelin gene expressions are also observed in these NP-treated rats or mice [4, 14].

14.4.2.1 Neuronal changes

The i.v., i.c.a., or i.c.v. NPs induced profound distortions of nerve cells within the cerebral cortex, hippocampus, cerebellum, thalamus, hypothalamus, and brain stem in rats or mice at light microscopy. These cell changes were most marked in Ag or Cu NP-treated animals. Nissl staining showed most pronounced neuronal damages in the mice brain as compared to rats. Neurons loss and/or nerve cell damage in the hippocampus were most severe, followed by the cerebellum, cerebral cortex, brain stem, thalamus, and hypothalamus [4, 14]. A very tight correlation between albumin-positive cells and neuronal damages are seen in NP-treated animals irrespective of the route of administration or metal type used. This suggests that BBB disruption is instrumental in neuronal damage.

14.4.2.2 Glial changes

NPs also influenced astrocytic activation as seen using overexpression of glial fibrillary acidic protein (GFAP) [33], a specific marker of astrocytes in the rat or mouse CNS. This activation of GFAP closely corresponds to extravasation of serum proteins in the brain. Most of the astrocytes in the NP-treated group were distorted in shape and are present in the edematous areas of the neuropil. Perivascular astrocytes also showed edematous swelling in NP-treated animals. This effect on astrocytes was most marked by Ag or Cu NPs (see Ref. [14]).

14.4.2.3 Myelin changes

Degeneration and/or loss of myelin following NP intoxication are seen using Luxol-fast blue (LFB) staining in rat or mouse brain. The loss of LFB staining was most marked in animals that received Ag or Cu NPs intravenously, intracarotidly, or through the i.c.v. route. This loss of myelinated fibers was further confirmed by immunohistochemistry of myelin basic protein (MBP) (see Ref. [14]).

14.4.2.4 Ultrastructural changes

TEM showed profound ultrastructural changes in the areas exhibiting La^{3+} leakage in NP-treated animals. Ag and Cu NPs caused most pronounced vesiculation of myelin in the brain or spinal cord. Disintegration of nerve cells with dissolution of the cell nucleus and nucleolus in the cortex is apparent in one Cu NP-treated rat. Collapse of microvessels and leakage of La^{3+} across the endothelial cells are also seen in the rat after Ag NP treatment. In these NP-treated animals perivascular edema, vacuolation, and damage of the neuropil are frequent. These ultrastructural damages are more pronounced in mice brain as compared to rats. The NPs induce selective damage to neurons, glia, or axons. This is apparent from the findings that one nerve cell or axon in a particular area of brain exhibits profound damage, whereas the adjacent neurons or axons are normal in appearance [14]. This suggests that NP-induced selective cellular or molecular mechanisms are responsible for brain damage rather than direct cellular toxicity by NPs [2–4, 13–15].

14.4.2.5 Heat shock protein expression

Cellular stress induced by NPs could play important roles in neurotoxicity [52–54]. This idea is supported by up-regulation of heat shock proteins (HSPs), which are also known as stress proteins, in the CNS of rats and mice after NP treatment [14]. The i.c.a., intravascular, or i.c.v. administration of Ag or Cu NPs induced profound up-regulation of inducible isoforms of HSP 72 kD expression in several CNS areas exhibiting BBB breakdown. The

HSP expression is confined within the distorted neurons and in glial cells in the neuropil [14]. Overexpression of HSP is seen in the cell cytoplasm of neurons in the cortex, hippocampus, cerebellum, thalamus, hypothalamus, brain stem, and spinal cord [4, 14]. A strong correlation between leakage of albumin and HSP expression in cells suggests that NP-induced BBB breakdown is instrumental in HSP gene expression.

14.5 Neurotoxicity of Other Nanoparticles

Apart from engineered NPs from metals other NPs from environmental or industrial sources, for example, Si, Mn, CNTs, and Ti, are also harmful to human health, depending on the duration of exposure [1, 7, 9]. However, effects of these NPs on brain function are still not well known. We evaluated their potential neurotoxic effects in animal models in vivo under different conditions. The salient findings of our investigations are described below.

14.5.1 Neurotoxicity of SiO_2 Nanoparticles

SiO_2 or dust particle exposure leads to various long-term diseases [7, 50]. Military personnel are exposed to SiO_2 NPs during combat operations or peacekeeping efforts in the Middle East [50]. These military or civilian personnel are often exposed to dust storms, and their continuous presence in these areas makes them vulnerable to SiO_2 exposure [7, 28]. Breathing air contaminated with SiO_2 could worsen the prevailing diseases, for example, cardiovascular or respiratory problems, including frequent hospitalization [1, 7]. However, the details of SiO_2 exposure on mental health functions are still not well known. The health problems of people living in desert areas are further complicated by a combination of high environmental heat and SiO_2 intoxication [19, 20, 50, 28]. Heat stress caused by high environmental heat alone induces BBB breakdown and brain dysfunction [29, 31, 32]. Thus, additional exposure of SiO_2 NPs may further exacerbate the problems of heat stress and associated brain pathology. Furthermore, any CNS injuries in people intoxicated with SiO_2 NPs in high environmental heat could also be further aggravated [50]. To test this hypothesis, experiments were carried out in our laboratory using chronic exposure of SiO_2 (50–60 nm) for one week (50 mg/kg, i.p.) in rats. Our results showed that SiO_2 exposure alone induced BBB breakdown to EBA and radioiodine on the eighth day at room temperature [50]. When these SiO_2-intoxicated rats were subjected to mild stress of immobilization for four hours or high environmental heat (33°C for four hours) the magnitude and intensity of the BBB breakdown were further exacerbated [51–53]. These animals also showed exacerbation of brain edema and brain pathologies [52, 53]. Subjection of normal rats to 33°C or immobilization stress for four hours did not induce BBB dysfunction. Likewise, when SiO_2-intoxicated rats were subjected to a focal spinal cord injury by making a longitudinal incision into the right dorsal horn at the T10–T11 segments, the outcome of injury in terms of breakdown of the BSCB, edema formation, and cord pathologies was significantly aggravated at five hours as compared to saline-treated traumatized rats [65].

These observations further confirm the idea that SiO_2 exposure alone induces BCSNB breakdown, resulting in CNS pathology, and subjection of them to additional stress or trauma exacerbates their brain pathology.

14.5.2 Neurotoxicity of Mn nanoparticles

Mn dust or MnO_2 exposure is quite common in people related to the welding or mining profession. The most common effects of MnO_2 exposure are development of Parkinson's disease (PD) [7–9, 66]. PD is noted in mining or welding personnel in their forties, whereas in the normal population PD could appear above the age of 60 [7, 9].

Although MnO_2 exposure is related to PD, systematic studies on Mn neurotoxicity in in vivo situations are still lacking. Our laboratory investigated MnO_2 neurotoxicity in a rat model. Rats were administered Mn NPs (30–40 nm size) in a dose of 10 mg or 20 mg/kg i.p. daily for seven days (Table 14.2). On the eighth day the BBB permeability to EBA and radioiodine, edema formation, and brain pathology were examined. Mn NP-treated rats exhibited profound BBB disruption, edema formation, and brain pathology in the sensory motor cortex, hippocampus, caudate putamen, cerebellum, and thalamus as compared to the hypothalamus, pons, medulla, and spinal cord [66]. A close correlation between BBB dysfunction and neuronal damage using Nissl staining was evident in Mn NP-treated animals. Interestingly these effects of Mn NPs are dose dependent. These results show that Mn NPs induce selective BBB breakdown and brain pathology [54–56].

14.5.3 Neurotoxicity of TiO_2 Nanoparticles

Ti exposure to humans is quite frequent nowadays from various natural, industrial, or household sources [7, 9]. Its use as an alloy in aerospace, military industries, aircraft, missiles, cosmetics, dental implants, mobile phones, and other transmission equipment causes routine exposure to humans [7]. However, the possible health effects of TiO_2 are still not well known. TiO_2 nanowires are used in electronics and also to deliver drugs in vivo [24, 67, 68]. Thus, effects of TiO_2 on neurotoxicity require further investigation.

Experiments carried out in our laboratory show that chronic exposure of TiO_2 NPs (40–50 nm) in a dose of 50 mg/kg i.p. once daily for seven days did not induce BBB dysfunction to protein tracers, for example, EBA or radioiodine (Table 14.2). In these animals brain edema formation was also absent and neuronal damage is seldom seen [57–60]. This suggests that TiO_2 in doses identical to other NPs could be safe and does not induce neurotoxicity. However, further studies are needed especially to determine whether brain pathology will be further exacerbated in TiO_2-treated animals if they are subjected to stress or trauma.

14.5.4 Neurotoxicity of Single-Walled Carbon Nanotubes

Motor vehicle exhaust, forest fires, coal burning, and breathing of polluted air filled with smoke are rich sources of carbon NP exposure [1, 7, 9]. On the other hand, CNTs are used for enhanced drug delivery for cancer therapy. Recently, functionalized single-walled (SW) or multiwalled (MW) CNTs are loaded with drugs or specific proteins that could be targeted specifically to receptors and tissues for therapeutic purposes [69]. However, in vivo administration of CNTs often leads to liver or kidney necrosis or damage [7–9, 69].

Interestingly, effects of CNTs on brain function are still not well investigated. We used an i.v. infusion of SWCNTs in a dose of 5 μg/kg/minute or 10 μg/kg/minute for 10 minutes in rats, and the BBB permeability to EBA and radioiodine, brain water content, and neuronal damages were examined 24 hours after administration (Table 14.2). Our observations showed that SWCNTs induced BBB disruption to EBA and radioiodine in the cerebral cortex, caudate nucleus, colliculi, and hypothalamus, together with increased brain water content 24 hours after infusion [60]. This effect was dose dependent. The magnitude and intensity of neuronal injury tightly correlated with BBB disruption in SWCNT-treated animals. This indicates that SWCNTs could induce brain damage, and thus their use in drug delivery, diagnostic purposes, or therapeutic purposes requires careful reconsideration. For a further discussion of CNTs and the propensity for biodegradation of these nanomaterials, see Chapter 10.

14.6 Nanoparticle Exacerbation of Brain Pathology

The possibility that NP intoxication alters the pathophysiological responses of CNS injuries following trauma, stress, or diseases was examined in our laboratory in rat models [1, 12]. In these experiments rats were chronically intoxicated with Ag, Cu, or Si NPs for one week in healthy or diabetic rats and their brain pathologies were examined [5, 12, 15]. The healthy animals intoxicated with these NPs were also subjected to hyperthermia, and neurotoxicity was evaluated using morphological techniques.

14.6.1 NPs Exacerbate Diabetes-Induced Brain Pathology

A large number of people suffer from diabetes mellitus, a disorder of endocrine metabolism [7, 9, 61]. Long-term diabetes is often uncontrolled and leads to brain dysfunction [13, 30]. Thus, NP exposure to people with diabetes may have devastating consequences on their brain functions. To explore this possibility, rats were made diabetic by streptozotocine treatment (50 mg/kg, i.p.

once daily for three days) and allowed to survive for four weeks [16, 18, 51]. These rats showed marked increase in plasma glucose (20±2 mM/L) as compared to saline-treated controls (5±1 nM/L) [59, 61]. The diabetic rats were intoxicated with Ag or Cu NPs (50.60 nm, 50 mg/kg, i.p. daily) after the third week of streptozotocine treatment for one week, and neurotoxicity was examined after completion of NP administration. The diabetic rats intoxicated with NPs showed about two- to threefold increase in the BBB permeability to EBA and radioiodine as compared to NP-treated healthy animals [16, 59, 61]. Brain edema formation was also enhanced in NP-treated diabetic rats by two- to fourfold as compared to healthy rats [59, 61]. Neuronal injuries and myelin vesiculation were much more aggravated in NP-treated diabetic rats as compared to healthy animals with identical NP treatment. Interestingly, Ag NPs are more neurotoxic in diabetic rats than Cu NPs. These observations suggest that neurotoxicity of NPs is aggravated in diabetes [61]. This indicates that patients with diabetes may be more vulnerable to the harmful effects of NPs.

14.6.2 NPs Exacerbate Hyperthermia-Induced Neurotoxicity

Heat-related illnesses or hyperthermia is a serious clinical problem in the military across the world [13, 22, 29, 31, 70]. Heat-related death or disability is quite frequent in the navy, air force, or army during combat operations in various parts of the world, particularly during summer months [22, 67, 68]. Common causes of heat illnesses in the military include fighter aircraft, exercise in summer months, fire in ship engines, missile explosions, and other event-related warfare [22]. Particularly, military missions in desert heat make military personnel more vulnerable to heat-related illnesses [13, 22, 29, 31]. Under these circumstances, if military personnel are exposed to various NPs from the environment their health could further deteriorate.

Thus, NP exposure could worsen heat-related illness with regard to brain function, which we examined in our laboratory in a rat model. Normal rats are exposed to whole-body hyperthermia (WBH) for four hours at 38°C in a biological oxygen demand (BOD) incubator, resulting in heat illnesses that are very similar to clinical situations [13, 32]. These heat-stressed animals show profound

BBB leakage to EBA and radioiodine, brain edema formation, and neuronal damages [13, 26, 29, 31, 40, 57, 71].

When Ag or Cu NP-treated (50–60 nm, 50 mg/kg, i.p. daily for one week) rats were exposed to WBH on the eighth day these animals showed about two- to fourfold increase in the BBB breakdown to EBA and radioiodine [15]. The brain edema was increased in these animals by four- to six-fold, and the magnitude and intensity of neuronal, glial, or myelin damages were enhanced by five- to six-fold as compared to saline-treated heat-stressed rats [41, 42]. Ag NPs exerted the most harmful effects on brain damage after WBH than Cu NPs [41]. These observations suggest that NP exposure exacerbates heat-induced brain pathology.

14.6.3 NP Intoxication Alters Pharmacology of Neuroprotection

Exacerbation of neurotoxicity in NP-treated groups following trauma or WBH suggests that any drug treatment for neuroprotection in these groups requires dose adjustment to counteract the aggravation of brain damage. To test this hypothesis we used a select group of drugs that are known to induce neuroprotection in different models of CNS trauma or WBH, for example, antioxidant compound H-290/51 (Astra Zeneca, Mölndal, Sweden) [68, 72], cerebrolysin (Ever NeuroPharma, Austria) [2, 17, 41–43, 45, 46], or SCI-1 to SCI-5 compounds (Acure Pharma, Uppsala, Sweden) [23–25].

Pretreatment with drugs, that is, H-290/51 (50 mg/kg, p.o.), cerebrolysin (2.5 ml/kg, i.v.), or SCI-1 to SCI-5 compounds (10 μg in 20 μL total dose), was able to thwart BBB or BSCB permeability, brain or spinal cord edema, and CNS pathology after WBH or spinal cord injury in normal rats [44]. However, these doses are insufficient when they are given to heat-stressed or spinal cord–traumatized rats pretreated with Ag, Cu, or Si NPs. Thus, a twofold increase in doses is required for these compounds to induce appreciable neuroprotection in NP-treated rats [42, 45]. This indicates that NP treatment alters the pharmacology of neuroprotective agents.

It appears that excessive oxidative stress and/or neuronal damage caused by the combination of NPs and trauma requires high doses of neuroprotective agents [45]. Alternatively, NP treatment may have altered the physiological mechanisms of drug bioavailability [23, 24]. Taken together, our observations suggest that NPs induce greater neurotoxicity following CNS injury and also reduce pharmacological effects of drugs.

14.7 Nanowired Drug Delivery for Neuroprotection

As mentioned above, NP intoxication exacerbates the pathophysiology of CNS injury and reduces drug effects. It appears that higher doses of drugs are able to penetrate the CNS more effectively to induce neuroprotection in NP-treated traumatized groups. However, enhanced drug delivery to the CNS could also be achieved using nanotechnology. Thus, drug formulation with NPs could effectively enhance drug delivery within the CNS to achieve greater neuroprotective ability. However, in this regard suitable NPs are needed that do not induce neurotoxicity when formulated for nanodrug delivery.

14.7.1 Nanowired Cerebrolysin Enhances Neuroprotection

We used TiO_2 nanowires as this nanomaterial is quite innocuous and does not induce neurotoxicity in the doses used for nanodrug delivery. The TiO_2 nanowired created in our laboratory was used to tag cerebrolysin and administered in a dose of 2.5 mL/kg (i.v.) in normal rats or animals chronically treated with Cu or Ag NPs (50 mg/kg, i.p. daily for one week) [17, 41, 42, 44, 46, 51–60]. When these NP-treated rats are exposed to four hours' WBH these animals showed exacerbation of brain pathology as compared to saline-treated rats. However, in these animals nanowired cerebrolysin (2.5 mL/kg dose) was able to significantly thwart brain damage after four hours' WBH as compared to the normal drug given in identical or in higher doses (5 mL/kg, i.v.) (Table 14.2). This suggests that nanowired cerebrolysin is far superior in enhancing neuroprotection NP-treated heat-stressed animals [44]. Nanowired cerebrolysin could cause a slow and sustained release of the drug within the CNS for longer periods, and the drug catabolism is also slowed down because of the drug–nanowire complex. Obviously, these factors will lead to greater beneficial effects than normal cerebrolysin [54–60].

14.7.2 Nanowired H-290/51 Enhances Neuroprotection

NP intoxication induces oxidative stress [15]. Thus, antioxidants compounds, for example, H-290/51, could be used for neuroprotection against NP-induced neurotoxicity. H-290/51 is a potent chain-breaking antioxidant that inhibits lipid peroxidation [15, 50, 68].

H-290/51 when given in spinal cord injury or WBH in rats treated with Cu or Ag NPs failed to achieve neuroprotection [15, 65]. However when H-290/51 was delivered using nanotechnology, that is, either tagged with TiO_2 nanowires or encapsulated within polyethylene glycol (PEG)ylated liposomes the nanowire–drug complex exerted profound neuroprotection in WBH or following spinal cord injury in NP-treated animals [15, 54, 55, 65, 68]. Interestingly, TiO_2-nanowired drug delivery is slightly but significantly better in inducing neuroprotection than a liposome-encapsulated compound in identical doses. The probable mechanisms of such differences are unclear and require further investigation.

14.7.3 Nanowired Acure Pharma Compounds Enhance Neuroprotection

Nanowired drugs are able to enhance neuroprotective efficacy of compounds in CNS injury [25]. However, whether nanowiring of different compounds with varying neuroprotective ability could also make them equally beneficial or the differences in their neuroprotective capability will still remain is unclear. This idea was examined by selecting three different drugs, AP 173 (SCI-1), AP 364 (SCI-5), and AP 713 (SCI-2), of varying neuroprotective ability for spinal cord injury [23–25]. These drugs were tagged with TiO_2 nanowires (50–60 nm) for drug delivery. Our observations showed that the compound AP 713 (10 μg dose) was the most superior in reducing the pathophysiology of spinal cord injury and motor functions in rats as compared to AP 173 or AP 364 given in identical doses [46]. Interestingly, when these drugs were given as nanowired delivery to the spinal cord after injury, AP 713 was still the most superior in attenuating spinal rod pathology or motor functions [23, 24] in normal or Cu or Ag NP-treated rats [55]. This observation confirms that nanowiring of a drug alone does not enhance the drug's capacity for neuroprotection. Nanowiring leads to a slow and sustained release of the drug within the CNS that also prevents it from catabolism [25]. Obviously, a slow and sustained release of drugs by nanowiring will achieve better neuroprotection.

14.8 Conclusions

In conclusion, NPs, depending on their inherent properties and concentration, induce neurotoxicity. This neurotoxicity of NPs depend on their ability to disrupt BBB function to large molecules, for example, proteins. NP intoxication also enhances brain pathology following trauma or hyperthermia, making them difficult to treat with standard doses of neurotherapeutic agents. Thus, NPs alter the course of brain pathology and also affect the pharmacology of drug responses following CNS insults. To overcome these difficulties in drug treatment in NP-intoxicated groups the drug dose requires adjustments. Alternatively, nanodrug delivery is required to contain NP-induced exacerbation of brain pathology. However, further studies are needed to understand the safety of these NPs that are used as tools for drug delivery.

Acknowledgments

These investigations were partially supported by the Air Force Office of Scientific Research (London), the Air Force Material Command, USAF, under grant number FA8655-05-1-3065. The US government is authorized to reproduce and distribute reprints for government purpose, notwithstanding any copyright notation thereon. The views and conclusions contained herein are those of the authors and should not be interpreted as necessarily representing the official policies or endorsements, either expressed or implied, of the Air Force Office of Scientific Research or the US government. We express sincere gratitude to several laboratories where a part of the work is performed or some data is recorded and evaluated. Financial support from the Swedish Medical Research Council (Grant Nr. 2710, HSS); Astra-Zeneca, Mölndal, Sweden (HSS), Alexander von Humboldt Foundation, Germany (HSS); IPSEN Medical, France (HSS); Ever NeuroPharma, Austria (AS), and the University Grants Commission, New Delhi, India (HSS); and the Indian Council of Medical Research, New Delhi, India (HSS/AS) is gratefully acknowledged. The authors have no conflict of interest with any financial agencies mentioned above.

References

1. Sharma, H. S. (2009). Nanoneuroscience and nanoneuropharmacology. *Prog. Brain Res.*, **180**, pp. 1–264 (Elsevier, Boston, MA).
2. Sharma, H. S., and Sharma, A. (2013). Silver neurotoxicity, in *Encyclopedia of Metalloproteins*, eds. Uversky, V. N., Kretsinger, R. H., and Permyakov, E. A. (Springer Verlag, New York), pp. 2031–2052.
3. Sharma, H. S., and Sharma, A. (2012). Recent perspectives on nanoneuroprotection & nanoneurotoxicity. *CNS Neurol. Disord. Drug Targets*, **11**(1), pp. 5–6.
4. Sharma, H. S., and Sharma, A. (2012). Neurotoxicity of engineered nanoparticles from metals. *CNS Neurol. Disord. Drug Targets*, **11**(1), pp. 65–80. Review.
5. Sharma, H. S., and Sharma, A. (2007). Nanoparticles aggravate heat stress induced cognitive deficits, blood-brain barrier disruption, edema formation and brain pathology. *Prog. Brain Res.*, **162**, pp. 245–273. Review.
6. Aitken, R. J., Chaudhry, M. Q., Boxall, A. B., and Hull, M. (2006). Manufacture and use of nanomaterials: current status in the UK and global trends. *Occup. Med. (Lond.)*, **56**(5), pp. 300–306. Review.
7. Buzea, C., Pacheco, I. I., and Robbie, K. (2007). Nanomaterials and nanoparticles: sources and toxicity. *Biointerphases*, **2**(4), MR17–M172.
8. Donaldson, K., and Seaton, A. (2012). A short history of the toxicology of inhaled particles. *Part. Fibre Toxicol.*, **9**(1), p. 13 [Epub ahead of print].
9. Handy, R. D., and Shaw, B. J. (2007). Toxic effects of nanoparticles and nanomaterials: implications for public health, risk assessment and the public perception of nanotechnology. *Health Risk Soc.*, **9**(2), pp. 125–144.

10. Kang, G. S., Gillespie, P. A., Gunnison, A., Moreira, A. L., Tchou-Wong, K. M., and Chen, L. C. (2011). Long-term inhalation exposure to nickel nanoparticles exacerbated atherosclerosis in a susceptible mouse model. *Environ. Health Perspect.*, **119**(2), pp. 176–181 [Epub 2010 Sep 22].

11. Sharma, H. S., and Sharma, A. (2010). Breakdown of the blood-brain barrier in stress alters cognitive dysfunction and induces brain pathology. New perspective for neuroprotective strategies, in *Brain Protection in Schizophrenia, Mood and Cognitive Disorders*, ed. Ritsner, M. (Springer-Verlag, Berlin, New York), pp. 243–304.

12. Sharma, H. S., and Sharma, A. (2012). Nanowired drug delivery for neuroprotection in central nervous system injuries: modulation by environmental temperature, intoxication of nanoparticles, and comorbidity factors. *WIREs Nanomed. Nanobiotechnol.*, **4**(2), pp. 184–203, doi: 10.1002/wnan.172 [Epub 2011 Dec 8].

13. Sharma, H. S., and Westman, J. (2004). *The Blood-Spinal Cord and Brain Barriers in Health and Disease* (Academic Press, San Diego), pp. 1–617.

14. Sharma, H. S., Ali, S. F., Hussain, S. M., Schlager, J. J., and Sharma, A. (2009). Influence of engineered nanoparticles from metals on the blood-brain barrier permeability, cerebral blood flow, brain edema and neurotoxicity. An experimental study in the rat and mice using biochemical and morphological approaches. *J. Nanosci. Nanotechnol.*, **9**(8), pp. 5055–5072.

15. Sharma, H. S., Ali, S. F., Tian, Z. R., Hussain, S. M., Schlager, J. J., Sjöquist, P. O., Sharma, A., and Muresanu, D. F. (2009). Chronic treatment with nanoparticles exacerbate hyperthermia induced blood-brain barrier breakdown, cognitive dysfunction and brain pathology in the rat. Neuroprotective effects of nanowired-antioxidant compound H-290/51. *J. Nanosci. Nanotechnol.*, **9**(8), pp. 5073–5090.

16. Lafuente, J. V., Sharma, A., Patnaik, R., Muresanu, D. F., and Sharma, H. S. (2012). Diabetes exacerbates nanoparticles induced brain pathology. *CNS Neurol. Disord. Drug Targets*, **11**(1), pp. 26–39. Review.

17. Menon, P. K., Muresanu, D. F., Sharma, A., Mössler, H., and Sharma, H. S. (2012). Cerebrolysin, a mixture of neurotrophic factors induces marked neuroprotection in spinal cord injury following intoxication of engineered nanoparticles from metals. *CNS Neurol. Disord. Drug Targets*, **11**(1), pp. 40–49. Review.

18. Muresanu, D. F., Sharma, A., and Sharma, H. S. (2010). Diabetes aggravates heat stress-induced blood-brain barrier breakdown, reduction in cerebral blood flow, edema formation, and brain

pathology: possible neuroprotection with growth hormone. *Ann. N. Y. Acad. Sci.*, **1199**, pp. 15–26.

19. (2012). Heat injuries, active component, US Armed Forces, 2011. *MSMR*, **19**(3), pp. 14–16.
20. (2012). Update: heat injuries, active component, US Armed Forces, 2011. *MSMR*, **19**(3), pp. 17–19.
21. Seaton, A., Tran, L., Aitken, R., and Donaldson, K. (2010). Nanoparticles, human health hazard and regulation. *J. R. Soc. Interfac.*, **6**(7 Suppl 1), pp. S119–S29 [Epub 2009 Sep 2].
22. Smalley, B., Janke, R. M., and Cole, D. (2003). Exertional heat illness in Air Force basic military trainees. *Mil. Med.*, **168**(4), pp. 298–303.
23. Sharma, H. S., Ali, S., Tian, Z. R., Patnaik, R., Patnaik, S., Lek, P., Sharma, A., and Lundstedt, T. (2009). Nano-drug delivery and neuroprotection in spinal cord injury. *J. Nanosci. Nanotechnol.*, **9**(8), pp. 5014–5037.
24. Sharma, H. S., Ali, S. F., Dong, W., Tian, Z. R., Patnaik, R., Patnaik, S., Sharma, A., Boman, A., Lek, P., Seifert, E., and Lundstedt, T. (2007). Drug delivery to the spinal cord tagged with nanowire enhances neuroprotective efficacy and functional recovery following trauma to the rat spinal cord. *Ann. N. Y. Acad. Sci.*, **1122**, pp. 197–218.
25. Tian, Z. R., Sharma, A., Nozari, A., Subramaniam, R., Lundstedt, T., and Sharma, H. S. (2012). Nanowired drug delivery to enhance neuroprotection in spinal cord injury. *CNS Neurol. Disord. Drug Targets*, **11**(1), pp. 86–95. Review.
26. Sharma, H. S. (2009). Blood–central nervous system barriers: the gateway to neurodegeneration, neuroprotection and neuroregeneration, in *Handbook of Neurochemistry and Molecular Neurobiology: Brain and Spinal Cord Trauma*, eds. Lajtha, A., Banik, N., and Ray, S. K. (Springer Verlag, Berlin, Heidelberg, New York), pp. 363–457.
27. Sharma, H. S. (2007). Nanoneuroscience: emerging concepts on nanoneurotoxicity and nanoneuroprotection. *Nanomed. (Lond.)*, **2**(6), pp. 753–758. Review.
28. Wu, J., Wang, C., Sun, J., and Xue, Y. (2011). Neurotoxicity of silica nanoparticles: brain localization and dopaminergic neurons damage pathways. *ACS Nano*, **5**(6), pp. 4476–4489 [Epub 2011 May 4].
29. Sharma, H. S. (1999). Pathophysiology of blood-brain barrier, brain edema and cell injury following hyperthermia: new role of heat shock protein, nitric oxide and carbon monoxide. an experimental study in

the rat using light and electron microscopy. *Acta Univ. Ups.*, **830**, pp. 1–94.

30. Rapoport, S. I. (1976). *Blood-Brain Barrier in Physiology and Medicine* (Raven, New York).
31. Sharma, H. S. (1982). *Blood-Brain Barrier in Stress.* PhD thesis. Banaras Hindu University (BHU Press, Varanasi, India), pp. 1–85.
32. Sharma, H. S., and Hoopes, P. J. (2003). Hyperthermia induced pathophysiology of the central nervous system. *Int. J. Hyperthermia*, **19**(3), pp. 325–354. Review.
33. Cervós-Navarro, J., Sharma, H. S., Westman, J., and Bongcam-Rudloff, E. (1998). Glial reactions in the central nervous system following heat stress. *Prog. Brain Res.*, **115**, pp. 241–274. Review.
34. Sharma, H. S. (2012). Recent perspectives on central nervous system injury and neuroprotection. *Int. Rev. Neurobiol.*, **102**, pp. 1–395 (Academic Press, London, New York).
35. Sharma, H. S. (2007). Neurotrophic factors in combination: a possible new therapeutic strategy to influence pathophysiology of spinal cord injury and repair mechanisms. *Curr. Pharm. Des.*, **13**(18), pp. 1841–1874. Review.
36. Sharma, H. S. (2011). Early microvascular reactions and blood-spinal cord barrier disruption are instrumental in pathophysiology of spinal cord injury and repair: novel therapeutic strategies including nanowired drug delivery to enhance neuroprotection. *J. Neural Transm.*, **118**(1), pp. 155–176.
37. Sharma, H. S., and Johanson, C. E. (2007). Sharma and Johanson 2007 Blood-cerebrospinal fluid barrier in hyperthermia. *Prog. Brain Res.*, **162**, pp. 459–478. Review.
38. Sharma, H. S., Miclescu, A., and Wiklund, L. (2011). Cardiac arrest-induced regional blood-brain barrier breakdown, edema formation and brain pathology: a light and electron microscopic study on a new model for neurodegeneration and neuroprotection in porcine brain. *J. Neural Transm.*, **118**(1), pp. 87–114.
39. Sharma, H. S., Olsson, Y., and Cervós-Navarro, J. (1993). p-Chlorophenylalanine, a serotonin synthesis inhibitor, reduces the response of glial fibrillary acidic protein induced by trauma to the spinal cord. An immunohistochemical investigation in the rat. *Acta Neuropathol.*, **86**(5), pp. 422–427.
40. Sharma, H. S., Westman, J., and Nyberg, F. (1998). Pathophysiology of brain edema and cell changes following hyperthermic brain injury. *Prog. Brain Res.*, **115**, pp. 351–412. Review.

41. Sharma, H. S., Muresanu, D. F., Patnaik, R., Stan, A. D., Vacaras, V., Perju-Dumbrav, L., Alexandru, B., Buzoianu, A., Opincariu, I., Menon, P. K., and Sharma, A. (2011). Superior neuroprotective effects of cerebrolysin in heat stroke following chronic intoxication of Cu or Ag engineered nanoparticles. A comparative study with other neuroprotective agents using biochemical and morphological approaches in the rat. *J. Nanosci. Nanotechnol.*, **11**(9), pp. 7549–7569.

42. Sharma, A., Muresanu, D. F., Mössler, H., and Sharma, H. S. (2012). Superior neuroprotective effects of cerebrolysin in nanoparticle-induced exacerbation of hyperthermia-induced brain pathology. *CNS Neurol. Disord. Drug Targets*, **11**(1), pp. 7–25. Review.

43. Sharma, H. S., Muresanu, D., Sharma, A., and Zimmermann-Meinzingen, S. (2010). Cerebrolysin treatment attenuates heat shock protein overexpression in the brain following heat stress: an experimental study using immunohistochemistry at light and electron microscopy in the rat. *Ann. N. Y. Acad. Sci.*, **1199**, pp. 138–148.

44. Sharma, H. S., Sharma, A., Mössler, H., and Muresanu, D. F. (2012). Neuroprotective effects of cerebrolysin, a combination of different active fragments of neurotrophic factors and peptides on the whole body hyperthermia-induced neurotoxicity: modulatory roles of co-morbidity factors and nanoparticle intoxication. *Int. Rev. Neurobiol.*, **102**, pp. 249–276.

45. Sharma, H. S., Muresanu, D. F., Patnaik, R., Stan, A. D., Vacaras, V., Perju-Dumbrav, L., Alexandru, B., Buzoianu, A., Opincariu, I., Menon, P. K., and Sharma, A. (2014). Superior antioxidant effects of cerebrolysin in heat stroke following chronic intoxication of Cu or Ag engineered nanoparticles. A comparative study with other antioxidant agents using biochemical and morphological approaches in the rat. *J. Nanosci. Nanotechnol.* (in press).

46. Sharma, H. S., Castellani, R. J., Smith, M. A., and Sharma, A. (2012). The blood-brain barrier in Alzheimer's disease: novel therapeutic targets and nanodrug delivery. *Int. Rev. Neurobiol.*, **102**, pp. 47–90.

47. Sharma, H. S. (2006). Hyperthermia influences excitatory and inhibitory amino acid neurotransmitters in the central nervous system. An experimental study in the rat using behavioural, biochemical, pharmacological, and morphological approaches. *J. Neural. Transm.*, **113**(4), pp. 497–519.

48. Sharma, H. S., and Kiyatkin, E. A. (2009). Rapid morphological brain abnormalities during acute methamphetamine intoxication in the rat:

an experimental study using light and electron microscopy. *J. Chem. Neuroanat.*, **37**(1), pp. 18–32.

49. Sharma, H. S., Gordh, T., Wiklund, L., Mohanty, S., and Sjöquist, P. O. (2006). Spinal cord injury induced heat shock protein expression is reduced by an antioxidant compound H-290/51. An experimental study using light and electron microscopy in the rat. *J. Neural Transm.*, **113**(4), pp. 521–536.
50. Sharma, H. S., Patnaik, R., Sharma, A., Sjöquist, P. O., and Lafuente, J. V. (2009). Silicon dioxide nanoparticles (SiO_2, 40–50 nm) exacerbate pathophysiology of traumatic spinal cord injury and deteriorate functional outcome in the rat. An experimental study using pharmacological and morphological approaches. *J. Nanosci. Nanotechnol.*, **9**(8), pp. 4970–4980.
51. Sharma, H. S., Muresanu, D. F., and Sharma A. (2013). Novel therapeutic strategies using nanodrug delivery, stem cells and combination therapy for CNS trauma and neurodegenerative disorders. *Expert Rev. Neurother.*, **13**(10), pp. 1085–1088, doi: 10.1586/14737175.2013.836297 [Epub 2013 Oct 3].
52. Sharma, H. S., Muresanu, D. F., Patnaik, R., and Sharma A. (2013). Exacerbation of brain pathology after partial restraint in hypertensive rats following SiO_2 nanoparticles exposure at high ambient temperature. *Mol. Neurobiol.*, **48**(2), pp. 368–379, doi: 10.1007/s12035-013-8502-y [Epub 2013 Jul 6].
53. Sharma, A., Muresanu, D. F., Patnaik, R., and Sharma, H. S. (2013). Size- and age-dependent neurotoxicity of engineered metal nanoparticles in rats. *Mol. Neurobiol.*, **48**(2), pp. 386–396, doi: 10.1007/s12035-013-8500-0 [Epub 2013 Jul 3].
54. Sharma, H. S., Menon, P. K., Lafuente, J. V., Aguilar, Z. P., Wang, Y. A., Muresanu, D. F., Mössler, H., Patnaik, R., and Sharma. A. (2014). The role of functionalized magnetic iron oxide nanoparticles in the central nervous system injury and repair: new potentials for neuroprotection with Cerebrolysin therapy. *J. Nanosci. Nanotechnol.*, **14**(1), pp. 577–595. Review.
55. Sharma, H. S., and Sharma, A. (2013). New perspectives of nanoneuroprotection, nanoneuropharmacology and nanoneurotoxicity: modulatory role of amino acid neurotransmitters, stress, trauma, and co-morbidity factors in nanomedicine. *Amino Acids*, **45**(5), 1055–1071, doi: 10.1007/s00726-013-1584-z. Review.
56. Sharma, H. S., Patnaik, R., Sharma, A., and Wiklund, L. (2013). The contribution of glial cells and water channel aquaporin-4 in the

neuropathology of cardiac arrest is still ignored. *CNS Neurol. Disord. Drug Targets*, **12**(1), p. 3.

57. Sharma, H. S., Muresanu, D. F., Lafuente, J. V., Patnaik, R., Tian, Z. R., Mössler, H., and Sharma, A. TiO2-nanowired cerebrolysin attenuated hyperthermia induced ubiquitin overexpression and brain pathology. Proceedings National Innovation and Showcase Summit 2014, TechConnect 2014; NSTI-Nanotech 2014, www.nsti.org, ISBN 978-1-4822-5827-1 Vol 2, 2014 pp. 303–306.
58. Sharma, A., Muresanu, D. F., Lafuente, J. V., Menon, P., Patnaik, R., Tian, Z. R., Mössler, H., and Sharma, H. S. Nanodelivery of cerebrolysin as adjunct therapy with functionalized magnetic iron oxide nanoparticles enhance neuroprotection following whole body hyperthermia. Proceedings National Innovation and Showcase Summit 2014, TechConnect 2014; NSTI-Nanotech 2014, www.nsti.org, ISBN 978-1-4822-5827-1 Vol 2, 2014 pp. 363–366.
59. Sharma, H. S., Feng, L., Lafuente, J. V., Muresanu, D. F., Tian, Z. R., Patnaik, R., and Sharma, A. Nanowired delivery of mesenchymal cell reduces diabetes induced aggravation of brain damage following heat stroke. Proceedings National Innovation and Showcase Summit 2014, TechConnect 2014; NSTI-Nanotech 2014, www.nsti.org, ISBN 978-1-4822-5827-1 Vol 2, 2014 pp. 246–249.
60. Sharma, H. S., Muresanu, D. F., Lafuente, J. V., Patnaik, R., Tian, Z. R., Mössler, H., and Sharma A. Nanodrug delivery by single-walled carbon nanotubes (SWCNTs) in the central nervous system induces neurotoxicity. Potenial neuroprotective effects of cerebrolsyin. Proceedings National Innovation and Showcase Summit 2014, TechConnect 2014; NSTI-Nanotech 2014, www.nsti.org, ISBN 978-1-4822-5827-1 Vol 2, 2014 pp. 339–342.
61. Sharma, H. S., Patnaik, R., and Sharma, A. (2010). Diabetes aggravates nanoparticles induced breakdown of the blood-brain barrier permeability, brain edema formation, alterations in cerebral blood flow and neuronal injury. An experimental study using physiological and morphological investigations in the rat. *J. Nanosci. Nanotechnol.*, **10**(12), pp. 7931–7945.
62. Kiyatkin, E. A., and Sharma, H. S. (2011). Expression of heat shock protein (HSP 72 kDa) during acute methamphetamine intoxication depends on brain hyperthermia: neurotoxicity or neuroprotection? *J. Neural Transm.*, **118**(1), pp. 47–60 [Epub 2010 Oct 8].
63. Kiyatkin, E. A., and Sharma, H. S. (2012). Environmental conditions modulate neurotoxic effects of psychomotor stimulant drugs of abuse. *Int. Rev. Neurobiol.*, **102**, pp. 147–171.

64. Westman, J., and Sharma, H. S. (1998). Heat shock protein response in the central nervous system following hyperthermia. *Prog. Brain Res.*, **115**, pp. 207–239.

65. Sharma, H. S., Muresanu, D. F., Sharma, A., Patnaik, R., and Lafuente, J. V. (2009). Nanoparticles influence pathophysiology of spinal cord injury and repair. *Prog. Brain Res.*, **180**, pp. 154–180.

66. Sharma, H. S., Patnaik, R., and Sharma, A. (2010). Exposure of manganese nanoparticles induce blood-brain barrier disruption, brain pathology and cognitive and motor dysfunctions in rats. *Eur. J. Neurol.*, **17**(Supplement: 3), pp. 469–469.

67. Moghimi, S. M., and Szebeni J. (2003). Stealth liposomes and long circulating nanoparticles: critical issues in pharmacokinetics, opsonization and protein-binding properties. *Prog. Lipid Res.*, **42**(6), pp. 463–478.

68. Muresanu, D. F., Sharma, A., Tian, Z. R., Smith, M. A., and Sharma, H. S. (2012). Nanowired drug delivery of antioxidant compound H-290/51 enhances neuroprotection in hyperthermia-induced neurotoxicity. *CNS Neurol. Disord. Drug Targets*, **11**(1), pp. 50–64. Review.

69. Donaldson, K., Aitken, R., Tran, L., Stone, V., Duffin, R., Forrest, G., and Alexander, A. (2006). Carbon nanotubes: a review of their properties in relation to pulmonary toxicology and workplace safety. *Toxicol. Sci.*, **92**(1), pp. 5–22 [Epub 2006 Feb 16]. Review.

70. Knochel, J. P. (1974). Environmental heat illness. An eclectic review. *Arch. Intern. Med.*, **133**(5), pp. 841–864.

71. Sharma, H. S., Westman, J., and Alm, P. (1998). Nitric oxide and carbon monoxide in the brain pathology of heat stress. *Prog. Brain Res.*, **115**, pp. 297–333.

72. Sharma, H. S., Sjöquist, P. O., and Ali, S. F. (2007). Drugs of abuse-induced hyperthermia, blood-brain barrier dysfunction and neurotoxicity: neuroprotective effects of a new antioxidant compound H-290/51. *Curr. Pharm. Des.*, **13**(18), pp. 1903–1923. Review.

Chapter 15

Dermatotoxicity of Nanomaterials

Nancy A. Monteiro-Riviere and Jim E. Riviere
Nanotechnology Innovation Center of Kansas State, Kansas State University, Manhattan, KS, USA
nmonteiro@ksu.edu

15.1 Introduction

Exposure of skin to nanomaterials is a primary route of exposure in many occupational, consumer, cosmetic, and environmental scenarios. A relatively large body of work has been conducted in these areas, although there are significant differences in the nature of materials studied as well as the quantity and mode of material exposure. Despite this increasing number of studies being reported, as will be seen below, there is still a paucity of safety and toxicology data conducted over longer exposure periods (days to months) that characterize many patterns of usage. As with many other studies in nanotoxicology, a lack of characterization as well as of consistency in types of materials studied (size, shape, surface coatings, vehicles/formulations) prevents generalizations to be easily made. These general factors and limitations have been adequately discussed in other chapters of the present text.

Handbook of Safety Assessment of Nanomaterials: From Toxicological Testing to Personalized Medicine
Edited by Bengt Fadeel

ISBN 978-981-4463-36-2 (Hardcover), 978-981-4463-37-9 (eBook)
www.panstanford.com

15.2 Why Is Skin Different to Other Routes of Exposure?

Compared to other routes of exposure, the skin is fundamentally different in that nanoparticles (NPs) are exposed on the skin's surface without having to first pass through an environment that could modify their properties. For example, when oral ingestion is considered, particles first encounter saliva and then other digestive fluids (gastric and intestinal juices) prior to reaching the absorptive surface of the gastrointestinal mucosa. In the inhalational scenario, aerosolized particles must traverse the complex bronchiolar pathways and depending on size be deposited onto either the mucous covering the bronchiolar epithelium or the surfactant-containing fluid lining the alveoli. In both oral and inhalational scenarios, deposited particles in a solution will then contact living cells, many of them with phagocytic and other particle transport characteristics related to physiological scavenging functions. In contrast, the exterior surface of the skin is the inert and acellular stratum corneum, which has evolved in mammals to be a water-impermeable barrier. In many species, the stratum corneum is perforated by invaginations of the skin that form hair shafts and follicles that are numerous in the furred species and with variable densities across body areas. It is only after a topically applied nanomaterial penetrates either the stratum corneum or the lining of the hair follicles would actual viable cells be encountered that could then elicit a toxicological response. Further penetration could potentially result in delivery to the capillary beds or lymphatics, which could result in systemic exposure, or interact with other cellular members of the immune system that could also elicit systemic or local effects (for an introduction to the immune system, see Chapter 8).

Topical exposure is very different in that the environment of the skin's surface is modulated by ambient conditions (temperature, humidity, ultraviolet [UV] sunlight exposure, sweat, compromised or diseased skin, presence or absence of clothing) that can affect both particle–particle and particle–skin interactions. In most topical exposure scenarios, NPs will be carried in some type of vehicle or formulation that may affect the state of individual NP aggregation. In other cases, pristine NPs could be deposited onto the skin and not be able to directly interact with the stratum corneum unless solvated by sweat or exogenous liquid sources often encountered in many pharmaceutical and consumer products. However, in the case of applications for nanomedicine, particle–formulation pairs may be specifically optimized to maintain activity of the skin surface without dermal penetration (e.g., sunscreens, surface antimicrobials), to penetrate hair shafts for sustained delivery to this region (e.g., antifungals etc.), or alternatively to be formulated to avoid aggregation and have surface chemistries compatible with partitioning into the intercellular lipids of the stratum corneum—the pathway for dermal absorption to deeper skin layers or the capillary beds and systemic delivery.

15.3 What Are the Biological Targets in the Skin?

As extensively reviewed elsewhere [1–4] the skin is the largest organ in the body and is the body's interface with the environment. Skin can be considered primarily an organ responsible for protecting the body from external insults that is reflected through its barrier, metabolic, UV light-shielding, and immunological functions, implemented, respectively, by the stratum corneum, metabolically active basal epidermal keratinocytes, UV-absorbing melanocytes, and the immune system components residing within the skin, including keratinocytes, resident mast cells, and dendritic Langerhans cells (LCs). A second primary function of mammalian skin is thermoregulation that is mediated by insulating fur/hair, heat-dissipating sweat, and a high density of heat-dissipating blood capillaries that far exceed metabolic demands. Skin is a crucial component of the nervous system involved in sensing pressure and heat and releasing neural-active mediators to produce pain in the affected areas. Skin is an endocrine organ involved in synthesis of vitamin D, androgens, estrogens, and insulin-like growth factor–binding protein-3, as well as serving as a target of numerous systemic hormones [3, 5]. Finally, skin also plays a major role in social interactions by determining the appearance of the species and individual to others—a function mediated by fur/hair, skin color determined by melanocyte density, skin texture controlled by the integrity and elasticity of the supporting subepidermal structure, efficacy of the stratum corneum barriers to produce a soft and pliable surface, and sexual attraction mediated by pheromone excretion.

This myriad of functions must be considered when assessing the nature of NP-induced dermatotoxicity, since abnormalities in these functions determine the nature of the dermal response to a toxic insult. Whether chemical or nanomaterial, these dermal functions can only be modulated if material can reach and interact with them. In contrast to much smaller organic chemicals and drugs, which may easily be able to partition into the intercellular lipids that form the stratum corneum barrier and interact with all cell types, nanomaterials are primarily restricted by their large sizes, which limit most nanomaterials from penetrating into the viable epidermis or dermis or being absorbed into the systemic circulation. If NPs can penetrate into and interact with skin elements, then the responses seen are characterized by the degree of modulation of normal skin functions.

The first level of chemical- or nanomaterial-induced toxicity will be an irritation response characterized by the skin's reaction to what it senses as a noxious threat, usually a damaging irritating poison or bacterial colonization, or loss of barrier function. The response is the typical one associated with the cardinal signs of inflammation, redness, heat, and pain. This response is mediated by increased blood flow to the affected site, movement of inflammatory cells from the vascular system through increased capillary permeability, epidermal proliferation, and ultimately barrier repair. This dermal irritation response is mediated by a complex cytokine cascade initiated by almost any variety of insults, be it UV light, chemical insult or irritation, contact allergen or allergic response, or microbial invasion. The cytokine cascade relative to skin structure is illustrated in Fig. 15.1. Relative to manufactured nanomaterials and topically applied nanomedicines, the following response could be theoretically initiated:

- NP alterations in barrier function through penetration and disruption of the stratum corneum layers
- Activation of the skin allergic response (e.g., poison ivy)
- Direct interaction with epidermal keratinocytes, resulting in irritation and cytokine release

- NP direct toxicity to keratinocytes, resulting in cell death and subsequent irritation
- Interaction and modulation of capillary functions

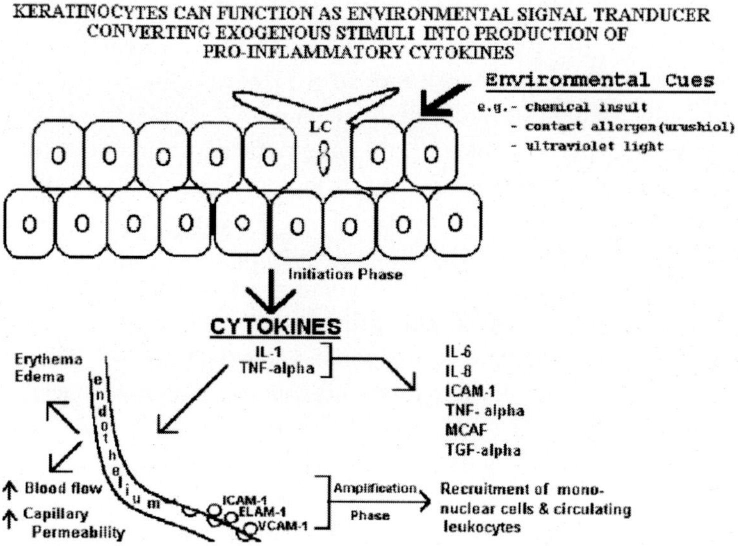

Figure 15.1 Schematic view of the immune response of the skin.

It should be pointed out that chemicals or nanomaterials do not need to fully penetrate the epidermis and interact directly with keratinocytes, nor do they need to be absorbed into the system circulation to initiate dermal irritation or an immune or sensitization response. Disruption of barrier function alone through penetration into or interaction with stratum corneum lipids will produce a perturbation that could result in dermal irritation [6, 7]. Similarly, NPs need only penetrate the deeper layers of the stratum corneum to reach the dendritic processes of LCs (Fig. 15.1) and initiate an immune response. In fact, from a pharmacological perspective, this is the target for delivering antigens using topical vaccination strategies.

The second level of potential nanomaterial-induced dermatotoxicity involves interactions that modulate epidermal and dermal cellular functions without causing massive cell death or cytokine release. These include keratinocyte or melanocyte proliferation or neoplastic transformation, resulting in more subtle changes and a chronic rather than acute response. Unfortunately, the ability of nanomaterials to induce these effects has not been extensively studied. Most in vivo experiments are short term and can only manifest as loss of barrier functions assessed by transepidermal water loss (TEWL) or as sensitization, inflammation, and/ or proliferation. In vitro experimental endpoints are geared toward irritation (cytokine release) and direct keratinocyte toxicity (loss of viability, visualization of cell death, markers of oxidative damage, etc.). Thus, where it exists at all, nanomaterial dermatotoxicology at the present is limited to assessment of acute and short-term responses, primarily using in vitro model systems.

A final consideration that has a major impact on experimentally assessing dermatotoxicity is the animal species employed since, as can be appreciated from the above discussion, differences in anatomy and physiology modulate the response potentially seen. Differences in hair density and skin thickness are the most obvious variables. Since most laboratory rodents used in experimental studies have a dense fur coat, the question often arises of their applicability to predict responses in humans who do not have a dense fur coat or very thin skin. Even within species, there are regional differences in hair density and skin thickness that may modulate responses [2, 8, 9]. Different species have different immunological systems, not all triggered by skin interactions. Finally, the molecular organization and biochemical composition of stratum corneum lipids differ between species and could affect results obtained [10]. The comparative biology of skin across different species has been extensively studied [1–3, 9] and should be consulted by the prudent nanomaterial researcher to ensure that responses seen in experimental studies are capable of occurring in humans.

15.4 Assessment of Nanomaterial Dermatotoxicity

On the basis of these considerations, what is the current paradigm for assessing cutaneous toxicity of nanomaterials? There are three very distinct avenues of investigation utilized: assess toxicity and irritation with in vitro epidermal keratinocyte cultures, assess the ability of NPs to penetrate skin using dermatomed skin mounted in static or flow-through in vitro diffusion cells, and conduct short-term (hours to a few days) in vivo topical exposure studies.

15.4.1 In vitro Studies

By far the largest number of "skin toxicity" studies with nanomaterials have been conducted using in vitro human epidermal keratinocyte (HEK) cells in culture [11–21]. In most cases, standard protocols developed for assessing chemical or drug toxicity have been employed. Toxicity is assessed morphologically by light or electron microscopy or by standard assays of cell viability with calcein AM, Live/Dead®, neutral red (3-amino-7-dimethylamino-2 methylphenazine hydrochloride), 3-[4,5-dimethyl-2-thiazol]-2,5-diphenyl-2H-tetrazolium bromide (MTT), Celltiter 96® AQueous one, alamar blue, Celltiter-Blue®, CytoTox One™ homogeneous membrane integrity, trypan blue and flow cytometry [12], or cytokine release (interleukin [IL]-8, in addition to IL-6, IL-10, IL-1b, and tumor necrosis factor alpha [TNF-α]). Our laboratory assesses this panel of biomarkers for HEK cells, which indicates a potential inflammatory response to a toxicological insult. The expression of IL-8 release in keratinocytes seems to play a major role with NP exposure. Specific examples of where these have been employed are discussed in all of the nanomaterial examples presented below. Additional studies have been conducted to determine the mechanism of HEK cellular uptake to define potential pathways. In a recent comprehensive study, different endocytic pathways (Fig. 15.2) were evaluated to understand cellular uptake of quantum dots (QDs) into HEK cells [22]. QD NPs were selected due to their intense photostable fluorescent properties and are available in various sizes, shapes, and surface coatings. This was one of the first studies that showed that QD NPs with a carboxylic acid surface coating were recognized by lipid rafts and not by clathrin or caveolae in HEK cells. Also, it showed that QDs were internalized into early endosomes and then transferred to late endosomes or lysosomes. Utilizing 24 endocytic interfering agents to investigate the mechanism by which QDs entered cells, depicted that QD endocytic pathways were primarily regulated by the G protein–coupled receptor-associated pathway and low-density lipoprotein receptor/scavenger receptor.

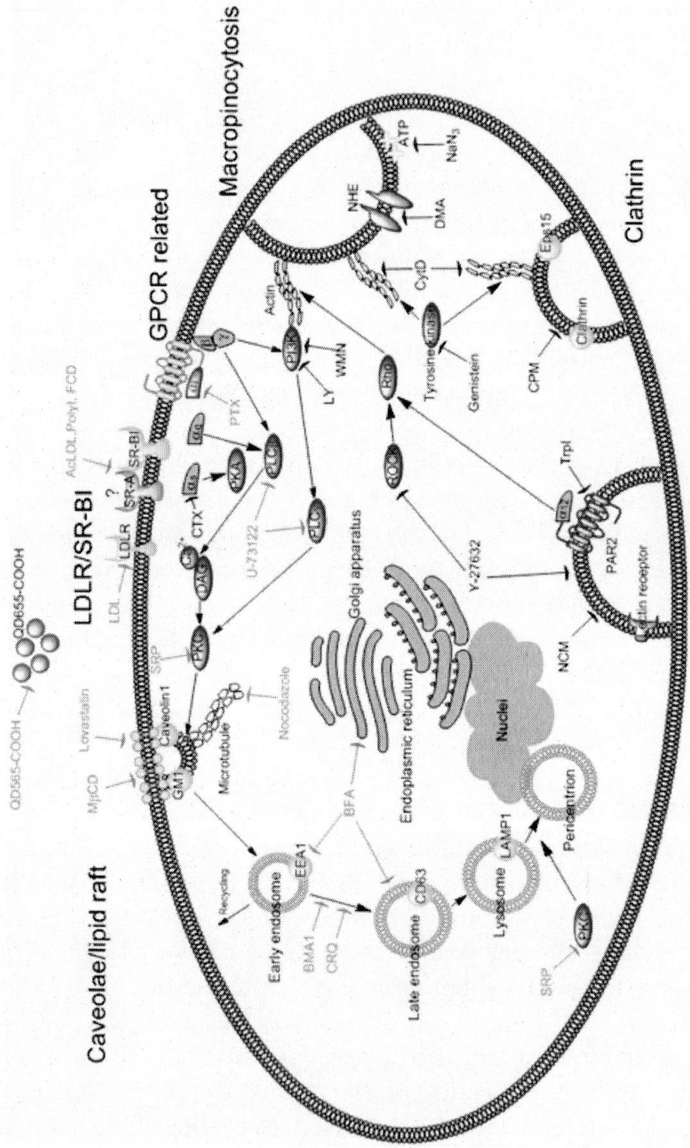

Figure 15.2 Endocytic pathways for QD uptake in human epidermal keratinocytes. From Ref. [22].

However, there are limitations of in vitro cell culture systems designed for small soluble molecules when applied to colloidal NP exposures [11, 12]. This comprehensive study emphasized the pitfalls of using classical toxicology assays to assess the toxicity of engineered NPs. The major limitation is that certain nanomaterials may interact with classic cell viability biomarkers and cause variable results with both false positive or false negative results. One of the hallmark properties of NPs that make them unique for many applications are their extremely large surface-area-to-mass ratios that result in extreme adsorptive properties, which provides a good sink for nonspecific adsorption of molecules. They may also adsorb proteins and nutrients in the culture medium [11].

The in vitro cell culture–based cytotoxicity-screening assays such as these are designed to minimize animal use in toxicity testing and have become the mainstay of modern toxicology testing strategies [23] (and see Chapter 4 on the use of in vitro tests in nanotoxicology). Here, we report on the use of several screening methods that can be used for high-throughput research and be less reliant on animals. Moreover, in vivo is not the most efficient or best method to screen thousands of new nanomaterials. It is not always feasible to use animals in testing in the United States or EU. Since the EU will not allow any in vivo testing on animals for cosmetics it makes in vitro testing even more important and puts a greater emphasis on in vitro testing. Therefore, these will make health assessments difficult and more perplexing to interpret when conducting health risk studies of engineered nanomaterials. While many of these new nanomaterials may or may not have a potential toxic effect, inaccurate screening could impede nanotechnology progress due to erroneous results.

A second concern is more theoretical and relates to how in vitro studies may be extrapolated to the in vivo setting. Under normal in vitro cell culture conditions, cells are bathed in a nutrient-rich protein-containing aqueous medium. For HEK cells, this may not reflect their in vivo environment where viable keratinocytes are sandwiched between the basement membrane and the nonviable stratum corneum and not in direct contact with protein-rich fluids. As will be discussed below, two properties of nanomaterials, which are very different from chemicals and drugs, are their propensity to aggregate (NP–NP interaction) and/or agglomerate (NP–protein–NP interaction) under appropriate conditions within the cells.

Figure 15.3 depicts aggregates of 20 nm silica-coated silver in HEK cells. These interactions, which may both increase the size of the nanomaterial and mask NP surface chemistries, are more likely to occur in the aqueous conditions seen in a cell culture medium rather than in the intercellular lipid domains of the stratum corneum layer through which NPs must pass to reach the viable HEK cells in vivo. In line with this limitation, in vitro cell culture systems do not possess the stratum corneum rate-limiting barrier to topical nanomaterial penetration, which in some ways could be considered a filter to the types of nanomaterials that could reach HEK cells in vivo. Another limitation is the relationship between the concentration in vitro that elicits a response and that which is seen in vivo under normal exposure conditions. Finally, in vitro studies of dendritic cells in culture have also been conducted to assess the NP cellular uptake into these immunologically critical cells [24]. This study showed dendritic cells that were differentiated from monocytes resulted in a six-fold increase in QD uptake compared to lymphocytes.

15.4.2 In vitro Skin Penetration Models

The second major line of investigation addresses the assumption that for a nanomaterial to be toxic to skin, it must penetrate the stratum corneum. If the particle cannot penetrate this barrier, it cannot be toxic to viable skin. To test this hypothesis, standard in vitro diffusion cell models employed extensively to assess chemical and drug dermal absorption have been used. These systems have been well described in the literature [25]. The primary differences between types of systems used in nanotoxicology studies are the type of diffusion cell, method of membrane preparation, species from which the skin membrane is harvested, composition of the receptor fluid, and method of topical nanomaterial application (neat, vehicle). The two primary types of diffusion cells used are static (e.g., Franz cells) [26] or flow-through (e.g., Bronaugh cells) [25, 27] systems that differ only in the nature of the perfusate bathing the dermal side of the skin. Static cells have fixed media from which samples are taken, while flow-through cells perfuse media to mimic blood flow under the skin. Membranes could include full-thickness skin, skin dermatomed to specific thicknesses, or epidermal membranes heat-separated from the underlying dermis.

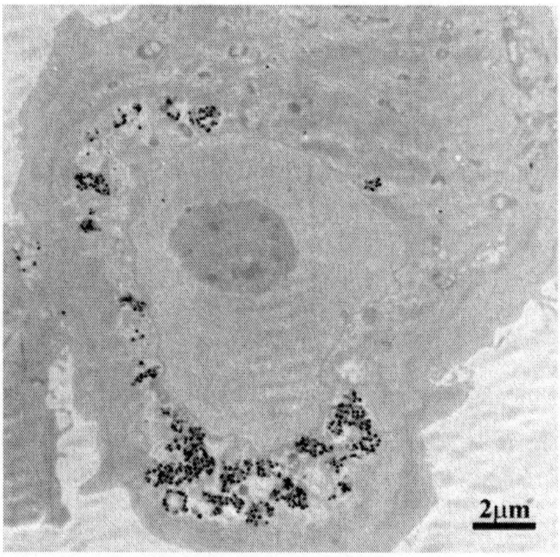

Figure 15.3 Silica-coated silver NP aggregates within the cytoplasmic vacuoles of human epidermal keratinocytes.

All sorts of animal species have been utilized, with human skin obtained from surgical procedures (viable) or cadavers (dead) and pig skin predominating in studies designed to assess absorption in humans. Haired laboratory rodents with very thin skin are often employed due to their easy availability, but extrapolation to humans may be problematic. A major concern with even clipped furry species is NP binding to the surface hairs. Many times it has been reported in the literature that the skin was shaved rather than clipped. This introduces nicks or small cuts in the skin that would allow for penetration of NPs. Another limitation of in vitro systems is differences in elastic modulus between the species. This is a concern when human skin is employed in vitro, since it contracts when collected with a biopsy punch. The membrane should be stretched when mounted onto a diffusion cell. This is not seen with other species such as pigs, and its effect on extrapolating NP penetration data from in vitro to in vivo human penetration is not known.

Penetration endpoints studied include the fraction of dose detected in the perfusate that reflects nanomaterial absorption or transdermal delivery, the amount of dose found deposited in various skin fractions (stratum corneum tape strips, sectioned skin biopsies), or the nonabsorbed amount left on the surface of the skin (surface wipes and washes). For fluorescent materials such as QDs, confocal laser microscopy is often used to visualize QD penetration in skin sections obtained from diffusion cells after exposure. More sophisticated quantitative imaging tools such as time-of-flight secondary ion mass spectrometry (TOF-SIMS) have also been employed recently [28]. Finally, biopsy sections may be examined by transmission electron microscopy/scanning electron microscopy (TEM/SEM) to specifically localize particles in specific microscopic regions.

Finally, the nature of the solvent or vehicle in which the nanomaterial is applied to the skin's surface is crucial, since most nanomaterials are exposed as solutions or colloid suspensions due to their particulate nature or for stability. The vehicle used to solubilize the nanomaterial may have profound effects on the amount of penetration seen. Such a phenomenon was seen when hydrophobic pristine fullerenes (C_{60}) were topically applied to in vitro porcine skin a series of organic solvents (toluene, cyclohexane, chloroform) or mineral oil [29]. Chloroform resulted in the deepest C_{60} stratum corneum penetration, while no penetration was detected when dosed in mineral oil. Vehicle modulations (oil in water versus water in oil) were also seen in the depth of stratum corneum penetration of titanium dioxide applied as sunscreen formulations in porcine skin [28]. Across all particles, such vehicle effects may be modulated by vehicle/formulation effects on solubility, NP aggregation state, effects of a vehicle's ionic strength and pH of the particle surface charge, and vehicle interactions with the stratum corneum lipid barrier (solvents, high pH).

Nanomaterial absorption into the perfusate has not been detected using all of the analytical methodologies commonly employed in any absorption study, including inductively coupled plasma mass spectroscopy (ICP-MS). The methods used are primarily a function of the material composition of the nanomaterial being studied. Methods employed have included all varieties of gas and liquid chromatography, fluorescence, and mass spectrometry. From the literature examined to date for all types of nanomaterials in all species, including humans, nanomaterials applied on the surface of the skin have not been detected in the perfusate at the limit of sensitivity for the analytical method employed.

Finally, there is another potential characteristic of NP interactions with the stratum corneum barrier lipid pathways that is fundamentally different than what is seen with most chemicals and small molecules. This is the potential for a particle that partitions into the stratum corneum to subsequently cause disruption and further penetration [30, 4]. The stratum corneum is not a spatially uniform structure but rather can be viewed as vertical stacks or columns of flattened keratinocytes, with uniform intercellular spaces filled with organized lipids within the columns but less uniform spaces between adjacent columns and clusters of columns and where adnexial structures such as hair or sweat ducts ascend up to the skin's surface. Intercellular spaces in these "edge" regions may be as wide as 20–30 nm, well within the size range to allow penetration of topically applied NPs. In fact, some NPs have been reported to preferentially locate in this region [31]. Some of these regions may be associated with hydrophilic transepidermal aqueous pathways that potentially are capable of generating a transepidermal water gradient under the proper conditions that could facilitate transport of small water-soluble nanomaterials. Once a nanomaterial, with a reactive surface, penetrates the intercellular spaces, disruption of the lipid barrier could also occur, further facilitating penetration. It must be stressed that to date, NP penetration into skin has been detected but not substantial transdermal flux across the epidermis.

Studies have demonstrated that mechanical action and perturbations of the skin barrier can affect the penetration of NPs. Tinkle et al. [32] reported that large 0.5 μm fluorescein isothiocyanate (FITC)-conjugated dextran beads could penetrate the stratum corneum of human skin and reach the epidermis following 30 minutes of flexing. Normally these particles would not penetrate the skin. Very small amino acid–derivatized fullerene NPs of 3.5 nm were able to penetrate the dermis of porcine skin that was flexed for 60 minutes and placed in flow-through diffusion cells for 8 hours. The nonflexed control skin showed penetration limited to the stratum granulosum layer of the epidermis [33]. Other studies have been conducted with QD655 and QD565 coated with carboxylic acid (hydrodynamic diameters of 18 nm and 14 nm, respectively) in flow-through diffusion cells with flexed, tape-stripped, and abraded rat skin for 8 and 24 hours. Penetration was not observed with the nonflexed, flexed, or tape-stripped skin, but penetration to the viable dermal layer occurred in the abraded skin. In some cases, retention of the QDs in hair follicles was noted in the flexed skin [34].

Currently, only a few penetration studies with NPs on compromised skin have been conducted. Titanium dioxide (TiO_2) and zinc oxide (ZnO) NPs in sunscreens were investigated to determine if NPs in different formulations can penetrate through ultraviolet-B (UVB)-damaged skin. Very little is known about their disposition in UVB-damaged skin. Pigs were used a models for human skin and were exposed to UVB that resulted in moderate sunburn. Both in vitro and in vivo studies were conducted with four sunscreen formulations: 10%-coated TiO_2 in oil/water (o/w), 10%-coated TiO_2 in water/oil (w/o), 5%-coated ZnO in o/w, and 5%-uncoated ZnO in o/w. Penetration occurred with the TiO_2 formulations more so than the ZnO formulations in both types of studies, but no absorption of TiO_2 or ZnO was noted by ICP-MS [28]. Very little research has been conducted with diseased skin. Healthy and psoriatic skin depicted Ti deeply penetrated the stratum corneum of psoriatic skin relative to normal skin but did not reach the viable layers of the epidermis [35].

15.4.3 In vivo Toxicity Studies

The final approach many would consider the "gold standard" of assessing nanomaterial dermatototoxic potential are in vivo studies. These model systems are identical to those used to assess chemical and drug skin reactions. Regulatory agencies worldwide have specific prescriptions for experimental design and interpretations. Markers of in vivo skin toxicity include standard visual scoring indices such as the classic Draize test, evaluation of the barrier integrity using TEWL, or classic histopathology to assess microscopic lesions. The advantage of such systems is that identical exposure scenarios that would be encountered in humans can often be replicated in vivo. The major problem is that few chronic exposure studies have ever been conducted for any nanomaterial in any animal model, certainly not in humans. In vivo studies were conducted daily for 14 days using 25 nm and 35 nm carbon-coated silver NPs (AgNPs), as well as to freshly synthesized or thoroughly washed 20 nm, 50 nm, and 80 nm AgNPs topically applied to the backs of weanling pigs. The penetration of AgNPs was only observed in the superficial layers of the stratum corneum, even after 14 days' TEM, and confirmed by energy-dispersive X-ray spectroscopy (EDS), which detected the localization of Ag [17].

15.4.4 Nanomaterial Properties in Relation to Skin Penetration and Dermatotoxicity

Before one can have a knowledgeable discussion of what nanomaterial properties cause skin damage, one must realize that only those properties that have actually been measured can enter into this assessment. The following physicochemical properties are most often used to characterize nanomaterials:

- Size distribution
- Agglomeration state
- Shape
- Crystal structure
- Chemical composition
- Surface area
- Surface chemistry
- Surface charge
- Porosity

The advantage of measuring these properties is that this provides a benchmark for what nanomaterials were actually studied relative to individual particle size, shape, and surface chemistry, including charge (zeta potential). The reader is referred to Chapter 3 for a more detailed discussion on characterization techniques. Of particular importance is the analysis for contaminants that may persist from manufacture that provide a toxicological signature, not of the NP, but of its synthesis method. The methods of synthesis is obviously of significance but can be misleading in understanding nanomaterial toxicology of the adverse effects seen, since toxicity may be related to the contaminant rather than the NP. However, the physical and chemical properties that are important in assessing biological interactions are those properties that reflect nanomaterial structure and function in the aqueous biological environment at physiologically compatible temperatures, ion concentrations, and pH. Ideally, NP properties should be measured in situ but are most often measured under controlled laboratory conditions that may not reflect the biological milieu. The major exception is the use of scanning and primarily TEM to image NPs in cells and in tissues. The major limitation here is that only the electron-dense components of NPs can be imaged. The problem with nanomaterial characterization is that their properties are sensitive to the environment they are in. Is aggregation or agglomeration facilitated, which dramatically alters particle size? When particle size is reported, is this the size of the individual nonaggregated particle or the size of the aggregated/agglomerated complex? These issues of characterization are now becoming more widely recognized and hopefully will result in the development of metrics and indices that reflect nanomaterial function in biologically relevant matrices.

One such approach developed in the author's laboratory is an attempt to measure a potential molecular interaction fingerprint of the surface of an NP measured in an aqueous system [36, 37]. We have defined the biological surface adsorption index (BSAI) as a five-component nanodescriptor metric that quantitates the surface adsorption forces (hydrophobicity, hydrogen bonding, polarity/polarizability, and lone par electrons) that govern nanomaterial interaction with biological components that lead to aggregation, protein corona formation, and interactions of nanomaterials with biological systems, such as cellular transport and trafficking domains. The properties assessed allow a broad classification of nanomaterials according to potential reactivity based upon descriptors that have been employed for generations to describe molecular interactions in the body (binding, partitioning, etc.). The advantage that BSAI-type indices provide is that these properties can be simultaneously assessed for each NP type.

15.4.5 Quantum Dot Penetration and Toxicity Studies

The properties associated with NP penetration and epidermal toxicity were studied under controlled conditions using a series of QDs of different sizes, shapes, and surface chemistries [4, 38, 34, 20]. Penetration of QD565 (spherical) and QD655 (ellipitical) with a CdSe (cadmium selenide) core and a ZnS (zinc sulfide) shell dispersed in borate buffer with diverse physicochemical properties were studied in porcine skin flow-through diffusion cells. QD565 coated with polyethylene glycol (PEG) (35 nm; neutral), PEG-amine (NH_2) (15 nm; cationic), or carboxylic acid (COOH) (14 nm; anionic) showed penetration into the stratum corneum layers and localization within the epidermal and dermal layers by eight hours. PEG- (45 nm) and PEG-NH_2- (20 nm) coated QD655 were localized within the epidermal layers by eight hours. The penetration of QD655-COOH (18 nm) into epidermal layers was evident only after 24 hours [38]. In addition, QD621 (nail shaped) (39–40 nm) with a CdSe core and a CdS shell dispersed in water were topically applied to assess penetration in the same systemn. After 24 hours, at a low concentration of 1 μM, QD621 were located primarily on the surface on the stratum corneum layers of the skin and no QD621 fluorescence was detected in the stratum granulosum, stratum spinosum, or stratum basale layers of the epidermis. However, by increasing the applied concentration to 10 μM, QD621 were detected in the upper stratum corneum bilipid layers or in between the stratum granulosum–stratum corneum interface, with only a small amount of fluorescence in the upper viable epidermal layers [20]. Transdermal flux of any of these diverse QDs was not detetced in the perfusate.

The penetration seen with QD565, QD655, and QD621 in flow-through diffusion porcine skin cells showed how differences in composition, size, configuration, surface charge, dispersing medium, and other physicochemical parameters could influence penetration. PEG-QD621 with a hydrodynamic diameter of 39±1 nm were capable of penetrating only the uppermost layers of the porcine stratum corneum after 24 hours of exposure [20], while confocal microscopy showed that all three surface coatings of the QD565 penetrated at 8 and 24 hours, but only the QD655 COOH took 24 hours to penetrate porcine skin [38]. QDs that are synthesized with the same core/shell and similar surface coatings and hydrodynamic diameters but different shapes and different penetration rates can penetrate intact skin. These studies showed that spherical QDs penetrated better than the ellipsoid, the cationic better than the anionic across shapes, and all better than the nail-shaped QD621 when compared to neutral QDs with similar hydrodynamic dimensions. These results are not unexpected if one considers the skin architecture and dimensions of the penetration paths. Nevertheless, one could argue that the vehicle-mediated alterations in the skin cannot be completely ruled out; the penetration effects reported still remain valid because the dosing solutions are representative of QDs as commercially supplied but at a twofold to fourfold lower concentration.

This study, using relatively easily traced QDs, clearly shows independent effects of size, surface chemistry, and shape on stratum corneum penetration into a defined in vitro skin model. Similar studies with other nanomaterials using the same skin penetration model need to be conducted.

15.5 Conclusions

Interaction of nanomaterials and the skin has turned out to be a fruitful area of investigation, both as a unique interface for studying NP–tissue interactions and because of the great potential for human exposure from occupational, medicinal, consumer product, and cosmetic perspectives. From a pharmacological and pharmaceutical perspective, it appears unlikely that transdermal delivery of NPs at pharmacologically significant fluxes into the systemic circulation is feasible. Use of topical dermatological medicinal preparations for surface activities (e.g., sunscreens), hair follicle delivery, or potential vaccinations seems feasible. From a toxicological perspective where large skin surfaces could be exposed to high exposure levels, skin becomes an important organ for assessment. In most cases, classic irritation might be the only sequela observed and this could be due to solvent or contaminant effects. Large transdermal fluxes are not required for toxicological effects to be initiated if the endpoint is sensitization or neoplastic transformation. This would be especially true if the skin barrier were disrupted by mechanical perturbation or disease.

References

1. Monteiro-Riviere, N. A. (1991). Comparative anatomy, physiology, and biochemistry of mammalian skin, in *Dermal and Ocular Toxicology Fundamentals and Methods*, ed. Hobson, D. W. (CRC Press, Boca Raton), pp. 3–71.
2. Monteiro-Riviere, N. A. (2006). The integument, in *Dellmann's Textbook of Veterinary Histology*, 6th ed., eds. Eurell, J., and Frappier, B. (Blackwell, Ames), pp. 320–349.
3. Monteiro-Riviere, N. A. (2010). Structure and function of skin, in *Toxicology of the Skin: Target Organ Series 29*, ed. Monteiro-Riviere, N. A. (Informa Healthcare, New York), pp. 1–18.

4. Monteiro-Riviere, N. A., and Baroli, B. (2010). Nanomaterial penetration, in *Toxicology of the Skin: Target Organ Series 29*, ed. Monteiro-Riviere, N. A. (Informa Healthcare, New York), pp. 333–346.

5. Zouboulis, C. C. (2009). The skin as an endocrine organ. *Dermato-Endocrinol.*, **1**, pp. 250–252.

6. Elias, P. M. (2006). Defensive functions of the stratum corneum: integrative aspects, in *Skin Barrier*, eds. Elias, P. M., and Feingold, K. R. (Taylor and Francis, New York), pp. 5–14.

7. Nickoloff, B. J. (1993). *Dermal Immune System* (CRC Press, Boca Raton).

8. Monteiro-Riviere, N. A. (2008). Anatomical factors that affect barrier function, in *Dermatotoxicology*, 7th ed., eds. Zhai, H., Wilhelm, K. P., and Maibach, H. I. (CRC Press, New York), pp. 39–50.

9. Monteiro-Riviere, N. A., Bristol, D. G., Manning, T. O., and Riviere, J. E. (1990). Interspecies and interregional analysis of the comparative histological thickness and laser Doppler blood flow measurements at five cutaneous sites in nine species. *J. Invest. Dermatol.*, **95**, pp. 582–586.

10. Monteiro-Riviere, N. A., Inman, A. O., Mak, V., Wertz, P., and Riviere, J. E. (2001). Effect of selective lipid extraction from different body regions on epidermal barrier function. *Pharm. Res.*, **18**, pp. 992–998.

11. Monteiro-Riviere, N. A., and Inman, A. O. (2006). Challenges for assessing carbon nanomaterial toxicity to the skin. *Carbon*, **44**, pp. 1070–1078.

12. Monteiro-Riviere, N. A., Inman, A. O., and Zhang, L. W. (2009). Limitations and relative utility of screening assays to assess engineered nanoparticle toxicity in a human cell line. *Toxicol. Appl. Pharmacol.*, **234**, pp. 222–235.

13. Monteiro-Riviere, N. A., Nemanich, R. J., Inman, A. O., Wang Y. Y., and Riviere, J. E. (2005). Multi-walled carbon nanotube interactions with human epidermal keratinocytes. *Toxicol. Lett.*, **155**, pp. 377–384

14. Monteiro-Riviere, N. A., Oldenburg, S. J., and Inman, A. O. (2010). Interactions of aluminum nanoparticles in human epidermal keratinocytes. *J. Appl. Toxicol.*, **30**, pp. 276–285.

15. Rouse, J. G., Yang, J., Barron, A. R., and Monteiro-Riviere, N. A. (2006). Fullerene based amino acid nanoparticle interactions with human epidermal keratinocytes. *Toxicol. In Vitro*, **20**, pp. 1313–1320.

16. Ryman-Rasmussen, J. P., Riviere, J. E., and Monteiro-Riviere, N. A. (2007). Surface coatings determine cytotoxicity and irritation

potential of quantum dot nanoparticles in epidermal keratinocytes. *J. Invest. Dermatol.*, **127**, pp. 143–153.

17. Samberg, M. E., Oldenburg, S. J., and Monteiro-Riviere, N. A. (2010). Evaluation of silver nanoparticle toxicity in skin in vivo and keratinocytes in vitro. *Environ. Health Perspect.*, **118**(3), pp. 407–413.
18. Saathoff, J. G., Inman, A. O., Xia, X. R., Riviere, J. E., and Monteiro-Riviere, N. A. (2011). In vitro toxicity assessment of three hydroxylated fullerenes in human skin cells. *Toxicol. In Vitro*, **25**, pp. 2105–2112.
19. Zhang, L. W., and Monteiro-Riviere, N. A. (2009a). Endocytic mechanisms and toxicity of a functionalized fullerene in human cells. *Toxicol. Lett.*, **191**, pp. 149–157.
20. Zhang, L. W., Yu, W. W., Colvin, V. L., and Monteiro-Riviere, N. A. (2008). Biological interactions of quantum dot nanoparticles in skin and in human epidermal keratinocytes. *Toxicol. Appl. Pharmacol.*, **228**, pp. 200–211.
21. Zhang, L. W., Zeng, L., Barron, A. R., and Monteiro-Riviere, N. A. (2007). Biological interactions of functionalized single-wall carbon nanotubes in human epidermal keratinocytes. *Int. J. Toxicol.*, **26**, pp. 103–113.
22. Zhang, L. W., and Monteiro-Riviere, N. A. (2009b). Mechanisms of quantum dot nanoparticle cellular uptake. *Toxicol. Sci.*, **110**, pp. 138–155.
23. National Research Council. (2007). *Toxicity Testing in the Twenty-First Century: A Vision and Strategy* (National Academy Press, Washington, DC).
24. Zhang, L. W., Baumer, W., and Monteiro-Riviere, N. A. (2011). Cellular uptake mechanisms and toxicity of quantum dots in dendritic cells. *Nanomedicine*, **6**(5), pp. 777–791.
25. Bronaugh, R. L. (2006). In vitro diffusion cell studies, in *Dermal Absorption Models in Toxicology and Pharmacology*, ed. Riviere, J. E., (CRC, Taylor and Francis, New York), pp. 21–28.
26. Franz, T. J. (1978). The finite dose technique as a valid in vitro model for the study of percutaneous absorption in man, in *Skin: Drug Application and Evaluation of Environmental Hazards*, eds., Simon, G. A., Paster, A., Klingberg, M. A., and Kaye, M. (Basel, Karger), *Curr. Probl. Dermatol.*, pp. 58–68.
27. Bronaugh, R. L., and Maibach, H. I. (1991). *In Vitro Percutaneous Absorption: Principles, Fundamentals, and Applications* (CRC Press, New York).

28. Monteiro-Riviere, N. A., Wiench, K., Landsiedel, R., Schulte, S., Inman, A. O., and Riviere, J. E. (2011). Safety evaluation of sunscreen formulations containing titanium dioxide and zinc oxide nanoparticles in UVB-exposed skin: An in vitro and in vivo study. *Toxicol. Sci.*, **123**, pp. 264–280.

29. Xia, X. R., Monteiro-Riviere, N. A., and Riviere, J. E. (2010a). Skin penetration and kinetics of pristine fullerenes (C_{60}) topically dosed in industrial organic solvents. *Toxicol. Appl. Pharmacol.*, **242**, pp. 29–37.

30. Cevc, G., and Vierl, U. (2010). Nanotechnology and the transdermal route. *J. Controlled Rel.*, **141**, pp. 277–299.

31. Baroli, B., Ennas, M. G., Loffredo, F., Isola, M., Pinna, R., and López-Quintela, M. A. (2007). Penetration of metallic nanoparticles in human full-thickness skin. *J. Invest. Dermatol.*, **127**, pp. 1701–1712.

32. Tinkle, S. S., Antonini, J. M., Rich, B. A., Roberts, J. R., Salmen, R., DePree, K., and Adkins, E. J. (2003). Skin as a route of exposure and sensitization in chronic beryllium disease. *Environ. Health Perspect.*, **111**, pp. 1202–1208.

33. Rouse, J. G., Yang, J., Ryman-Rasmussen, J. P., Barron, A. R., and Monteiro-Riviere, N. A. (2007). Effects of mechanical flexion on the penetration of fullerene amino acid-derivatized peptide nanoparticles through skin. *Nano Lett.*, **7**, pp. 155–160.

34. Zhang, L. W., and Monteiro-Riviere, N. A. (2008). Assessment of quantum dot penetration into intact, tape stripped, abraded and flexed rat skin. *Skin Pharmacol. Physiol.*, **21**, pp. 166–180.

35. Pinheiro, T., Pallon, J., Alves, L. C., Veríssimo, A., Filipe, P., Silva, J. N., and Silva, R. (2007). The influence of corneocyte structure on the interpretation of permeation profiles of nanoparticles across skin. *Nucl. Instrum. Methods Phys. Res. B*, **260**, pp. 119–123.

36. Xia, X. R., Monteiro-Riviere, N. A., Mathur, S., Oldenberg, S., Fadell, B., and Riviere, J. E. (2011). Mapping the surface adsorption forces of nanomaterials in biological systems using the biological surface adsorption index (BSAI) approach. *ACS Nano*, **5**, pp. 9074–9081.

37. Xia, X. R., Monteiro-Riviere, N. A., and Riviere, J. E. (2010b). An index for characterization of nanomaterials in biological systems. *Nat. Nanotechnol.*, **5**, pp. 671–675.

38. Ryman-Rasmussen, J. P., Riviere, J. E., and Monteiro-Riviere, N. A. (2006). Penetration of intact skin by quantum dots with diverse physicochemical properties. *Toxicol. Sci.*, **91**, pp. 159–165.

Chapter 16

Reproductive Toxicity of Nanomaterials

Margaret Saunders,[a] **Gary Hutchison,**[b] **and Sara Correia Carreira**[c]
[a]*Bioengineering, Research and Innovation Hub (BIRCH), Department of Medical Physics & Bioengineering, University Hospitals Bristol NHS Trust, St Michael's Hospital, Bristol, UK*
[b]*Centre for Nano Safety, School of Life, Sport and Social Science, Edinburgh Napier University, Edinburgh, UK*
[c]*Bristol Centre for Functional Nanomaterials (BCFN), University of Bristol, Bristol, UK*
m.saunders@bristol.ac.uk

16.1 Introduction

The field of nanomedicine is still in its infancy, but important lessons can be learned from studies of nanomaterials and nanoparticles (NPs) developed for other purposes. Application of such knowledge will enable the safe development of nanomedicine, whilst minimizing any potential health or environmental risks. Needless to say, toxicity assessment is vital, particularly when nanomaterials may have medical applications. The study of nanotoxicology has expanded greatly in recent years, and recent reviews have addressed the toxicology of a range of nanomaterials used specifically

Handbook of Safety Assessment of Nanomaterials: From Toxicological Testing to Personalized Medicine
Edited by Bengt Fadeel

ISBN 978-981-4463-36-2 (Hardcover), 978-981-4463-37-9 (eBook)
www.panstanford.com

in nanomedicine [1, 2] or that are suitable for applications in nanomedicine [3–5]. These highlight the potential for toxicity of many medically relevant nanomaterials such as silica, gold, silver, carbon nanotubes (CNTs), liposomes, dendrimers, quantum dots, superparamagnetic iron oxide nanoparticles (SPIONs), and titanium dioxide, dependent upon specific characteristics such as chemistry, solubility, extent of agglomeration, size, shape, and charge [1, 2, 3, 5]. However, an area that has still received little attention is that of reproductive and developmental nanotoxicity [1, 4, 6]. This area is important because of the greatly enhanced susceptibility of the developing embryo/fetus to the effects of exposure to external agents or xenobiotics compared to the adult as a result of the rapid development during gestation and the altered metabolism [7] and the impact of adverse effects on fertility and normal reproduction. In this chapter, we focus on the potential impact of nanomaterials on male and female reproductive health.

A selection of engineered nanomaterials and their medical applications are summarized below in Table 16.1.

16.2 The Reproductive System

Due to the complexity of the human reproductive system there is potential for exposure effects on different aspects with varying outcome severity. For example, impaired function of the placenta during gestation can result in effects ranging from fetal growth retardation to fetal or neonatal death and some effects may not manifest until later in life [47]. Therefore there may be opportunities for nanotoxicological insult through occupational or environmental exposure as well as through intentional (i.e., medical) administration.

16.2.1 The Female Reproductive System

The female reproductive system includes the ovaries, the oviducts, the uterus and cervix, and the vagina, as shown in Fig. 16.1A. The function of these organs is regulated by hormones excreted from the hypothalamus and the pituitary. The production of female sex hormones and female gametes (oocytes) occurs in the ovaries, and

Table 16.1 Applications of medically relevant engineered nanomaterials

Nanomaterial	Application
Gold	Gold is used in catalytic, optical, electronic components of probes, sensors, and optical devices, promising bioimaging probes, and photothermal agents [8–13].
Silver	Silver possesses antimicrobial properties enabling use in wound dressings, antimicrobial coatings on surgical instruments, and prostheses [14–19].
Magnetic iron oxide (Fe_3O_4)	The magnetic properties of iron oxide NPs bear an immense potential for the development of new biomedical applications such as imaging, in vivo cell tracking, targeted drug delivery, and hyperthermia treatment [20–26].
Silica (SiO_2)	Silica products are commonly used in the electronics industry and as a food additive, but nanosized amorphous silica may be used in medical applications such as photodynamic therapy, drug delivery, and biomedical imaging [27–30].
Quantum dots	Quantum dots are semiconductor nanocrystals that possess high luminosity and are therefore considered to be promising bioimaging probes [29]. However, the vast majority of quantum dots contain toxic heavy metals, such as cadmium. Capping of the quantum dot core may reduce toxic effects of cadmium [31], and cadmium-free alternatives with similar optical properties are currently being explored [32].
Polymeric micelles	Nanosized polymeric micelles are most commonly used as drug carriers and are composed of two distinct domains, a drug-loading core and a hydrophilic shell [33–37]. Polymeric particles are believed to be nontoxic and biocompatible [38, 39].
Titanium dioxide (TiO_2)	Although best known as a component in sunscreens, pigments, and paints, titanium dioxide NPs also have applications as drug delivery vehicles and excipient formulations and potential for use in photodynamic therapy [40–44].
CNTs	These hold great promise for biomedical use as carriers for drug delivery, DNA/protein sensors, biocatalysts, tissue scaffolds, and imaging agents with the potential to deliver large amounts of therapeutic agents to the target site [1, 45, 46].

transport of oocytes is mediated through the fallopian tubes (or oviducts), where sperm that enter the female reproductive tract through the vagina may fertilize the oocytes. Upon fertilization, implantation of the fertilized oocyte into the uterine wall and development of the placenta occur. The uterus provides an adequate environment for the development and eventual delivery of the fetus, and the placenta provides its life support. Therefore, proper regulation of every organ comprising the female reproductive system is essential for its correct function.

- *Ovarian structure and function.* The ovaries are the female gonads, analogous to the male testes. Apart from being gonads, they also function as endocrine glands. The ovaries are located on either side of the uterus, on the lateral wall of the pelvis. The surface of the ovaries is covered by a single layer of cells called the ovarian surface epithelium (OSE) [48]. OSE cells secrete cytokines, growths factors, and steroids that play an important role in ovarian physiology [49]. The ovary itself consists of a number of ovarian follicles, which contain an occyte each, surrounded by granulosa cells. Ovarian granulosa cells are involved in steroidgenesis and play an important role in maintaining female fertility by regulating ovarian function and oocyte development [50].

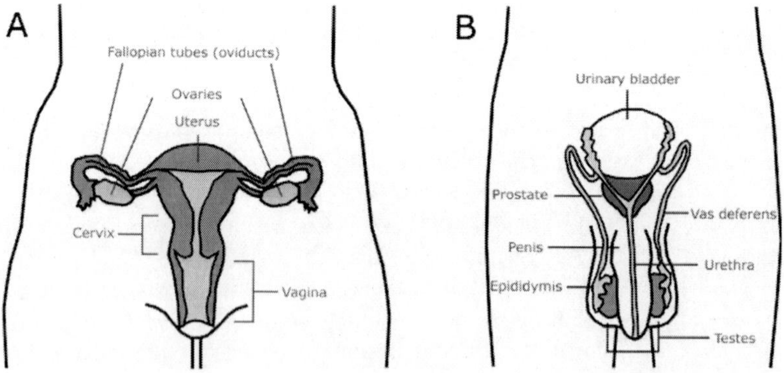

Figure 16.1 Schematic of the female (A) and the male (B) reproductive system.

Uterine structure and function. The lumen of the uterus is lined by the endometrium, which is composed of glandular and luminal epithelia that are supported by a connective tissue stroma. Mediated by steroid hormones during the menstrual cycle, the phenotypes of endometrial luminal and glandular epithelium undergo dramatic changes in preparation for implantation of the fertilized oocyte [51]. Therefore, the endometrium represents an important fertility-determining factor. The uterus is joined up to the vagina via the cervix. The ectocervix is the portion of the cervix facing the vaginal canal, whereas the endocervix is the internal portion that opens into the uterus. Endocervical epithelial cells produce innate immune mediators that likely aid in the protection from pathogens and therefore play an important role in maintaining the health of the female reproductive tract [52].

Vaginal structure and function. The vagina extends from the exterior of the body to the uterus. The lining along the vagina is composed of a multilayered stratified epithelium [53], which provides the first line of defense against pathogens [54] and secretes mucous, which provides both a barrier and a target to drug delivery [55].

Placenta and the blood–placenta barrier. Upon fertilization, an oocyte develops into a blastocyst before implantation into the uterine wall. The outer layer of the blastocyst consists of trophoblasts, which provide the early embryo with nutrients and continue to develop into a large part of the placenta. Therefore, trophoblast growth and migration are critical events during placental development and embryogenesis. Cytotrophoblast stem cells within the placenta may differentiate into extravillous trophoblasts (EVTs), which invade the endometrium and remodel the uterine blood vessels, or into syncytiotrophoblasts (STBs), which form a syncytial layer that secretes human chorionic gonadotropin (hCG) to sustain the pregnancy and form the placental barrier. Transfer between the maternal and the fetal circulation through the placental barrier occurs across the endothelial-syncytial membrane and depends upon a number of factors, for example, thickness of the membrane, blood flow, and pressure [56]. The placenta plays a vital role in nurturing the developing fetus, as well as forming a barrier to protect it against potentially toxic components.

16.2.2 The Male Reproductive System

The male reproductive system consists of two testes and accessory reproductive organs (i.e., a duct system, glands, and a penis) with the primary function of sperm production and delivery of sperm to ensure fertilization can occur (Fig. 16.1B). The testes are externally located in the scrotum and protected by a surrounding sac [57].

Testicular structure and function. The testes' internal physiology can be subdivided into two distinct compartments, seminiferous tubules and an interstitial space. The interstitial space consists of Leydig cells, which produce testosterone, blood vessels, macrophages, lymphocytes, connective tissue, and lymphatic vessels [58]. Seminiferious tubules have peritubular myeloid cells on the external surface, and the internal basement membrane is lined with Sertoli and germ cells. Each Sertoli cell extends from the basement membrane to the lumen, forming the blood–testes barrier, with the function of maintaining and nurturing germ cell maturation into spermatozoa in an immune-privileged environment. A complex communication between all cell types is crucial for normal testicular function [59]. The convoluted network of seminiferous tubules converges together to form an interconnected complex called the rete testis where efferent ductules enter the epididymis, a structure loosely attached to the testes. The efferent ductules empty into the vas deferens, responsible for transportation of sperm to the urethra in seminal fluid generated by seminal vesicles, the prostate gland, and bulbourethral glands [57].

The primary function of the testes is to produce sperm by means of spermatogenesis. Alongside spermatogenesis, androgen biosynthesis is a main function of mammalian testis. A wide range of physiological and developmental mechanisms are controlled by androgen production in the testes; however, other major actions include male phenotype development and maintenance of spermatogenesis and blood–testes barrier integrity [60]. Androgens such as testosterone are key mediators of spermatogenesis, and their production is interrelated with functional spermatogenesis; therefore, both processes need to be highly regulated. This regulation is achieved through endocrine, paracrine, and autocrine signaling. Gonadotropin releasing hormone (GnRH) initiates endocrine regulation and is produced by specialized neurons in the hypothalamus. GnRH signals the pituitary gland to secrete both follicle-stimulating hormone (FSH) and luteinizing hormone (LH), which binds to specific receptors on Leydig and Sertoli cells (luteinizing hormone receptor [LHR] and follicle-stimulating hormone receptor [FSHR], respectively). Testes also produce other important molecules such as cytokines, inhibin (a and b), activin, MIH, Ins13, and oxytocin [61].

Blood–testes barrier. The blood–testes barrier, unlike the blood–brain and retina barriers, is made up of adherens and coexisting tight junctions. The junctions are not fixed and have the ability to disassemble to accommodate the migration of spermatocytes across the barrier during the process of spermatogenesis, while maintaining its integrity [62]. The tight junction component of the barrier forms a physical barricade that restricts molecules and cells from entering or exiting the lumen. To ensure the proper development and maturation of germ cells it is critical that the correct microenvironment be maintained by the second component, a physiological barrier that consists of transporters that regulate the movement of substances in or out of the lumen. The final component of the blood–testes barrier is the immunological barrier that prevents access by the immune system. Individually the three components on their own do not formulate a complete barrier; it is the interaction between all three components that creates a barrier of maximum competence [58].

16.3 Reproductive Health

Both environmental pollution and lifestyle (e.g., smoking, obesity) may have adverse effects on female reproductive health. Regarding environmental pollution, compounds that influence hormone function, like pesticides and persistent pollutants, as well as heavy metals, such as lead, are particularly associated with adverse reproductive health effects in human females [63]. Male reproductive health has been in decline over the past decade, with reports of falling sperm counts, reduced semen quality, and increased incidence of disease. It is unclear what is driving this decline, but it has been suggested that it may be a combination of environmental factors and lifestyle. Studies have demonstrated that reproductive health, in particular semen quality, is affected by air pollution [64] and smoking [65] and that reproductive development can be compromised by a number of environmental chemicals [66]. Because both male and female reproductive systems have shown sensitivity to xenobiotic exposure, they are likely to be a target for NP-driven side effects.

16.4 Reproductive Toxicity Testing

Few studies are currently available to determine the possible impact of exposure to nanomaterials upon reproductive and developmental outcomes despite the social and clinical relevance [67]. Whilst the direct impact on the developing embryo/fetus is of obvious concern, it is important to also take into account possible effects on both male and female fertility, a complex area where there are still many unanswered questions. Reproductive toxicity includes adverse effects on sexual function and fertility in adult males and females, as well as developmental toxicity in the offspring [68]. This covers any effect of chemicals that would interfere with sexual function and fertility. This may include, but not be limited to, alterations to the female and male reproductive systems, adverse effects on the onset of puberty, gamete production and transport, reproductive cycle normality, sexual behavior, fertility, parturition, premature reproductive senescence, or modifications in other functions that are dependent on the integrity of the reproductive systems. Adverse effects on or via lactation are also included in reproductive toxicity and are also included in reproductive toxicity, and for pragmatic purposes of classification, developmental toxicity essentially means adverse effects induced during pregnancy or as a result of parental exposure. These effects can be manifested at any point in the life span of the organism. The major manifestations of developmental toxicity include death of the developing organism, structural abnormality, altered growth, and functional deficiency [68]. Thus, this is a highly complex area requiring a broad range of models.

16.5 In vitro and in vivo Models for Reproductive Nanotoxicology

Harmonized guidelines have been developed for reproduction toxicity testing of pharmaceuticals and are regulated by the US Food and Drug Administration (FDA) and the European Medicines Agency and delineated in the International Committee on Harmonization (ICH) document *S5(R2)* [69, 70]. Screening and testing for developmental and reproductive toxicities of chemicals have been harmonized under guidelines produced for the US Environmental Protection Agency (EPA) and the Organisation for Economic Co-operation and Development (OECD) [70] and cover prenatal developmental toxicity, reproduction and fertility effects, a one-generation reproductive toxicity study, a reproduction/developmental toxicity screening test, a combined repeat-dose toxicity study with the reproduction/ developmental screening test, and developmental neurotoxicity. Specific guidelines for reproductive nanotoxicology do not yet exist.

16.5.1 In vitro Models to Study Effects on the Female Reproductive Tract

Since the female reproductive tract comprises many organs, several different models for each component have been developed. In vitro models of ovarian tissues, the uterine endometrium, and trophoblasts have mainly been used to study fertility and embryo implantation [71], as well as tumor development in ovarian and cervical cancer [72, 73]. However, these models may also useful to study the impact of engineered nanomaterials on the function of female reproductive organs, as well as to explore targeted diagnosis and therapy of cancerous cells.

Models to study steroidgenesis and ovarian function. Primary granulosa cells extracted from follicles are frequently used to study ovarian hormone secretion. Stelzer et al. [74] studied the effects of gold NPs on primary rat granulosa cells and found that the particles entered granulosa cells and altered estradiol secretion. Similarly, Liu et al. [75] found that calcium phosphate NPs increased apoptosis by interfering with the cell cycle of cultured primary human ovarian granulosa cells. The particles were able to enter the cells, but no change in the levels of either progesterone or estradiol was measured in the cell culture medium. Reproductive toxicity has also been show for CdTe/ZnTe quantum dots, which interfered with the development of primary mouse follicles [76].

A model that is used to study estrogen-dependent growth of tumors is the ovarian cancer cell line BG-1 [72]. These cells are also used to study targeted diagnostics and therapy for ovarian cancer using NPs [77]. Core-shell hydrogel NPs functionalized with peptides that specifically target ovarian carcinoma cell lines have proved to efficiently deliver short interfering RNA (siRNA) to BG-1 cells and knock down specific receptors without impairing cell viability [77].

Models for OSE are also frequently used to study tumor development, as it is believed to be the source of most ovarian tumors [78]. Primary OSE cells obtained from patients undergoing hysterectomy are a common model. These cells are difficult to culture and are usually more stable when cultured as a more in vivo–like 3D culture [79] or after immortalization [80]. Cell lines derived from tumors of the human OSE, like NIH:OVCAR-3 and SK-OV-3, have been used to study targeted molecular magnetic resonance imaging (MRI) using magnetic NPs [81].

Models to study uterine function. Most endometrial epithelial cell lines are derived from endometrial adenocarcinomas and have a range of characteristics. The first readily available endometrial cell line was HEC-1 [82]. Because of their low adhesive properties, these cells have been used widely in implantation research as a model of nonreceptive epithelial cells, but they have also been used to assess cytotoxicity of polymeric nanodrug carriers [83].

The Ishikawa cell line is a very well-characterized human endometrial cell line that has mixed characteristics of glandular and luminal epithelium [84] and is considered a good model for studying normal endometrial function. The cell line was derived from a well-differentiated adenocarcinoma of the human endometrial epithelium and expresses functional steroid receptors for estrogen, progesterone, and androgen [85–87]. Ishikawa cells have been used to assess toxicity of a variety of chemicals [88], as well as the effect of endocrine disruptors like bisphenol-A on the female reproductive system [89].

Different models are available for the distinct parts of the cervix. For example, End/E6E7 and Ect1/E6E7 are derived from normal endo- and ectocervical epithelium, respectively, and immortalized through transduction with the human papilloma virus genes [90]. Because the endocervix is susceptible to sexually transmitted pathogens, endocervical cell lines are often used to study sexually transmitted diseases [91]. The human endocervical epithelial cell lines ME-180, SiHa, and CaSki originate from a cervical carcinoma and have been used as a screening tool for epithelial toxicity of the female genital tract [92] and also to test efficacy of antitumor drugs [93]. Recently, a new, immortalized cervical epithelial cell line named A2EN has been derived from an endocervical tissue explant [94]. This cell line is thought to be particularly useful to study immune responses mediated by the endocervix.

The most commonly studied cervical tumor cell line is undoubtedly HeLa. It is an immortalized epithelial cell line derived from a cervical adenocarcinoma. However, it is seldom used to study specific reproductive toxicity or targeted delivery of NPs. Rather, it is used as a general model to assess cytotoxic effects and cellular uptake of NPs [95, 96].

Models to study placental development and barrier function. A large number of trophoblast cell lines are available that were derived from a variety of sources, including normal placenta, malignant tissue, embryonal carcinomas, and placental choriocarcinomas [97]. Depending on their properties, the different cell lines may be more or less adequate models for the different functions of the placenta. For example, the primary EVT cell lines SGHPL-4 and HTR-8/SVneo have generally been used to study a variety of events during placental development, including adhesion to the uterus, invasion of the uterine wall, and proliferation [98–101]. BeWo, JEG-3, and JAR cells derived from a human placental choriocarcinoma are frequently used as models for the placental barrier, although the most established cell line for this purpose is BeWo [102]. BeWo cells have been used to assess placental transport of environmental pollutants [103], as well as polystyrene (PS) NPs [104].

In addition to the available in vitro models, ex vivo placental perfusion is a powerful tool to assess placental transfer of NPs in a setup that resembles the in vivo human placenta very closely. Human placentas from cesarean sections or natural births can be obtained relatively easily after informed consent of the patient. Preterm placentas may also be obtained from legal pregnancy terminations. In a placental perfusion study investigating the maternal–fetal transfer of polyethylene glycol (PEG)ylated gold NPs, no transplacental transport of the particles was observed and gold aggregates were only located in the STBs and trophoblasts but not in fetal capillaries [105]. This contrasts with a study utilizing fluorescently labelled PS beads in a recirculating system, where 50 nm, 80 nm, or 240 nm diameter beads appeared in the fetal perfusate within 15 minutes but 500 nm particles were retained in the placental tissue and maternal circulation [106]. In another study using the human perfused placenta system, fluorescently tagged polyamidoamine (PAMAM) dendrimers (5–6 nm size range) exhibited a low but consistent rate of transfer from maternal to fetal circulation [107].

Models to study the vaginal barrier. 3D models of the vaginal lining using the immortalized vaginal epithelial cell line V19I have been developed for high-throughput toxicity testing of microbicides [108], and cervicovaginal mucous has been used to study NP uptake kinetics for vaginal drug delivery [55].

16.5.2 In vivo Models to Study Impacts on the Female Reproductive Tract

Most animal studies that are aimed at assessing reproductive toxicity of nanomaterials on females involve pregnant rats or mice. Endpoint measurements are usually biodistribution of NPs in the reproductive organs, the placenta, and the fetus, as well as birth outcomes.

Exposure of pregnant rats to silica NPs reduced weight gain in dams and their offspring but had no effect on other parameters of physical development of the newborn rats and caused no teratogenic effects [109]. Silver NPs intravenously injected into pregnant mice accumulated to some extent in the visceral yolk sac and the endometrium but did not accumulate significantly in embryos [110]. Similarly, intravenously injected gold NPs (40 nm) were not detected in the ovaries, placentas, and fetuses of pregnant mice [111]. These findings suggest that silver and gold NPs do not penetrate the placental barrier, at least in mice. This is a promising first result, but it is difficult to extrapolate these results directly to humans due to the different structure of the rodent compared to the human placenta [112, 113] and different nanomaterials may not behave in the same way.

CNTs and fullerenes (e.g., C_{60}) are a group of nanomaterials that have attracted a lot of attention due to their versatility in terms of applications but also regarding their potential toxicity. CNT effects on the respiratory tract have been studied to a large extent because of the resemblance of CNTs to asbestos fibers and the high potential for inhalational exposure [114] (see Chapter 5). Their potential toxicity to the reproductive system has only been addressed in recent years. Multiwalled carbon nanotubes (MWCNTs) in 2% carboxymethyl cellulose were administered intraperitoneally (i.p.) or intratracheally (i.t.) to pregnant mice on day 9 of gestation [115]. Fetuses were removed on day 18 of gestation and examined for anomalies. Malformations were observed in all groups treated with MWCNTs administered i.p. (2–5 mg/kg), and early resorption of fetuses increased with dose. In contrast, malformations and increased early resorptions were only observed in the groups treated with the higher doses (4 or 5 mg/kg) but not the lower dose (3 mg/kg) administered i.t. In both groups the proportion of malformations increased with dose. Whilst some maternal MWCNT-induced toxicity was observed in the i.p. group this was not the case for the i.t. group, indicating a teratogenic effect [115]. In contrast, a study by Lim et al. [116] demonstrated that administration of MWCNTs to pregnant rats by oral gavage of 0, 40, 100 and 1000 mg/kg/day from gestational day 6–19 failed to show any embryo-fetal toxicity when examined after euthanasia on gestational day 20.

In another study [117], pristine and oxidized single-walled carbon nanotubes (SWCNTs) were assessed for embryotoxicity in mice. SWCNTs at 10 ng–30 μg per mouse were administered by retrobulbar injection on gestation day 5.5. The animals were sacrificed 10 days later, and the uteri, placentas, and fetuses were examined. Exposure to oxidized SWCNTs resulted in a high percentage of early miscarriages and malformations, whereas exposure to pristine SWCNTs resulted in a lower percentage, with the lowest effective dose being 100 ng/mouse. Morphological alterations and increased reactive oxygen species (ROS) were observed in placentas of malformed fetuses only, and similarly, increased ROS were detected in the malformed fetuses but not maternal tissues. The embryonic stem cell test (EST) predicted the embryotoxicity of the SWCNTs in vitro, indicating the oxidized SWCNTs as more toxic.

A small study of effects of exposure to C_{60} fullerenes in pregnant mice [118] demonstrated that i.p. exposure to a very high dose of C_{60} (137 mg/kg) at day 10 of gestation resulted in embryonic death with severe malformation by 18 hours after exposure. Exposure to lower doses (50–25 mg/kg) still caused malformations with clear evidence of C_{60} distribution into the embryo, whereas the polyvinylpyrrolidone (PVP) vehicle and water controls caused no malformations. A study in pregnant rats clearly demonstrated that radiolabeled [^{14}C] C_{60} (0.3 mg/kg) given by tail vein injection on gestation day 15 was distributed to the reproductive tract, placenta, and fetuses within 24–48 hours of administration [119]. Exposure of lactating rats on postnatal day 8 resulted in distribution of the C_{60} to the milk and subsequently to the gastrointestinal tract of suckling pups.

Engineered TiO_2 NPs have been shown to cross the murine placenta and accumulate in the fetus [120, 121]. Shimizu et al. [120] administered a subcutaneous injection of TiO_2 NPs (1 mg/mL) to pregnant mice on gestational days 6, 9, 12, and 15. Brain tissue from male fetuses (embryonic day 16) and pups (postnatal days 2, 7, 14, and 21) showed altered gene expression compared to unexposed controls. As well as demonstrating placental transfer, a study in pregnant mice in which they received subcutaneous injections of 1 mg/mL TiO_2 NPs on days 3, 7, 10, and 14 postcoitum showed evidence of apoptotic effects in the olfactory bulb of six-week old gestationally exposed male offspring [121]. In another murine study, pregnant females received a single oral dose of 10 mg/kg, 100 mg/kg, or 1000 mg/kg of either TiO_2 or silver NPs or the vehicle control on gestational day 9 [122]. Fetuses were evaluated on gestational day 19 (one day prior to spontaneous delivery). A single oral dose of 100 mg/kg or 1000 mg/kg of TiO_2 NPs in pregnant mice resulted in an increase in fetal deformities and mortality. In contrast, whilst silver NPs, at a 10 mg/kg dose only, also increased fetal mortality, there was no increase in fetal deformities at any exposure dose.

Further studies have demonstrated that gestational exposure can impact neurodevelopment. Mice were administered 42 mg/m^3 aerosolized TiO_2 NPs formulated as a powder for one hour/day via inhalation on gestational days 8–18, and gestationally exposed offspring showed moderate neurobehavioural alterations [123].

Studies on nonpregnant females show very little effect on the reproductive organs. Inhalational exposure of rats to nanosized silver did not result in histopathological changes of reproductive organs, even in animals exposed to high concentrations (1.2×10^6 particles/cm^3, 1.8×10^9 nm^2/cm^2) of silver NPs [124]. Similarly, no histopathological changes were observed in the endometrium and myometrium of rats exposed orally to gellan gum–capped gold NPs [125].

16.5.3 In vitro Models to Study Effects on the Male Reproductive System

As a result of positive biodistribution studies greater significance has been given to earlier in vitro testicular toxicity/mechanistic studies and as such has resulted in a number of more recent studies conducted using NPs. These in vitro models provide a useful platform to investigate the cellular interactions with NPs and individual cell responses, something not easily assessed in vivo. Predominantly this research has relied on, and has been investigated using, immortalized cell lines, usually with cells identified as important in ensuring normal testicular function, for example, Leydig, Sertoli, and germ cells. A number of studies have utilized primary cells; however, these to date are more limited in number and have tended to focus on the impact of nanomaterials on human spermatozoa.

Models to study steroidgenesis. An important function of normal reproductive health in both males and females is the regulation of steroidogenic activity. This area of research into understanding the impact of NPs on hormone regulation has so far been ignored. Only one in vitro study has assessed cells producing steroids, in particular using a murine Leydig cell line model (TM3) to assess the impact of NPs on androgens synthesis. The study concluded that NPs could affect steroidogenesis through the overexpression of some genes such as the steroidogenic acute regulatory gene (*StAR*) [126].

Models to study sperm development. Many in vitro studies have focused their attention on assessing the impact of nanomaterials on germ/sperm cells. An immortalized spermatogonial stem cell model (C18-4) has shown that regardless of the cell viability marker assessed (lactate dehydrogenase leakage, mitochondrial function, apoptosis, or necrosis assay), the NP form of an element (15 nm silver, 30 nm aluminum, and 30 nm molybdenum) displayed greater toxicity than the soluble form (silver carbonate, sodium molybdate, and aluminum chloride) [127]. Another study by Braydich-Stolle et al. [128] using the same spermatagonial stem cell line (C18-4) set out to examine the impact of silver NP size and coating, concluding that a significant decline in cell proliferation was dependent on these two factors. The study analyzed the traditional toxicity assays to assess the mechanism driving this effect. However, ROS production, a known particle-driven effect, and/or apoptosis did not seem to play a major role. Using this cell line, the authors were able to deduce that the silver NPs were interacting with cellular pathways, specifically Fyn kinase, a downstream part of Ret and glial cell line–derived neurotrophic factor (GDNF)-mediated signaling linked to proliferation.

Another study by Asare et al. [129] utilized cells from a human testicular embryonic carcinoma NT2 and isolated primary testicular cells from two strains of C57BL6 mice, one WT and the other genetically modified to serve as a representative model for human male reproductive toxicity. The cells were exposed to 21 nm titanium dioxide particles and 20 nm and 200 nm silver particles in a time and dose response study. Both sizes of silver NPs were shown to inhibit normal cell function and cause greater cell death than the titanium dioxide. These cell-based studies allow nanotoxicologists to investigate the mechanisms that underlie the physiological response, and in this particular study, DNA damage in the human cells was shown to have increased in a concentration-dependent manner after exposure to the 200 nm silver particles.

Models to study male infertility. A few in vitro studies have been conducted to assess the impact of nanomaterials on male fertility. These have focused on assessing human spermatozoa samples. One study assessed 9 nm gold NPs at a concentration of 44 ppm on healthy human sperm. The control sperm motility rate of 95% decreased postexposure to gold NPs to 75%. The study did not investigate toxicity, except for motility; however, a morphological assessment of samples was conducted, highlighting fragmentation of spermatazoa and spermatazoa with gold NPs in the head and tail sections [130]. An earlier study using human spermatozoa found contradictory results, showing that NPs had no effect on the ability of spermatozoa to initiate an acrosomic reaction and did not affect their motility [131]. The later study did have some biases and is difficult to interpret [130].

A study by Makhluf et al. [132] assessed whether iron oxide NPs coated with polyvinyl alcohol (PVA) were able to penetrate sperm cells without affecting motility or the spermatozoa's ability to undergo an acrosome reaction, two functions that are crucial for successful fertilization. Using bovine sperm and transmission electron microscopy (TEM) to assess and validate findings the acrosome reaction was assessed to determine ability of the sperm to fertilize the egg. Analysis showed the particles crossed the sperm plasma membrane and bound to the acrosome, and a cross section of the tail revealed high particle binding to the sperm mitochondria. They concluded that particle-treated cells were not damaged and that sperm function was normal after treatment. This study offers the potential opportunity to control movement and target these magnetic sperm cells in an animal body. The importance of showing this is to be able to target these magnetic sperm cells for biomedical and diagnostic applications. This research shows great promise; however, little toxicology data is available on the impact of NPs on fertilization and their presence in the fertilized oocyte.

16.5.4 In vivo Models to Study Effects on the Male Reproductive System

With respect to reproductive physiology, a number of studies have attempted to assess biodistribution of administered NPs to the testes, utilizing imaging techniques such as TEM, confocal microscopy, and mass spectrometry. To date these studies have only been conducted in rodent models [133–141]. This research has raised expectations for the use of nanomaterials in medical applications specifically targeting male reproductive physiology. It should be noted that potential fears about their ability to impair human reproduction, especially male fertility, may be an unwanted side effect of deliberate exposure [4, 142].

Amorphous silica NPs are being utilized in an increasing number of applications such as medicine, cosmetics, and foods. In a study conducted by Morishita et al. [143] mice were injected intravenously on two consecutive days with 70 nm silica particles and compared to exposures of 300 nm silica particles. The intratesticular distribution of these particles 24 hours after the second injection was analyzed via TEM. The silica NPs were detected within Sertoli cells and spermatocytes, including in the nuclei of spermatozoa. The larger silica particles were not observed in the testes. The study concluded that silica NPs were able to deposit in the testes and to penetrate the blood–testes barrier and the nuclear membranes of spermatozoa without producing noticeable testicular injury. Additionally, repeated intravenous injections of CNTs, used in medical applications such as drug delivery and biosensing, in mice caused testicular damage that was reversible after injections were discontinued, and did not affect fertility [134].

The impact of inhaled NPs on the reproductive system has been investigated in a number of animal studies. In mice, i.t. administered carbon black NPs seem to impair spermatogenesis while increasing testosterone synthesis. A study by Yoshida et al. [144] investigated the effect of i.t. administration of three sizes of carbon black NPs—14 nm, 56 nm, and 95 nm (0.1 mg/mouse, 10 times every week)—in ICR male mice to investigate their adverse effects on reproductive function. After exposure testosterone levels were shown to increase above normal levels with the possibility of altered endocrine regulation. In rats, inhalation of a specific amount of diesel exhaust NPs disrupted the endocrine activity of the male reproductive system by increasing plasma and testicular testosterone and inhibin levels [145]. This could be explained by increased expression of the StAR protein and cytochrome P450 [146]. The study by Yoshida et al. [144] carried out histological examination of the mice testes and identified vacuolation in the Sertoli cells, suggesting that NPs were able to alter the morphology of the Sertoli cells and possibly interfere with their function, primarily as an endocrine regulator and as the blood—testes barrier. On the basis of the knowledge that NPs can translocate from the lung and deposit in the testes studies have been conducted examining inhalation of a more medically relevant element, cerium oxide (CeO_2). CeO_2 NPs have been shown to mimic the properties of superoxide dismutase (SOD), an endogenous cellular defense molecule effective against the buildup of excess free radicals. CeO_2 is therefore thought to have the ability to provide protection from ROS more efficiently than SOD itself. In a 28-day study in which rats were exposed to CeO_2 NPs, the NPs distributed to tissues other than the lungs, such the testes and the epididymis, after a single and after repeated six-hour exposure [147].

Biodistribution of NPs to the testes has not just been established in adult models. A number of studies have demonstrated that blood-borne NPs can reach the fetuses of pregnant rats [148] and that there is evidence of health effects in the offspring exposed to diesel exhaust particles via the mother [123, 149, 150], which include effects on fetal testes and Sertoli cell development, as well as more long-term effects such as impairment of spermatogenesis in adulthood [151–153]. On the basis of this knowledge, a study conducted by Takeda et al. [121] exposed pregnant mice via subcutaneous injection to TiO_2 NPs (25–70 nm in particle size) at 100 g/mouse/day on gestational days 3, 7, 10, and 14. Male offspring were autopsied at postnatal day 4 or during week 6. Lower body weights were found among offspring of dams exposed to TiO_2. Aggregates of TiO_2 NPs (100–200 nm) were detected in Leydig cells, Sertoli cells, and spermatids in the testes of pups on postnatal day 4 and week 6. The study reports a decreased number of Sertoli cells in disorganized and disrupted seminiferous tubules and tubule lumens with few mature sperm, and decreases in daily sperm production and epididymal sperm motility were observed at postnatal week 6 in pups of the TiO_2-treated group.

Many of the studies conducted to date have attempted to either localize nanomaterials in the testes or assess their impact on normal cellular function. More studies are arising that look to harness nanotechnology as a medical implement to alter reproductive health. For example, contraceptive techniques that target the vas deferens have received much attention due to good safety, efficacy, and reversibility. Studies conducted by Chen et al. [154] demonstrated that a filtering-type nano–copper complex/polymer composite placed in the vas deferens of both 20 dogs and 40 rabbits was efficacious, with no sperm in the semen 12 months postimplantation, which was comparable to the vasectomized controls. The procedure was deemed safe, with no structural changes or increased cell death in the testes, epididymis, or vas deferens and no significant changes in the quantities of copper and zinc ions in semen. It was also concluded that this technology could be used as a reversible male contraceptive device, as a 60%–80% pregnancy rate was observed after removal of the nanocomposite from the vas deferens.

Although the studies highlighted here have generally focused on animals (mice and rats), with different exposure routes with NPs of varied physicochemical properties, all of the studies conducted have concluded that NPs are able to reach the testes. This raises the requirement for more work to be conducted to assess the impact of nanotechnology, in particular nanomedicines, on male reproductive health.

16.6 Parameters That Influence Nanoparticle Effects on the Reproductive System

16.6.1 The Female Reproductive System

The most important parameters that influence the effects that NPs have on a biological system are size and surface chemistry. Very often these parameters also affect each other, for example, the surface chemistry of the particle influences its hydrodynamic radius and so ultimately its size in a biological medium [155]. The effect of particle size on placental transport has been demonstrated in pregnant mice injected with fluorescently labeled silica NPs. Small (70 nm) but not large (300 nm) particles penetrated the placenta and accumulated in trophoblast cells and in the fetal liver and brain [156]. Also, the combined effect of NP size and surface chemistry on transfer of NPs across extraembryonal tissues has been demonstrated in mouse embryos isolated from pregnant mice. After removal of the embryos from the womb, different sizes of fluorescently labeled amine- and carboxyl-modified PS beads were injected via the extraembryonic tissues. It was observed that larger (200 nm) amine-modified PS beads translocated to a greater extent to the embryo than smaller (20 nm, 100 nm) carboxyl-modified particles [157]. Not only the effects on placental transfer but also the effects on other endpoints of reproductive nanotoxicity have been studied. For example, short-term exposure of mouse oocytes to CdSe-core quantum dots resulted in long-term toxic effects on embryogenesis, including oocyte maturation, fertilization, and subsequent embryonic development. However, this cytotoxic effect was prevented by surface modification with coatings such as ZnS [158], highlighting the influence of surface chemistry on NP toxicity.

16.6.2 The Male Reproductive System

As in the case with the female system the male is susceptible to the influence of the effects of NP size and surface chemistry. NPs hardly contribute to particle mass and are more accurately quantified by their number or surface (size, chemistry, and charge). It is well documented that smaller nanosized materials are more likely to penetrate cells and tissues at primary exposure sites, such as the lung or gastrointestinal tract. They may also enter the bloodstream from where they may translocate into other organs, allowing interaction with secondary organ-specific cells and tissues. The testis has a good vascular network, predominantly to maintain a low testicular temperature. However, this increases the chance of deposition of NPs to the testes via the blood circulation. Park et al. [159] exemplified the significance of this in a study in which mice were treated with 1 mg/kg silver NPs for 14 days via oral administration. The smaller particles (22 nm, 42 nm, and 71 nm) distributed to the testis and other organs, including the brain, lung, liver, and kidney, while the larger (323 nm) particles were not detected in any of the other tissues assessed.

16.7 Conclusions

Undoubtedly, NPs have tremendous potential in the biomedical field, especially regarding improved diagnosis and therapy. However, possible adverse effects of NPs on human health and on the environment need to be carefully assessed before large-scale application is implemented. The potential of NPs to interact with cells of the male and female reproductive systems and to disrupt their normal function is particularly poorly understood. It seems that particle size and surface chemistry are major factors that determine biological effects of NPs, both beneficial as well as toxic. Important endpoint measurements for reproductive toxicity of NPs are effects on fertility (e.g., hormone secretion, oocyte maturation, sperm motility), implantation, and placental transport and toxicity. Robust and relevant in vitro models are needed, particularly to study placental transport, because there are very few animal models that mimic the structure of the human placenta.

Relevant exposure routes of women to medically relevant NPs are likely to be during therapeutic intervention (e.g., oral, intravenous, and/or vaginal application of NP drugs) or for diagnostic purposes (e.g., injection of magnetic NPs as contrast agents). Knowledge of biodistribution and bioaccumulation of NPs is important, especially in women of childbearing age. NPs may persist in tissues, and delayed exposure during pregnancy after NP application may be possible. However, it is more likely that pregnant women may be exposed to medical NPs during the early stages of pregnancy, when they may not be aware of it, which is why careful assessment of transplacental passage of NPs is important. Crossing of the placental barrier and interference with its function are a major concern, but they may also bear great potential. There is good evidence that NPs can be tailored to avoid toxicity and maximize positive treatment outcomes. Therefore, it may be possible to design NPs especially tailored for use in pregnant women, for example, by encapsulating drugs in particles that do not cross the placental barrier or impair its function. This means that a variety of essential drugs (e.g., antiepileptic drugs) could be made available to pregnant women without introducing risks to the unborn child.

To date, only very few studies focus on the potential impact on male reproductive health. The testis provides researchers an opportunity to explore and study one of the few physiological human barriers and assess the impact of NP exposure from an aspect of both toxicology and hazard assessment to the opposite question of efficacy and distribution of new nanomedicines. Mass dose quantification of NPs correlated to an observed biological response has traditionally been used as a useful toxicological metric. The challenges to assess the mass deposited within testicular tissues and relate this to an observed effect on key cells and functions, such as the blood–testes barrier, may be difficult due to the cost and specialist nature of present technologies. The focus of future male reproductive risk assessment will require more focus on particle clearance rates and longer exposures because to date most studies have addressed acute exposures only. With regard to NP characteristics, size, elemental composition, and charge have been shown to each influence the response in the testes, highlighting the importance of considering these parameters in the development of nanomedicines.

Finally, to achieve optimal targeted design and minimal toxic side effects, systematic characterization of NPs used in in vitro and in vivo studies is essential so that observed effects can be directly related to particular NP properties or combinations of properties. Further development of in vivo–like 3D cell culture and coculture models as opposed to standard 2D models is also very much recommended. The studies available to date provide an ever-growing body of evidence. However, due to their low number and lack of standardization across them, it is difficult to draw concrete conclusions, with some studies even leading to contradictory results. Development of good practice or standard operating procedures for NP testing would be very beneficial, so results from different labs may be directly compared and collated.

References

1. Zhao, J., and Castranova, V. (2011). Toxicology of nanomaterials used in nanomedicine. *J. Toxicol. Environ. Health B Crit. Rev.*, **14**, pp. 593–632, doi: 10.1080/10937404.2011.615113.
2. Maurer-Jones, M. A., Bantz, K. C., Love, S. A., Marquis, B. J., and Haynes, C. L. (2009). Toxicity of therapeutic nanoparticles. *Nanomedicine (London)*, **4**, pp. 219–241, doi: 10.2217/17435889.4.2.219.
3. Kunzmann, A., Andersson, B., Thurnherr, T., Krug, H., Scheynius, A., and Fadeel, B. (2011). Toxicology of engineered nanomaterials: focus on biocompatibility, biodistribution and biodegradation. *Biochim. Biophys. Acta*, **1810**, pp. 361–373, doi: 10.1016/j.bbagen.2010.04.007.
4. Ema, M., Kobayashi, N., Naya, M., Hanai, S., and Nakanishi, J. (2010). Reproductive and developmental toxicity studies of manufactured nanomaterials. *Reprod. Toxicol.*, **30**, pp. 343–352, doi: 10.1016/j.reprotox.2010.06.002.
5. Kumar, V., Kumari, A., Guleria, P., and Yadav, S. K. (2012). Evaluating the toxicity of selected types of nanochemicals. *Rev. Environ. Contam. Toxicol.*, **215**, pp. 39–121, doi: 10.1007/978-1-4614-1463-6_2.
6. Park, M. V., Annema, W., Salvati, A., Lesniak, A., Elsaesser, A., Barnes, C., McKerr, G., Howard, C. V., Lynch, I., Dawson, K. A., Piersma, A. H., and de Jong, W. H. (2009). In vitro developmental toxicity test detects inhibition of stem cell differentiation by silica nanoparticles. *Toxicol. Appl. Pharmacol.*, **240**, pp. 108–116, doi: 10.1016/j.taap.2009.07.019.

7. Saunders, M. (2007). Placental biological barrier models for evaluation of nanoparticle transfer, in *Nanotoxicology: Characterization, Dosing and Health Effects*, eds. Monteiro-Riviere, N. A., and Tran, C. L. (Informa Healthcare USA, New York), pp. 115–126.
8. Bhattacharya, R., and Mukherjee, P. (2008). Biological properties of "naked" metal nanoparticles. *Adv. Drug Delivery Rev.*, **60**, pp. 1289–1306.
9. Boisselier, E., and Astruc, D. (2009). Gold nanoparticles in nanomedicine: preparations, imaging, diagnostics, therapies and toxicity. *Chem. Soc. Rev.*, **38**, pp. 1759–1782, doi: 10.1039/b806051g.
10. Daniel, M. C., and Astruc, D. (2004). Gold nanoparticles: assembly, supramolecular chemistry, quantum-size-related properties, and applications toward biology, catalysis, and nanotechnology. *Chem. Rev.*, **104**, pp. 293–346, doi: 10.1021/cr030698+.
11. Kisailus, D., Najarian, M., Weaver, J. C., and Morse, D. E. (2005). Functionalized gold nanoparticles mimic catalytic activity of a polysiloxane-synthesizing enzyme. *Adv. Mater.*, **17**, pp. 1234–1239.
12. Murphy, C. J., Gole, A. M., Stone, J. W., Sisco, P. N., Alkilany, A. M., Goldsmith, E. C., and Baxter, S. C. (2008). Gold nanoparticles in biology: beyond toxicity to cellular imaging. *Acc. Chem. Res.*, **41**, pp. 1721–1730, doi: 10.1021/ar800035u.
13. Yu, A., Liang, Z., Cho, J., and Caruso, F. (2003). Nanostructured electrochemical sensor based on dense gold nanoparticle films. *Nano Lett.*, **3**, pp. 1203–1207.
14. Chen, X., and Schluesener, H. J. (2008). Nanosilver: a nanoproduct in medical application. *Toxicol. Lett.*, **176**, pp. 1–12, doi: 10.1016/j.toxlet.2007.10.004.
15. Jain, J., Arora, S., Rajwade, J. M., Omray, P., Khandelwal, S., and Paknikar, K. M. (2009). Silver nanoparticles in therapeutics: development of an antimicrobial gel formulation for topical use. *Mol. Pharm.*, **6**, pp. 1388–1401.
16. Kim, J. S., Kuk, E., Yu, K. N., Kim, J. H., Park, S. J., Lee, H. J., Kim, S. H., Park, Y. K., Park, Y. H., and Hwang, C. Y. (2007). Antimicrobial effects of silver nanoparticles. *Nanomed.: Nanotechnol., Biol. Med.*, **3**, pp. 95–101.
17. Monteiro, D. R., Gorup, L. F., Takamiya, A. S., Ruvollo-Filho, A. C., Camargo, E. R., and Barbosa, D. B. (2009). The growing importance of materials that prevent microbial adhesion: antimicrobial effect of medical devices containing silver. *Int. J. Antimicrob. Agents*, **34**, pp. 103–110.

18. Rai, M., Yadav, A., and Gade, A. (2009). Silver nanoparticles as a new generation of antimicrobials. *Biotechnol. Adv.*, **27**, pp. 76–83.

19. Singh, M., Singh, S., Prasad, S., and Gambhir, I. (2008). Nanotechnology in medicine and antibacterial effect of silver nanoparticles. *Digest J. Nanomater. Biostruct.*, **3**, pp. 115–122.

20. Chen, Y. C., Lee, W. F., Tsai, H. H., and Hsieh, W. Y. (2012). Paclitaxel and iron oxide loaded multifunctional nanoparticles for chemotherapy, fluorescence properties, and magnetic resonance imaging. *J. Biomed. Mater. Res. A*, **100**, pp. 1279–1292.

21. Chertok, B., Moffat, B. A., David, A. E., Yu, F., Bergemann, C., Ross, B. D., and Yang, V. C. (2008). Iron oxide nanoparticles as a drug delivery vehicle for MRI monitored magnetic targeting of brain tumors. *Biomaterials*, **29**, pp. 487–496.

22. Gupta, A. K., and Gupta, M. (2005). Synthesis and surface engineering of iron oxide nanoparticles for biomedical applications. *Biomaterials*, **26**, pp. 3995–4021.

23. Laurent, S., Forge, D., Port, M., Roch, A., Robic, C., Vander Elst, L., and Muller, R. N. (2008). Magnetic iron oxide nanoparticles: synthesis, stabilization, vectorization, physicochemical characterizations, and biological applications. *Chem. Rev.*, **108**, pp. 2064.

24. Lee, H., Lee, E., Do Kyung, K., Jang, N. K., Jeong, Y. Y., and Jon, S. (2006). Antibiofouling polymer-coated superparamagnetic iron oxide nanoparticles as potential magnetic resonance contrast agents for in vivo cancer imaging. *J. Am. Chem. Soc.*, **128**, pp. 7383–7389.

25. Pankhurst, Q. A., Connolly, J., Jones, S., and Dobson, J. (2003). Applications of magnetic nanoparticles in biomedicine. *J. Phys. D: Appl. Phys.*, **36**, p. R167.

26. Peng, X. H., Qian, X., Mao, H., and Wang, A. Y. (2008). Targeted magnetic iron oxide nanoparticles for tumor imaging and therapy. *Int. J. Nanomed.*, **3**, p. 311.

27. Chen, A. M., Zhang, M., Wei, D., Stueber, D., Taratula, O., Minko, T., and He, H. (2009). Co-delivery of doxorubicin and Bcl-2 siRNA by mesoporous silica nanoparticles enhances the efficacy of chemotherapy in multidrug-resistant cancer cells. *Small*, **5**, pp. 2673–2677.

28. Cheng, S. H., Lee, C. H., Yang, C. S., Tseng, F. G., Mou, C. Y., and Lo, L. W. (2009). Mesoporous silica nanoparticles functionalized with an oxygen-sensing probe for cell photodynamic therapy: potential cancer theranostics. *J. Mater. Chem.*, **19**, pp. 1252–1257.

29. De Jong, W. H., and Borm, P. J. A. (2008). Drug delivery and nanoparticles: applications and hazards. *Int. J. Nanomed.*, **3**, p. 133.

30. Lee, J. E., Lee, N., Kim, T., Kim, J., and Hyeon, T. (2011). Multifunctional mesoporous silica nanocomposite nanoparticles for theranostic applications. *Acc. Chem. Res.*, **44**, pp. 893–902.

31. Choi, A. O., Cho, S. J., Desbarats, J., Lovrić, J., and Maysinger, D. (2007). Quantum dot-induced cell death involves Fas upregulation and lipid peroxidation in human neuroblastoma cells. *J. Nanobiotechnol.*, **5**, p. 1, doi: 10.1186/1477-3155-5-1.

32. Pons, T., Pic, E., Lequeux, N., Cassette, E., Bezdetnaya, L., Guillemin, F., Marchal, F., and Dubertret, B. (2010). Cadmium-free CuInS2/ZnS quantum dots for sentinel lymph node imaging with reduced toxicity. *ACS Nano*, **4**, pp. 2531–2538, doi: 10.1021/nn901421v.

33. Brewer, E., Coleman, J., and Lowman, A. (2011). Emerging technologies of polymeric nanoparticles in cancer drug delivery. *J. Nanomater.*, **2011**, p. 1.

34. Chan, J. M., Zhang, L., Yuet, K. P., Liao, G., Rhee, J. W., Langer, R., and Farokhzad, O. C. (2009). PLGA-lecithin-PEG core-shell nanoparticles for controlled drug delivery. *Biomaterials*, **30**, pp. 1627–1634.

35. Cho, K., Wang, X., Nie, S., and Shin, D. M. (2008). Therapeutic nanoparticles for drug delivery in cancer. *Clin. Cancer Res.*, **14**, pp. 1310–1316.

36. Kamaly, N., Xiao, Z., Valencia, P. M., Radovic-Moreno, A. F., and Farokhzad, O. C. (2012). Targeted polymeric therapeutic nanoparticles: design, development and clinical translation. *Chem. Soc. Rev.*, **41**, pp. 2971–3010, doi: 10.1039/c2cs15344k.

37. Patil, Y. B., Toti, U. S., Khdair, A., Ma, L., and Panyam, J. (2009). Single-step surface functionalization of polymeric nanoparticles for targeted drug delivery. *Biomaterials*, **30**, pp. 859–866.

38. des Rieux, A., Fievez, V., Garinot, M., Schneider, Y. J., and Preat, V. (2006). Nanoparticles as potential oral delivery systems of proteins and vaccines: a mechanistic approach. *J. Controlled Rel.*, **116**, pp. 1–27, doi: 10.1016/j.jconrel.2006.08.013.

39. Patil, Y., and Panyam, J. (2009). Polymeric nanoparticles for siRNA delivery and gene silencing. *Int. J. Pharm.*, **367**, pp. 195–203.

40. Chen, Y., Wan, Y., Wang, Y., Zhang, H., and Jiao, Z. (2011). Anticancer efficacy enhancement and attenuation of side effects of doxorubicin with titanium dioxide nanoparticles. *Int. J. Pharm.*, **6**, p. 2321.

41. Dissette, V., Bozzi, P., Bignozzi, C. A., Dalpiaz, A., Ferraro, L., Beggiato, S., Leo, E., Vighi, E., and Pasti, L. (2010). Particulate adducts based on sodium risedronate and titanium dioxide for the bioavailability enhancement of oral administered bisphosphonates. *Eur. J. Pharm. Sci.*, **41**, pp. 328–336.

42. Fadeel, B., and Garcia-Bennett, A. E. (2010). Better safe than sorry: understanding the toxicological properties of inorganic nanoparticles manufactured for biomedical applications. *Adv. Drug Delivery Rev.*, **62**, pp. 362–374.

43. Li, Q., Wang, X., Lu, X., Tian, H., Jiang, H., Lv, G., Guo, D., Wu, C., and Chen, B. (2009). The incorporation of daunorubicin in cancer cells through the use of titanium dioxide whiskers. *Biomaterials*, **30**, pp. 4708–4715.

44. Zhang, H., Wang, C., Chen, B., and Wang, X. (2012). Daunorubicin-TiO2 nanocomposites as a "smart" pH-responsive drug delivery system. *Int. J. Nanomed.*, **7**, p. 235.

45. Koyama, S., Haniu, H., Osaka, K., Koyama, H., Kuroiwa, N., Endo, M., Kim, Y. A., and Hayashi, T. (2006). Medical application of carbon-nanotube-filled nanocomposites: the microcatheter. *Small*, **2**, pp. 1406–1411, doi: 10.1002/smll.200500416.

46. Liang, F., and Chen, B. (2010). A review on biomedical applications of single-walled carbon nanotubes. *Curr. Med. Chem.*, **17**, pp. 10–24.

47. Saunders, M. (2009). Transplacental transport of nanomaterials. *Nanomed. Nanobiotechnol.*, **1**, pp. 671–684, doi: 10.1002/wnan.053.

48. Okamura, H., Katabuchi, H., and Nagai, R. (1997). Ultrastructure of human ovulation: histofunctional parameters, in *Microscopy of Reproduction and Development*, ed. Motta, P. M. (Antonio Delfino Editore, Rome, Italy), pp. 155–161.

49. Okamura, H., and Katabuchi, H. (2004). Pathophysiological dynamics of human ovarian surface epithelial cells in epithelial ovarian carcinogenesis. *Int. Rev. Cytol.*, **242**, pp. 1–54.

50. Matzuk, M. M., Burns, K. H., Viveiros, M. M., and Eppig, J. J. (2002). Intercellular communication in the mammalian ovary: oocytes carry the conversation. *Sci. Signaling*, **296**, p. 2178.

51. Jabbour, H. N., Kelly, R. W., Fraser, H. M., and Critchley, H. O. (2006). Endocrine regulation of menstruation. *Endocr. Rev.*, **27**, pp. 17–46.

52. Quayle, A. (2002). The innate and early immune response to pathogen challenge in the female genital tract and the pivotal role of epithelial cells. *J. Reprod. Immunol.*, **57**, pp. 61–79.

53. Pudney, J., Quayle, A. J., and Anderson, D. J. (2005). Immunological microenvironments in the human vagina and cervix: mediators of cellular immunity are concentrated in the cervical transformation zone. *Biol. Reprod.*, **73**, pp. 1253–63, doi: 10.1095/biolreprod.105.043133.
54. Herbst-Kralovetz, M. M., Quayle, A. J., Ficarra, M., Greene, S., Rose, W. A., Chesson, R., Spagnuolo, R. A., and Pyles, R. B. (2008). Quantification and comparison of toll-like receptor expression and responsiveness in primary and immortalized human female lower genital tract epithelia. *Am. J. Reprod. Immunol.*, **59**, pp. 212–224, doi: 10.1111/j.1600-0897.2007.00566.x.
55. Tang, B. C., Dawson, M., Lai, S. K., Wang, Y. Y., Suk, J. S., Yang, M., Zeitlin, P., Boyle, M. P., Fu, J., and Hanes, J. (2009). Biodegradable polymer nanoparticles that rapidly penetrate the human mucus barrier. *Proc. Natl. Acad. Sci. U S A*, **106**, pp. 19268–19273, doi: 10.1073/pnas.0905998106.
56. Bourget, P., Roulotm C., and Fernandez, H. (1995). Models for placental transfer studies of drugs. *Clin. Pharmacokinet.*, **28**, pp. 161–180.
57. Widmaier, E. P., Raff, H., and Strang, K. T. (2004). *Vander, Sherman & Luciano's Human Physiology: The Mechanisms of Body Functions* (McGraw Hill, New York).
58. Mital, P., Hinton, B. T., and Dufour, J. M. (2011). The blood-testis and blood-epididymis barriers are more than just their tight junctions. *Biol. Reprod.*, **84**, pp. 851–858.
59. Wang, R. S., Yeh, S., Tzeng, C. R., and Chang, C. (2009). Androgen receptor roles in spermatogenesis and fertility: lessons from testicular cell-specific androgen receptor knockout mice. *Endocr. Rev.*, **30**, pp. 119–132, doi: 10.1210/er.2008-0025.
60. Collins, L. L., and Chang, C. (2002). Androgens and the androgen receptor in male sex development and fertility, in *Androgens and Androgen Receptor: Mechanisms, Functions and Clinical Applications*, ed. Chang, C. (Kluwer Academic, Norwell), pp. 299–323.
61. Johnson, M. H., and Everitt, B. J. (2007). *Essential Reproduction* (Wiley-Blackwell, UK).
62. Yan, H. H., and Cheng, C. Y. (2005). Blood-testis barrier dynamics are regulated by an engagement/disengagement mechanism between tight and adherens junctions via peripheral adaptors. *Proc. Natl. Acad. Sci. U S A*, **102**, pp. 11722–11728, doi: 10.1073/pnas.0503855102.
63. Mendola, P., Messer, L. C., and Rappazzo, K. (2008). Science linking environmental contaminant exposures with fertility and reproductive health impacts in the adult female. *Fertil. Steril.*, **89**, pp. e81–e94.

64. Selevan, S. G., Borkovec, L., Slott, V. L., Zudova, Z., Rubes, J., Evenson, D. P., and Perreault, S. D. (2000). Semen quality and reproductive health of young Czech men exposed to seasonal air pollution. *Environ. Health Perspect.*, **108**, pp. 887–894.

65. Pincock, S. (2004). BMA says smoking harms reproductive capability. *Lancet*, **363**, pp. 628, doi: 10.1016/S0140-6736(04)15634-1.

66. Sharpe, R. M., and Irvine, D. S. (2004). How strong is the evidence of a link between environmental chemicals and adverse effects on human reproductive health? *BMJ*, **328**, pp. 447–451, doi: 10.1136/bmj.328.7437.447.

67. Pietroiusti, A. (2012). Health implications of engineered nanomaterials. *Nanoscale*, **4**, pp. 1231–1247, doi: 10.1039/c2nr11688j.

68. United Nations Economic Commission for Europe (UNECE) (2011). Reproductive toxicity, in *Globally Harmonized System of Classification and Labelling of Chemicals (GHS). Part 3. Health and Environmental Hazards* (ed. UNECE, New York/Geneva).

69. International Committee on Harmonization (ICH) (2005). *Detection of Toxicity to Reproduction for Medicinal Products & Toxicity to Male Fertility S5(R2). ICH Harmonized Tripartite Guildeline*, (ed. International Committee on Harmonization of Technical Requirements for Registration of Pharmaceuticals for Human Use, Geneva, Switzerland).

70. Makris, S. L., Kim, J. H., Ellis, A., Faber, W., Harrouk, W., Lewis, J. M., Paule, M. G., Seed, J., Tassinari, M., and Tyl, R. (2011). Current and future needs for developmental toxicity testing. *Birth Defects Res. B Dev. Reprod. Toxicol.*, **92**, pp. 384–394, doi: 10.1002/bdrb.20335.

71. Hannan, N. J., Paiva, P., Dimitriadis, E., and Salamonsen, La. (2010). Models for study of human embryo implantation: choice of cell lines? *Biol. Reprod.*, **82**, pp. 235–45, doi: 10.1095/biolreprod.109.077800.

72. Baldwin, W. S., Curtis, S. W., Cauthen, C. A., Risinger, J. I., Korach, K. S., and Barrett, J. C. (1998). BG-1 ovarian cell line: an alternative model for examining estrogen-dependent growth in vitro. *In Vitro Cell. Dev. Biol. Animal*, **34**, pp. 649–654.

73. Narisawa-Saito, M., Yoshimatsu, Y., Ohno, S.-i., Yugawa, T., Egawa, N., Fujita, M., Hirohashi, S., and Kiyono, T. (2008). An in vitro multistep carcinogenesis model for human cervical cancer. *Cancer Res.*, **68**, pp. 5699–705, doi: 10.1158/0008-5472.CAN-07-6862.

74. Stelzer, R., and Hutz, R. J. (2009). Gold nanoparticles enter rat ovarian granulosa cells and subcellular organelles, and alter in-vitro estrogen accumulation. *J. Reprod. Dev.*, **55**, pp. 685–690.

75. Liu, X., Qin, D., Cui, Y., Chen, L., Li, H., Chen, Z., Gao, L., Li, Y., and Liu, J. (2010). The effect of calcium phosphate nanoparticles on hormone production and apoptosis in human granulosa cells. *Reprod. Biol. Endocrinol.*, **8**, p. 32, doi: 10.1186/1477-7827-8-32.

76. Xu, G., Lin, S., Law, W.-C., Roy, I., Lin, X., Mei, S., Ma, H., Chen, S., Niu, H., and Wang, X. (2012). The invasion and reproductive toxicity of QDs-transferrin bioconjugates on preantral follicle in vitro. *Theranostics*, **2**, pp. 734–745.

77. Blackburn, W. H., Dickerson, E. B., Smith, M. H., McDonald, J. F., and Lyon, L. A. (2009). Peptide-functionalized nanogels for targeted siRNA delivery. *Bioconjug. Chem.*, **20**, pp. 960–968, doi: 10.1021/bc800547c.

78. Scully, R. E. (1995). Pathology of ovarian cancer precursors. *J. Cell Biochem. Suppl.*, **23**, pp. 208–218.

79. Lawrenson, K., Benjamin, E., Turmaine, M., Jacobs, I., Gayther, S., and Dafou, D. (2009). In vitro three-dimensional modelling of human ovarian surface epithelial cells. *Cell Proliferat.*, **42**, pp. 385–393, doi: 10.1111/j.1365-2184.2009.00604.x.

80. Sasaki, R., Narisawa-Saito, M., Yugawa, T., Fujita, M., Tashiro, H., Katabuchi, H., and Kiyono, T. (2009). Oncogenic transformation of human ovarian surface epithelial cells with defined cellular oncogenes. *Carcinogenesis*, **30**, pp. 423–431, doi: 10.1093/carcin/bgp007.

81. Larsen, B., Haag, M., Serkova, N., Shroyer, K., and Stoldt, C. (2008). Controlled aggregation of superparamagnetic iron oxide nanoparticles for the development of molecular magnetic resonance imaging probes. *Nanotechnology*, **19**, p. 265102.

82. Kuramoto, H., Tamura, S., and Notake, Y. (1972). Establishment of a cell line of human endometrial adenocarcinoma in vitro. *Am. J. Obstet. Gynecol.*, **114**, pp. 1012–1019.

83. das Neves, J., Michiels, J., Ariën, K. K., Vanham, G., Amiji, M., Bahia, M. F., and Sarmento, B. (2012). Polymeric nanoparticles affect the intracellular delivery, antiretroviral activity and cytotoxicity of the microbicide drug candidate dapivirine. *Pharmaceut. Res.*, **29**, pp. 1468–1484, doi: 10.1007/s11095-011-0622-3.

84. Mo, B., Vendrov, A. E., Palomino, W. A., DuPont, B. R., Apparao, K., and Lessey, B. A. (2006). ECC-1 cells: a well-differentiated steroid-responsive endometrial cell line with characteristics of luminal epithelium. *Biol. Reprod.*, **75**, pp. 387–394.

85. Lessey, B. A., Ilesanmi, A. O., Castelbaum, A. J., Yuan, L., Somkuti, S. G., Satyaswaroop, P. G., and Chwalisz, K. (1996). Characterization of the functional progesterone receptor in an endometrial adenocarcinoma

cell line (Ishikawa): Progesterone-induced expression of the [alpha] 1 integrin. *J. Steroid Biochem. Mol. Biol.*, **59**, pp. 31–39.

86. Lovely, L. P., Appa Rao, K. B., Gui, Y., and Lessey, Ba. (2000). Characterization of androgen receptors in a well-differentiated endometrial adenocarcinoma cell line (Ishikawa). *J. Steroid Biochem. Mol. Biol.*, **74**, pp. 235–241.

87. Nishida, M., Kasahara, K., Kaneko, M., Iwasaki, H., and Hayashi, K. (1985). Establishment of a new human endometrial adenocarcinoma cell line, Ishikawa cells, containing estrogen and progesterone receptors. *Nihon Sanka Fujinka Gakkai Zasshi*, **37**, pp. 1103–1111.

88. Schaefer, W. R., Fischer, L., Deppert, W. R., Hanjalic-Beck, A., Seebacher, L., Weimer, M., and Zahradnik, H. P. (2010). In vitro-Ishikawa cell test for assessing tissue-specific chemical effects on human endometrium. *Reprod. Toxicol. (Elmsford, N.Y.)*, **30**, pp. 89–93, doi: 10.1016/j.reprotox.2010.02.002.

89. Naciff, J. M., Khambatta, Z. S., Reichling, T. D., Carr, G. J., Tiesman, J. P., Singleton, D. W., Khan, S., and Daston, G. P. (2010). The genomic response of Ishikawa cells to bisphenol A exposure is dose- and time-dependent. *Toxicology*, **270**, pp. 137–149, doi: 10.1016/j.tox.2010.02.008.

90. Fichorova, R. N., Rheinwald, J. G., and Anderson, D. J. (1997). Generation of papillomavirus-immortalized cell lines from normal human ectocervical, endocervical, and vaginal epithelium that maintain expression of tissue-specific differentiation proteins. *Biol. Reprod.*, **57**, pp. 847–855.

91. Follows, S., Murlidharan, J., Massari, P., Wetzler, L., and Genco, C. (2009). Neisseria gonorrhoeae infection protects human endocervical epithelial cells from apoptosis via expression of host antiapoptotic proteins. *Infect. Immun.*, **77**, pp. 3602–3610.

92. Gali, Y., Ariën, K. K., Praet, M., Van den Bergh, R., Temmerman, M., Delezay, O., and Vanham, G. (2010). Development of an in vitro dual-chamber model of the female genital tract as a screening tool for epithelial toxicity. *J. Virol. Methods*, **165**, pp. 186–197, doi: 10.1016/j.jviromet.2010.01.018.

93. Chang, C.-H., Gupta, P., Michel, R., Loo, M., Wang, Y., Cardillo, T. M., and Goldenberg, D. M. (2010). Ranpirnase (frog RNase) targeted with a humanized, internalizing, anti-Trop-2 antibody has potent cytotoxicity against diverse epithelial cancer cells. *Mol. Cancer Ther.*, **9**, pp. 2276–2286, doi: 10.1158/1535-7163.MCT-10-0338.

94. Buckner, L. R., Schust, D. J., Ding, J., Nagamatsu, T., Beatty, W., Chang, T. L., Greene, S. J., Lewis, M. E., Ruiz, B., and Holman, S. L. (2011). Innate

immune mediator profiles and their regulation in a novel polarized immortalized epithelial cell model derived from human endocervix. *J. Reprod. Immunol.*, **92**(1–2), pp. 8–20.

95. Khan, J. A., Pillai, B., Das, T. K., Singh, Y., and Maiti, S. (2007). Molecular effects of uptake of gold nanoparticles in HeLa cells. *Chembiochem.: Eur. J. Chem. Biol.*, **8**, pp. 1237–1240, doi: 10.1002/cbic.200700165.

96. Pan, Y., Leifert, A., Ruau, D., Neuss, S., Bornemann, J., Schmid, G., Brandau, W., Simon, U., and Jahnen-Dechent, W. (2009). Gold nanoparticles of diameter 1.4 nm trigger necrosis by oxidative stress and mitochondrial damage. *Small*, **5**, pp. 2067–2076, doi: 10.1002/smll.200900466.

97. King, A., Thomas, L., and Bischof, P. (2000). Cell culture models of trophoblast II: trophoblast cell lines; a workshop report. *Placenta*, **21**, pp. S113–S119, doi: 10.1053/plac.1999.0526.

98. Cartwright, J. E., Holden, D. P., and Whitley, G. S. (1999). Hepatocyte growth factor regulates human trophoblast motility and invasion: a role for nitric oxide. *Br. J. Pharmacol.*, **128**, pp. 181–189, doi: 10.1038/sj.bjp.0702757.

99. Iacob, D., Cai, J., Tsonis, M., Babwah, A., Chakraborty, C., Bhattacharjee, R. N., and Lala, P. K. (2008). Decorin-mediated inhibition of proliferation and migration of the human trophoblast via different tyrosine kinase receptors. *Endocrinology*, **149**, pp. 6187–6197, doi: 10.1210/en.2008-0780.

100. Paiva, P., Salamonsen, L. A., Manuelpillai, U., and Dimitriadis, E. (2009). Interleukin 11 inhibits human trophoblast invasion indicating a likely role in the decidual restraint of trophoblast invasion during placentation. *Biol. Reprod.*, **80**, pp. 302–310, doi: 10.1095/biolreprod.108.071415.

101. Svinarich, D. M., Bitonti, O. M., Romero, R., and Gonik, B. (1996). Induction and posttranslational expression of cytokines in a first-trimester trophoblast cell line by lipopolysaccharide. *Am. J. Obstet. Gynecol.*, **175**, pp. 970–973.

102. Correia Carreira, S., Cartwright, L., Mathiesen, L., Knudsen, L. E., and Saunders, M. (2011). Studying placental transfer of highly purified non-dioxin-like PCBs in two models of the placental barrier. *Placenta*, **32**, pp. 283–291, doi: 10.1016/j.placenta.2010.12.024.

103. Saunders, M. (2009). Transplacental transport of nanomaterials. *Nanomed. Nanobiotechnol.*, **1**, pp. 671–684, doi: 10.002/wnan.053.

104. Cartwright, L., Poulsen, M. S., Nielsen, H. M., Pojana, G., Knudsen, L. E., Saunders, M., and Rytting, E. (2012). In vitro placental model

optimization for nanoparticle transport studies. *Int. J. Nanomed.*, **7**, pp. 497–510, doi: 10.2147/IJN.S26601.

105. Myllynen, P. K., Loughran, M. J., Howard, C. V., Walsh, A. A., and Vähäkangas, K. H. (2008). Kinetics of gold nanoparticles in the human placenta. *Reprod. Toxicol. (Elmsford, N.Y.)*, **26**, pp. 130–137, doi: 10.1016/j.reprotox.2008.06.008.

106. Wick, P., Malek, A., Manser, P., Meili, D., Maeder-Althaus, X., Diener, L., Diener, P. A., Zisch, A., Krug, H. F., and von Mandach, U. (2010). Barrier capacity of human placenta for nanosized materials. *Environ. Health Perspect.*, **118**, pp. 432–436, doi: 10.1289/ehp.0901200.

107. Menjoge, A. R., Rinderknecht, A. L., Navath, R. S., Faridnia, M., Kim, C. J., Romero, R., Miller, R. K., and Kannan, R. M. (2011). Transfer of PAMAM dendrimers across human placenta: prospects of its use as drug carrier during pregnancy. *J. Controlled Rel.*, **150**, pp. 326–338, doi: 10.1016/j.jconrel.2010.11.023.

108. Hjelm, B. E., Berta, A. N., Nickerson, C. A., Arntzen, C. J., and Herbst-Kralovetz, M. M. (2010). Development and characterization of a three-dimensional organotypic human vaginal epithelial cell model. *Biol. Reprod.*, **82**, pp. 617–627, doi: 10.1095/biolreprod.109.080408.

109. Durnev, A. D., Solomina, A. S., Daugel-Dauge, N. O., Zhanataev, A. K., Shreder, E. D., Nemova, E. P., Shreder, O. V., Veligura, V. A., Osminkina, L. A., Timoshenko, V. Y., and Seredenin, S. B. (2010). Evaluation of genotoxicity and reproductive toxicity of silicon nanocrystals. *Bull. Exp. Biol. Med.*, **149**, pp. 445–449.

110. Austin, C. A., Umbreit, T. H., Brown, K. M., Barber, D. S., Dair, B. J., Francke-Carroll, S., Feswick, A., Saint-Louis, M. A., Hikawa, H., Siebein, K. N., and Goering, P. L. (2011). Distribution of silver nanoparticles in pregnant mice and developing embryos. *Nanotoxicology*, pp. 1–11, doi: 10.3109/17435390.2011.626539.

111. Sadauskas, E., Wallin, H., Stoltenberg, M., Vogel, U., Doering, P., Larsen, A., and Danscher, G. (2007). Kupffer cells are central in the removal of nanoparticles from the organism. *Part. Fibre Toxicol.*, **4**, p. 10, doi: 10.1186/1743-8977-4-10.

112. Enders, A. C., and Blankenship, T. N. (1999). Comparative placental structure. *Adv. Drug Delivery Rev.*, **38**, pp. 3–15.

113. King, B. F. (1992). Comparative studies of structure and function in mammalian placentas with special reference to maternal-fetal transfer of iron. *Am. Zool.*, **32**, pp. 331–342.

114. Donaldson, K., Aitken, R., Tran, L., Stone, V., Duffin, R., Forrest, G., and Alexander, A. (2006). Carbon nanotubes: a review of their properties

in relation to pulmonary toxicology and workplace safety. *Toxicol. Sci.*, **92**, pp. 5–22.

115. Fujitani, T., Ohyama, K., Hirose, A., Nishimura, T., Nakae, D., and Ogata, A. (2012). Teratogenicity of multi-wall carbon nanotube (MWCNT) in ICR mice. *J. Toxicol. Sci.*, **37**, pp. 81–89.
116. Lim, J. H., Kim, S. H., Shin, I. S., Park, N. H., Moon, C., Kang, S. S., Park, S. C., and Kim, J. C. (2011). Maternal exposure to multi-wall carbon nanotubes does not induce embryo-fetal developmental toxicity in rats. *Birth Defects Res. B Dev. Reprod. Toxicol.*, **92**, pp. 69–76, doi: 10.1002/bdrb.20283.
117. Pietroiusti, A., Massimiani, M., Fenoglio, I., Colonna, M., Valentini, F., Palleschi, G., Camaioni, A., Magrini, A., Siracusa, G., Bergamaschi, A., Sgambato, A., and Campagnolo, L. (2011). Low doses of pristine and oxidized single-wall carbon nanotubes affect mammalian embryonic development. *ACS Nano*, **5**, pp. 4624–4633, doi: 10.1021/nn200372g.
118. Tsuchiya, T., Oguri, I., Yamakoshi, Y. N., and Miyata, N. (1996). Novel harmful effects of [60]fullerene on mouse embryos in vitro and in vivo. *FEBS Lett.*, **393**, pp. 139–145.
119. Sumner, S. C., Fennell, T. R., Snyder, R. W., Taylor, G. F., and Lewin, A. H. (2010). Distribution of carbon-14 labeled C60 ([14C]C60) in the pregnant and in the lactating dam and the effect of C60 exposure on the biochemical profile of urine. *J. Appl. Toxicol.*, **30**, pp. 354–360, doi: 10.1002/jat.1503.
120. Shimizu, M., Tainaka, H., Oba, T., Mizuo, K., Umezawa, M., and Takeda, K. (2009). Maternal exposure to nanoparticulate titanium dioxide during the prenatal period alters gene expression related to brain development in the mouse. *Part. Fibre Toxicol.*, **6**, p. 20, doi: 10.1186/1743-8977-6-20.
121. Takeda, K., Suzuki, K., Ishihara, A., Kubo-Irie, M., Fujimoto, R., Tabata, M., Oshio, S., Nihei, Y., Ihara, T., and Sugamata, M. (2009). Nanoparticles transferred from pregnant mice to their offspring can damage the genital and cranial nerve systems. *J. Health Sci.*, **55**, pp. 95–102.
122. Philbrook, N. A., Winn, L. M., Afrooz, A. R., Saleh, N. B., and Walker, V. K. (2011). The effect of TiO(2) and Ag nanoparticles on reproduction and development of Drosophila melanogaster and CD-1 mice. *Toxicol. Appl. Pharmacol.*, **257**, pp. 429–436, doi: 10.1016/j.taap.2011.09.027.
123. Hougaard, K. S., Jackson, P., Jensen, K. A., Sloth, J. J., Loschner, K., Larsen, E. H., Birkedal, R. K., Vibenholt, A., Boisen, A. M., Wallin, H., and Vogel, U. (2010). Effects of prenatal exposure to surface-coated nanosized

titanium dioxide (UV-Titan). A study in mice. *Part. Fibre Toxicol.*, **7**, p. 16, doi: 1743-8977-7-16 [pii]10.1186/1743-8977-7-16.

124. Ji, J. H., Jung, J. H., Kim, S. S., Yoon, J.-U., Park, J. D., Choi, B. S., Chung, Y. H., Kwon, I. H., Jeong, J., Han, B. S., Shin, J. H., Sung, J. H., Song, K. S., and Yu, I. J. (2007). Twenty-eight-day inhalation toxicity study of silver nanoparticles in Sprague-Dawley rats. *Inhal. Toxicol.*, **19**, pp. 857–871, doi: 10.1080/08958370701432108.

125. Dhar, S., Mali, V., Bodhankar, S., Shiras, A., Prasad, B. L. V., and Pokharkar, V. (2011). Biocompatible gellan gum-reduced gold nanoparticles: cellular uptake and subacute oral toxicity studies. *J. Appl. Toxicol.*, **31**, pp. 411–420, doi: 10.1002/jat.1595.

126. Komatsu, T., Tabata, M., Kubo-Irie, M., Shimizu, T., Suzuki, K., Nihei, Y., and Takeda, K. (2008). The effects of nanoparticles on mouse testis Leydig cells in vitro. *Toxicol. In Vitro*, **22**, pp. 1825–1831, doi: 10.1016/j.tiv.2008.08.009.

127. Braydich-Stolle, L., Hussain, S., Schlager, J. J., and Hofmann, M. C. (2005). In vitro cytotoxicity of nanoparticles in mammalian germline stem cells. *Toxicol. Sci.*, **88**, pp. 412–419, doi: 10.1093/toxsci/kfi256.

128. Braydich-Stolle, L. K., Lucas, B., Schrand, A., Murdock, R. C., Lee, T., Schlager, J. J., Hussain, S. M., and Hofmann, M. C. (2010). Silver nanoparticles disrupt GDNF/Fyn kinase signaling in spermatogonial stem cells. *Toxicol. Sci.*, **116**, pp. 577–589, doi: 10.1093/toxsci/kfq148.

129. Asare, N., Instanes, C., Sandberg, W. J., Refsnes, M., Schwarze, P., Kruszewski, M., and Brunborg, G. (2012). Cytotoxic and genotoxic effects of silver nanoparticles in testicular cells. *Toxicology*, **291**, pp. 65–72, doi: 10.1016/j.tox.2011.10.022.

130. Wiwanitkit, V., Sereemaspun, A., and Rojanathanes, R. (2009). Effect of gold nanoparticles on spermatozoa: the first world report. *Fertil. Steril.*, **91**, pp. e7–e8, doi: 10.1016/j.fertnstert.2007.08.021.

131. Ben-David Makhluf, S., Qasem, R., Rubinstein, S., Gedanken, A., and Breitbart, H. (2006). Loading magnetic nanoparticles into sperm cells does not affect their functionality. *Langmuir*, **22**, pp. 9480–9482, doi: 10.1021/la061988z.

132. Makhluf, S. B., Abu-Mukh, R., Rubinstein, S., Breitbart, H., and Gedanken, A. (2008). Modified PVA-Fe3O4 nanoparticles as protein carriers into sperm cells. *Small*, **4**, pp. 1453–1458, doi: 10.1002/smll.200701308.

133. Araujo, L., Sheppard, M., Löbenberg, R., and Kreuter, J. (1999). Uptake of PMMA nanoparticles from the gastrointestinal tract after oral administration to rats: modification of the body distribution after

suspension in surfactant solutions and in oil vehicles. *Int. J. Pharm.*, **176**, pp. 209–224.

134. Bai, Y., Zhang, Y., Zhang, J., Mu, Q., Zhang, W., Butch, E. R., Snyder, S. E., and Yan, B. (2010). Repeated administrations of carbon nanotubes in male mice cause reversible testis damage without affecting fertility. *Nat. Nanotechnol.*, **5**, pp. 683–689, doi: 10.1038/nnano.2010.153.

135. Balasubramanian, S. K., Jittiwat, J., Manikandan, J., Ong, C. N., Yu, L. E., and Ong, W. Y. (2010). Biodistribution of gold nanoparticles and gene expression changes in the liver and spleen after intravenous administration in rats. *Biomaterials*, **31**, pp. 2034–2042, doi: 10.1016/j.biomaterials.2009.11.079.

136. De Jong, W. H., Hagens, W. I., Krystek, P., Burger, M. C., Sips, A. J., and Geertsma, R. E. (2008). Particle size-dependent organ distribution of gold nanoparticles after intravenous administration. *Biomaterials*, **29**, pp. 1912–1919, doi: 10.1016/j.biomaterials.2007.12.037.

137. Kim, J. S., Yoon, T. J., Yu, K. N., Kim, B. G., Park, S. J., Kim, H. W., Lee, K. H., Park, S. B., Lee, J. K., and Cho, M. H. (2006). Toxicity and tissue distribution of magnetic nanoparticles in mice. *Toxicol. Sci.*, **89**, pp. 338–347, doi: 10.1093/toxsci/kfj027.

138. Kwon, J. T., Hwang, S. K., Jin, H., Kim, D. S., Minai-Tehrani, A., Yoon, H. J., Choi, M., Yoon, T. J., Han, D. Y., Kang, Y. W., Yoon, B. I., Lee, J. K., and Cho, M. H. (2008). Body distribution of inhaled fluorescent magnetic nanoparticles in the mice. *J. Occup. Health*, **50**, pp. 1–6.

139. Lankveld, D. P., Oomen, A. G., Krystek, P., Neigh, A., Troost-de Jong, A., Noorlander, C. W., Van Eijkeren, J. C., Geertsma, R. E., and De Jong, W. H. (2010). The kinetics of the tissue distribution of silver nanoparticles of different sizes. *Biomaterials*, **31**, pp. 8350–8361, doi: 10.1016/j.biomaterials.2010.07.045.

140. Lee, C. M., Jeong, H. J., Kim, D. W., Sohn, M. H., and Lim, S. T. (2012). The effect of fluorination of zinc oxide nanoparticles on evaluation of their biodistribution after oral administration. *Nanotechnology*, **23**, p. 205102, doi: 10.1088/0957-4484/23/20/205102.

141. Park, J. H., von Maltzahn, G., Ong, L. L., Centrone, A., Hatton, T. A., Ruoslahti, E., Bhatia, S. N., and Sailor, M. J. (2010). Cooperative nanoparticles for tumor detection and photothermally triggered drug delivery. *Adv. Mater.*, **22**, pp. 880–885, doi: 10.1002/adma.200902895.

142. Lan, Z., and Yang, W. X. (2012). Nanoparticles and spermatogenesis: how do nanoparticles affect spermatogenesis and penetrate the blood-testis barrier. *Nanomedicine (London)*, **7**, pp. 579–596, doi: 10.2217/nnm.12.20.

143. Morishita, Y., Yoshioka, Y., Satoh, H., Nojiri, N., Nagano, K., Abe, Y., Kamada, H., Tsunoda, S., Nabeshi, H., Yoshikawa, T., and Tsutsumi, Y. (2012). Distribution and histologic effects of intravenously administered amorphous nanosilica particles in the testes of mice. *Biochem. Biophys. Res. Commun.*, **420**, pp. 297–301, doi: 10.1016/j.bbrc.2012.02.153.

144. Yoshida, S., Hiyoshi, K., Ichinose, T., Takano, H., Oshio, S., Sugawara, I., Takeda, K., and Shibamoto, T. (2009). Effect of nanoparticles on the male reproductive system of mice. *Int. J. Androl.*, **32**, pp. 337–342.

145. Li, C., Taneda, S., Taya, K., Watanabe, G., Li, X., Fujitani, Y., Ito, Y., Nakajima T and Suzuki, A. K. (2009). Effects of inhaled nanoparticle-rich diesel exhaust on regulation of testicular function in adult male rats. *Inhal. Toxicol.*, **21**, pp. 803–811, doi: 10.1080/08958370802524381.

146. Ramdhan, D. H., Ito, Y., Yanagiba, Y., Yamagishi, N., Hayashi, Y., Li, C., Taneda, S., Suzuki, A. K., Watanabe, G., Taya, K., Kamijima, M., and Nakajima, T. (2009). Nanoparticle-rich diesel exhaust may disrupt testosterone biosynthesis and metabolism via growth hormone. *Toxicol. Lett.*, **191**, pp. 103–108, doi: 10.1016/j.toxlet.2009.08.013.

147. Geraets, L., Oomen, A. G., Schroeter, J. D., Coleman, V. A., and Cassee, F. R. (2012). Tissue distribution of inhaled micro- and nano-sized cerium oxide particles in rats: results from a 28-day exposure study. *Toxicol. Sci.*, **127**, pp. 463–473, doi: 10.1093/toxsci/kfs113.

148. Semmler-Behnke, M., Fertsch, S., Schmid, G., Wenk, A., and Kreyling, W. G. (2007). *Uptake of 1.4 nm versus 18 nm Gold Particles by Secondary Target Organs Is Size Dependent in Control and Pregnant Rats after Intratracheal or Intravenous Application*, (ed. EuroNanoForum 2007, European Commission, Düsseldorf), pp. 102–104.

149. Yoshida, M., Yoshida, S., Sugawara, I., and Takeda, K. (2002). Maternal exposure to diesel exhaust decreases expression of steroidogenic factor-1 and Müllerian inhibiting substance in the murine fetus. *J. Health Sci.*, **48**, pp. 317–324.

150. Yoshida, S., Ono, N., Tsukue, N., Oshio, S., Umeda, T., Takano, H., and Takeda, K. (2006). In utero exposure to diesel exhaust increased accessory reproductive gland weight and serum testosterone concentration in male mice. *Environ. Sci.*, **13**, pp. 139–147.

151. Izawa, H., Kohara, M., Watanabe, G., Taya, K., and Sagai, M. (2007). Diesel exhaust particle toxicity on spermatogenesis in the mouse is aryl hydrocarbon receptor dependent. *J. Reprod. Dev.*, **53**, pp. 1069–1078.

152. Ono, N., Oshio, S., Niwata, Y., Yoshida, S., Tsukue, N., Sugawara, I., Takano, H., and Takeda, K. (2007). Prenatal exposure to diesel exhaust impairs mouse spermatogenesis. *Inhal. Toxicol.*, **19**, pp. 275–281, doi: 10.1080/08958370601069257.

153. Ono, N., Oshio, S., Niwata, Y., Yoshida, S., Tsukue, N., Sugawara, I., Takano, H., and Takeda, K. (2008). Detrimental effects of prenatal exposure to filtered diesel exhaust on mouse spermatogenesis. *Arch. Toxicol.*, **82**, pp. 851–859, doi: 10.1007/s00204-008-0302-1.

154. Chen, Z. L., Huang, X. B., Suo, J. P., Li, J., and Sun, L. (2010). The contraceptive effect of a novel filtering-type nano-copper complex/polymer composites intra-vas device on male animals. *Int. J. Androl.*, **33**, pp. 810–817, doi: 10.1111/j.1365-2605.2009.01028.x.

155. Chen, Z., Chen, H., Meng, H., Xing, G., Gao, X., Sun, B., Shi, X., Yuan, H., Zhang, C., Liu, R., Zhao, F., Zhao, Y., and Fang, X. (2008). Bio-distribution and metabolic paths of silica coated CdSeS quantum dots. *Toxicol. Appl. Pharmacol.*, **230**, pp. 364–371, doi: 10.1016/j.taap.2008.03.022.

156. Yamashita, K., Yoshioka, Y., Higashisaka, K., Mimura, K., Morishita, Y., Nozaki, M., Yoshida, T., Ogura, T., Nabeshi, H., Nagano, K., Abe, Y., Kamada, H., Monobe, Y., Imazawa, T., Aoshima, H., Shishido, K., Kawai, Y., Mayumi, T., Tsunoda S-I, Itoh, N., Yoshikawa, T., Yanagihara, I., Saito, S., and Tsutsumi, Y. (2011). Silica and titanium dioxide nanoparticles cause pregnancy complications in mice. *Nat. Nanotechnol.*, **6**, pp. 321–328, doi: 10.1038/nnano.2011.41.

157. Tian, F., Razansky, D., Estrada, G. G., Semmler-Behnke, M., Beyerle, A., Kreyling, W., Ntziachristos, V., and Stoeger, T. (2009). Surface modification and size dependence in particle translocation during early embryonic development. *Inhal. Toxicol.*, **21**(Suppl, 1), pp. 92–96, doi: 10.1080/08958370902942624.

158. Hsieh, M.-S., Shiao, N.-H., and Chan, W.-H. (2009). Cytotoxic effects of CdSe quantum dots on maturation of mouse oocytes, fertilization, and fetal development. *Int. J. Mol. Sci.*, **10**, pp. 2122–2135, doi: 10.3390/ijms10052122.

159. Park, E.-J., Bae, E., Yi, J., Kim, Y., Choi, K., Lee, S. H., Yoon, J., Lee, B. C., and Park, K. (2010). Repeated-dose toxicity and inflammatory responses in mice by oral administration of silver nanoparticles. *Environ. Toxicol. Pharmacol.*, **30**, pp. 162–168.

Chapter 17

Nanomedicine: Ethical Considerations

Todd Kuiken
Woodrow Wilson International Center for Scholars, Washington, DC, USA
todd.kuiken@wilsoncenter.org

17.1 Introduction

Medicine is moving toward being able to predict what you will die from and when. As nanomedicine, genetics, and diagnostics improve, they promise the ability of healthcare professionals to diagnose, treat, and share medical information nearly instantaneously. As nanomedicine improves, the question, what does it mean to be human? arises. Exactly what will be the defining characteristics of a human? How long should a person live? Do people have the right to die? How do humans and society deal with this expanded life span? How will we determine when a person is dying? Nanomedicine raises all of these ethical questions, none of which are new or necessarily unique and would hold true for any new medical device or medicine that was being evaluated in a clinical trial. The issue becomes how much risk society is willing to accept with a new technology before it is proven effective and "safe." This will become even more evident

Handbook of Safety Assessment of Nanomaterials: From Toxicological Testing to Personalized Medicine
Edited by Bengt Fadeel

ISBN 978-981-4463-36-2 (Hardcover), 978-981-4463-37-9 (eBook)
www.panstanford.com

as personalized medicine is realized and the rise of do-it-yourself biology enables individuals to diagnose and potentially treat themselves outside the traditional medical establishment.

17.1.1 The Technology Landscape

New developments in science and technology are breaking down the walls between information and communication technology, micronano systems, and biology through miniaturization and the ability to manipulate matter at the atomic scale and to interface those with biological systems [1]. Chow-White and Garcia-Sancho suggest that biology and computing have been converging over time and are intimately linked and the lines between the two are now indivisible, producing new data practices and a new scientific approach to understanding genetic code and the human body [2]. They conclude that "biology, computing, and social orders interact and are reciprocally shaped around spaces of convergence, but none of them fully determines the sequencer, the database, or other genomic technologies" [3]; however, they all need each other in order to exist.

According to the US National Institutes of Health, "Nanomedicine, an off shoot of nanotechnology, refers to highly specific medical intervention at the molecular scale for curing disease or repairing damaged tissues, such as bone, muscle, or nerve" [3]. In a recent review of nanosized drug synthesis methods and manufacturing technology, it's believed that nanosized colloidal carriers of drugs "represent one of the approaches to discovering new methods of pharmacotherapy" [4]. The authors go on to say that colloidal delivery systems can increase the effectiveness of drug delivery by "optimizing the biodistribution and toxicodynamics of drugs." Nazarov et al. state that the idea of delivering a drug directly to the site of concern and overcoming histohematic barriers is beginning to move from a theoretical prospect to a practical reality. Nanosized drugs will be highly individualized, and their effectiveness will depend on many factors, including structure and the chemical nature of the carrier. This will create an individualized approach to medicine and in particular designing nanosized forms of drugs. One of the main goals that nanomedicine is attempting to reach is to change the distribution of drugs in an organism so that they can reach only their site of action (targeted delivery of drugs). The theory behind this lies in the nanoparticles' ability to overcome low solubility and unsatisfactory absorption of contemporary drugs because of their highly developed surface. This enables nanoparticles to be used not only for therapeutic treatment but also for diagnosis of diseases [4]. For a comprehensive review of nanomedicine applications see Riehemann et al. [5], Sandhiya et al. [6], and Moghimi et al. [7] (also refer to Chapter 1 for a historical perspective on nanomedicine).

Another subset of nanotechnology is referred to as bionanotechnology and DNA nanotechnology in particular, which utilizes biomolecular self-assembly to construct artificial structures and devices based on the properties of DNA or proteins. Advances in nanotechnology are providing new tools and materials, creating the ability to match the scale and interface with biological systems and components, in addition to exhibiting electrical, magnetic, optical, thermal, and chemical properties aiding in the construction of complex networks of functional parts [8]. This was envisioned in the early developments of nanotechnology research programs [9].

17.1.2 Informatics/Databases

Over the past 20 years we have seen the rise of e-science [10], which utilizes massive computer networks, information science tools, and social networking technologies to combine and analyze large-scale datasets. The Human Genome Project is one example where the success of the project was in part due to the success of the informatics. Informatics is a method of using information and computer science for collecting, analyzing, and applying information. "X-informatics" has been used as a general descriptor for such applications within a specific field, discipline, or science, that is, nanoinormatics for nanotechnology, bioinformatics for biology, and ecoinformatics for ecology [11].

Nanoinformatics "is necessary for intelligent development and comparative characterization of nanomaterials, for design and use of optimized nanodevices and nanosystems, for development of advanced instrumentation and manufacturing processes, and for assurance of occupational and environmental safety and health" [11]. The difficulty with successfully utilizing such large-scale databases and informatics tools is being able to verify the accuracy of the data, setting up metrics for inputting data into a database, determining who controls such a database, and determining who "owns" the data. Current and future convergence technologies will rely even more so on x-informatics to process the massive amounts of data that are being generated and utilized.

Convergence technologies are more and more requiring data, components, and knowledge from a variety of disciplines that will draw on metrics from a host of places, all of which will need to be verified and stored and ownership issues will need to be established. Access to such large-scale databases and computer networks will either inhibit or foster greater innovation and development in fields such as nanotechnology that are more and more relying on "large-scale infrastructure and systematic curation" of data.

One critical aspect of digital databases and how effective they are is who is responsible for them and who has access to them? There are vast numbers of laboratories, companies, and communities that are involved in the creation and preservation of scientific data, each with its own legal and ethical protocols and with varying degrees of openness to outside users.

17.1.3 Proteomics

Proteomics is the study of proteins expressed by a genome and will require access to the databases previously discussed. It provides methods for correlating the vast amounts of genomic data with protein information that is being produced through analysis of cells under normal and altered states. It consists of high-throughput identification and characterization of proteins and integrating it with genomic data. Characterization of novel catalysts using proteomics will allow synthetic biologists to expand their protein toolbox in order to design bioprocesses for biopharmaceuticals and bioproducts. One of the challenges is being able to identify the function of all the various protein parts within cellular systems; better identification of these parts will allow a better description of the entire biological system that could be used as a cell factory [12].

Johnson et al. suggest that the solution to many of the technical challenges in proteomics and protein-based molecular diagnostics is nanomaterials and that nanoproteomics will provide platforms for the discovery of next-generation biomarkers, which will enable the molecular diagnostics field to take off [13]. Advances in nanotechnology have allowed the identification of low-abundance proteins, for example [14]. Johnson et al. conclude that in the near term nanotechnology will have a major impact on proteomics and diagnostics, leading to new devices and sensing modes and augmenting existing methods. As bioinformatics and proteomics advance the support of computational biologists, mathematicians, and statisticians will be required to be able to analyze and interpret the vast amounts of data the field is generating [15].

17.2 Personalized Medicine

Goldstein predicted that by 2016 doctors will be able to scan their patients' entire genome in a matter of minutes [16]. While that prediction was a bit optimistic the technology and medicine in general are moving toward genetic-based personalized therapies. Weston and Hood predict that in the next 20 years there will be a fundamental shift in healthcare, moving from a reactive model to one that can predict and prevent cellular dysfunction, capable of determining a probabilistic, individualized future health history [15]. Gonzalez-Nilo et al. conclude that implementing the tools of nanoinformatics, which they define as the intersection of bioinformatics, computational chemistry, and nanobiotechnology, will open the door toward personalized medicine by accelerating the design of highly specific biomedical treatments, increasing the efficacy, bioavailability, and specificity of nanomaterials and reducing side effects [17]. As the "omics" technologies (i.e., genomics and proteomics) improve they are exposing the weaknesses of traditional therapeutics, which do not allow for different patients' "physiologic idiosyncrasies." Doctors no longer rely primarily on images such as a mammogram or a patient's pathology. They now include the molecular profile or genetic map of a patient [18]. Nanotechnology, along with systems biology and the convergence of the other "omics" technologies, is predicted to play a critical role in being able to develop patient-specific therapies.

Nanotechnology is predicted to "provide access to previously inaccessible data as related to 'omic' technology components" and "enable innovative therapeutic modalities that leverage the validated systems biology outputs for exquisitely specific individualized therapy" [18]. Areas of nanotechnology that are predicted to aid personalized medicine are early detection diagnostics, implantable drug delivery devices, nanobased injectable therapeutics, nanobased contrast agents, and tissue engineering.

The ability of patients to control the most important aspects of their care (a diabetes patient to measure his or her glucose level) is very appealing and is why self-care testing is the majority of the point-of-care testing market [19]. Nanotechnology has led to significant technological advances in the miniaturization of devices, particularly in point-of-care testing. These devices encompass a variety of applications from in vivo testing to handheld glucose tests. The market was worth $11.3 billion in 2007 and was predicted to grow to $18.7 billion in 2013 [20]. New lab-on-a-chip manufacturing techniques are enabling them to be connected to and embedded in smartphones and USB drives, turning your phone into a portable diagnostic device capable of communicating directly with your doctor's office [21]. Macklis and Sharma predict that within the next few years nearly all medical devices will be controlled or monitored remotely or via Internet-accessible interfaces, making the electronic medical records system a much more important monitor and response network with the potential for automated responses [22]. Intrabody communication and personal network security are needed to evaluate and act upon the continuous test results that come with continuous monitoring, potentially leading to less aggressive and less expensive therapeutics [23]. Beyond point-of-care diagnostics, the nanomedicine market is estimated to reach $1 trillion by 2015 [19].

Personalized medicine requires the analysis of massive amounts of genomic data using miniaturized biomedical devices and lab-on-a-chip technologies that have been influenced by nanotechnology and nanofabrication [24]. Personalized medicine is also based on a systems biology approach utilizing modern molecular medicine and the analysis of large datasets for complex diseases [25]. However, personalized medicine will only be successful if medicines with subtle differences designed for individuals on the basis of their DNA can be manufactured reliably and at small scales [26].

As these technologies take hold, a new governance structure may be needed to regulate the transmission of medical data, the sharing of devices between users, the potential of "homemade" or hacked instruments, and the standardization of results and analysis methods amongst various countries. Lymberis suggests that micro- and nanotechnologies will change the way that healthcare is organized and priced and ultimately remove the barriers between medical and pharmaceutical practices and home care and test laboratories [27]. New ethical issues may arise with personalized medicine and home diagnostic tests. For instance, would it be ethical to make available a point-of-care cancer diagnostic test? Many doctors believe that counseling is vitally important for such diagnoses and wonder if making such tests available is ethical [1].

17.2.1 Size/Scope of the Nanomedicine Market

According to BCC Research the global nanomedicine market reached $63.8 billion in 2010 and $72.8 billion in 2011 and is expected to grow to $130.9 billion by 2016 [28]. In 2006, Wagner et al. analyzed the emerging nanomedicine landscape and found more than 150 start-ups and small and medium-sized enterprises focused on nanomedicine research and development projects: 38 nanoenabled products with sales totaling $6.8 billion. Twenty-three of those products were drug delivery systems, with three of them accounting for $3.2 billion. They found that three-quarters of the research market was focused on drug delivery applications [29]. Resnik and Tinkle reported in 2007 that drug delivery accounted for 78% of sales and 58% of patent filings worldwide [30]. Moving a new nanosized drug or product to market requires support from large pharmaceutical companies. Small start-ups are pursuing a host of ideas utilizing nanotechnology in order to improve treatment and diagnosis. Despite this, Wagner et al. found that these small companies and industry R&D managers agreed that large pharmaceutical companies are skittish on emphasizing nanotechnology in their business strategy [29]. The Project on Emerging Nanotechnologies (PEN) maintains an interactive map showing the locations of companies, universities, and government labs across the U.S. that are involved in nanotechnology research, development, and commercialization, currently with over 200 entries in the medicine and health fields (http://www.nanotechproject.org/inventories/map/).

17.3 Ethical and Policy Implications Surrounding Nanomedicine

Medicine is moving toward being able to predict what you will die from and when. As nanomedicine, genetics, and diagnostics improve they promise the ability of healthcare professionals to diagnose, treat, and share medical information nearly instantaneously. The potential development of a lab on a chip that could be implanted into a human that monitors the person's health raises concerns about who monitors the information, who has access to the information, and whether or not the chip can be used for purposes other than medical surveillance [31]. This chip can not only diagnose current disease but also analyze your DNA to determine which diseases you may be susceptible to in the future. This raises ethical concerns of the patients' right to know, right not to know, and duty to know. This has been referred to as the principle of autonomy [32]. The European Group on Ethics in Science and New Technologies forecasted that the long-term ethical issues will surround civil liberties in regard to implants that will potentially make life safer but will also create privacy debates [33].

Spagnolo and Daloiso suggest that ethical issues should take a back seat to determining the scientific merit of a medicine if the potential benefit justifies the use of nanoparticles, even if the design and effects of the nanomedicine are not completely understood [31]. The shift to personalized medicine not only in terms of targeted drug deliver but also in terms of personalized diagnosis and treatment creates a potential for patients becoming autonomous, increased technology use, and dependence causing reduced personal interactions and will inevitably change the way medicine is currently practiced. While technology advances reduce the invasiveness of medical treatments and improve efficiency, they have the potential of shifting the responsibility away from doctors to patients [31].

"Good facts of course are essential to the development of good ethics but the same set of facts may result in very different ethical arguments and conclusions for different people" [34]. White argues that from an ethical perspective, considerations of benefits and harms would supersede issues of profitability in terms of importance [34]. While ethically this should be how things work, the reality is that profitability in the eyes of pharmaceutical companies is priority number one, albeit it goes hand in hand with weighing the risks versus the benefits of a particular drug, medical device, or diagnostic tool.

17.3.1 Clinical Trials

A search using "nanoparticles" of ClinicalTrials.gov produces 116 clinical trials that have been approved and are completed or are currently being conducted. The European Group on Ethics in Science and New Technologies proposed a series of questions surrounding nanomedicine [35]:

(1) How should the dignity of people participating in nanomedicine research trials be respected?
(2) How can we protect the fundamental rights of citizens who may be exposed to free particles in the environment?
(3) How can we promote responsible use of nanomedicine that protects both human health and the environment?
(4) What are the specific ethics issues, such as justice, solidarity, and autonomy, that have to be considered in this scientific domain?

The ethical issues surrounding nanomedicine are not necessarily about the technology but about the implications of the technology and how we look, think, and feel about our bodies (assuming the applications nanomedicine promises come to fruition). Bawa and Johnson ask whether nanobased therapies have the potential to further marginalize those individuals who are perceived as disabled [36].

The institutional review board is charged with determining whether (1) risks will be minimized, (2) risks will be reasonable in relation to expected benefits to the subjects or society, (3) provisions for data and safety monitoring will be adequate, (4) informed consent will be properly sought and documented, (5) selection of subjects will be equitable, (6) protections for vulnerable populations will be adequate, and (7) privacy and confidentiality will be protected before approving a clinical trial [37, 38]. These questions will still be assessed with nanomedicine. The only issue that may be new or unique to nanomedicine is whether or not the subjects are told about the "nano" portion of the medicine and what that means.

Resnik and Tinkle conclude that the potential risk of nanomaterials is the biggest issue surrounding nanomedicine and its use in clinical trials. They go on to say that additional research on the physicochemical properties and the potential impacts on humans and the environment is essential for nanomedicine to develop responsibly [37]. Resnik and Tinkle argue that the Food and Drug Administration (FDA) should conduct phase IV, or postmarket, studies for some drugs using nanoparticles and mandate this research for some nanomedicine products. They argue that since this type of postmarket study is not legally required of the companies the FDA should conduct the research [37]. I would argue, however, that the FDA should mandate that the companies conduct this research or impose fees on companies in order for the FDA to conduct the studies using independent laboratories.

There is a risk in overanalyzing the ethical questions surrounding nanomedicine, as concluded by Ferrari, Philibert, and Sanhai (2009) [39]. The authors state that "the greatest risk in nanomedicine may well be in letting our concerns paralyze our action and not taking advantage of the full, revolutionary potential that nanotechnology in medicine can offer humankind" [39]. This argument is countered by Khushf (2007) [40], who suggests that ethical considerations seek to anticipate/mitigate potential risks and disruptive developments. He goes on to say that ethics is too often looked upon as a constraint on research and mistakenly is considered relevant only when human subjects are involved [40].

17.3.2 Is Hype Driving the Ethics Debate?

One of the issues surrounding nanotechnology, and in particular nanomedicine, is the hype (both good and bad) that inevitably arises with a new technology that is fighting for limited research dollars. Gordijn (2005) [41] expresses this view as follows:

> Optimistic visionaries predict truly utopian states of affairs. Pessimistic thinkers present all manner of apocalyptic visions. Whereas the utopian views follow from one-sidedly focusing on the potential benefits of nanotechnology, the apocalyptic perspectives result from giving exclusive attention to possible worst-case scenarios.

Some of the hype surrounding nanomedicine insists it will provide low-priced and superior equipment for medical research that will be available far and wide, nanomachines will be programmed to travel through our bloodstream to clean out fatty deposits, medical diagnosis and drug delivery will be transformed, and preventive medicine will include nanorobots that would provide a defense against invading viruses in our bodies. In short nanomedicine has proclaimed to be the solution to many, if not all, medical problems [41].

While many futurists and ethicists point to nanotechnology applications for active implants, experts believe this area of medicine will have the least impact from nanotechnology. Nanotechnology will be used only as an enabling technology for optimizing certain components, and a majority of the components will be driven by advances in microtechnology [42]. The enabling component of nanotechnology in relation to implants carries over to drug delivery and diagnostics. This enabling capability has the potential to add innovative functionality and is highly versatile.

Nanomedicine is expected to enhance human capabilities and properties. This surrounds the idea that nanotechnology will enable the construction of stronger and enhanced tissues and organs and improve our mental capacity. The latter of these potential implications of nanomedicine is where a lot of the ethical debates are and will take place. However, according to Moghimi et al., 2005 [7], the future of nanomedicine will depend largely on the rational design of nanomaterials on the basis of a thorough understanding of their biological processes rather than forcing applications based on hype.

17.3.3 Public Acceptance

Godman [43] asks the question, Do we want to find criteria for risk acceptance, or do we wish to open the discussion so that the core values of technology development are called into question? Grunwald echoes this idea stating that the unique ethical issues arise from the public discourse about nanotechnology and not anything unique in the actual ethical tenants [44]. Similarly, Petersen suggests that it is more important to create a "conversation about" nanotechnology rather than creating a new field of nanoethics [45].

This is also put forth by Sparrow [46]. These questions follow with the argument that there is nothing inherently "new" about the ethical considerations surrounding nanotechnology/medicine [47] but rather a new technology raising the same ethical issues surrounding medicine.

PEN has conducted polling on US citizens for the past four years tracking the public's understanding and acceptance of nanotechnology. The results have shown that the public is willing to accept certain risks associated with nanotechnology, with the understanding there will be some sort of public benefit (Fig. 17.1a). Their overall understanding and exposure to nanotechnology, however, has been limited and has not changed significantly (Fig. 17.1b). Cameron argues that the scope of the public's potential unease with nanotechnology and nanomedicine could be greater than early skepticism of recombinant DNA because of the hype surrounding nano and its ability to manipulate and change the human condition [48]. He goes on to say that a "clear repudiation of the 'trans-humanist' visions of nanotechnology should occur because," according to Cameron, "they have the effect of raising the element of risk associated with nanotechnology without any countervailing economic or other benefit."

Cameron advocates that a clear distinction be made between government-funded research and research that is associated with "fringe philosophers and social thinkers." This has strong implications to nanomedicine in the realm of human enhancement beyond that obtained from traditional drug delivery. Creating an atmosphere for public discourse on nanomedicine enables public participation in decision making and avoids social fears that are often based on a superficial knowledge, which can help plan new regulations and avoid past political mistakes [49].

17.4 Ethical Dilemma: Is Anything New or Unique to Nanomedicine?

The 2012 Olympics raised some interesting ethical questions [50] that bear resemblance to our discussion here on nanomedicine and were raised by Bawa and Johnson in 2009: "Is it morally acceptable to us as a society that athletes or military personnel have significant

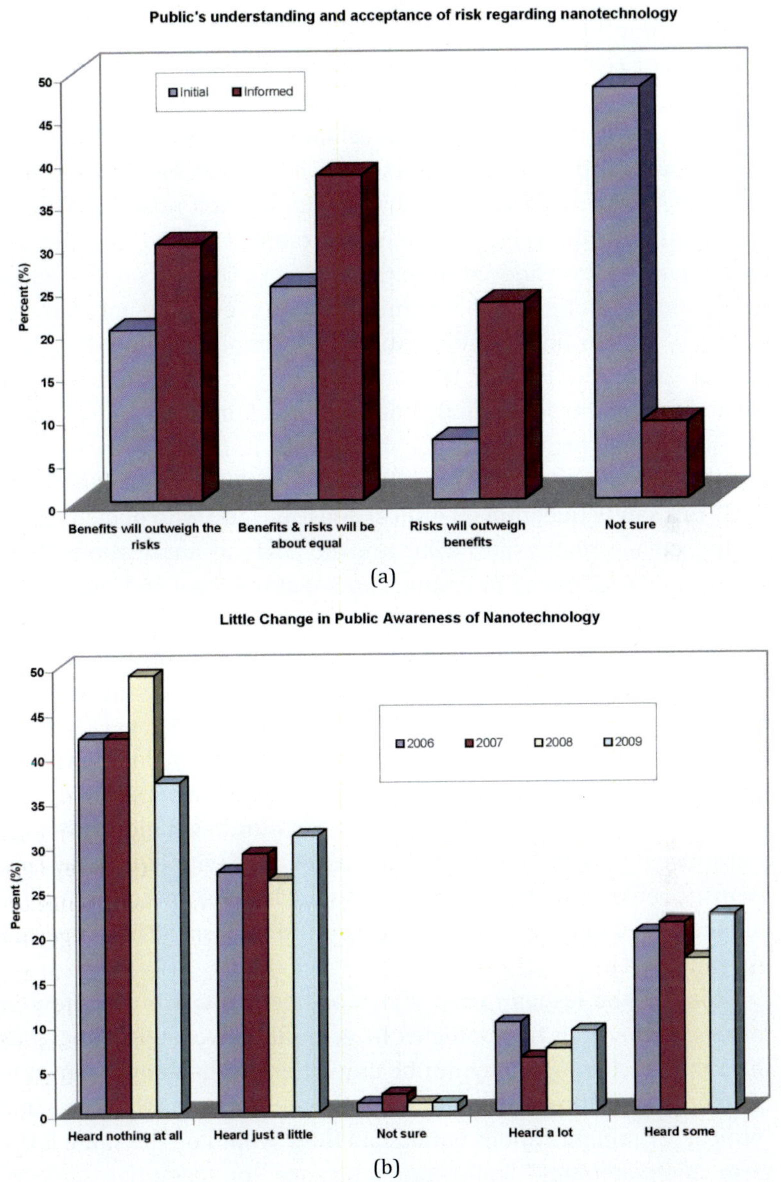

Figure 17.1 (a) Public understanding of risk and benefits of nanotechnology. Participants were asked before and after being provided information about nanotechnology. Survey conducted in 2008 with 1,003 participants [66]. (b) Public awareness of nanotechnology. Survey conducted in 2006, 2007, 2008, and 2009 [66].

parts of their bodies altered to enhance performance in competitive or combat situations?" While the Olympic sprinter was determined to not have a competitive advantage, born out by the fact he lost the race, the overall ethical question remains, how many implantable devices would it take for a person to no longer be considered a person? A better question that Bawa and Johnson raise is, When is a medical procedure/intervention/treatment regarded as a therapy, and when is it considered an enhancement? Or will society adapt and progress alongside the technology and view it as natural? For instance, will an implantable nanodevice simply become the norm similar to contact lenses? If nanotechnology is revolutionary, then we are right to be concerned about it. If it is familiar, then we know all too well the problems it may bring with it. If nanotechnology is inevitable, then why should we go out of our way to make it possible [51]? Is a safety question an ethical question, and who decides [52]? Is the science really safe, what are the societal implications, how will individuals' behavior change, how will the legal and regulatory systems react, and what fundamental moral and ethical beliefs are challenged [53]? As nanomedicine improves, the question, what does it mean to be human? arises. Exactly what will be the defining characteristics of a human? How long should a person live? Do people have the right to die? How do humans and society deal with this expanded life span? How will we determine when a person is dying? Put more succinctly, the ethical issues surrounding nanotechnology and subsequently nanomedicine are associated with equity, privacy, security, environment, and metaphysical questions relating to enhanced human capabilities and the interaction of humans and machines [54].

None of the issues raised above are new, nor are they new in terms of emerging technologies in general. Lewenstein concludes that "nanotechnology may not be any different than any other area of emerging science and technology." He goes on to say that the same "ethical" arguments about nanotechnology would be the same if the term "biotechnology," "information science," or "cognitive science" were used. Others have argued that these are re-emerging ethical issues rather than emerging ones [55]. Preston makes a similar argument stating that the ethical issues brought up in regard to nanotechnology are the same issues that have been raised previously with other technologies and there is nothing new or "distinctly nano" [56]. He highlights three themes that focus particularly on nanomedicine but could be replicated for any "new" medicine or medical treatment/device:

- The introduction of radically new materials into human and natural environments
- Technology that consciously seek to replicate the process of evolution by natural selection
- Use of nanotechnologies to enhance humans away from their inherited evolutionary identity

Sandler [57] argues that there is more to the ethics surrounding nanomedicine than risks (as argued in part by Allhoff [58]) and that "nanomedicine" itself must be separated into technology sectors—information, diagnostic, therapeutic, and enhancement—for the ethical evaluations to be effective. He goes on to say that the enhancement issues raise a broad range of ethical issues that are unique to nanomedicine. Johnson [59] shares similar views, although her thesis is derived from the belief that nanoparticles utilized in nanomedicine will not be required to go through the typical clinical trial process as new drugs and therefore require a more focused ethical debate. Others have argued that the ethical issues surrounding nanomedicine are so great that "new agencies need to be formed to consider and regulate the social and policy questions raised by the future of nanomedicine" [60].

Bawa and Johnson argue that it's difficult for ethicists to predict what the issues surrounding nanomedicine will be, considering that much of what has been "hyped" has yet to occur [61]. They go on to say that this hype surrounding nanomedicine should not overshadow the societal and ethical implications. However, according to Lenk and Biller-Andorno, society will have to decide anew (ethically, morally, and personally) whether it chooses to accept nanotechnology and under what conditions [55].

From a 10,000-foot view, the ethical questions surrounding nanotechnology in a sense have either been ignored or been resolved, depending on the pervasiveness of nanotechnology in the marketplace and the billions of dollars invested by governments across the globe. The question becomes, who will receive access to these technologies? While access is an ethical question, similar questions were asked and are still being debated in terms of computers and access to the Internet. It's a continuing pattern of the rich getting access to newly developed technologies/medicines, while the poor wait for the economics to dictate when they get access. This pattern has not changed, nor will it change, and while it certainly is an ethical question, it is not new or unique to nanomedicine.

Nanomedicine with all its promise (or hype) raises numerous ethical questions, none of which are new or unique and would hold true for any new medical device or medicine that was being evaluated in a clinical trial. After all the reason they are being tested is because there are unknowns. While there are clearly structured ways to evaluate ethical questions, ethics at its core is a personal decision-making process that an individual comes to, albeit society can dictate ethics onto the public by way of public policy. The issue becomes how much risk as a society we are willing to accept with a new technology before it is proven effective and "safe" ("safe" being a relative term as all medicines have the risk of causing side effects). The same questions being raised about nanomedicine were asked with the advent of other emerging medical breakthroughs, that is, chemotherapy and genetics, and will be asked when the next technological breakthrough occurs. The more pertinent "ethical" questions that need to be addressed are the access and distribution of these technologies to those who would otherwise not be able to afford them.

I have argued that there is nothing inherently new about the ethical questions being raised in regard to nanomedicine. Others have argued the opposite and conclude that extra oversight is needed because of "heightened uncertainty regarding risks, fast-evolving science yielding complex and increasingly active materials, likelihood of research on vulnerable participants including cancer patients, and potential risks to others beyond the research participant" [62]. They go on to say that nanomedicine research requires additional oversight because new science and technology poses uncertain but significant risks and that human subject research challenges the current oversight system by raising issues not addressed by the current structure. However, I do not see how nanomedicine raises any new or significantly different concerns in terms of a new drug that would have to go through the FDA approval process, if the argument is simply because it uses nanotechnology. However, if the argument is that the FDA does not have the expertise or scientific data to evaluate such a drug or device, as Stratmeyer seems to suggest—"Quality data generated in a consistent manner is needed by FDA to evaluate benefits versus risks when making product approval decisions" [63]—then I could see an argument for additional funding in order to acquire such expertise and information. However, this should be the norm for all new drugs and medical devices that the FDA evaluates.

Why should we concern ourselves with the ethics surrounding nanotechnology and in particular nanomedicine? One argument suggests that studying the ethical issues and consequences of nanotechnology enables the introduction of the technology to modern societies, but failure to examine these issues may lead to public resistance and diminish any expected benefits society may gain from the technology [64]. Others have argued that the ethical debates about potential applications using nanotechnology "could lead to completely unfounded and unjustified ethical problems which then could become issues for nanotechnology in the real societal debate" [65]. Grunwald further suggests that "speculative handling of the ethical questions of nanotechnology would produce artificial problems, whose communication to the public might cause damage to innovation policy in the field of nanotechnology and could prevent exploiting its chances" [66].

As nanomedicine advances does access to the technology create a disadvantage for those unable to acquire the technology for economic, social, or geographical reasons? Again this is a broader ethical question that is certainly not new or unique to nanomedicine or medicine in general. For instance, access to healthcare and/or cutting-edge medical treatments is not universal. In fact, even where physical location is not a barrier, economic access is still a barrier. The US healthcare system is a perfect example where the ability to gain access to cutting-edge treatments is based solely on one's ability to pay for it, not the absence of the technology itself. Access to healthcare in general may be changing, however, with the advent of personalized medicine and access to increasingly cheaper gene sequencing technologies. Self-diagnostics and treatments may still be a futuristic notion, but with the advent of synthetic biology, DIYbio, and nanotechnology the ethical debate surrounding this should be discussed. For instance, do the same ethical considerations researchers, doctors, and drug companies have to address when conducting a clinical trial or drug approval hold, or are they even necessary when a "person" is diagnosing or testing on himself or herself? How do the ethical considerations change when designing or approving a drug that has been tailored to an individual on the basis of his or her genetic makeup? Patients potentially will be giving up their genetic privacy in order for drug companies and doctors to design a treatment specifically for them. One ethical and intellectual property (IP) question that arises is whether or not the drug company would, could, or should "own" your genetic makeup in order to design a drug based on a your DNA. How would such a drug be priced, and do patients have the same legal recourse against a drug company if the drug was designed specifically for them?

17.5 Conclusions

Technology moves at a rapid pace, yet our governance structures that are supposed to enable technology development, prevent unintended risks, and provide economic development are struggling to keep pace. If you take nanotechnology as an example, the governance structure is a mishmash of regulations that varies from country to country and, within those countries, agency to agency. Too often special working groups that are formed to evaluate and develop governance structures for the technology of the day get lost in certain aspects of definitions and fail to recognize the convergence and reoccurring issues surrounding new technologies. Instead it seems, we continue to rehash the same issues (environmental health and safety, ethics, IP, etc.) for each and every new technology platform that is developed. The questions and sometimes the answers don't change but simply the technology platform.

Similar to politics where the extremes on both sides create hype and hysteria, nanotechnology and its impact on nanomedicine will probably be somewhere in the middle of these two diverging views. Ethical debates, however, tend to focus on the extremes and cause diverging viewpoints. In fact it may be that ethics is not capable of evaluating a technology that has yet to prove itself or has yet to show either of these two extremes. This is not to say that ethical considerations should not be examined in regard to nanomedicine, but the ethical issues are not necessarily new (issues such as a patient's rights, clinical trials, human enhancement, safety, etc.), but rather the techniques and science to achieve these improvements are new.

Acknowledgments

This chapter is an updated and adapted version of the article published in 2010, "Nanomedicine and Ethics: Is There Anything New or Unique?" [67].

References

1. Lymberis, A. (2010). Micro-nano-biosystems: an overview of European research. *Min. Invasive Ther.*, **19**, pp. 136–143.
2. Chow-White, P. A., and Garcia-Sancho, M. (2012). Bidirectional shaping of convergence: interactions between biology and computing from the first DNA sequencer to global genome datadases. *Sci. Technol. Hum. Val.*, **37**(1), pp. 124–164.
3. National Institutes of Health (2009), http://commonfund.nih.gov/nanomedicine/, accessed December 11, 2009.
4. Nazarov, G. V., Galan, S. E., Nazarova, E. V., Karkishchenko, N. N., Muradov, M. M., and Stepanov, V. A. (2009). Drug synthesis methods and manufacturing technology: nanosized forms of drugs (a review). *Pharm. Chem. J.*, **43**(3), pp. 163–170.
5. Riehemann, K, Schneider, S. W., Luger, T. A., Godin, B., Ferrari, M., and Fuchs, H. (2009). Nanomedicine-challenges and perspectives. *Angew. Chem., Int. Ed.*, **48**, pp. 872–897.
6. Sandhiya, S., Aibor Dkhar, S., and Surendiran, A. (2009). Emerging trends of nanomedicine: an overview. *Fundam. Clin. Pharmacol.*, **23**, pp. 263–269.
7. Moghimi, S. M., Hunter, A. C., and Murray, J. C. (2005). Nanomedicine: current status and future prospects. *FASEB J.*, **19**, pp. 311–330.
8. Doktycz, M., and Simpson, M. L. (2007). Nano-enabled synthetic biology. *Mol. Syst. Biol.*, **3**, pp. 1–10.
9. Roco, M. (2003). Nanotechnology: convergence with modern biology. *Curr. Opin. Biotechnol.*, **14**, pp. 337–346.
10. Hey, T., Tansley, S., and Tolle, K. eds. (2009). *The Fourth Paradigm: Data-Intensive Scientific Discovery* (Redmond, WA: Microsoft Research).

11. *Nanoinformatics 2020 Roadmap. National Science Foundation*, http://nanotechinformatics.org/nanoinformatics/index.php/Main_Page.

12. Armengaud, J. (2010). Proteogenomics and systems biology: quest for the ultimate missing parts. *Expert Rev. Proteomics*, **7**(1), pp. 65–77.

13. Johnson, C. J., Zhukovsky, N., Cass, A. E. G., and Nagy, J. M. (2008). Proteomics, nanotechnology and molecular diagnostics. *Proteomics*, **8**, pp. 715–730.

14. Marko, N. F., Weil, R. J., and Toms, S. A. (2007). Nanotechnology in proteomics. *Future Drugs*, **4**(5), pp. 617–626.

15. Weston, A. D., and Hood, L. (2004). Systems biology, proteomics, and the future of health care: toward predictive, preventative, and personalized medicine. *J. Proteome Res.*, **3**, pp. 179–196.

16. Goldstein, A. H. (2005). *Nanomedicine's Brave New World*, http://www.salon.com/2005/11/28/nanomedicine/.

17. González-Nilo, F., Pérez-Acle, T., Guínez-Molinos, S., Geraldo, D. A., Sandoval, C., Yévenes, A., Santos, L. S., Laurie, V. F., Mendoza, H., Cachau, R. E. (2011). Nanoinformatics: an emerging area of information technology at the intersection of bioinformatics, computational chemistry and nanobiotechnology. *Biol. Res.*, **44**, pp. 43–51.

18. Sakamoto, J. H., van de Ven, A. L., Godin, B., Blanco, E., Serda, R. E., Grattoni, A., Ziemys, A., Bouamrani, A., Hu, T., Ranganathan, S. I., De Rosa, E., Martinez, J. O., Smid, C. A., Buchanan, R. M., Lee, S. Y., Srinivasan, S., Landry, M., Meyn, A., Tasciotti, E., Liu, X., Decuzzi, P., and Ferrari, M. (2010). Enabling individual therapy through nanotechnology. *Pharmacol. Res.*, **62**, pp. 57–89.

19. Melo, M. R., Clark, S., and Barrio, D. (2011). Miniaturization and globalization of clinical laboratory activities. *Clin. Chem. Lab. Med.*, **49**(4), pp. 581–586.

20. Kricka, L. J., and Thorpe, G. H. (2010). Technology of handheld devices for point of care testing, in *Point of Care Testing: Needs, Opportunity, and Innovation*, eds., Price, C. P., St. John, A., and Kricka, L. J. (AAAS Press, Washington, DC).

21. Ruano-López, J. M., Agirregabiria, M., Olabarria, G., Verdoy, D., Bang, D. D., Bu, M., Wolff, A., Voigt, A., Dziuban, J. A., Walczak, R., and Berganzo, J. (2009). The SmartBioPhone, a point of care vision under development through two European projects: OPTOLABCARD and LABONFOIL. *Lab. Chip.*, **9**, pp. 1495–1499.

22. Macklis, R. M., and Sharma, N. (2011). Convergence technology in cancer medicine. *Expert Rev. Med. Devices.*, **8**(2), pp. 263–273.

23. Zhu, H., Xu, R., and Yuan, J. (2009). High speed intra-body communication for personal health care. *Conf. Proc. IEEE Eng. Med. Biol. Soc.*, pp. 709–712.

24. Mohamadi, M. R., Mahmoudian, L., Kaji, N., Tokeshi, M., Chuman, H., and Baba, Y. (2006). Nanotechnology for genomics and proteomics. *Nanotoday*, **1**(1), pp. 38–45.

25. Louridas, G. E., and Lourida, K. G. (2012). The new biology: a bridge to clinical cardiology. *Hippokratia*, **16**(2), pp. 106–112.

26. International Risk Governance Council. (2010). *Guidelines for the Appropriate Risk Governance of Synthetic Biology* (International Risk Governance Council).

27. Lymberis, A. (2010). Micro-nano-biosystems: an overview of European research. *Min. Invasive Ther.*, **19**, pp. 136–143.

28. http://www.giiresearch.com/report/bc229027-nanotechnology-medical-applications-global-market.html, last accessed August 15, 2012.

29. Wagner, V., Dullaart, A., Bock, A. K., and Zweck, A. (2006). The emerging nanomedicine landscape. *Nat. Biotechnol.*, **24**(10), pp. 1211–1217.

30. Resnik, D. B., and Tinkle, S. S. (2007). Ethics in nanomedicine. *Nanomedicine*, **2**(3), pp. 345–350.

31. Spagnolo, A. G., and Daloiso, V. (2008). Outlining ethical issues in nanotechnologies. *Bioethics*, **23**(7), pp. 394–402.

32. Evers, J., Aerts, S., and De Tavernier, J. (2008). An ethical argument in favor of nano-enabled diagnostics in livestock disease control. *Nanoethics*, **2**, pp. 163–178.

33. European Group on Ethics in Science and New Technologies (2005). *Ethical Aspects of ICT Implants in the Human Body. Opinion No. 20*, http://ec.europa.eu/european_group_ethics/docs/avis20_en.pdf, accessed December 9, 2009.

34. White, G. B. (2009). Missing the boat on nanoethics. *Am. J. Bioeth.*, **9**(10), pp. 18–19.

35. European Group on Ethics in Science and New Technologies to the European Commission (2007). *Opinion on the Ethical Issues Aspects of nanomedicine. Opinion No. 21*, http://ec.europa.eu/european_group_ethics/activities/docs/opinion_21_nano_en.pdf.

36. Bawa, R., and Johnson, S. (2009). Emerging issues in nanomedicine and ethics, in *Nanotechnology and Society: Current and Emerging Ethical Issues*, eds. Allhoff, F., Lin, P. (Springer).

37. Resnik, D. B., and Tinkle, S. S. (2007). Ethical issues in clinical trials involving nanomedicine. *Contemp. Clin. Trials*, **28**, pp. 433–441.
38. 21 C. F. R. 56.111 (1998). See also 45 C.F.R. 46.111 (2005).
39. Ferrari, M., Philibert, M. A., and Sanhai, W. R. (2009). Nanomedicine and Society. *Clin. Pharmacol. Ther.*, **85**(5), 466–467.
40. Khushf, G. (2007). Upstream ethics in nanomedicine: a call for research. *Nanomedicine*, **2**(4), pp. 511–521.
41. Gordijn, B. (2005). Nanoethics: from Utopian dreams and apocalyptic nightmares towards a more balanced view. *Sci. Eng. Ethics*, **11**, pp. 521–533.
42. Wagner, V., Dullaart, A., Bock, A. K., and Zweck, A. (2006). The emerging nanomedicine landscape. *Nat. Biotechnol.*, **24**(10), pp. 1211–1217.
43. Godman, M. (2008). But is it unique to nanotechnology? Reframing nanoethics. *Sci. Eng. Ethics*, **14**, pp. 391–403.
44. Grunwald, A. (2005). Nanotechnology: a new field of ethical inquiry? *Sci. Eng. Ethics*, **11**(2), pp. 187–201.
45. Petersen, A. (2009). Introduction: the ethical challenges of nanotechnologies. *Bioeth. Inq.*, **6**, pp. 9–12.
46. Sparrow, R. (2009). The social impacts of nanotechnology: an ethical and political analysis. *Bioeth. Inq.*, **6**, pp. 13–23.
47. MacDonald, C. (2004). Nanotech is novel; the ethical issues are not. *Scientist*, **18**, p. 3.
48. Cameron, N. M. S. (2006). Nanotechnology and the human future: policy, ethics, and risk. *Ann. N. Y. Acad. Sci.*, **1093**, pp. 280–300.
49. Guerra, G. (2008). European regulatory issues in nanomedicine. *Nanoethics*, **2**, pp. 87–97.
50. http://www.huffingtonpost.com/2012/07/26/prosthetic-limbs-unfair-advantage-oscar-pistorius_n_1706146.html.
51. Sparrow, R. (2007). Revolutionary and familiar, inevitable and precarious: rhetorical contradictions in enthusiasm for nanotechnology. *Nanoethics*, **1**, pp. 57–68.
52. Lewenstein, B. V. (2005). What counts as a "social and ethical issue" in nanotechnology? *HYLE-Int. J. Philos. Chem.*, **11**(1), pp. 5–18.
53. Satava, R. M. (2002). Disruptive visions: moral and ethical challenges from advanced technology and issues for the new generation of surgeons. *Surg. Endosc.*, **16**, pp. 1403–1408.
54. Mnyusiwalla, A., Daar, A. S., and Singer, P. A. (2003). Mind the gap: science and ethics in nanotechnology. *Nanotechnology*, **14**, R9–R13.

55. Lenk, C., and Biller-Andorno, N. (2007). Nanomedicine-emerging or re-emerging ethical issues? A discussion of four ethical themes. *Med. Health Care Philos.*, **10**, pp. 173–184.

56. Preston, C. J. (2005). The promise and threat of nanotechnology. *HYLE-Int. J. Philos. Chem.*, **11**(1), pp. 19–44.

57. Sandler, R. (2009). Nanomedicine and nanomedical ethics. *Am. J. Bioeth.*, **9**(10), pp. 16–17.

58. Allhoff, F. (2009). The coming era of nanomedicine. *Am. J. Bioeth.*, **9**(10), pp. 3–11.

59. Johnson, S. (2009). The era of nanomedicine and nanoethics: has it come, is it still coming, or will it pass us by? *Am. J. Bioeth.*, **9**(10), pp. 1–2.

60. McGee, E. M. (2009). Nanomedicine: ethical concerns beyond diagnostics, drugs, and techniques. *Am. J. Bioeth.*, **9**(10), pp. 14–15.

61. Bawa, R., and Johnson, S. (2007). The ethical dimensions of nanomedicine. *Med. Clin. North Am.*, **91**, pp. 881–887.

62. Wolf, S., and Jones, C. M. (2011). Designing oversight for nanomedicine research in human subjects: systematic analysis of exceptional oversight for emerging technologies. *J. Nanopart. Res.*, **13**, pp. 1449–1465.

63. Stratmeyer, M. E., Goering, P. L., Hitchins, V. M., and Umbreit, T. H. (2010). What we know and don't know about the bioeffects of nanoparticles: developing experimental approaches for safety assessment. *Biomed. Microdevices*, **12**, pp. 569–573.

64. Grunwald, A. (2010). From speculative nanoethics to explorative philosophy of nanotechnology. *Nanoetchics*, **4**, pp. 91–101.

65. Nordman, A., and Rip, A. (2009). Mind the gap revisited. *Nat. Nanotechnol.*, **4**, pp. 273–274.

66. Grunwald, A. (2010). From speculative nanoethics to explorative philosophy of nanotechnology. *Nanoethics*, **4**, pp. 91–101.

67. Kuiken, T. (2010). Nanomedicine and ethics: is there anything new or unique? *WIREs Nanomed. Nanobiotechnol.*, [Epub ahead of print].

Epilogue

Toward Personalized and Curative Medicine

Patrick Hunziker
University Hospital Basel, Switzerland
CLINAM Foundation for Nanomedicine, Basel, Switzerland
patrick.hunziker@usb.ch

E.1 Today's Medicine

E.1.1 The Achievements of Modern Medicine

The development of modern medicine has led to seemingly miraculous advances compared to a century ago: The poxvirus has been eradicated, and measles, a once devastating epidemic disease, has largely been eliminated from those communities able and willing to be vaccinated. The mortality from myocardial infarction has decreased from 30% to about 5% in those patients who reach hospital care, thanks to intensive care units, coronary artery balloon angioplasty with stenting, and drugs. Some cancers can be cured, even in advanced stages. Despite further increase in human immunodeficiency virus (HIV) infections, the development

Handbook of Safety Assessment of Nanomaterials: From Toxicological Testing to Personalized Medicine
Edited by Bengt Fadeel

ISBN 978-981-4463-36-2 (Hardcover), 978-981-4463-37-9 (eBook)
www.panstanford.com

of full-blown acquired immunodeficiency syndrome (AIDS) has significantly dropped in developed countries. Osteosynthesis allows repairing hip fractures and permits the patient to regain ambulation quickly instead of remaining bedridden and succumbing to pneumonia or pulmonary embolism. In some patients, even multiorgan system failure that includes severe infection and systemic inflammation, profound circulatory shock despite pharmacologic support, pulmonary failure despite artificial ventilation, renal failure, and coma can be overcome by combination of the full spectrum of medical devices and drugs available (Fig. E.1).

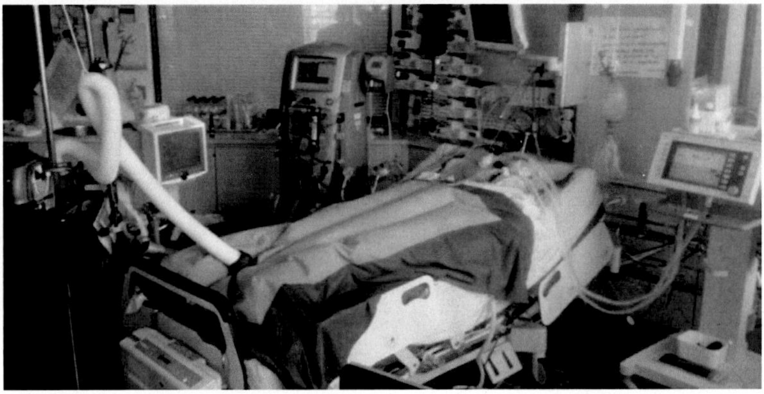

Figure E.1 Range of advanced therapies and organ support. The patient with infection, multiorgan failure, septic shock, heart failure, lung failure, and kidney failure treated (from left to right) by external thermoregulation, external heart/lung assist, kidney replacement therapy, tube feeding, and a ventilator and by drugs inducing artificial coma, antibiotics, anti-inflammatory drugs, catecholamines, and sedation. The patient fully recovered. *Source*: University Hospital Basel, Switzerland.

Despite a significant increase in *prevalence* of diseases like diabetes associated with a Western lifestyle, the loss of *length* and *quality of life* by these diseases is limited (Fig. E.2) in developed countries. In contrast, despite a much lower prevalence of diabetes in developing countries, the *disease burden*, measured as the loss of quality-adjusted life years, is significantly higher in less developed countries. A key clue to explain this striking difference is in the large structural and financial effort for healthcare enabled by the economies in developed countries.

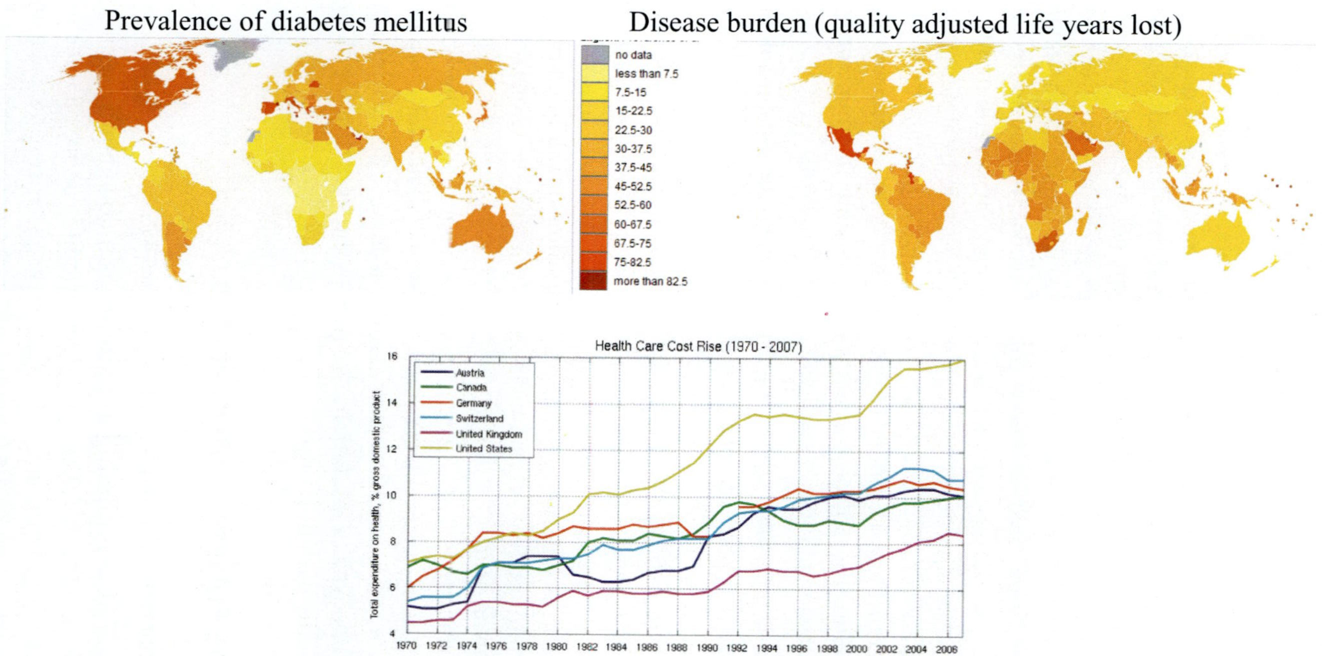

Figure E.2 Top: High diabetes prevalence but limited number of life years lost in developed countries; limited diabetes prevalence but high disease burden in developing countries. Bottom: High expenditure in healthcare in developed countries as one factor explaining the discrepancy. *Source*: Wikipedia.

E.1.2 The Limitations of Modern Medicine: Efficacy, Toxicity, and Cost Triangle

The counterexamples to the apparent omnipotence of medicine are numerous: Patients with *atherosclerosis*, the main cause of death in developed countries, are not cured by current therapies but, despite interventional therapies and lifelong expensive medications, will almost invariably come back with the same problem—progression of atherosclerosis. Successes in *cancer*, the number two killer disease, are often limited: Many cancer trials are considered a success because average survival was prolonged by several months in a statistically significant manner; more effective treatment can often not be given because of systemic toxicity. *HIV infection* no longer kills quickly with modern treatment, but disease suppression requires lifelong therapy with a combination regime. In other words, the key limitation of today's medicine is the triangle of:

- ***limited efficacy***: While unable to cure many prevalent diseases, disease suppression through long-term medication is done instead.
- ***limiting toxicity***: The inability to cure certain diseases, like cancer, is in part related to toxicity and includes both generic toxicity (e.g., bleeding with platelet inhibitor treatment) as well as patient-specific toxicity rooted in the genome of the patient (e.g., rhabdomyolysis with certain cholesterol-lowering drugs).
- ***limiting cost***: The large costs arising from regular physician visits and hospitalizations, long-term medication, repeat interventions, and loss of autonomy in chronic disease lead in a subtle way to limitation of access to the best care for all or for some patient groups.

E.2 The Future of Medicine

The future of medicine requires new paradigms:

- **A new understanding of patients** as unique human beings, characterized by a specific demographic context, a specific age, a unique genome, and a specific gene expression. Such an understanding of the human body as a complex system with

individual characteristics is the first step toward *personalized medicine.*

- **A new understanding of disease** as a health disorder with a potentially unique genetic setup, an individual gene expression profile, a specific disease localization, a specific disease state, and, in the case of infection, a specific biological and resistance profile of the microorganism.
- **A new understanding of drug therapy** that aims at optimal drug effects at the disease site through targeting (instead of just achieving high levels everywhere in the body), while minimizing undesired effects in remote organs and bystander cells, thereby widening the therapeutic index.
- **New therapeutic strategies** that transcend the concept of "standard drugs for standard indications, given in standard dosage for a standard duration, and tested in megatrials" (often referred to as "evidence-based medicine"). Such new strategies aim at finding patient-specific answers: "Given a specific patient with a unique genome, a disease-specific gene expression profile, a specific disease location, and a specific disease state, which drug or combination therapy, which individualized dosage schedule (potentially time varying), and for which duration will bring what benefit (in quantitative terms) and lead to which individual toxicity (in quantitative terms) at which absolute cost and cost-effectiveness?"
- **A new understanding of toxicity** as a biological process in individuals that is based on a quantitative understanding of interaction of drugs with specific biological pathways and that can be reliably predicted on the basis of the genomic, proteomic, and metabolomic setup of a person.
- **A new understanding of the cost of healthcare** that does not limit its scope on "expenditure per disease case" but takes into account the indirect impact to the individual, the indirect cost to the patient and to society, and the benefit of healthcare expenditures that flow back into the economy of a country.
- **A focus on cure.** "Cure" might be arbitrarily defined as "disease-specific freedom from adverse health events and from the need for physician visits, medications, and interventions for an extend time period." Some of the most widespread diseases in developed countries, namely, diabetes,

atherosclerosis, chronic pulmonary disease, hypertension, and HIV infection, are not cured by therapy but rather transformed into chronic, subclinical disease states that require long-term drug therapy and lead to regular physician visits and repeated hospitalization. The ability to "cure" even one of these diseases, even at a substantial cost, might free enormous resources for society.

E.3 Strategic Issues for Nanomedicine

The development of the field of nanomedicine was provided in Chapter 1. Instead of giving an exhaustive list of potential pieces of work, some strategic issues will be mentioned here:

- **Nanomedicine will promote interdisciplinarity in medical research and development.** On the basis of the history of the nanosciences as a field incorporating knowledge of physics, chemistry, pharmacology, biology, medicine, mathematical modeling, and engineering, the nanomedicine community is well trained in communication across the boundaries of scientific fields, and this higher-level perspective of cross fertilization may be as important as a contribution to the future of medicine as any detailed technical development.
- **Nanomedicine enhances efficacy and reduces toxicity of drug therapies by combining organ-, cell-, and receptor-specific delivery with arbitrary payloads.** The combination of targeted delivery with conventional pathway-targeting drugs has shown to greatly increase efficacy at the target cell, while largely abolishing toxicity in bystander cells and tissues [1]. For the near future, progress in synthesis of highly reproducible, highly pure, well-characterized materials with excellent stealth characteristics, proven capability for receptor-specific binding, and well-understood degradation/elimination is required to accelerate the translation from proof of concept to industrial manufacture and clinical trial.

Other issues:

- **Understanding nanomaterial interactions with living systems**: Despite significant efforts in the past years, the goal of a full understanding of nanomaterial interaction with biologic fluids, cells and tissues is not yet achieved. One aspect of particular importance is the adsorption of plasma proteins to nanomaterials. This appears to be the critical step for opsonization (leading to premature elimination in liver and spleen) and for complement activation (leading to immediate immunotoxicity reactions). Likewise, the interaction of nanomaterials with cell membranes is a critical mechanistic step in the activity of nanomedicines but is not fully understood. Certain nanomaterials with and without adsorbed plasma proteins are known to interact with circulating blood cells, an effect that might be exploited therapeutically or should be avoided by suited nanomaterial design. Better, a detailed mechanistic understanding of all these processes [2] is required, and tools to predict such processes in a generic fashion for new materials are desired.

- **Understanding the targets:** Typical development of new drugs uses standardized in vitro tests, cell culture models, and animal models for preclinical development. Unfortunately, the models used most often, for example, in research in cancer, atherosclerosis, and inflammatory diseases, show major differences from the physiological situation in humans. An important example is the differential activity of several esterases in human versus animal plasma [3]. Thus, esterase-cleavable ligands on nanomaterials risk to have a largely different fate in humans compared to any laboratory animal, limiting the value of preclinical in vivo experiments. Likewise, tumor vascularization and tumor biology of human xenotransplant models in mice, a standard model in cancer research, do not well replicate the situation in humans [4]. Thus, the existing and evolving methods that allow gene expression profiling in microscopic tissue domains from human tissue, or in the future even of single cells, may help our search for targets and biomarkers and deepen our understanding of undesired events in the disease courses, like evasion of a few cancer cells from destruction by targeted therapies, followed by tumor recurrence.
- **Standardized and predictive testing:** The available literature shows a vast array of test setups for nanomaterial

characterization and for nano–bio interactions. Most studies are therefore not easily comparable. To render the broad ongoing academic research comparable and to facilitate translation to real-world applications, there is an urgent need for standardization. While for some tests, industrial standard protocols for small molecules or pharmaceutical "biologicals" are sufficient, other issues, like immunotoxicity of nanomaterials or nanomaterial–cell interactions need new standards. Finding a simple and rapid in vitro test (possibly even applicable in a point-of-care setting) that is able to reliably predict complement activation-related immunotoxicity for individual patients would be a significant single contribution to the field.

- **Predictive modeling:** From 1990 to 2000, the term "nanomedicine" was, for the most part, not in use (but nanomedicines were developed and have reached the clinic; see Chapter 1), while in 2000–2010, the field was thoroughly shaped by academic experimental research. In view of the huge variability in the human genome, the large number of diseases and disease variants, potentiated by the combinatorial freedom in combining metallic, polymeric, and lipid nanomaterials with ligands like small molecules, antibodies, and aptamers, and the large choice of potential payloads from the domain of chemical and biological substances, it appears that the search for optimal nanosystems is much worse than looking for a needle in a haystack, in particular if a study of the interaction of those materials with the vast number of biological substrates is intended. To master this huge level of complexity, novel computer-assisted models and strategies are required: beyond nice graphical representations, predictive models are required that allow to build composite nanosystems in silico and to screen their interaction with biological substrates thoroughly. Calibration of such models with systematically acquired experimental data will be of particular importance to the plausibilization of results and may also produce discrepancies between models and experiments that improve our understanding of the scientific fundaments and open the door for new discoveries.

Modeling for personalized toxicity and for personalized medicine: A particular largely unexplored domain is the use of modeling to predict and to optimize treatment regimens in personalized medicine, with the goal of maximal efficacy, minimal toxicity, and optimal cost effectiveness. How can we expand established concepts of in silico toxicity screening [5] to render them useful in clinical practice, to an individual patient in a physician's office, and in a next step to the nanomedical toxicity prediction in the same setting? There is ample room for promising collaborations between nanoscientists, computer scientists, and clinicians.

Understanding the process of translation from academia to industry: The concept of open innovation [6] implies a strong influx of academic concepts into industrial development. For the young field of nanomedicine, which has not yet been fully adopted by the industry, this aspect of knowledge transfer is of particular importance. To develop new concepts for medical therapies, a deeper knowledge in the academic community of the industrial development process, of industrial assessment of innovation and risk, of "no-gos," and of the industrial value chain is important.

- **Cure—the grand goal of nanomedicine and medicine in general**: The basic concepts of nanomedicine (e.g., to exploit the supramolecular structure and to use physical, biological, and chemical effects that only arise in assemblies of molecules or metal atoms but not in single molecules or atoms) show clearly that nanomedicine is more than just a slight extension of conventional small-molecular drugs or pharmaceutical biologicals. This broader range of achievable functionalities opens new doors for radical innovation, like delivering synthetic nanosized organelles to cells [7], a new therapeutic paradigm for regenerative medicine, and other nonpharmaceutical concepts. We should grasp these possibilities to fulfill the great dream of the healing profession, namely, to cure and to slow down not only disease progression at a high cost on a long term but ultimately failing treatments.

E.4 The Ultimate Goal

The final goal of the research described in this book is to use science to contribute to the benefit of human health and society. To achieve this, even basic and technical research and developments profit from moments of critical reflection on the larger goal of our endeavors in this fascinating field of science.

References

1. Pavel, B., Benito, S. M., Saw, C., Burger, P., Heider, H., Pfisterer, M., Marsch, S. Meier, W., and Hunziker, P. (2005). Cell targeting by a generic receptor-targeted polymer nanocontainer platform. *J. Controlled Rel.*, **102**, pp. 475–488.
2. Fadeel, B., and Garcia-Bennett, A. E. (2010). Better safe than sorry: understanding the toxicological properties of inorganic nanoparticles manufactured for biomedical applications. *Adv. Drug Delivery Rev.*, **62**, pp. 362–374.
3. Bin, L., Sedlacek, M., Manoharan, I., Boopathy, R., Duysen, E. G., Masson, P., and Lockridge, O. (2005). Butyrylcholinesterase, paraoxonase, and albumin esterase, but not carboxylesterase, are present in human plasma. *Biochem. Pharmacol.*, **70**, pp. 1673–1684.
4. Annapoorni, R., and Weinberg, R. A. (2003). Comparative biology of mouse versus human cells: modelling human cancer in mice. *Nat. Rev. Cancer*, **3**, pp. 952–959, http://www.stanford.edu/class/archive/cbio/cbio241/cbio241.1072/coursework/rangarajan2003.pdf.
5. Angelo, V., Dobler, M., and Smiesko, M. (2012). VirtualToxLab: a platform for estimating the toxic potential of drugs, chemicals and natural products. *Toxicol. Appl. Pharmacol.*, **261**, pp. 142–153
6. http://en.wikipedia.org/wiki/Open_innovation.
7. Ben-Haim, N., Broz, P., Marsch, S., Meier, W., and Hunziker, P. (2008). Cell-specific integration of artificial organelles based on functionalized polymer vesicles. *Nano Lett.*, **8**(5), pp. 1368–1373.

Index